Essentials of Materials Science and Technology

Essentials of Materials Science and Technology

Edited by **Lily Chen**

NY RESEARCH
P R E S S

New York

Published by NY Research Press,
23 West, 55th Street, Suite 816,
New York, NY 10019, USA
www.nyresearchpress.com

Essentials of Materials Science and Technology
Edited by Lily Chen

International Standard Book Number: 978-1-63238-474-4 (Hardback)

The publisher's policy is to use permanent paper from mills that operate a sustainable forestry policy. Furthermore, the publisher ensures that the text paper and cover boards used have met acceptable environmental accreditation standards.

Trademark Notice: Registered trademark of products or corporate names are used only for explanation and identification without intent to infringe.

Printed in the United States of America.

Contents

Preface

The purpose of the book is to provide a glimpse into the dynamics and to present opinions and studies of some of the scientists engaged in the development of new ideas in the field from very different standpoints. This book will prove useful to students and researchers owing to its high content quality.

Materials science is an interdisciplinary field formed by combining the principles of physics, engineering, chemistry, mineralogy, and metallurgy. While materials science studies the properties and structure of substances like ceramics, plastics and metals, materials technology is concerned with the manufacturing and design of new and improved materials. There has been rapid progress in this discipline and its applications are finding their way across multiple industries. This book is a valuable compilation of topics, ranging from the basic to the most complex advancements in this field. It will help the readers in keeping pace with the rapid changes in these areas of study. It provides comprehensive insights into the field of materials science and technology by presenting the most up-to-date researches from around the world.

At the end, I would like to appreciate all the efforts made by the authors in completing their chapters professionally. I express my deepest gratitude to all of them for contributing to this book by sharing their valuable works. A special thanks to my family and friends for their constant support in this journey.

Editor

Silicon-nitride-based integrated optofluidic biochemical sensors using a coupled-resonator optical waveguide

Jiawei Wang, Zhanshi Yao and Andrew W. Poon *

Photonic Device Laboratory, Department of Electronic and Computer Engineering, The Hong Kong University of Science and Technology, Hong Kong, China

Edited by:
Dan-Xia Xu,
National Research Council Canada,
Canada

Reviewed by:
Koji Yamada,
Nippon Telegraph and Telephone
Corporation, Japan
Weidong Zhou,
University of Texas at Arlington, USA
Robert Halir,
Universidad de Málaga, Spain

*Correspondence:
Andrew W. Poon,
Photonic Device Laboratory,
Department of Electronic and
Computer Engineering, The Hong
Kong University of Science and
Technology, Clear Water Bay,
Kowloon, Hong Kong, China
eeawpoon@ust.hk

Silicon nitride (SiN) is a promising material platform for integrating photonic components and microfluidic channels on a chip for label-free, optical biochemical sensing applications in the visible to near-infrared wavelengths. The chip-scale SiN-based optofluidic sensors can be compact due to a relatively high refractive index contrast between SiN and the fluidic medium, and low-cost due to the complementary metal-oxide-semiconductor (CMOS)-compatible fabrication process. Here, we demonstrate SiN-based integrated optofluidic biochemical sensors using a coupled-resonator optical waveguide (CROW) in the visible wavelengths. The working principle is based on imaging in the far field the out-of-plane elastic-light-scattering patterns of the CROW sensor at a fixed probe wavelength. We correlate the imaged pattern with reference patterns at the CROW eigenstates. Our sensing algorithm maps the correlation coefficients of the imaged pattern with a library of calibrated correlation coefficients to extract a minute change in the cladding refractive index. Given a calibrated CROW, our sensing mechanism in the spatial domain only requires a fixed-wavelength laser in the visible wavelengths as a light source, with the probe wavelength located within the CROW transmission band, and a silicon digital charge-coupled device/CMOS camera for recording the light scattering patterns. This is in sharp contrast with the conventional optical microcavity-based sensing methods that impose a strict requirement of spectral alignment with a high-quality cavity resonance using a wavelength-tunable laser. Our experimental results using a SiN CROW sensor with eight coupled microrings in the 680 nm wavelength reveal a cladding refractive index change of ~1.3×10^{-4} refractive index unit (RIU), with an average sensitivity of ~281 ± 271 RIU^{-1} and a noise-equivalent detection limit of 1.8×10^{-8} ~ 1.0×10^{-4} RIU across the CROW bandwidth of ~1 nm.

Keywords: silicon nitride, biochemical sensor, integrated optofluidics, coupled-resonator optical waveguide, microring resonators, CMOS-compatible, elastic light scattering, visible wavelengths

Introduction

In recent years, the increasing demands of medical diagnostics outside a clinic or a laboratory and self-monitoring for personal healthcare have highly motivated the rapid research and development of portable, low-cost biochemical sensors (Estevez et al., 2012). Particularly, miniaturized, label-free biochemical sensors are highly desired in order to be readily deployed at or carried to the sensing environment and to read-out in real-time, quantitative biochemical information about the environment (Vollmer et al., 2008). Among various demonstrated chip-scale photonic biochemical sensors, optical microresonator-based biosensors featuring optical resonances with a high quality (Q) factor ($10^3 \sim 10^4$) promise a high sensitivity [few tens to hundreds of nanometer resonance shift per refractive index unit (RIU)], a low detection limit ($10^{-7} \sim 10^{-4}$ RIU) and a compact footprint (few to hundreds of micrometer square) (De Vos et al., 2007; Ciminelli et al., 2013; Sedlmeir et al., 2014). However, such high-Q microcavity-based sensors working in the spectral domain are constrained by a narrow resonance bandwidth as the sensing window, which requires a strict resonance alignment and thus may compromise the reliability of the sensor system. Besides, the sensing implementation typically requires a precision wavelength-scanning setup, such as a wavelength-tunable laser, which may limit the portability of the sensor system.

Other than microcavity-based biochemical sensors, integrated interferometric optical biochemical sensors also attract increasing attentions. Various kinds of interferometer structures, including Mach–Zehnder interferometers (MZI) (Densmore et al., 2008; Kozma et al., 2009; Duval et al., 2013; Halir et al., 2013; Dante et al., 2015), Young interferometers (Ymeti et al., 2007), and Hartman interferometers (Xu et al., 2007) have been adopted as integrated interferometric biochemical sensors, demonstrating a high sensitivity ($10^2 \sim 10^4$ rad/RIU) along with a low detection limit ($10^{-7} \sim 10^{-5}$ RIU). One key merit of such integrated inter-ferometric sensors is that they require a relatively simple configuration, which typically comprises a fixed-wavelength laser source and a photodetector. However, these interferometric sensors are not tolerant to equipment noises that cause output intensity variations, such as laser intensity variations.

Previously, our research group has proposed a coupled-resonator optical waveguide (CROW)-based biochemical sensing scheme using what we termed "pixelized pattern detection" in the spatial domain (Lei and Poon, 2011). The scheme employs the discrete transition of the CROW eigenstate excited at a fixed laser wavelength upon a small change in the cladding refrac-tive index, Δn, and detects the resulting change in mode-field-intensity distribution by far-field measurement of the out-of-plane elastic-light-scattering intensity patterns. Such a sensing scheme in principle only requires relatively simple optical sources and imaging systems including a fixed-wavelength laser and a cam-era. Recently, we have experimentally demonstrated a proof of concept of such a chip-scale CROW-based sensor on the silicon-on-insulator (SOI) platform in the 1550 nm telecommunication wavelengths (Wang et al., 2014). We have extended the scheme by detecting the continuous modulation of the CROW mode-field-intensity distribution at a fixed wavelength upon a Δn by correlating the elastic-light-scattering patterns with reference pat-terns at the CROW eigenstates. Compared with interferometric sensors, the correlation analysis allows our sensing scheme to be more tolerant to equipment noises that are common to all pixels of the CROW sensor yet do not cause a spectral shift, including laser intensity variations. Our previous experiment demonstrated a Δn of $\sim 1.5 \times 10^{-4}$ RIU and a noise-equivalent detection limit (NEDL) of $2 \times 10^{-7} \sim 9 \times 10^{-4}$ RIU. However, the choice of the SOI platform and the experimental setup config-uration (including a 1550 nm laser, an optical amplifier and an InGaAs camera) render our previous work not practical for point-of-care optical biochemical sensing applications. Particularly, in order to leverage the wide availability of smartphones for bio-chemical sensing (Lakshminarayanan et al., 2015), it would be advantageous to switch the operational wavelength of the sensor from the telecommunication wavelengths to the visible or near-infrared wavelengths that can be readily recorded using high-resolution silicon charge-coupled device (CCD)/complementary metal-oxide-semiconductor (CMOS) cameras.

In this paper, we report our experimental demonstration of the CROW-based biochemical sensors in the visible wavelengths in the silicon-nitride (SiN) platform. The SiN platform is trans-parent to the visible and near-infrared wavelengths (Gorin et al., 2008; Subramanian et al., 2013) and its fabrication process is CMOS-compatible. After the CROW calibration steps, our sens-ing scheme in principle only requires a fixed-wavelength, low-output-power, visible laser source, and a silicon CCD/CMOS camera for recording out-of-plane light-scattering patterns from the top-view. This offers a promising opportunity to integrate the CROW sensor with a smartphone that is equipped with a compact laser source and a high-resolution camera with a properly designed optical interface for future smartphone-based point-of-care applications.

Principle and Methods

Principle and the Sensing Algorithm

Figure 1 illustrates the principle of the CROW-based biochemical sensor following our previous work (Wang et al., 2014). Here, we outline the key concepts of the principle for understanding this work. **Figure 1A** schematically shows a SiN CROW sen-sor comprising eight coupled microring resonators with identi-cal design, coupled to input and output bus waveguides in an add-drop filter configuration. For a perfect CROW compris-ing C coupled identical single-mode resonators, the inhomoge-neously broadened transmission spectrum features a combination of split mode resonances, with each mode slightly shifted from the original resonance frequency due to inter-cavity-coupling effect. Therefore, the eigenstate number N within each transmission band always equals to the resonator number C. While a perfect CROW exhibits distinctive mode-field-amplitude distributions at eigenstates, the pair of symmetric and anti-symmetric split-modes at different eigenfrequencies have non-distinctive mode-field-intensity distributions. In practice, a CROW inevitably suf-fers from fabrication imperfections. The coupled resonators are no longer identical nor are identically coupled. The symme-try breaking between the pair of symmetric and anti-symmetric split-modes therefore results in distinctive mode-field-intensity distributions at all discernable eigenstates. The resulting phase disorders and coupling disorders can result in the split mode resonances to be spectrally overlapped. Therefore, in the presence of structural non-uniformity, N could be equal to or smaller than C ($N \leq C$).

Figure 1B schematically illustrates the inhomogeneously broadened transmission bands upon applying cladding refrac-tive indices n_0 and $n_0 + \Delta n$, for an imperfect eight-microring

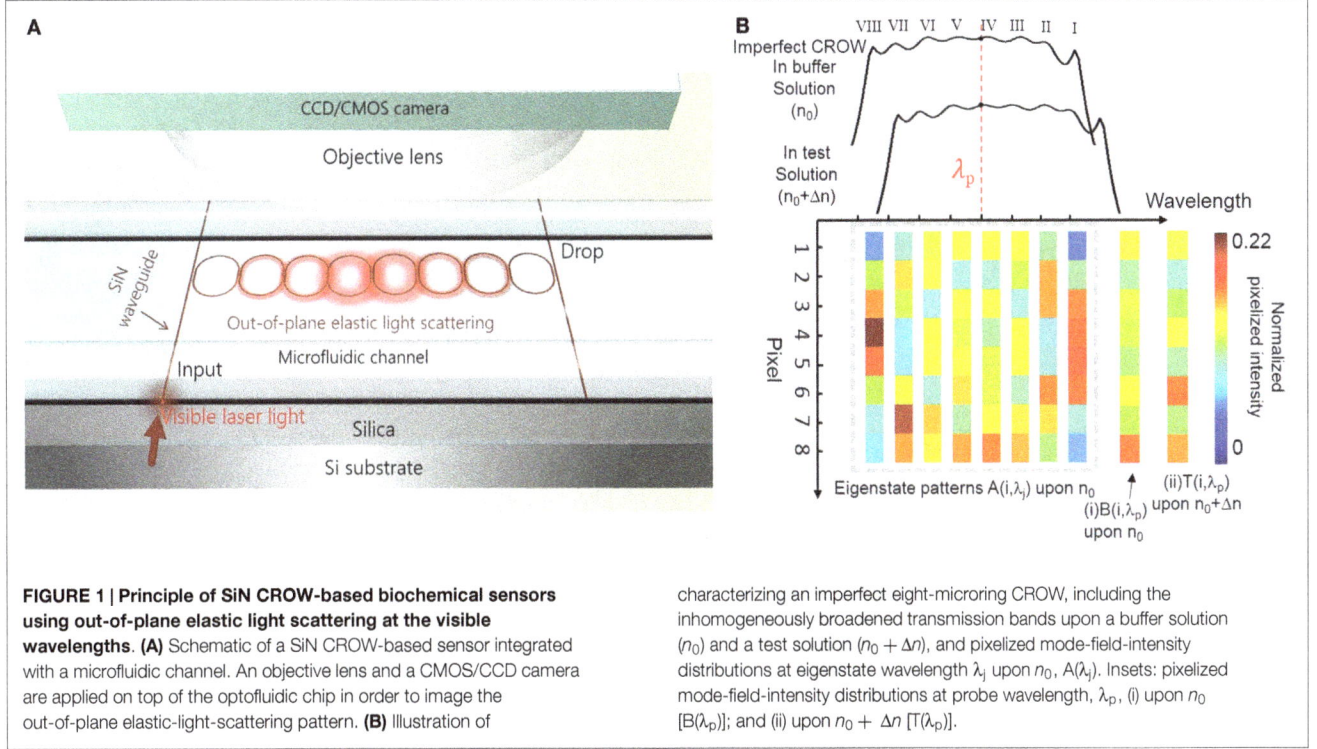

FIGURE 1 | Principle of SiN CROW-based biochemical sensors using out-of-plane elastic light scattering at the visible wavelengths. (A) Schematic of a SiN CROW-based sensor integrated with a microfluidic channel. An objective lens and a CMOS/CCD camera are applied on top of the optofluidic chip in order to image the out-of-plane elastic-light-scattering pattern. **(B)** Illustration of characterizing an imperfect eight-microring CROW, including the inhomogeneously broadened transmission bands upon a buffer solution (n_0) and a test solution ($n_0 + \Delta n$), and pixelized mode-field-intensity distributions at eigenstate wavelength λ_j upon n_0, A(λ_j). Insets: pixelized mode-field-intensity distributions at probe wavelength, λ_p, (i) upon n_0 [B(λ_p)]; and (ii) upon $n_0 + \Delta n$ [T(λ_p)].

CROW exhibiting a complete set of eight distinctive eigenstate mode-field-intensity distributions. With the mode-field intensity of each microring integrated as a pixel, we denote the pixelized one-dimensional pattern at the eigenstate as {A_j}, with j indexing the eigenstate. Any mode-field-amplitude distribution at an arbitrary wavelength, λ_p, within the CROW transmission band upon n_0 can be expressed by a linear superposition of the complete set of eigenstate mode-field-amplitude distributions upon n_0. Therefore, we are able to uniquely identify any pixelized mode-field-intensity profile at λ_p upon n_0, B(λ_p), as shown in inset (i), with {A_j} by a correlation analysis. Upon a small Δn applied homogenously to the cladding, we can uniquely identify by the correlation analysis any pixelized mode-field-intensity distribution at λ_p upon $n_0 + \Delta n$, T(λ_p), as shown in inset (ii), with {A_j}.

As in our previous work (Wang et al., 2014), we adopt the Pearson's correlation coefficient, ρ, in order to analyze the degree of correlation between a pixelized pattern at an arbitrary probe wavelength λ_p, B(λ_p), and the pixelized patterns at the eigenstate wavelengths λ_j, A(λ_j). For a CROW with a number of coupled single-mode cavities, C, and a number of discernable eigenstates, N ($\leq C$), we define ρ at λ_p for A(λ_j) as follows:

$$\rho_j(\lambda_p) = \frac{\sum_{i=1}^{C} (A(i, \lambda_j) - \overline{A(\lambda_j)})(B(i, \lambda_p) - \overline{B(\lambda_p)})}{\sqrt{\sum_{i=1}^{C} (A(i, \lambda_j) - \overline{A(\lambda_j)})^2} \sqrt{\sum_{i=1}^{C} (B(i, \lambda_p) - \overline{B(\lambda_p)})^2}}$$

(1)

where j = 1, 2, ..., N is the eigenstate number, and i = 1, 2, ..., C is the cavity (pixel) number. A(i, λ_j) and B(i, λ_p) are the pixel values normalized to the total intensity of the entire patterns,

respectively. The bar sign denotes the mean of the entire pixelized pattern over C pixels.

We adopt the Pearson's correlation coefficient approach to describe the linear dependence of the measured and calibrated intensity distributions. The Pearson's correlation approach is insensitive to both level and scale variations of the intensity distributions. Therefore, the approach is tolerant to equipment noise sources, such as uniform background light imaged onto the camera and the intensity variation of the laser source, which are common to all pixels and do not cause a spectral shift. However, this approach still suffers from the noises that cause a spectral shift, such as a wavelength drift of the laser source and thermal variations in the test environment.

Here, we detail our sensing algorithm following our previous work (Wang et al., 2014). **Figure 2** shows a flow chart illustrating our sensing algorithm including calibration. We first generate a library of correlation coefficients $\left\{ \rho'_j (\lambda_0) \right\}$, defined at a fixed reference wavelength λ_0 centered at the CROW transmission band. The library is calibrated over a range of Δn values, Δn_d, given by an integer multiple of a minimum refractive index change interval Δn_i. The $\left\{ \rho'_j (\lambda_0) \right\}$ thus comprises a library of data array of N (rows) $\times M$ (columns), where M is given by $\Delta n_d / \Delta n_i$.

For sensing, we first measure the pixelized mode-field-intensity pattern in a buffer solution at a fixed probe wavelength λ_p (which is generally offset from λ_0) as B(λ_p) (**Figure 1B**). We correlate B(λ_p) with the eigenstate patterns {A_j} in order to extract {$\rho_j(\lambda_p)$}. We look for the closest match of {$\rho_j(\lambda_p)$} with the library $\left\{ \rho'_j (\lambda_0) \right\}$, using only the principal (largest) component, ρ^p, and the second-principal (second-largest) component, ρ^s, of {$\rho_j(\lambda_p)$} in order to streamline the pattern recognition process

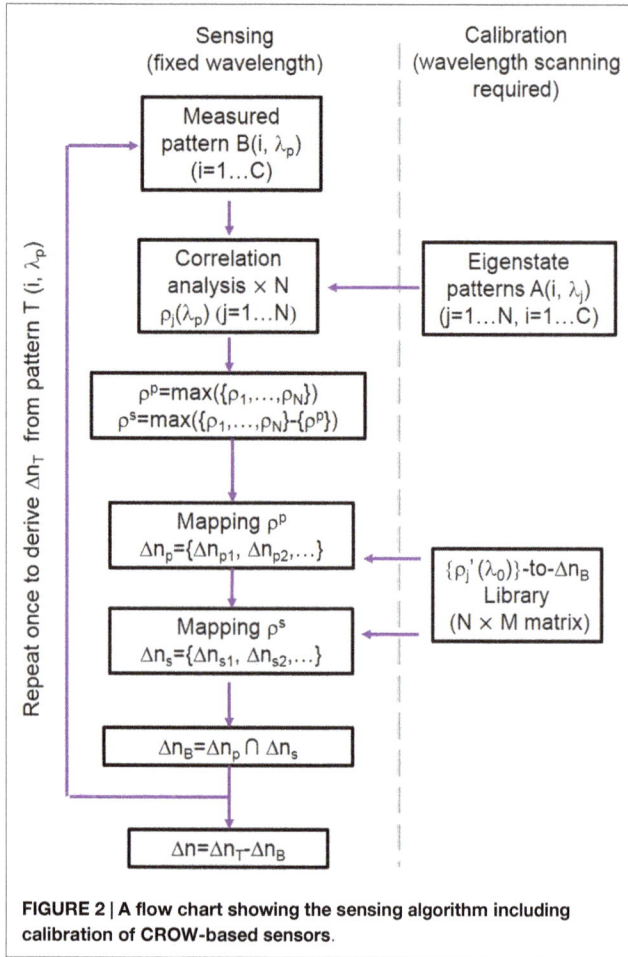

FIGURE 2 | A flow chart showing the sensing algorithm including calibration of CROW-based sensors.

(Wang et al., 2014). We thus obtain a unique equivalent refractive index change for the buffer solution, Δn_B, which is only due to the offset between λ_p and λ_0. We repeat the same procedure for measuring the pattern at λ_p upon the test solution, $T(\lambda_p)$, and obtain another unique equivalent refractive index change Δn_T. Finally, we obtain $\Delta n = \Delta n_T - \Delta n_B$.

Transfer-Matrix Modeling of Imperfect CROW Sensors

We model imperfect SiN CROWs in 680 nm wavelengths using transfer-matrix method with empirical inputs (Wang et al., 2014) (see Supplementary Materials S1 and S2). We measure and accumulate statistics of the measured waveguide widths and coupling gap widths from scanning-electron microscope (SEM) characterization of our fabricated devices. We sample six waveguide widths and three coupling gap width in one coupling region, and measure a total of eighteen coupling regions in two representative eight-microring CROW devices (see Supplementary Material S3). The statistics of the waveguide widths and the coupling gap widths approximately follow two Gaussian distributions. We extract the fabricated waveguide width of 427.5 ± 1.1 nm and coupling gap spacing of 129.1 ± 1.0 nm. In the modeling, we assume that the two Gaussian distributions are independent, and we generate a set of varied waveguide widths and coupling gap spacing randomly distributed across the CROW using the Gaussian number generator in Matlab.

We study the effects of these empirical inputs on the device parameters, including the waveguide effective refractive index, n_{eff}, and the inter-cavity coupling coefficient, κ. We calculate using the numerical finite-element method (FEM) (COMSOL RF module) the n_{eff} of a SiN channel ridge waveguide for the transverse-magnetic (TM)-polarized mode, as a function of waveguide width around 427.5 nm at a fixed waveguide height of 300 nm upon a water upper-cladding. We adopt the measured material refractive index of the deposited 300 nm-thick SiN film as a function of wavelength using ellipsometry. The mean value of the calculated n_{eff} is 1.5994 ± 0.0003 at 686 nm. We choose the TM polarization mode in order to obtain a large evanescent field exposure near the waveguide top surface for better light–analyte interaction. We calculate the coupling coefficient in each directional coupling region as a function of the coupling gap spacing, assuming the waveguide width is fixed at 427.5 nm. We estimate the waveguide propagation loss upon a water upper-cladding to be relatively high at ~17 dB/cm based on our measurements. We attribute this primarily to surface-roughness-induced scattering losses from the waveguide sidewall. We apply the designed racetrack arc radius and interaction length into the modeled CROW. We find from our FEM calculations a linear relationship between Δn and the resulting effective refractive index change Δn_{eff}, which we apply to our transfer-matrix modeling (see Supplementary Materials S1 and S2).

Device Fabrication

We fabricate the CROW devices in a 4″ silicon wafer. The silicon wafer is first grown with a ~2 μm-thick thermal oxide. We grow nitrogen-rich SiN by plasma-enhanced chemical vapor deposition (PECVD) ($SiH_4:NH_3 = 25:40$ (sccm), 300°C, 13.56 MHz). The thickness of SiN layer is ~300 nm. We fabricate the CROW device pattern by electron-beam lithography (JEOL JBX-6300FS) using a positive electron-beam resist ZEP-520A. We transfer the device pattern to the SiN layer by inductively coupled plasma etching with C_4F_8 and SF_6 gases (STS ICP DRIE Silicon Etcher). **Figure 3A** shows the optical micrograph of the fabricated SiN eight-microring CROW device. The racetrack microring comprises two half circles with a radius of 20 μm and two straight waveguides with an interaction length (L_c) of 4 μm. We design the waveguide width to be 450 nm and the coupling gap spacing to be 100 nm. **Figure 3B** shows a zoom-in-view optical microscope image of the CROW. **Figure 3C** shows a SEM picture of the coupling region.

We fabricate a microfluidic chamber on a polydimethylsiloxane (PDMS) layer. We pattern a SU8 film by contact photolithography as a mold in order to form the PDMS microfluidic channel by imprinting. The designed dimension of microfluidic channel is 8 mm × 2 mm × 50 μm (length, width, and height). We use a puncher to make two holes, each with a diameter of 1 mm, as an inlet and outlet for solution delivery. The diced silicon chip and the PDMS microfluidic layer are treated with oxygen plasma and directly bonded, with the microfluidic channel encompassing the CROW sensor. The bonded PDMS–SiN interface is stable enough for repeating the sensing experiments for many times under a relatively high fluidic pump pressure. **Figure 3D** schematically shows the cross-sectional view of the optofluidic chip.

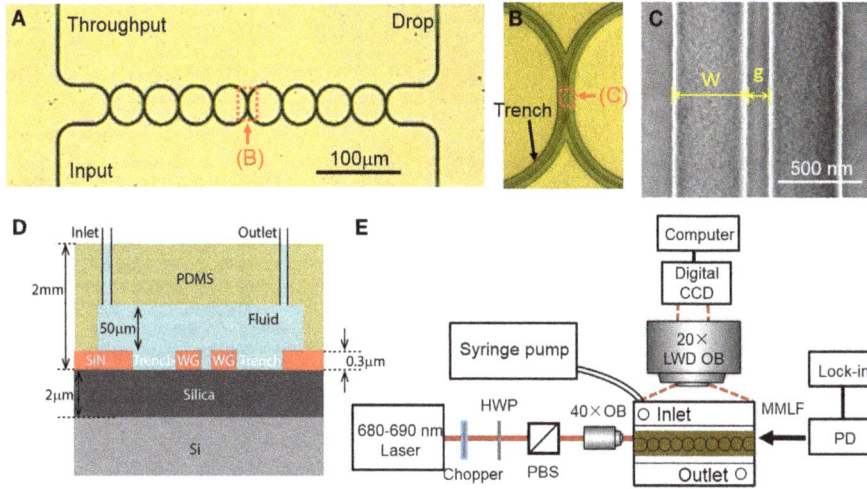

FIGURE 3 | (A) Optical micrograph of the fabricated eight-microring CROW. **(B)** Zoom-in-view picture of the CROW. **(C)** Scanning-electron microscope image of an inter-cavity coupling region of the CROW. **(D)** A cross-sectional view of the SiN chip integrated with a microfluidic channel. **(E)** Schematic of the experimental setup. HWP, half-wave plate; PBS, polarizing beam splitter; LWD OB, long-working-distance objective lens; OB, objective lens; PD, photodetector; MMLF, multimode lensed fiber.

Experimental Method

Figure 3E schematically shows the experimental setup. The wavelength-tunable laser light in the 680 nm wavelengths is end-fired into a tapered 3 μm-wide SiN waveguide through an objective lens (NA = 0.65). The laser power before coupling into the chip is ~2 mW. The polarization is controlled by a half-wave plate before a polarizing beam splitter. The output light from the throughput- or drop-port is collected using a multimode lensed fiber to a silicon power meter and a lock-in amplifier.

For elastic-light-scattering pattern imaging from the top view, we use a long-working-distance microscope objective lens (20× Mitutoyo Plan Apo, NA = 0.42) and a CCD camera (Diagnostic Instruments, Inc., RT3) with 1600 × 1200 pixels (7.4 μm-sized pixels). The camera has an effective differential cooling of −43°C and an 8-bit analog-to-digital conversion in data readout. We fix the exposure time as 60 ms and the gain of ~1. For background subtraction, we set the probe wavelength in between the CROW transmission bands in order to obtain a background image.

In order to acquire the library of calibrated correlation coefficients, we scan the laser wavelength in steps of 0.02 nm over ~2 free spectral ranges (FSRs) of the CROW sensor. We record at each wavelength eight successive images over a time period of 4 s (at 2 frames/s). We take average of these successive images in order to reduce the systematic equipment noise contribution. In the sensing tests, we inject the buffer and test solutions, and start recording the images after the scattering pattern is stabilized upon an essentially static fluidic medium. We record over 50 successive images during a time period of 25 s at a fixed probe wavelength.

In order to calibrate the spectral sensitivity of the CROW, we prepare NaCl solutions with mass concentrations from 1 to 5% (in steps of 1%) and test the transmission band spectral shifts upon a Δn. Between each measurement, we rinse the chip by injecting deionized (DI) water using a fluidic pump. We obtain the resonance spectral shifts by fitting the throughput-transmission spectra with a sum of multiple inverted Lorentzian lineshapes, each centered at the resonance (eigenstate) wavelength. The overall transmission band shift is taken as the average value of the spectral shifts of all the eigenstates.

Results

Modeling Results

Figure 4 shows the modeling results for $N = C$ (see Supplementary Material S4 for modeling results corresponding to the case $N < C$). **Figure 4A** schematically shows an imperfect SiN CROW with varied waveguide width and coupling gap width of each microring. Inset shows the numerically calculated waveguide mode-field-amplitude profile in the TM mode at 686 nm wavelength. **Figure 4B** shows the modeled throughput- and drop-transmission spectra of an imperfect eight-microring CROW. We define the CROW transmission bandwidth, $\Delta\lambda_{BW}$, as the spectral range between the first and last discernable eigenstates within the transmission band. **Figure 4C** shows the modeled pixelized patterns at the eight eigenstates. **Figure 4D** shows the calculated library $\left\{\rho_j'\left(\lambda_0\right)\right\}$ as a function of Δn, with $\Delta n_d = 2.523 \times 10^{-2}$ and $\Delta n_i = 3.6 \times 10^{-4}$ RIU. **Figure 4E** shows the calculated differential correlation coefficients per unit Δn, given as $|d\left(\rho_j'\left(\lambda_0\right)\right)/d\left(\Delta n\right)|$.

We define the CROW sensitivity (in units of RIU^{-1}) at an arbitrary λ_p within the transmission band as the larger $|d\left(\rho_j'\left(\lambda_p\right)\right)/d\left(\Delta n\right)|$ of the ρ^p and ρ^s. **Figure 4F** shows the modeled sensitivity as a non-linear function of λ_p. The sensitivity in the transmission band spans a range from ~73 to ~1440 RIU^{-1}, with an average sensitivity of ~553 ± 290 RIU^{-1}. We quantify the non-uniformity of the sensitivity by the ratio of SD value to average sensitivity value. A lower ratio value suggests a more uniform sensitivity. The extracted non-uniformity ratio from

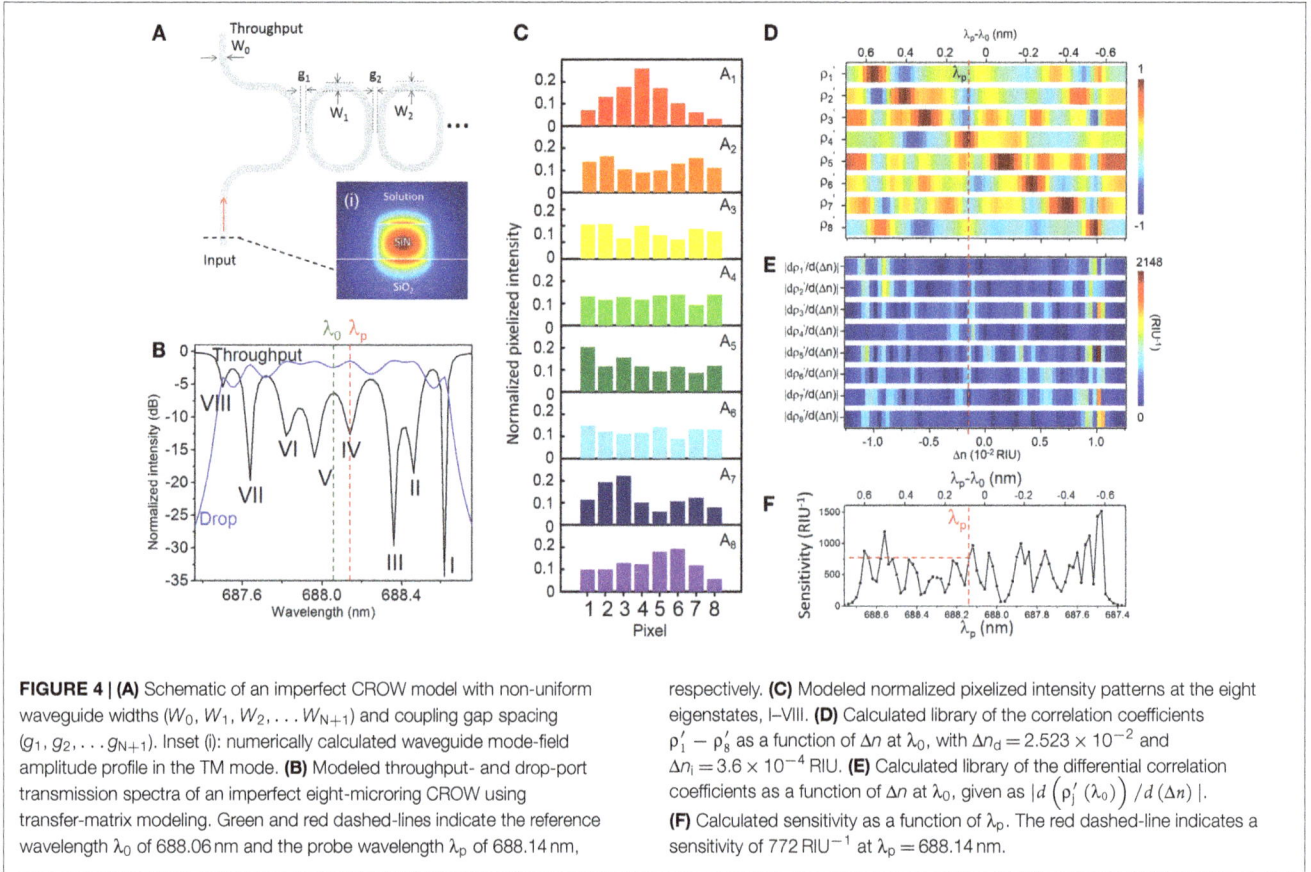

FIGURE 4 | (A) Schematic of an imperfect CROW model with non-uniform waveguide widths (W_0, W_1, W_2, ... W_{N+1}) and coupling gap spacing (g_1, g_2, ... g_{N+1}). Inset (i): numerically calculated waveguide mode-field amplitude profile in the TM mode. (B) Modeled throughput- and drop-port transmission spectra of an imperfect eight-microring CROW using transfer-matrix modeling. Green and red dashed-lines indicate the reference wavelength λ_0 of 688.06 nm and the probe wavelength λ_p of 688.14 nm, respectively. (C) Modeled normalized pixelized intensity patterns at the eight eigenstates, I–VIII. (D) Calculated library of the correlation coefficients $\rho'_1 - \rho'_8$ as a function of Δn at λ_0, with $\Delta n_d = 2.523 \times 10^{-2}$ and $\Delta n_i = 3.6 \times 10^{-4}$ RIU. (E) Calculated library of the differential correlation coefficients as a function of Δn at λ_0, given as $|d\left(\rho'_j(\lambda_0)\right)/d(\Delta n)|$. (F) Calculated sensitivity as a function of λ_p. The red dashed-line indicates a sensitivity of 772 RIU^{-1} at $\lambda_p = 688.14$ nm.

Figure 4F is ~0.52. Although such a sensitivity variation is not ideal, we can obtain a practical sensitivity within a wide enough wavelength window without fine-tuning the probe wavelength. As an example, we can set a practical sensitivity of ~100 RIU^{-1} in order to sense a Δn down to 10^{-5} RIU (assuming a noise-induced uncertainty of correlation coefficients of ~$\pm 10^{-3}$). From Figure 4F, the width of the probe wavelength window with a sensitivity >100 RIU^{-1} is 1.1 nm. We consider this sufficiently wide for sensing with a practical sensitivity at an arbitrarily set probe wavelength. If a higher practical sensitivity of, say, 200 RIU^{-1} is desired, the width of the probe wavelength window with a sensitivity >200 RIU^{-1} narrows to ~1.06 nm.

Here, we arbitrarily choose λ_p at 688.14 nm near the center of the CROW transmission band (Figure 4B) in order to model the sensing test. The sensitivity at λ_p is ~772 RIU^{-1}. Figure 5 illustrates the modeled sensing results. Figure 5A shows the modeled pixelized patterns at λ_p, B(λ_p) and T(λ_p), assuming a water buffer ($n_0 = 1.331$) and an arbitrarily chosen Δn value of 2.50×10^{-3} RIU, respectively. Figure 5B shows the two sets of correlation coefficients extracted from the two modeled pixelized patterns without and with Δn. The ρ^p and ρ^s without Δn are ρ_4 and ρ_5, respectively. The ρ^p and ρ^s with Δn are ρ_5 and ρ_3, respectively. Figures 5C,D show the zoom-in view of the calculated library $\left\{\rho'_j(\lambda_0)\right\}$ as a function of Δn. Insets show the detailed mappings of ρ^p and ρ^s with the library. We extract using linear interpolation from the library $\Delta n = \Delta n_T - \Delta n_B = 2.52 \times 10^{-3}$ RIU, which

agrees with the arbitrarily chosen Δn value. We attribute the deviation of 2×10^{-5} RIU to the interpolation error. In principle, the maximum error upon the sampling interval in the library is given by $\pm \Delta n_i/2$, which is ~1.8×10^{-4} RIU given the assumed Δn_i value.

Calibrating the CROW Sensor in a Buffer Solution

Figure 6 summarizes the characterization results upon a buffer solution (DI water). Figure 6A shows the measured TM-polarized transmission spectra with DI water upper-cladding. The measured FSR of ~1.80 nm is consistent with the microring circumference. The CROW exhibits an inhomogeneously broadened transmission band, with a $\Delta\lambda_{BW}$ of ~1.10 nm. We discern eight eigenstates within each transmission band (labeled by I to VIII for the first transmission band, and I′–VIII′ for the second transmission band in Figure 6A).

Figure 6B shows the measured elastic-light-scattering images at eigenstates I–VIII. We observe a non-uniform scattering image profile across each microring. We attribute this to the extra modulation of the surface roughness and local defects to the intrinsic mode-field-intensity distributions. We notice an obvious "local hotspot" in the coupling region between microring 3 and microring 4 in all the light-scattering images. We attribute that to the larger surface roughness localized in the coupling region between microring 3 and microring 4. We integrate within a certain window the elastic-light-scattering intensity of each microring to form a single pixel. The window excludes the coupling region in order to avoid scattering-induced crosstalks between

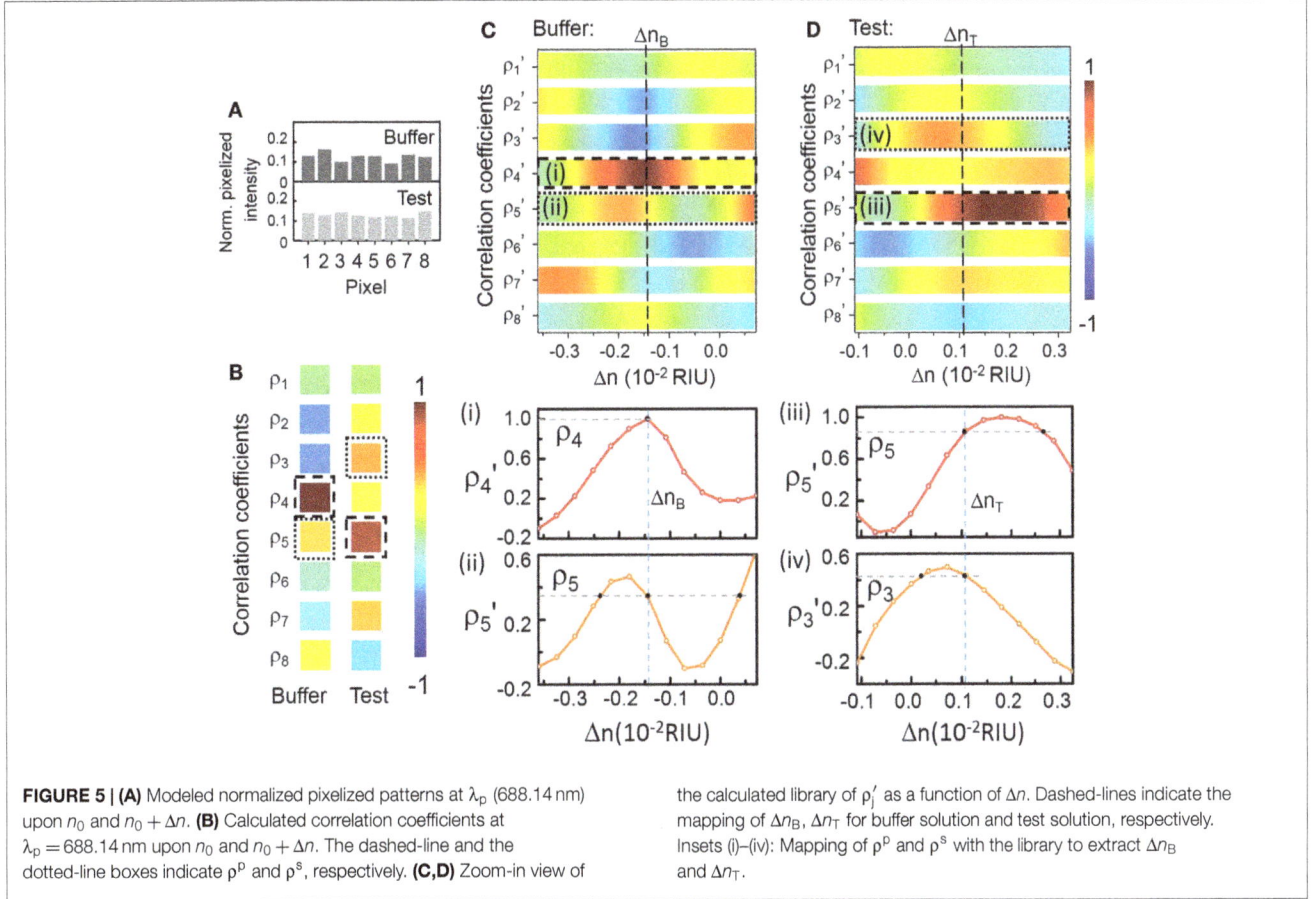

FIGURE 5 | **(A)** Modeled normalized pixelized patterns at λ_p (688.14 nm) upon n_0 and $n_0 + \Delta n$. **(B)** Calculated correlation coefficients at $\lambda_p = 688.14$ nm upon n_0 and $n_0 + \Delta n$. The dashed-line and the dotted-line boxes indicate ρ^p and ρ^s, respectively. **(C,D)** Zoom-in view of the calculated library of ρ_i' as a function of Δn. Dashed-lines indicate the mapping of Δn_B, Δn_T for buffer solution and test solution, respectively. Insets (i)–(iv): Mapping of ρ^p and ρ^s with the library to extract Δn_B and Δn_T.

the coupled waveguides and local hotspots. Here we normalize the patterns with the estimated contributions of the surface-roughness-induced scattering as a step for pattern correction (see Supplementary Material S5). **Figure 6C** shows the corrected pixelized mode-field-intensity patterns at the eight eigenstates. We use the corrected pixelized patterns for sensing.

Figure 6D shows the measured library of the calibrated correlation coefficients as a function of Δn. Here, we calibrate the sensor by scanning the laser wavelength over $\pm \Delta \lambda$ ($\Delta \lambda = 0.7$ nm) about the center of the CROW transmission band spanning a FSR upon a fixed buffer solution (DI water), with a minimum wavelength step of 0.02 nm. This interval corresponds to a Δn_i of $\sim 3.5 \times 10^{-4}$ RIU, based on the calibrated linear spectral sensitivity of ~ 57.30 nm/RIU of the CROW sensor (see Supplementary Material S6). We also convert $\Delta \lambda$ back to Δn using the calibrated linear spectral sensitivity. The corresponding range of Δn_d is $\sim \pm 1.2 \times 10^{-2}$ RIU.

Figure 6E shows the calculated $|d\rho_j'/d(\Delta n)|$ as a function of Δn. **Figure 6F** shows the calculated sensitivity as a function of λ_p over the $\lambda_0 \pm \Delta \lambda$ range. The calculated sensitivity value shows highly non-uniform profiles. The sensitivity ranges from ~ 15 to ~ 1420 RIU^{-1}, with an average value of $\sim 281 \pm 271$ RIU^{-1}. The extracted non-uniformity ratio from **Figure 6F** is ~ 0.96. The width of the probe wavelength window with a sensitivity > 100 RIU^{-1} is 0.88 nm. Whereas, the width of the probe wavelength window with a sensitivity > 200 RIU^{-1} narrows to ~ 0.48 nm, which is

still relatively tolerant to set a probe wavelength. In conventional microcavity-based sensing methods, the sensitivity is only applicable within the high-Q transmission band (~ 0.1 nm in De Vos et al., 2007), which is generally much narrower than our probe wavelength window.

We define the NEDL at λ_p as the uncertainty of extracted Δn. We repeat the extraction of Δn values based on ρ^p and ρ^s at each λ_p for eight times and calculate the SD of the eight extracted Δn values. **Figure 6G** shows the extracted NEDL values as a function of λ_p, which shows a high dependence on the choice of λ_p. The NEDL values range from $\sim 2 \times 10^{-8}$ to $\sim 1 \times 10^{-4}$ RIU. We observe particularly low NEDL values ($\sim 10^{-8}$ RIU) at λ_p aligning with the eigenstate wavelengths. We attribute the low NEDL at each eigenstate to the particularly low uncertainty of ρ^p ($\sim \pm 10^{-6} - 10^{-4}$) close to 1 at each eigenstate. Upon eight repeated tests at a fixed probe wavelength at each eigenstate, the measured pixelized patterns only slightly deviate from the calibrated eigenstate distributions due to the low noise in the cooled silicon CCD camera and the low thermo-optic coefficient of SiN. The low uncertainties of ρ^p at each eigenstate are converted into particularly low NEDL values.

In order to quantify the sensing resolution, here we define the resolution of the CROW sensor as the lowest refractive index change that can be sensed reliably and repeatedly. In practice, there are two main limiting factors to the resolution. One is the interpolation error in extracting Δn. The other is

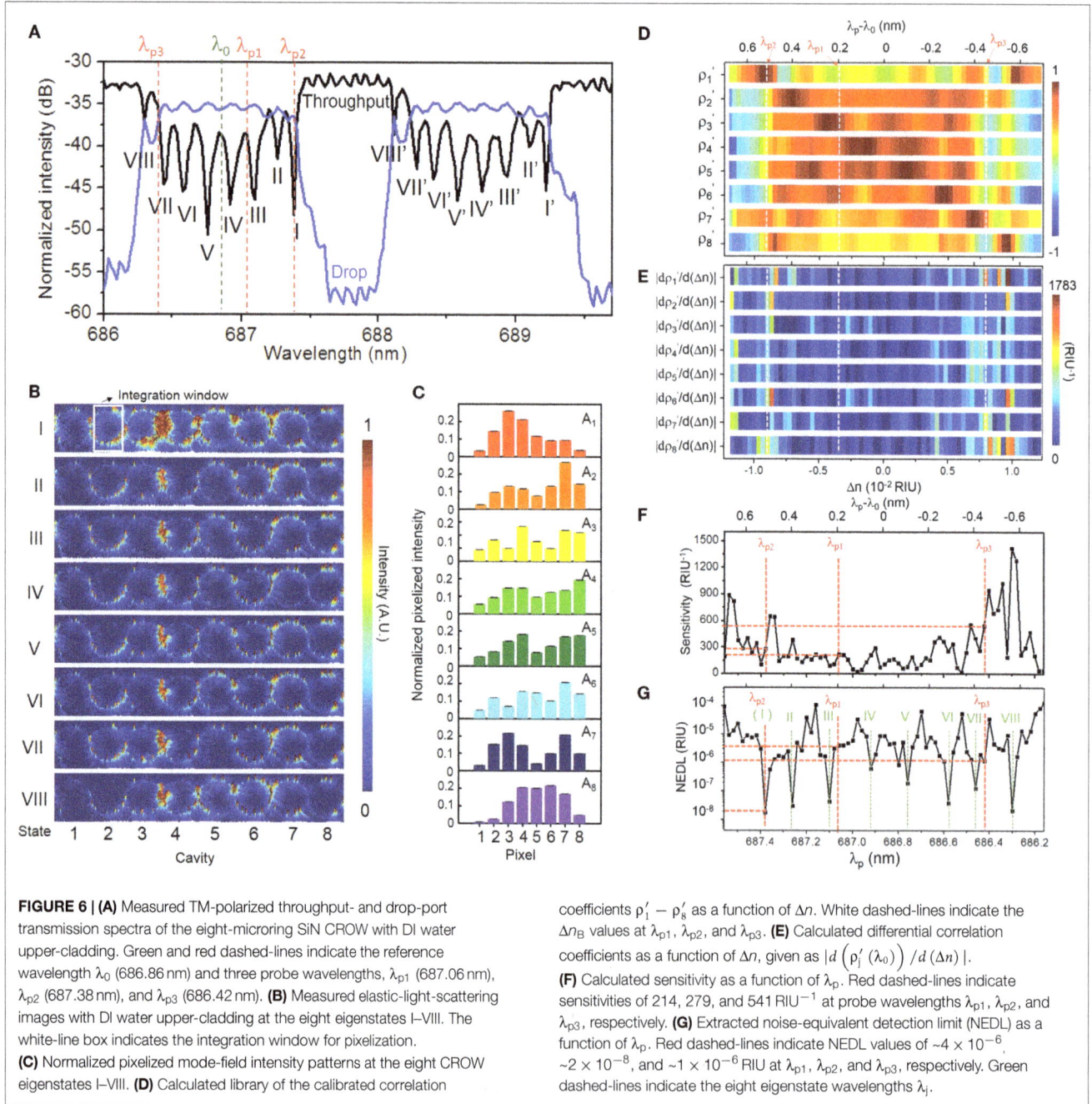

FIGURE 6 | (A) Measured TM-polarized throughput- and drop-port transmission spectra of the eight-microring SiN CROW with DI water upper-cladding. Green and red dashed-lines indicate the reference wavelength λ_0 (686.86 nm) and three probe wavelengths, λ_{p1} (687.06 nm), λ_{p2} (687.38 nm), and λ_{p3} (686.42 nm). **(B)** Measured elastic-light-scattering images with DI water upper-cladding at the eight eigenstates I–VIII. The white-line box indicates the integration window for pixelization. **(C)** Normalized pixelized mode-field intensity patterns at the eight CROW eigenstates I–VIII. **(D)** Calculated library of the calibrated correlation coefficients $\rho'_1 - \rho'_8$ as a function of Δn. White dashed-lines indicate the Δn_B values at λ_{p1}, λ_{p2}, and λ_{p3}. **(E)** Calculated differential correlation coefficients as a function of Δn, given as $|d\left(\rho'_j(\lambda_0)\right)/d(\Delta n)|$. **(F)** Calculated sensitivity as a function of λ_p. Red dashed-lines indicate sensitivities of 214, 279, and 541 RIU^{-1} at probe wavelengths λ_{p1}, λ_{p2}, and λ_{p3}, respectively. **(G)** Extracted noise-equivalent detection limit (NEDL) as a function of λ_p. Red dashed-lines indicate NEDL values of ~4×10^{-6}, ~2×10^{-8}, and ~1×10^{-6} RIU at λ_{p1}, λ_{p2}, and λ_{p3}, respectively. Green dashed-lines indicate the eight eigenstate wavelengths λ_j.

the NEDL taking into account all the noise sources that our correlation approach is not tolerant to. Therefore, given a calibration interval of Δn_i (3.5×10^{-4} RIU), the worst resolution is ~1.8×10^{-4} RIU given $\pm \Delta n_i/2$ (1.8×10^{-4} RIU) and the NEDL (~$1.8 \times 10^{-8} - 1.0 \times 10^{-4}$ RIU in **Figure 6G**). The interpolation-error-limited resolution ($\pm \Delta n_i/2$) suggests that a Δn below $\Delta n_i/2$ may not be tested reliably or repeatedly. The resolution can be improved by adopting a finer Δn_i.

We also calibrate the CROW sensor in the adjacent transmission band (see Supplementary Material S7). The pixelized mode-field intensity patterns at eigenstates I′–VIII′ show a high similarity with the corresponding patterns at eigenstates I–VIII, respectively. The extracted sensitivity and NEDL range are both close to the calibrated results of the first transmission band.

Blind Sensing Test Results

We implement blind sensing tests at three different probe wavelengths (λ_{p1}, λ_{p2}, and λ_{p3}) within the CROW transmission band. We prepare one buffer solution (DI water) and three NaCl solutions, X, Y, and Z, with different mass concentration values unknown to the researcher conducting the sensing tests. We study the images upon the buffer solution at the initial stage and upon rinsing after each sensing test. We confirm that the pixelized pattern returns to the baseline pattern (see Supplementary Material S8). **Table 1** summarizes the experimental sensing results.

TABLE 1 | Sensing results at the three probe wavelengths upon the buffer solution and the three test solutions.

λ_p	Solution	ρ^p	ρ^s	Δn_B or Δn_T ($\times 10^{-3}$ RIU)	Δn ($\times 10^{-3}$ RIU)	Sensed concentration (%)
λ_{p1} (687.06 nm)	Buffer (DI water)	ρ_3 (0.927 ± 0.003)	ρ_6 (0.866 ± 0.002)	~−3.54 ± 0.02	–	–
	X (NaCl)	ρ_6 (0.924 ± 0.003)	ρ_2 (0.838 ± 0.003)	~4.21 ± 0.01	~7.75 ± 0.02	~4.35 ± 0.01
	Y (NaCl)	ρ_4 (0.946 ± 0.002)	ρ_5 (0.854 ± 0.003)	~−2.46 ± 0.03	~1.08 ± 0.04	~0.61 ± 0.03
	Z (NaCl)	ρ_3 (0.903 ± 0.002)	ρ_6 (0.854 ± 0.008)	~−3.41 ± 0.01	~0.13 ± 0.03	~0.073 ± 0.014
λ_{p2} (687.38 nm)	Buffer (DI water)	ρ_1 (0.99996 ± 0.00003)	ρ_7 (0.622 ± 0.001)	~−9.1800 ± 0.0001	–	–
	X (NaCl)	ρ_4 (0.941 ± 0.007)	ρ_5 (0.875 ± 0.005)	~−0.79 ± 0.02	~8.39 ± 0.02	~4.70 ± 0.01
	Y (NaCl)	ρ_2 (0.841 ± 0.012)	ρ_6 (0.807 ± 0.012)	~−8.16 ± 0.02	~1.03 ± 0.02	~0.58 ± 0.01
	Z (NaCl)	ρ_1 (0.989 ± 0.001)	ρ_7 (0.622 ± 0.003)	~−9.14 ± 0.01	~0.05 ± 0.01	~0.03 ± 0.01
λ_{p3} (686.42 nm)	Buffer (DI water)	ρ_7 (0.851 ± 0.001)	ρ_2 (0.485 ± 0.002)	~7.775 ± 0.003	–	–
	X (NaCl)	ρ_1 (0.205 ± 0.015)	ρ_7 (−0.013 ± 0.014)	–	–	–
	Y (NaCl)	ρ_2 (0.519 ± 0.006)	ρ_8 (0.305 ± 0.009)	~8.81 ± 0.01	~1.04 ± 0.01	~0.58 ± 0.01
	Z (NaCl)	ρ_7 (0.783 ± 0.005)	ρ_2 (0.477 ± 0.005)	~7.90 ± 0.01	~0.13 ± 0.01	~0.069 ± 0.005

Prepared concentration values: X: (4.5 ± 0.1)%, Y: (0.60 ± 0.02)%, Z: (0.070 ± 0.002)%.

Sensing at an Arbitrarily Set Probe Wavelength λ_{p1}

Figure 7 shows the sensing results at an arbitrarily set probe wavelength λ_{p1} (687.06 nm) near the center of the CROW transmission band. The sensitivity at λ_{p1} is ~214 RIU^{-1} (see **Figure 6F**). The NEDL at λ_{p1} is ~4 × 10^{-6} RIU (see **Figure 6G**). **Figure 7A** shows the measured elastic-light-scattering images of the CROW upon the buffer solution and the three test solutions at λ_{p1}. **Figure 7B** shows the corresponding pixelized patterns. **Figure 7C** shows the corresponding calculated correlation coefficients. **Figures 7D–G** show the mapping of ρ^p and ρ^s in the buffer solution and the three test solutions with the library. Insets (i)–(viii) show the mapping ρ^p and ρ^s to the corresponding Δn_B or Δn_T using linear interpolations in between Δn_i.

We acquire for solution X a Δn_X of ~(7.75 ± 0.02) × 10^{-3} RIU and for solution Y a Δn_Y of ~(1.08 ± 0.04) × 10^{-3} RIU, both corresponding to a relatively large Δn but still within Δn_d. We acquire for solution Z (**Figure 7G**) a Δn_Z of ~(1.3 ± 0.3) × 10^{-4} RIU. For all three solutions, we convert from the measured Δn values the sensed concentration values (see **Table 1**), which show a good agreement with the prepared values.

Sensing at λ_{p2} Aligned with Eigenstate I

Figure 8 shows the sensing results at a specifically chosen probe wavelength λ_{p2} (687.38 nm) aligned with eigenstate I. The sensitivity at λ_{p2} is ~279 RIU^{-1} (see **Figure 6F**). The NEDL at λ_{p2} is ~2 × 10^{-8} RIU (see **Figure 6G**), which is much lower compared with that at λ_{p2}. **Figure 8A** shows the measured elastic-light-scattering images upon the buffer solution and the three test solutions. **Figure 8B** shows the corresponding pixelized patterns. **Figure 8C** shows the corresponding calculated correlation coefficients. **Figures 8D–G** show the mapping of ρ^p and ρ^s values in the buffer solution and the three test solutions with the library (see Supplementary Material S9 for detailed mappings).

We acquire for solution X a Δn_X of ~(8.40 ± 0.02) × 10^{-3} RIU and for solution Y a Δn_Y of ~(1.03 ± 0.02) × 10^{-3} RIU. Both sensing results agree with the prepared concentrations of solutions X and Y. For solution Z (**Figure 8G**), we acquire a Δn_Z of ~(0.5 ± 0.1) × 10^{-4} RIU, corresponding to a mass concentration of ~(0.03 ± 0.01)%. This, however, shows a significant deviation from the prepared concentration [~(0.070 ± 0.002)%]. We

attribute this deviation to a not sufficiently fine calibration of the library and the error from linear interpolation. The calibrated response of ρ^p around the eigenstate is in the proximity to the maximum (unity). The limited sampling resolution of Δn_i may not be sufficient to describe the response around an extremum.

Sensing at λ_{p3} Near Eigenstate VII

Figure 9 shows the sensing results at another specifically chosen probe wavelength λ_{p3} (686.42 nm). We specifically set λ_{p3} at the blue-edge of the transmission band near eigenstate VII. The sensitivity at λ_{p3} is ~541 RIU^{-1} (see **Figure 6F**). The NEDL at λ_{p3} is ~1 × 10^{-6} RIU (see **Figure 6G**). We consider λ_{p3} as a near optimized choice with a relatively high sensitivity and a low NEDL. **Figure 9A** shows the measured elastic-light-scattering images upon the buffer solution and the three test solutions. **Figure 9B** shows the corresponding pixelized patterns. **Figure 9C** shows the corresponding calculated correlation coefficients. **Figures 9D–F** show the mapping of ρ^p and ρ^s values with the library (see Supplementary Material S9 for detailed mappings).

For solution X, however, we observe an almost dark scattering pattern, which suggests that λ_{p3} upon solution X is relatively shifted out of the transmission band. Both the extracted ρ^p and ρ^s values out of $\rho_j(\lambda_{p3})$ upon solution X are particularly low. By mapping the extracted $\rho_j(\lambda_{p3})$ values with the library, we find no match to indicate the corresponding Δn_x. Therefore, in the case that there is a chance to measure a large Δn near Δn_d (in the order of 10^{-2} ~ 10^{-3} RIU in this case), it is better to position λ_p close to the red-side of the transmission band in order to leverage the dynamic range given by $\Delta \lambda_{BW}$ in full.

For solution Y, we acquire a Δn_Y of ~(1.04 ± 0.02) × 10^{-3} RIU. For solution Z, we acquire a Δn_Z of ~(1.24 ± 0.1) × 10^{-4} RIU. Both sensing results agree with the prepared concentrations of solutions Y and Z. Compared with the sensing result of solution Z at λ_{p1}, we obtain a more accurate value of Δn_Z with a much improved uncertainty. We attribute this to a higher sensitivity and a lower NEDL at λ_{p3} than those at λ_{p1}.

Discussion

Here, we benchmark our work with other silicon- and SiN-based on-chip optical biochemical sensors that have been demonstrated

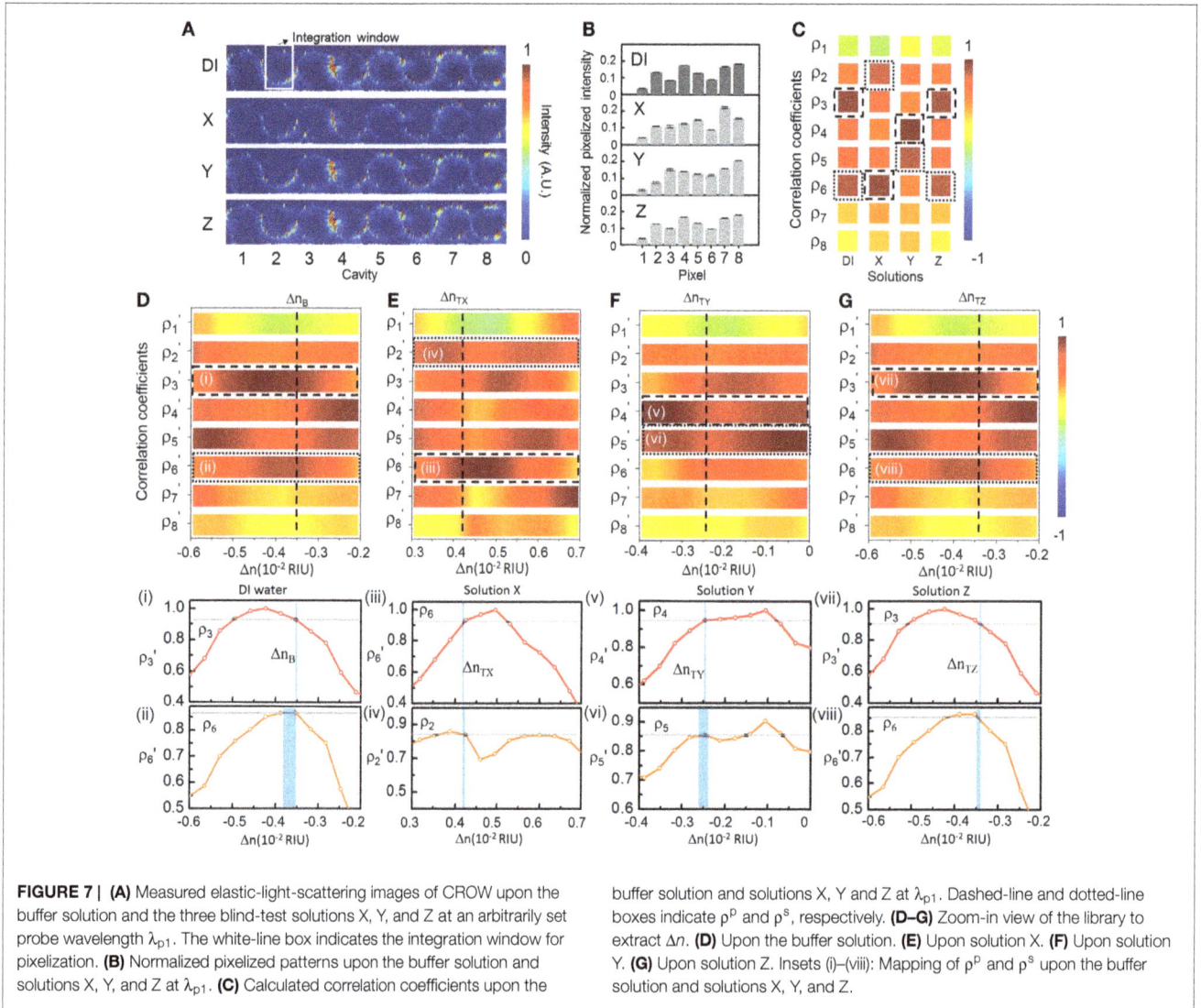

FIGURE 7 | **(A)** Measured elastic-light-scattering images of CROW upon the buffer solution and the three blind-test solutions X, Y, and Z at an arbitrarily set probe wavelength λ_{p1}. The white-line box indicates the integration window for pixelization. **(B)** Normalized pixelized patterns upon the buffer solution and solutions X, Y, and Z at λ_{p1}. **(C)** Calculated correlation coefficients upon the buffer solution and solutions X, Y and Z at λ_{p1}. Dashed-line and dotted-line boxes indicate ρ^p and ρ^s, respectively. **(D–G)** Zoom-in view of the library to extract Δn. **(D)** Upon the buffer solution. **(E)** Upon solution X. **(F)** Upon solution Y. **(G)** Upon solution Z. Insets (i)–(viii): Mapping of ρ^p and ρ^s upon the buffer solution and solutions X, Y, and Z.

in recent years, including our previous work (Wang et al., 2014), as summarized in **Table 2**. All of the work including this work have attained a detection limit of $10^{-7} \sim 10^{-4}$ RIU. Two of the microcavity-based sensors (Ghasemi et al., 2013; Doolin et al., 2015) and three of the MZI-based sensors (Duval et al., 2013; Misiakos et al., 2014; Dante et al., 2015) operate on the SiN-based platform in the visible wavelengths.

Most of the reported microcavity-based sensors in the literature (except Ghasemi et al., 2013; Doolin et al., 2015) operate in the telecommunication wavelengths (1.3/1.55 μm) and require a wavelength-tunable laser and a non-silicon photodetector. Whereas, our CROW sensor operating in the visible wavelengths only requires in principle a fixed-wavelength visible laser diode and a silicon CCD/CMOS camera after the library preparation.

In terms of the sensor calibration, the main difference between our library preparation and the conventional calibration process for a microcavity-based sensor is the recording of the pixelized patterns instead of single intensity values. A typical calibration for a conventional microcavity-based sensor [e.g., De Vos et al. (2007) and Iqbal et al. (2010)] involves scanning laser wavelength across a narrow transmission band. As an example, in the work of

De Vos et al., calibrating the spectral sensitivity of a microring sensor of Q ~ 20,000 involved measuring the microring transmission spectrum three times for each of the four given NaCl solutions with different concentrations (De Vos et al., 2007). In contrast, our library preparation involves scanning laser wavelength across the CROW transmission band, recording the pixelized patterns at each wavelength step corresponding to the refractive index interval Δn_i and deriving the corresponding correlation coefficients with the eigenstate patterns. The pattern recording and additional computation of the correlation coefficients render our library preparation more reliable and tolerant to the equipment noises that are common to all pixels compared with recording single intensity values multiple times.

A major issue requiring further developments is the significant variation of sensitivity values upon different probe wavelengths. We can modify the CROW design in order to attain a more uniform sensitivity (see Supplementary Material S10). Our modeling results suggest that an imperfect CROW with a reduced cavity size along with an enhanced inter-cavity coupling coefficient offers a more uniform sensitivity. Upon a small cavity radius $R = 10$ μm and a strong inter-cavity coupling coefficient

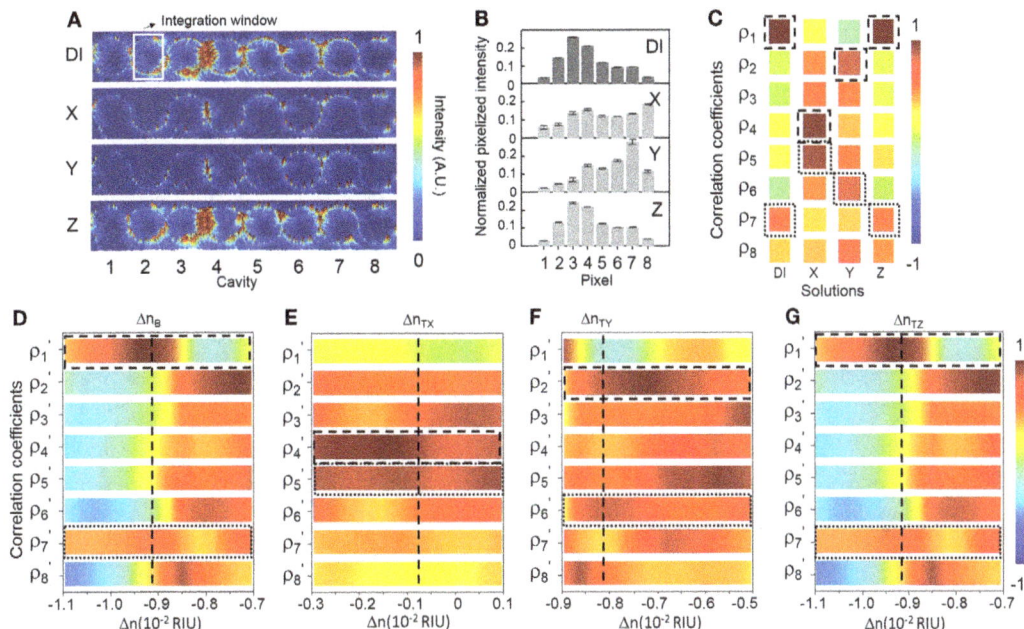

FIGURE 8 | (A) Measured elastic-light-scattering images of CROW upon the buffer solution and the three blind-test solutions X, Y, and Z at a specifically chosen probe wavelength λ_{p2} at eigenstate I. The white-line box indicates the integration window for pixelization. **(B)** Normalized pixelized patterns upon the buffer solution and solutions X, Y, and Z at λ_{p2}. **(C)** Calculated correlation coefficients upon the buffer solution and solutions X, Y, and Z at λ_{p2}. Dashed-line and dotted-line boxes indicate ρ^p and ρ^s, respectively. **(D–G)** Zoom-in view of the library to extract Δn. **(D)** Upon the buffer solution. **(E)** Upon solution X. **(F)** Upon solution Y. **(G)** Upon solution Z.

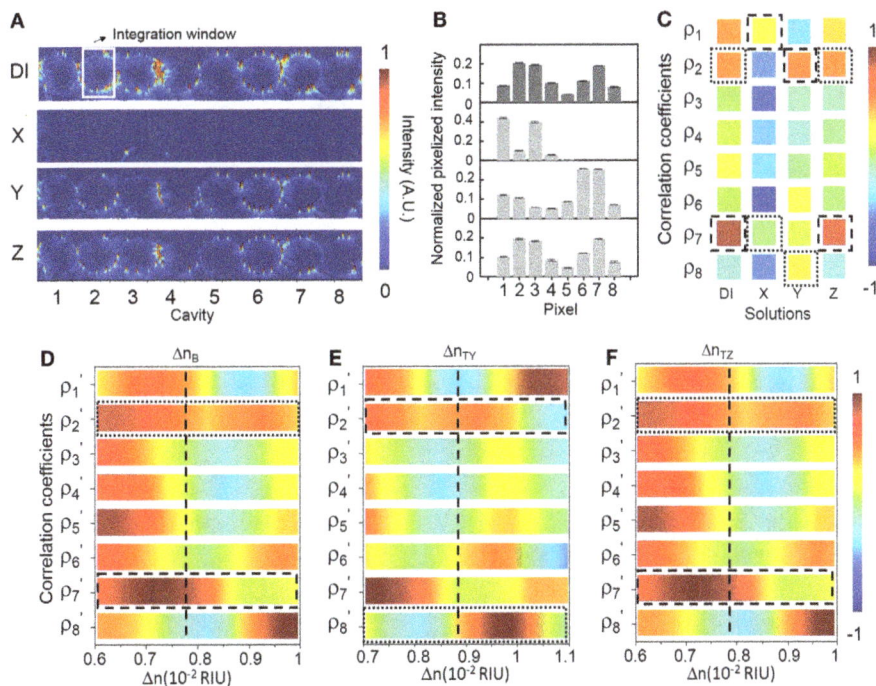

FIGURE 9 | (A) Measured elastic-light-scattering images of CROW upon the buffer solution and the three blind-test solutions X, Y, and Z at a specifically chosen probe wavelength λ_{p3} near eigenstate VII. The white-line box indicates the integration window for pixelization. **(B)** Normalized pixelized patterns upon the buffer solution and solutions X, Y and Z at λ_{p3}. **(C)** Calculated correlation coefficients upon the buffer solution and solutions X, Y, and Z at λ_{p3}. Dashed-line and dotted-line boxes indicate ρ^p and ρ^s, respectively. **(D–F)** Zoom-in view of the library to extract Δn. **(D)** Upon the buffer solution. **(E)** Upon solution Y. **(F)** Upon solution Z.

TABLE 2 | Summary of silicon- and silicon-nitride-based on-chip optical biochemical sensors.

Device config.	Reference	Material platform	Operational wavelength (nm)	Footprint (μm^2)	Q-factor	Sensitivity (RIU^{-1})	Detected Δn (RIU)	Detection limit (RIU)
MZI	Densmore et al. (2008)	SOI	~1550	~40,000	N/A	$920\,\pi$ rad	~7×10^{-4}	~1×10^{-5}
	Duval et al. (2013)	Si_3N_4	658	~10^8	N/A	$4950\,\pi$ rad	~3×10^{-4}	~2×10^{-7}
	Misiakos et al. (2014)	SiN	~600–900	~10^7	N/A	581 rad	~4×10^{-5}	~1×10^{-5}
	Dante et al. (2015)	Si_3N_4	660	~60,000	N/A	$6000\,\pi$ rad	~2×10^{-4}	~4×10^{-7}
Microdisk	Wang et al. (2013)	SOI	~1550	~3	~100	130 nm	~9×10^{-3}	~8×10^{-4}
	Doolin et al. (2015)	Si_3N_4	~770	~900	10000	200 nm	~4×10^{-4}	~10^{-6}
Microring with slot-waveguide	Barrios et al. (2007)	Si_3N_4	~1300	~20,000	1800	212 nm	~10^{-3}	~2×10^{-4}
	Claes et al. (2009)	SOI	~1550	~240	~450	298 nm	~4×10^{-3}	~4.2×10^{-5}
	Carlborg et al. (2010)	Si_3N_4	~1300	~20,000	–	248 nm	~3×10^{-4}	~5×10^{-6}
Microring	De Vos et al. (2007)	SOI	~1550	~110	20,000	70 nm	~9×10^{-4}	~10^{-5}
	Iqbal et al. (2010)	SOI	~1550	~900	43,000	163 nm	~10^{-6}	–
	Ghasemi et al. (2013)	SiN	~656	~400	–	48 nm	–	–
	Liu et al. (2014)	SOI	~1550	~1600	15000	6000 rad	~4×10^{-4}	~2.5×10^{-6}
Eight-microring CROW in the spatial domain	Wang et al. (2014)	SOI	~1550	~1716	N/A	~199	~1.5×10^{-4}	2×10^{-7} ~ 9×10^{-4}
	(This work)	SiN	~680	~14080	N/A	~281 ± 271	~1.3×10^{-4}	2×10^{-8} ~ 1×10^{-4}

$\kappa \sim 0.9$, we obtain for an imperfect eight-microring CROW a modeled sensitivity of ~$384 \pm 153\,RIU^{-1}$, with an improved non-uniformity ratio of ~0.40 compared to the modeled ratio of ~0.52 following our experimental device parameters. Assuming a practical sensitivity of ~$100\,RIU^{-1}$, the width of the modeled probe wavelength window with a sensitivity >$100\,RIU^{-1}$ is 2.2 nm, which is much improved compared to the modeled width of 1.1 nm following the experimental device parameters. If a higher practical sensitivity of $300\,RIU^{-1}$ is desired, the modeled probe wavelength window width with a sensitivity >$300\,RIU^{-1}$ is ~1.56 nm, which is still sufficiently wide for practical applications. Based on our current imperfect CROW model, we can further design the CROW with tailored non-uniform parameters to optimize the sensitivity and sensitivity variation.

In summary, we demonstrated a SiN CROW-based sensing scheme in the spatial domain in the visible wavelengths. Given a calibrated CROW sensor, this sensing scheme in principle only requires a low-power, fixed-wavelength laser source in the visible wavelengths and a silicon CCD or CMOS camera to image the elastic-light-scattering patterns in the far field. Our proof-of-concept experiment using an eight-microring CROW on the SiN-on-silica platform showed an average sensitivity of ~$281 \pm 271\,RIU^{-1}$ and a NEDL of $2 \times 10^{-8} \sim 1 \times 10^{-4}\,RIU$. Our blind sensing tests using NaCl solutions showed a detection of ~$1.26 \times 10^{-4}\,RIU$. Therefore, we have shown that such a chip-scale, microresonator-based SiN CROW sensor operating in the visible wavelengths is promising as a potentially high-performance, portable, and low-cost optical biochemical sensor for applications such as point-of-care biochemical analyses and self-monitoring of personal healthcare using smartphones.

Acknowledgments

This work is supported by grants from the Research Grants Council of the Hong Kong Special Administrative Region, China (Project No. 618010, 617612, and 617913). The authors acknowledge the HKUST Nanoelectronics Fabrication Facility for fabricating the optofluidic chip.

References

Barrios, C. A., Gylfason, K. B., Sánchez, B., Griol, A., Sohlström, H., Holgado, M., et al. (2007). Slot-waveguide biochemical sensor. *Opt. Lett.* 32, 3080–3082. doi:10.1364/ol.32.003080

Carlborg, C. F., Gylfason, K. B., Kazmierczak, A., Dortu, F., Polo, M. B., Catala, A. M., et al. (2010). A packaged optical slot-waveguide ring resonator sensor array for multiplex label-free assays in labs-on-chips. *Lab. Chip* 10, 281–290. doi:10.1039/b914183a

Ciminelli, C., Campanella, C. M., Dell'olio, F., Campanella, C. E., and Armenise, M. N. (2013). Label-free optical resonant sensors for biochemical applications. *Prog. Quant. Electron.* 37, 51–107. doi:10.1016/j.pquantelec.2013.02.001

Claes, T., Molera, J. G., De Vos, K., Schacht, E., Baets, R., and Bienstman, P. (2009). Label-free biosensing with a slot-waveguide-based ring resonator in silicon on insulator. *Photonics J. IEEE* 1, 197–204. doi:10.1109/jphot.2009.2031596

Dante, S., Duval, D., Fariña, D., González-Guerrero, A. B., and Lechuga, L. M. (2015). Linear readout of integrated interferometric biosensors using a periodic wavelength modulation. *Laser Photon. Rev.* 9, 248–255. doi:10.1002/lpor.201400216

De Vos, K., Bartolozzi, I., Schacht, E., Bienstman, P., and Baets, R. (2007). Silicon-on-insulator microring resonator for sensitive and label-free biosensing. *Opt. Express* 15, 7610–7615. doi:10.1364/oe.15.007610

Densmore, A., Xu, D.-X., Janz, S., Waldron, P., Mischki, T., Lopinski, G., et al. (2008). Spiral-path high-sensitivity silicon photonic wire molecular sensor with temperature-independent response. *Opt. Lett.* 33, 596–598. doi:10.1364/ol.33.000596

Doolin, C., Doolin, P., Lewis, B. C., and Davis, J. P. (2015). Refractometric sensing of Li salt with visible-light Si3N4 microdisk resonators. *Appl. Phys. Lett.* 106, 081104. doi:10.1063/1.4913618

Duval, D., Osmond, J., Dante, S., Dominguez, C., and Lechuga, L. M. (2013). Grating couplers integrated on Mach-Zehnder interferometric biosensors operating in the visible range. *Photonics J. IEEE* 5, 3700108. doi:10.1109/jphot.2013. 2251873

Estevez, M.-C., Alvarez, M., and Lechuga, L. M. (2012). Integrated optical devices for lab-on-a-chip biosensing applications. *Laser Photon. Rev.* 6, 463–487. doi:10. 1002/lpor.201100025

Ghasemi, F., Eftekhar, A. A., Gottfried, D. S., Song, X., Cummings, R. D., and Adibi, A. (2013). "Self-referenced silicon nitride array microring biosensor for toxin detection using glycans at visible wavelength," in *SPIE BiOS* (San Francisco, CA: International Society for Optics and Photonics).

Gorin, A., Jaouad, A., Grondin, E., Aimez, V., and Charette, P. (2008). Fabrication of silicon nitride waveguides for visible-light using PECVD: a study of the effect of plasma frequency on optical properties. *Opt. Express* 16, 13509–13516. doi:10.1364/oe.16.013509

Halir, R., Vivien, L., Le Roux, X., Xu, D.-X., and Cheben, P. (2013). Direct and sensitive phase readout for integrated waveguide sensors. *Photonics J. IEEE* 5, 6800906. doi:10.1109/jphot.2013.2276747

Iqbal, M., Gleeson, M. A., Spaugh, B., Tybor, F., Gunn, W. G., Hochberg, M., et al. (2010). Label-free biosensor arrays based on silicon ring resonators and high-speed optical scanning instrumentation. *IEEE J. Sel. Top. Quantum Electron.* 16, 654–661. doi:10.1109/jstqe.2009.2032510

Kozma, P., Hamori, A., Cottier, K., Kurunczi, S., and Horvath, R. (2009). Grating coupled interferometry for optical sensing. *Appl. Phys. B* 97, 5–8. doi:10.1007/s00340-009-3719-1

Lakshminarayanan, V., Zelek, J., and Mcbride, A. (2015). "Smartphone Science" in eye care and medicine. *Opt. Photonics News* 26, 44–51. doi:10.1364/opn.26.1. 000044

Lei, T., and Poon, A. W. (2011). Modeling of coupled-resonator optical waveguide (CROW) based refractive index sensors using pixelized spatial detection at a single wavelength. *Opt. Express* 19, 22227–22241. doi:10.1364/oe.19. 022227

Liu, J., Zhou, X., Qiao, Z., Zhang, J., Zhang, C., Xiang, T., et al. (2014). Integrated optical chemical sensor based on an SOI ring resonator using phase-interrogation. *Photonics J. IEEE* 6, 1–7. doi:10.1109/jphot.2014.2352973

Misiakos, K., Raptis, I., Makarona, E., Botsialas, A., Salapatas, A., Oikonomou, P., et al. (2014). All-silicon monolithic Mach-Zehnder interferometer as a refractive index and bio-chemical sensor. *Opt. Express* 22, 26803–26813. doi:10.1364/OE. 22.026803

Sedlmeir, F., Zeltner, R., Leuchs, G., and Schwefel, H. G. (2014). High-Q MgF^2 whispering gallery mode resonators for refractometric sensing in aqueous environment. *Opt. Express* 22, 30934–30942. doi:10.1364/oe.22.030934

Subramanian, A., Dhakal, A., Peyskens, F., Selvaraja, S., Baets, R., Neutens, P., et al. (2013). Low-loss singlemode PECVD silicon nitride photonic wire waveguides for 532–900 nm wavelength window fabricated within a CMOS pilot line. *Photonics J. IEEE* 5, 2202809. doi:10.1109/jphot.2013.2292698

Vollmer, F., Arnold, S., and Keng, D. (2008). Single virus detection from the reactive shift of a whispering-gallery mode. *Proc. Natl. Acad. Sci. U.S.A.* 105, 20701–20704. doi:10.1073/pnas.0808988106

Wang, J., Yao, Z., Lei, T., and Poon, A. W. (2014). Silicon coupled-resonator optical-waveguide-based biosensors using light-scattering pattern recognition with pixelized mode-field-intensity distributions. *Sci. Rep.* 4, 07528. doi:10.1038/srep07528

Wang, X., Guan, X., Huang, Q., Zheng, J., Shi, Y., and Dai, D. (2013). Suspended ultra-small disk resonator on silicon for optical sensing. *Opt. Lett.* 38, 5405–5408. doi:10.1364/ol.38.005405

Xu, J., Suarez, D., and Gottfried, D. S. (2007). Detection of avian influenza virus using an interferometric biosensor. *Anal. Bioanal. Chem.* 389, 1193–1199. doi:10.1007/s00216-007-1525-3

Ymeti, A., Greve, J., Lambeck, P. V., Wink, T., Van Hövell, S. W., Beumer, T. A., et al. (2007). Fast, ultrasensitive virus detection using a Young interferometer sensor. *Nano Lett.* 7, 394–397. doi:10.1021/nl062595n

Conflict of Interest Statement: The authors declare that the research was conducted in the absence of any commercial or financial relationships that could be construed as a potential conflict of interest.

2

A facile route to metal oxides/single-walled carbon nanotube macrofilm nanocomposites for energy storage

Zeyuan Cao and Bingqing Wei*

Department of Mechanical Engineering, University of Delaware, Newark, DE, USA

Nanocomposites consisting of transition-metal oxides and carbon nanomaterials with a desired size and structure are highly demanded for high-performance energy storage devices. Here, a facile two-step and cost-efficient approach relying on directly thermal treatment of chemical vapor deposition products is developed as a general synthetic method to prepare a family of metal oxides [M_xO_y (M = Fe, Co, Ni)]/single-walled carbon nanotube (SWNT) macrofilm nanocomposites. The M_xO_y nanoparticles obtained are of 3–17 nm in diameter and homogeneously anchor on the free-standing SWNT macrofilms. NiO/SWNT also exhibits a high-specific capacitance of 400 F g^{-1} and fast charge-transfer Faradaic redox reactions to achieve asymmetric supercapacitors with a high power and energy density. All M_xO_y/SWNT nanocomposites could deliver a high capacity beyond 1000 mAh g^{-1} and show excellent cycling stability for lithium-ion batteries. The impressive results demonstrate the promise for energy storage devices and the general approach may pave the way to synthesize other functional nanocomposites.

Keywords: carbon nanotube macrofilms, metal oxides, nanocomposites, asymmetric supercapacitors, lithium-ion batteries

Edited by:
Mihri Ozkan,
University of California Riverside, USA

Reviewed by:
Zhengjun Zhang,
Tsinghua University, China
Dexian Ye,
Virginia Commonwealth University,
USA

Correspondence:
Bingqing Wei,
Department of Mechanical
Engineering, University of Delaware,
Newark, DE 19716, USA
weib@udel.edu

Introduction

Nanocomposites, consisting of a set of first-row transition-metal oxides such as Fe_2O_3 (Wang et al., 2012; Cao and Wei, 2013a), Fe_3O_4 (Su et al., 2011; Cao and Wei, 2013b), CoO (Guo et al., 2013), Co_3O_4 (Wu et al., 2010; Liang et al., 2012), NiO (Lee et al., 2005; Zhou et al., 2012), V_2O_5 (Cao and Wei, 2013c), and MnO_2 (Yu et al., 2011; Qin et al., 2013) in a variety of structures at nanoscale, and carbon nanomaterials such as carbon nanotubes (CNTs) and graphene, have been extensively explored and investigated for electrochemical applications, in particular, energy storage (Cao and Wei, 2013d; Xu et al., 2013). One common strategy in designing such materials is coupling size-controllable nanoparticles on the supporting carbon scaffolds to tailor surface area, conductivity, and charge-transfer interaction in order to obtain the desired characteristics such as a high energy and power density as well as a long-cycle stability (Poizot et al., 2000; Grugeon et al., 2001; Li et al., 2007). However, current synthetic methods based on such a strategy vary with specific cases and are complicated with high cost, which impedes the practical production and application. Hence, efficient, general, and robust synthesis methods are sought to address these challenges. In our previous work, we have successfully demonstrated a facile heat treatment process to convert single-walled CNT (SWNT) macrofilms, containing Fe catalysts and obtained via chemical vapor deposition (CVD), to alpha-Fe_2O_3/SWNT nanocomposites (Cao and Wei, 2013a). The resulting Fe_2O_3/SWNT nanocomposite films exhibited good electric conductivity and strain accommodation for lithium-ion batteries with a high-specific capacity (Cao and Wei, 2013a).

Herein, following the analogous formation mechanism of the Fe_2O_3/SWNT nanocomposite, we generalize this facile route to the family of M_xO_y (M = Fe, Co, Ni) with SWNT macrofilm nanocomposites for electrochemical applications. The size of M_xO_y nanoparticles is successfully controlled within 3–17 nm and the desired morphology of electrically interconnected networks that associate metal oxides with SWNT is formed. We particularly discussed an optimized asymmetric supercapacitor that consists of NiO/SWNT as the positive electrode and a pure CNT macrofilm as the negative electrode, which has shown a high specific capacitance of 130 F g^{-1} after a long-term discharge–charge duration of 5000 cycles. In addition, both of the NiO/SWNT and Co_3O_4/SWNT nanocomposites exhibit an exceptionally high capacity beyond 1000 mAh g^{-1} in Li-ion batteries.

The general approach is composed of two steps to synthesize M_xO_y/SWNT macrofilms as illustrated in **Scheme 1**. First, bis(cyclopentadienyl)metal (MCp2, M = Fe, Co, Ni) and sulfur (with an atomic ratio of M/S = 10) were mixed as precursors to produce SWNT macrofilms by CVD. Taking NiO as an example, nickelocene (NiCp2) is decomposed during the CVD reaction to Ni ions as catalyst sources and two parallel Cp rings as carbon source for SWNT formation. Sulfur is a predominant additive to control the wall number of CNTs so that to promote SWNT growth as well as to enhance the growth rate. The deposition lasted for half an hour at 1100°C and the as-prepared SWNT macrofilms were not achieved until cooled down to room temperature followed by the second step, calcination in air. The heat treatment resulted in the oxidation of the metal catalysts to metal oxides and also the growth of the oxides. The detailed synthetic procedures are described in the following section.

Materials and Methods

Synthesis of M_xO_y (M = Fe, Co, Ni)/SWNT Macrofilm Nanocomposites

Single-walled carbon nanotube (SWNT) macrofilms were synthesized by the modified floating CVD method reported in the previously published work (Zhu and Wei, 2007). In brief, a mixture of bis(cyclopentadienyl)metal (MCp2, M = Fe, Co, Ni) and sulfur (atomic ratio M:S = 1:10, both from Sigma Aldrich) as

precursors in a crucible boat was placed at the inlet of a ceramic tube of an electric furnace. The furnace was pre-heated to 1100–1150°C with an argon gas flow of 500 mL min^{-1}. Meanwhile, the MCp2 with a relatively low sublimation point (~100°C) was introduced by the Ar carrier gas flow into the central reaction zone of the furnace. During the deposition, a mixed gas flow of Ar (1500 mL min^{-1}) and H_2 (150 mL min^{-1}) was delivered and the MCp2 then started pyrolysis at such a high temperature to generate carbon source and metal catalysts for CNT growth. It was widely recognized that sulfur is indispensable in controlling the wall number of CNTs. Sulfur acts as an additive to promote SWNT growth as well as to enhance the growth rate of SWNTs. After a 30 min reaction, large-area SWNT macrofilms [M-CNT (M = Fe, Co, Ni)] containing metal catalysts with uniformity (randomly homogeneous entanglement of SWNT bundles) and conformability (distribution of the nanotube diameter ranges from 0.8 to 1.3 nm) could be collected from the entire furnace tube. It is the result of a reverse gas flow generated from the central hot zone to the cold ends caused by the pressure difference (Cao and Wei, 2013d). The thickness of the films can be controlled by mainly adjusting the amount of the precursor mixtures and reaction time. The M-CNT macrofilms were peeled off from the tube walls when the furnace was cooled down. Subsequently, the direct thermal treatment was performed to the M-CNT by calcination in air for 30 min below the critical temperature that CNT began to burn out (about 450°C). During the heat treatment, metal catalysts were oxidized to metal oxide nanoparticles with fine sizes. Three samples of Co_3O_4/SWNT were prepared at different annealing temperatures: Co-CNT (100°C), Co_3O_4/SWNT (200°C), and Co_3O_4/SWNT (400°C) because of the more complicated phases than NiO/SWNT, which was obtained by annealing only at 400°C.

Structural Characterization

Morphological and structural characterizations were performed using transmission electron microscope (TEM) and high-resolution TEM (HRTEM) on JEOL JEM-2010F field emission TEM operating at 200 kV. The crystalline phase of the samples was determined using X-ray diffraction (XRD) collected on a Philips X'Pert powder diffractometer with Cu

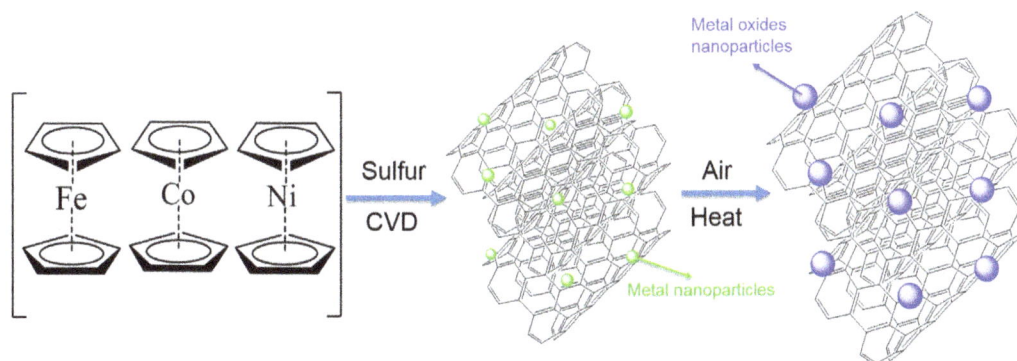

SCHEME 1 | Two-step synthetic route to M_xO_y (M = Fe, Co, Ni)/SWNT macrofilm nanocomposites.

Kα radiation (λ = 0.15418 nm) operating at 45 kV and 40 mA. Raman spectroscopy (Bruker SENTERRA with 532 nm laser excitation) was employed to verify the transformation of metal catalysts to metal oxides. The thermogravimetric analysis (TGA) was carried out on a Mettler Toledo TGA/DSA 1 STARe System under air flow (20 mL min⁻¹) with a heating rate of 5°C min⁻¹.

Lithium-Ion Battery Electrochemical Measurements

The total mass of free-standing M_xO_y/SWNT films as working electrodes were weighed by a micro/ultramicro balance (Mettler Toledo XP6) with 0.001 mg accuracy. CR2032 coin cells were assembled under a half-cell system with lithium ribbons (0.38 mm thick, 99.9%, Sigma Aldrich) as both counter and reference electrodes in an argon-filled glovebox (MBRAUN UNIlab). Celgard 2500 were used as separators. Electrolyte is 1 M LiPF₆ dissolved in 1:1 (v/v) ethylene carbonate (EC): diethyl carbonate (DEC) (Ferro Co.). The CV curves were collected by PARSTAT 2273 (Princeton Applied Research) potentiostat/galvanostat. The galvanostatic discharge–charge tests were carried out using BT-4 four-channel battery test equipment (Arbin Instrument, Ltd.).

Asymmetric Supercapacitor Electrochemical Measurements

Similarly, the free-standing NiO/SWNT macrofilms were employed as working electrodes. The CV tests of the NiO/SWNT electrodes were performed on the potentiostat PARSTAT 2273 (Princeton Applied Research) in a three-electrode cell. Pt wire was used as the counter electrode and Ag/AgCl electrode as reference electrode. The specific capacitance of the electrode can be calculated from the CV curves according to the following equation:

$$C = \frac{\int i \, dV}{\nu \Delta V} \tag{1}$$

where C is the specific capacitance (faraday per gram) based on the mass of the active materials, i is the current density response (with the unit of ampere per gram. $i = I/m$, I is the current response in ampere), ΔV is the voltage window of CV, ν is the scan rate. To assemble an asymmetric supercapacitor, the loading mass ratio between NiO/SWNT (+) as positive electrode and purified SWNT (−) as negative electrode was estimated to be 0.25 from the equation:

$$\frac{m_+}{m_-} = \frac{C_- \Delta E_-}{C_+ \Delta E_+} \tag{2}$$

where m is the mass, C is the specific capacitance and ΔE is the potential window for electrodes on each side. The subscripts of + and − represent the positive and negative electrodes, respectively. The electrochemical measurements of the asymmetric supercapacitor were carried out at room temperature under a two-electrode configuration in a CR2032 coin cell with Whatman˚ glass microfiber filter as separator in 1 M KOH aqueous electrolyte. The cycling performance was measured by galvanostatic discharge–charge tests

on BT-4 four-channel battery test equipment (Arbin Instrument, Ltd.). The specific capacitance could be also calculated by:

$$C = \frac{i \Delta t}{\Delta V} \tag{3}$$

where i is the galvanostatic charge/discharge current density (ampere per gram), Δt is the time for a full discharge in one cycle, ΔV is the voltage window delimited for charge/discharge cycling.

Results and Discussion

The size, morphology, and structure of the resulting materials were characterized using TEM. TEM images show size distribution of the Co_3O_4 nanoparticles in the range of 5–17 nm (**Figure 1A**) and the NiO nanoparticles with a diameter of 3–10 nm (**Figure 1D**). It is noted that the isolated-dispersed tiny particles in **Figure 1A** are Co catalysts, implying an incomplete oxidization at the low temperature of 200°C. In contrast, NiO has a better uniformity with a narrow size distribution due to complete transformation to NiO nanoparticles at a higher temperature of 400°C. It was confirmed that the Co residues in the Co_3O_4/SWNT sample (200°C) could be further oxidized when they were heated at 400°C in air. **Figures 1B,E** show similar morphologies of the M_xO_y/SWNT macrofilm nanocomposites. Metal oxide nanoparticles are coupled with the entangled serpentine CNT bundles. HRTEM images (**Figures 1C,F**) reveal the well-defined lattice fringes corresponding to face-centered cubic Co_3O_4 and NiO, respectively. TGA (shown in **Figure 2A**) demonstrates 23 wt. % CNT in the Co_3O_4/SWNT nanocomposite (200°C) while the NiO/SWNT nanocomposite contains only 12 wt. % carbon scaffold.

X-ray diffraction and Raman spectroscopy further confirm the transformation from metal catalysts to metal oxides during calcination. The XRD patterns before and after heat treatments are shown in **Figures 2B,C**. The results indicate that the crystal structure of Co_3O_4 is cubic phase with *Fd-3m* space group according to JCPDS No. 42-1467 and NiO is of the natural Brunsenite phase with *Fm-3m* space group (JCPDS No. 47-1049). This is in a good agreement with the HRTEM observations. Consistently, two distinct phases, Co_3O_4 (marked by asteroid) and pure Co (marked by spade) are present in the XRD pattern of the Co_3O_4/SWNT sample (200°C). The absence of Co diffraction peaks in the Co_3O_4/SWNT sample annealed at 400°C confirms the complete oxidation of the Co catalysts. Raman spectra (**Figure 2D**) show the RBM features of SWNT for all the samples at the wavenumber of ~100 cm⁻¹ as well as the typical D and G modes at 1350 and 1587 cm⁻¹, consistent with the TEM results (Jorio et al., 2003). The presence of two small peaks, E_g (488 cm⁻¹) and F_{2g} (522 cm⁻¹) separate at each side of 500 cm⁻¹ with F_{2g} (618 cm⁻¹) and the strongest peak A_{1g} (691 cm⁻¹) are typically indicative of the Co_3O_4 formation after the heat treatment (Hadjiev et al., 1988). Likewise, the first-order transverse optical (TO, 404 cm⁻¹), longitudinal optical (LO, 564 cm⁻¹), and two two-photon (2TO at 705 cm⁻¹, 2LO at 1080 cm⁻¹) peaks belong to NiO nanoparticles (Wang et al., 2002; Zhou et al., 2012).

The electrochemical performance was studied in a half cell that consists of the free-standing M_xO_y/SWNT macrofilms as working

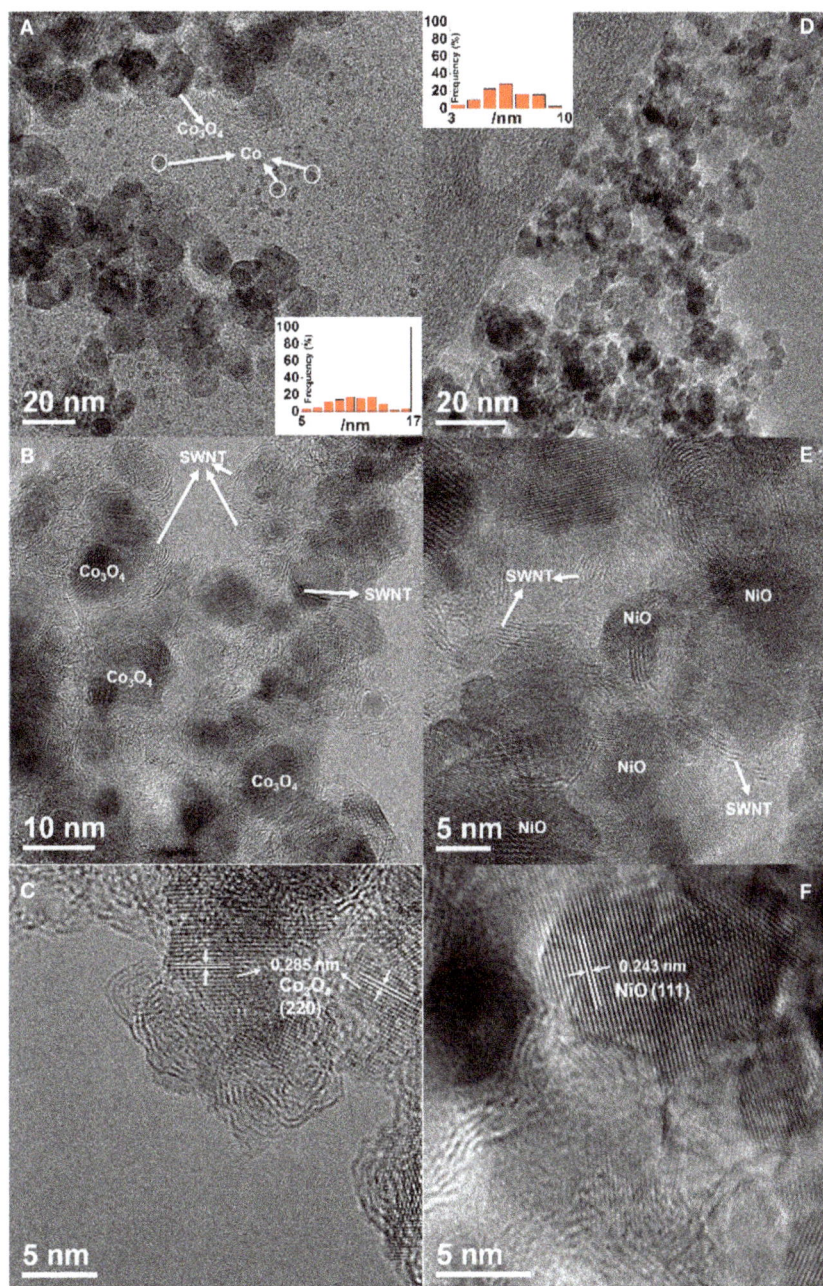

FIGURE 1 | (A,B,D,E) TEM and **(C,F)** HRTEM images of **(A–C)** Co_3O_4/SWNT (200°C) and **(D–F)** NiO/SWNT. Insets: histograms of the nanoparticle size distribution.

electrodes and lithium metal as both counter and reference electrode under a two-electrode configuration. Cyclic voltammetry (CV) was performed at an extremely slow scan rate of 0.1 mV s^{-1} to adapt to the sluggish kinetics of lithiation/delithiation. The CV plot (**Figure 3A**) exhibits the first two cycles for M_xO_y/SWNT in fresh cells. The cathodic peaks at the potential higher than 1 V vs. Li$^+$/Li on the first-cycle curves for both samples are ascribed to the formation of a solid electrolyte interface (SEI) film (Varghese et al., 2008; Zhou et al., 2012). The disappearance of them in the second cycle indicates the irreversible processes of SEI (Guo et al.,

2013). The peaks with a maximum current at 0.45 and 0.87 V for NiO/SWNT and Co_3O_4/SWNT, respectively, are assigned to the reduction of M ions to M^0 according to conversion reactions $(2y\text{Li} + M_xO_y \leftrightarrow x\text{M} + y\text{Li}_2\text{O})$ (Taberna et al., 2006). Both the reactions shift to a more positive potential on the CV curves in the following cycles. The difference in the number of the cathodic peaks (one at 1 V for NiO/SWNT, two at 0.9 and 1.36 V for Co_3O_4/SWNT) suggests the different reduction steps from the single oxidized state of Ni (II) in NiO and the multi-oxidized states of Co (II or III) in Co_3O_4, which agrees with the previously reported

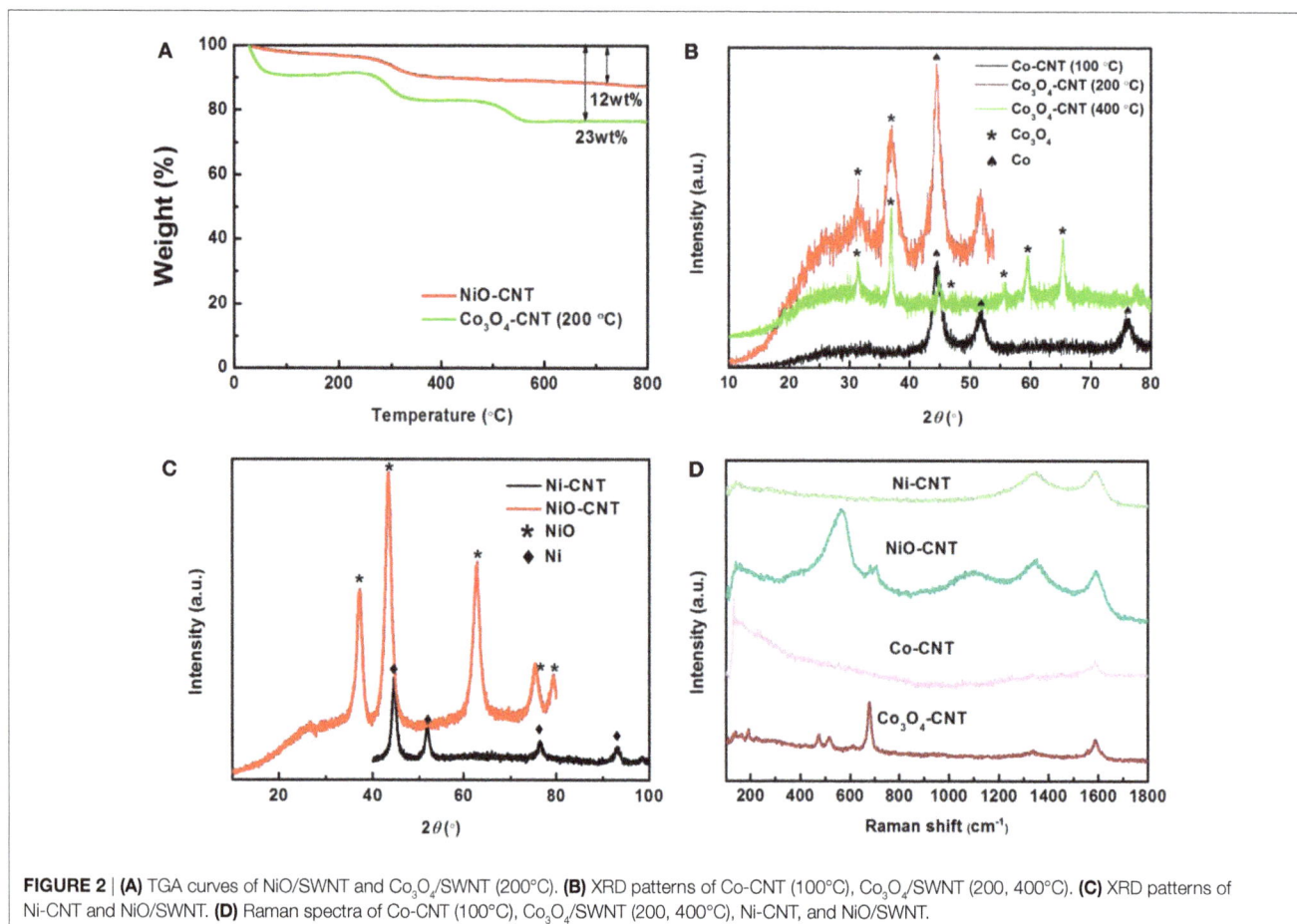

FIGURE 2 | (A) TGA curves of NiO/SWNT and Co$_3$O$_4$/SWNT (200°C). **(B)** XRD patterns of Co-CNT (100°C), Co$_3$O$_4$/SWNT (200, 400°C). **(C)** XRD patterns of Ni-CNT and NiO/SWNT. **(D)** Raman spectra of Co-CNT (100°C), Co$_3$O$_4$/SWNT (200, 400°C), Ni-CNT, and NiO/SWNT.

results of Fe$_2$O$_3$ (Cao and Wei, 2013a). Afterwards, the subsequent galvanostatic discharge/charge cycling measurements were carried out between 0.001 and 3 V at a constant current density of 50 mA g^{-1} with the same samples. **Figure 3B** shows their discharge/charge voltage profiles. Consistent with CV analysis, two plateaus at ~1.5 and ~1 V on the discharge curve of Co$_3$O$_4$/SWNT were observed corresponding to the two cathodic peaks, while NiO/SWNT has one plateau at ~1 V during discharge. A common discharge tail in the low voltage region between 0.8 and 0.001 V with more inclined slope is present for M$_x$O$_y$/SWNT, which is attributed to the partial capacity contribution from the SWNT macrofilms. Upon charging, there are two plateaus for both samples, which originate from the reverse processes of the conversion reactions from M^0 to M$_x$O$_y$ (Varghese et al., 2008). For Co$_3$O$_4$/SWNT, they (1.26 and 2 V) correspond to the two anodic peaks at 1.28 and 2 V on the CV curve. By contrast, NiO/SWNT has the higher ones at 1.32 and 2.2 V corresponding to peaks (1.39 and 2.22 V) in the CV plot. The continuous cycling results of NiO/SWNT over 80 cycles are selectively recorded in every 10 cycles as shown in **Figures 3C,D**. Basically, the discharge/charge curves are almost overlapped (**Figure 3C**), indicating a good cyclic stability. The discharge capacity maintains at 894 mAh g^{-1} after 80 cycles with a retention as high as 88%. Compared with non-free-standing NiO/graphene (883 mAh g^{-1} after 50 cycles) (Zhou et al., 2012)

and Co$_3$O$_4$/graphene (~935 mAh g^{-1} after 30 cycles) (Wu et al., 2010) nanocomposites prepared by the cumbersome synthesis, to the best of our knowledge, the M$_x$O$_y$/SWNT nanocomposites (over 1000 mAh g^{-1}) demonstrate one of the best electrochemical performance with an additional binder-free benefit among the reported hybrid anode materials.

In spite of the higher energy density, lithium-ion batteries usually suffer from a lower power density than supercapacitors. Taking advantage of both virtues of these two energy storage devices, we turn to an alternative approach to compose the asymmetric supercapacitors with a battery-type Faradaic electrode (as the energy source) and a capacitor-type electrode (as the power source) (Yan et al., 2012). They can extend operation voltage in the cell system with aqueous electrolyte beyond the water splitting limit (~1.23 V) by making full use of the two electrodes at different potential windows, consequently resulting in an enhanced specific capacitance and significantly improved energy density (Yan et al., 2012). **Figure 4A** shows the typical CV curves of the NiO/SWNT nanocomposite at different scan rates in 1 M KOH aqueous solution. All the CV curves have a pair of strong redox peaks, which are representative of the pseudo-capacitive behavior due to Faradaic redox reactions. This is distinct from an approximately ideal rectangular shape of CV curves for electric double layer capacitors. The redox couple, for example, located at around 0.27 V (cathodic)/0.35 V (anodic) vs.

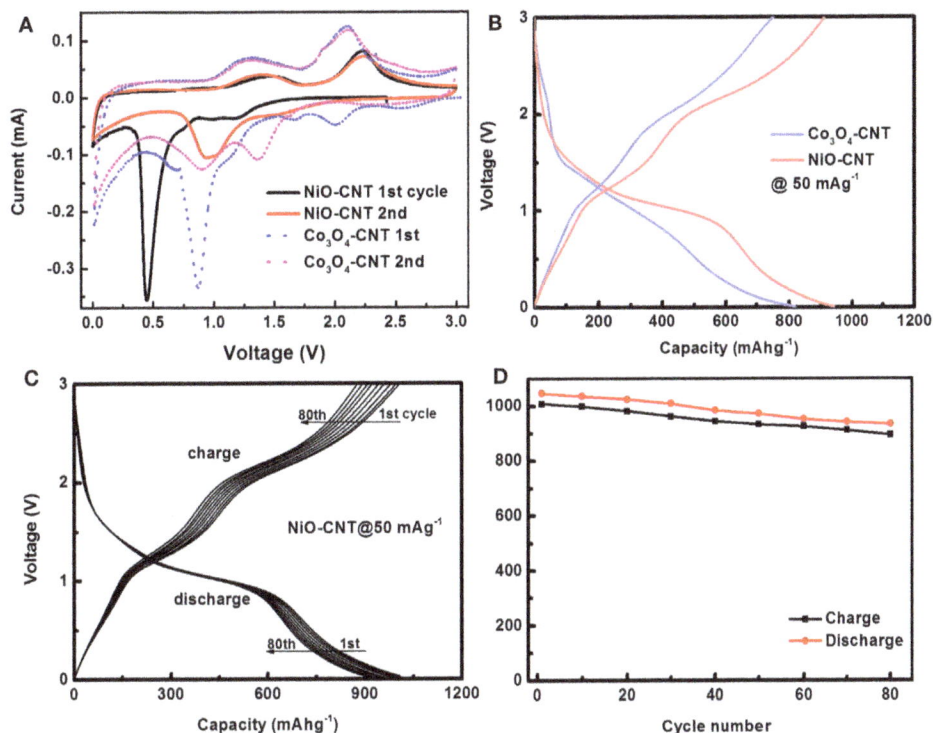

FIGURE 3 | Lithium-ion battery performance of MxOy/SWNT in half cells. (A) CV curves at a slow scan rate of 0.1 mV s⁻¹, and (B) galvanostatic discharge/charge voltage profiles at a constant current density of 50 mA g⁻¹ after CV tests of Co_3O_4/SWNT and NiO/SWNT. (C) Discharge/charge curves at the selected cycles, (D) cycling performance of NiO/SWNT electrodes.

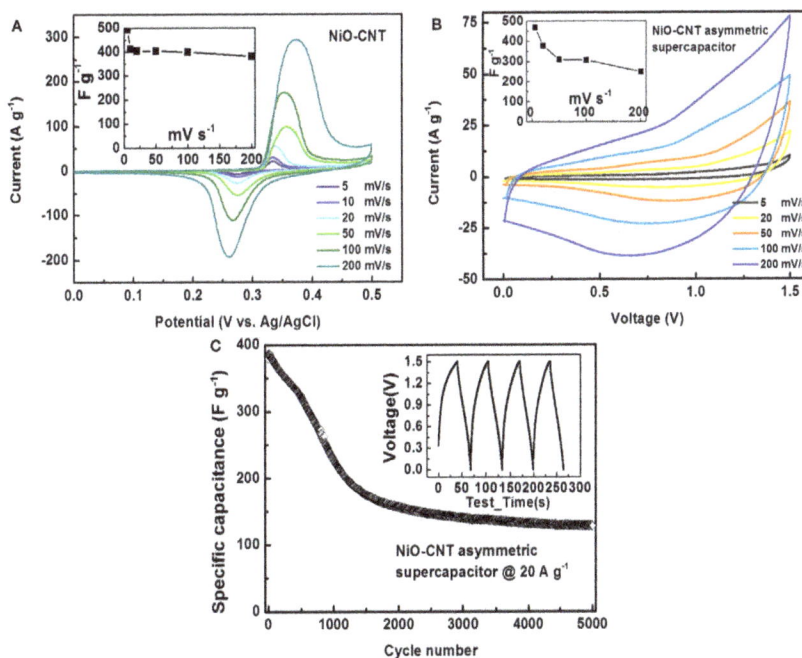

FIGURE 4 | (A) CV curves of NiO/SWNT at various scan rates. Inset: specific capacitance as a function of the scan rates calculated from CV. **(B)** CV curves of an asymmetric supercapacitor composed of (+) NiO/SWNT-SWNT (−) at different scan rates between 5 and 200 mV s⁻¹. Inset: specific capacitance vs. scan rates. **(C)** Galvanostatic charge/discharge cyclic performance of the asymmetric supercapacitor within a voltage window of 1.5 V at a current density of 20 A g⁻¹. Inset: charge/discharge curves.

Ag/AgCl at the scan rate of 50 mV s^{-1}, is attributed to the reversible Faradaic redox reactions between NiO and NiOOH according to the equation: $NiO + OH^- - e^- \leftrightarrow NiOOH$ (Lu et al., 2011). The specific capacitance of the NiO/SWNT electrodes as a function of scan rates calculated from the CV curves (inset of **Figure 4A**) is stable at ~400 F g^{-1}, suggesting the fast charge transfer in the Faradaic processes (Yan et al., 2012). Purified SWNT macrofilms without metal oxide nanoparticles, which deliver a specific capacitance of ~40 F g^{-1} as described in previous work (Yu et al., 2009), are employed as negative electrodes to assemble the asymmetric supercapacitors in 1 M KOH aqueous electrolyte with the NiO/SWNT positive electrodes by carefully matching their mass ratio (see Section "Materials and Methods" for details). **Figure 4B** exhibits the CV curves of the asymmetric supercapacitor at various scan rates from 5 to 200 mV s^{-1} within an operation voltage window of 1.5 V. The specific capacitance of the asymmetric cell (based on the total mass of the two electrodes) maintains as high as 249 F g^{-1} at 200 mV s^{-1} from 468 F g^{-1} at 5 mV s^{-1} as the scan rate increases. The galvanostatic charge/discharge curves at a current density of 20 A g^{-1} (inset of **Figure 4C**) show a good linear correlation of voltage with time, confirming a rapid *I–V* response of NiO/SWNT (Fan et al., 2011). The nearly equilateral triangular shapes demonstrate the excellent electrochemical reversibility. It is worth noting that at such a high-current density, the specific capacitance approaches a steady state during long-life cycling up to 5000 cycles. The performance above 130 F g^{-1} is sufficiently high to enable the asymmetric supercapacitor with qualified high energy and power density for durable and practical applications.

Conclusion

In summary, we have developed a general and facile strategy for synthesizing M_xO_y (M = Fe, Co, Ni)/SWNT macrofilm nanocomposites via CVD growth followed by a thermal treatment. We have successfully demonstrated their promise in energy storage. NiO/SWNT exhibits an excellent electrochemical performance in asymmetric supercapacitors with a high power and energy density. All M_xO_y/SWNT have shown a high-specific capacity and cycling stability for lithium-ion batteries. This work proposed an important family of material candidates to serve lithium-ion battery industry. It may also stimulate the evolution of new technique involving production of many advanced, low-cost transitional metal oxide nanocomposites for electrochemical applications.

Acknowledgments

The authors gratefully acknowledge the financial supports from the US National Science Foundation (NSF) under the contract of 1067947 and AFOSR MURI-FA9550-12-1-0035.

References

Cao, Z., and Wei, B. Q. (2013a). α-Fe$_2$O$_3$/single-walled carbon nanotube hybrid films as high-performance anodes for rechargeable lithium-ion batteries. *J. Power Sources* **241**, 330. doi:10.1016/j.jpowsour.2013.04.101

Cao, Z., and Wei, B. Q. (2013b). High rate capability of hydrogen annealed iron oxide-single walled carbon nanotube hybrid films for lithium-ion batteries. *ACS Appl. Mater. Interfaces* **5**, 10246. doi:10.1021/am403028z

Cao, Z., and Wei, B. Q. (2013c). V$_2$O$_5$/single-walled carbon nanotube hybrid mesoporous films as cathodes with high-rate capacities for rechargeable lithium ion batteries. *Nano Energy* **2**, 481. doi:10.1016/j.nanoen.2012.11.013

Cao, Z., and Wei, B. Q. (2013d). A perspective: carbon nanotube macro-films for energy storage. *Energy Environ. Sci.* **6**, 3183. doi:10.1039/C3EE42261E

Fan, Z., Yan, J., Wei, T., Zhi, L., Ning, G., Li, T., et al. (2011). Asymmetric supercapacitors based on graphene/MnO$_2$ and activated carbon nanofiber electrodes with high power and energy density. *Adv. Funct. Mater.* **21**, 2366. doi:10.1002/adfm.201100058

Grugeon, S., Laruelle, S., Herrera-Urbina, R., Dupont, L., Poizot, P., and Tarascon, J.-M. (2001). Particle size effects on the electrochemical performance of copper oxides toward lithium. *J. Electrochem. Soc.* **148**, A285. doi:10.1149/1.1353566

Guo, S., Zhang, S., Wu, L., and Sun, S. (2013). Co/CoO nanoparticles assembled on graphene for electrochemical reduction of oxygen. *Angew. Chem. Int. Ed. Engl.* **124**, 11940. doi:10.1002/ange.201206152

Hadjiev, V., Iliev, M., and Vergilov, I. (1988). The Raman spectra of Co$_3$O$_4$. *J. Phys. C Solid State Phys.* **21**, L199. doi:10.1088/0022-3719/21/7/007

Jorio, A., Pimenta, M., Souza Filho, A., Saito, R., Dresselhaus, G., and Dresselhaus, M. (2003). Characterizing carbon nanotube samples with resonance Raman scattering. *New J. Phys.* **5**, 139. doi:10.1088/1367-2630/5/1/139

Lee, J. Y., Liang, K., An, K. H., and Lee, Y. H. (2005). Nickel oxide/carbon nanotubes nanocomposite for electrochemical capacitance. *Synth. Met.* **150**, 153. doi:10.1016/j.synthmet.2005.01.016

Li, J., Tang, S., Lu, L., and Zeng, H. C. (2007). Preparation of nanocomposites of metals, metal oxides, and carbon nanotubes via self-assembly. *J. Am. Chem. Soc.* **129**, 9401. doi:10.1021/ja071122v

Liang, Y., Wang, H., Diao, P., Chang, W., Hong, G., Li, Y., et al. (2012). Oxygen reduction electrocatalyst based on strongly coupled cobalt oxide nanocrystals and carbon nanotubes. *J. Am. Chem. Soc.* **134**, 15849. doi:10.1021/ja305623m

Lu, Q., Lattanzi, M. W., Chen, Y., Kou, X., Li, W., Fan, X., et al. (2011). Supercapacitor electrodes with high-energy and power densities prepared from monolithic NiO/Ni nanocomposites. *Angew. Chem. Int. Ed. Engl.* **123**, 6979. doi:10.1002/ange.201103449

Poizot, P., Laruelle, S., Grugeon, S., Dupont, L., and Tarascon, J. M. (2000). Nano-sized transition-metal oxides as negative-electrode materials for lithium-ion batteries. *Nature* **407**, 496. doi:10.1038/35035045

Qin, J., Zhang, Q., Cao, Z., Li, X., Hu, C., and Wei, B. Q. (2013). MnO$_2$/SWCNT macro-films as flexible binder-free anodes for high-performance Li-ion batteries. *Nano Energy* **2**, 733. doi:10.1016/j.nanoen.2012.12.009

Su, J., Cao, M., Ren, L., and Hu, C. (2011). Fe$_3$O$_4$–graphene nanocomposites with improved lithium storage and magnetism properties. *J. Phys. Chem. C* **115**, 14469. doi:10.1021/jp201666s

Taberna, P. L., Mitra, S., Poizot, P., Simon, P., and Tarascon, J. M. (2006). High rate capabilities Fe$_3$O$_4$-based Cu nano-architectured electrodes for lithium-ion battery applications. *Nat. Mater.* **5**, 567. doi:10.1038/nmat1672

Varghese, B., Reddy, M., Yanwu, Z., Lit, C. S., Hoong, T. C., Subba Rao, G., et al. (2008). Fabrication of NiO nanowall electrodes for high performance lithium ion battery. *Chem. Mater.* **20**, 3360. doi:10.1021/am100791z

Wang, W., Liu, Y., Xu, C., Zheng, C., and Wang, G. (2002). Synthesis of NiO nanorods by a novel simple precursor thermal decomposition approach. *Chem. Phys. Lett.* **362**, 119. doi:10.1016/S0009-2614(02)00996-X

Wang, Z., Luan, D., Madhavi, S., Hu, Y., and Lou, X. W. (2012). Assembling carbon-coated α-Fe$_2$O$_3$ hollow nanohorns on the CNT backbone for superior lithium storage capability. *Energy Environ. Sci.* **5**, 5252. doi:10.1039/C1EE02831F

Wu, Z.-S., Ren, W., Wen, L., Gao, L., Zhao, J., Chen, Z., et al. (2010). Graphene anchored with Co$_3$O$_4$ nanoparticles as anode of lithium ion batteries with enhanced reversible capacity and cyclic performance. *ACS Nano* **4**, 3187. doi:10.1021/nn100740x

Xu, C., Xu, B., Gu, Y., Xiong, Z., Sun, J., and Zhao, X. (2013). Graphene-based electrodes for electrochemical energy storage. *Energy Environ. Sci.* **6**, 1388. doi:10.1039/c3ee23870a

Yan, J., Fan, Z., Sun, W., Ning, G., Wei, T., Zhang, Q., et al. (2012). Advanced asymmetric supercapacitors based on Ni(OH)$_2$/graphene and porous graphene electrodes with high energy density. *Adv. Funct. Mater.* **22**, 2632. doi:10.1002/adfm.201102839

Yu, C., Masarapu, C., Rong, J., Wei, B., and Jiang, H. (2009). Stretchable supercapacitors based on buckled single-walled carbon nanotube macrofilms. *Adv. Mater.* **21**, 4793–4797. doi:10.1002/adma.200901775

Yu, G., Hu, L., Liu, N., Wang, H., Vosgueritchian, M., Yang, Y., et al. (2011). Enhancing the supercapacitor performance of graphene/MnO$_2$ nanostructured electrodes by conductive wrapping. *Nano Lett.* **11**, 4438. doi:10.1021/nl2026635

Zhou, G., Wang, D.-W., Yin, L.-C., Li, N., Li, F., and Cheng, H.-M. (2012). Oxygen bridges between NiO nanosheets and graphene for improvement of lithium storage. *ACS Nano* **6**, 3214. doi:10.1021/nn300098m

Zhu, H., and Wei, B. (2007). Direct fabrication of single-walled carbon nanotube macro-films on flexible substrates. *Chem. Commun. (Camb.)* **29**, 3042–3044. doi:10.1039/b702523h

Conflict of Interest Statement: The authors declare that the research was conducted in the absence of any commercial or financial relationships that could be construed as a potential conflict of interest.

3

Optical characterization and lasing in three-dimensional opal-structures

Yoshiaki Nishijima[1]* and Saulius Juodkazis[2,3,4]*

[1] Department of Electrical and Computer Engineering, Graduate School of Engineering, Yokohama National University, Yokohama, Japan, [2] Centre for Micro-Photonics, Faculty of Science, Engineering and Technology, Swinburne University of Technology, Melbourne, VIC, Australia, [3] Melbourne Centre for Nanofabrication (MCN), Australian National Fabrication Facility, Clayton, VIC, Australia, [4] Center of Nanotechnology, King Abdulaziz University, Jeddah, Saudi Arabia

Edited by:
P. Davide Cozzoli,
University of Salento, Italy

Reviewed by:
Carola Kryschi,
Friedrich-Alexander Universität
Erlangen-Nürnberg, Germany
Francesco Scotognella,
Politecnico di Milano, Italy

*Correspondence:
Yoshiaki Nishijima,
Department of Electrical and
Computer Engineering, Graduate
School of Engineering, Yokohama
National University, 79-5
Tokiwadai, Hodogaya-ku, Yokohama
240-8501, Japan
nishijima@ynu.ac.jp;
Saulius Juodkazis,
Centre for Micro-Photonics, Faculty of
Science, Engineering and Technology,
Swinburne University of Technology,
3122 Hawthorn, VIC, Australia
sjuodkazis@swin.edu.au

The lasing properties of dye-permeated opal pyramidal structures are compared with the lasing properties of opal films. The opal-structures studied were made by sedimentation of microspheres and by sol–gel inversion of the direct-opals. Forced-sedimentation by centrifugation inside wet-etched pyramidal pits on silicon surfaces was used to improve the structural quality of the direct-opal structures. Single crystalline pyramids with the base length of ~100 μm were formed by centrifuged sedimentation. The lasing of dyes in the well-ordered crystalline and poly-crystalline structures showed a distinct multi-modal spectrum. Gain via a distributed feedback was responsible for the lasing since the photonic band gap was negligible in a low refractive index contrast medium; the indices of silica and ethylene glycol are 1.46 and 1.42, respectively. A disordered lasing spectrum was observed from opal films with structural defects and multi-domain regions. The three-dimensional structural quality of the structures was assessed by *in situ* optical diffraction and confocal fluorescence. A correlation between the lasing spectrum and the three-dimensional structural quality was established. Lasing threshold of a sulforhodamine dye in a silica opal was controlled via Förster mechanism by addition of a donor rhodamine 6G dye. The lasing spectrum had a well-ordered modal structure, which was spectrally stable at different excitation powers. The sharp lasing threshold characterized by a spontaneous emission coupling ratio, $\beta \simeq 10^{-2}$, was obtained.

Keywords: colloidal crystals, opals, lasing, micro-fabrication, Förster energy transfer, microscopy, diffraction

1. Introduction

Photonic crystals (PhCs) (John, 1987; Yablonovitch, 1987) fabricated by top-down or bottom-up methods are being used for a growing number of applications, such as spontaneous emission control (Li et al., 2007), micro lasing (Matsubara et al., 2008; Nishijima et al., 2008), and refractive index sensing (Nishijima et al., 2007; Kita et al., 2008; Wu et al., 2008). PhCs can also be combined with plasmonics to yield new methods for controlling photo-excited processes, which for example, can result in an enhancement of fluorescence (Tawa et al., 2008). Opals are a class of PhCs, which could potentially be used for many applications, because opal structures covering macroscopically large areas can be easily fabricated by self-assembly of (sub-)micrometer sized spheres. Opal and inverse-opal structures with a well-controlled three-dimensional (3D) morphology could find various uses relating to PhCs, micro lasers, structured waveguides, and magneto-optical applications (Conti and Fratalocchi, 2008; Choueikani et al., 2009).

One of the problems associated with the fabrication of large opal structures is the difficulty to control the crystalline quality of the face-centered-cubic opals even when the microspheres used for sedimentation have a small, ~1%, size distribution. It was recently demonstrated that a narrow, ~1%, size distribution of microspheres is essential for obtaining high-fidelity photonic stopgaps and wave guiding in Si-inverse opals (Lavrinenko et al., 2009). Developing methods for controlling the structure of the opal sedimentation lattices and selecting between face-centered-cubic (fcc) and hexagonal-closest-packing (hcp) symmetry is a key area of research concerning the growth of artificial opals (Velasco et al., 1998; Auer and Frenkel, 2001). The fcc and hcp packings have nearly equal Helmholtz formation energies (Bolhuis et al., 1997; Bruce et al., 1997; Pronk and Frenkel, 1999, 2003). Therefore, in order to control the colloidal crystal morphology, colloid epitaxy (Santamaria et al., 2002), deposition on a micro patterned surface by sedimentation (Mizeikis et al., 2001), or dip-coating using Langmuir–Blodgett technique (Hur and Won, 2008) are used. Recently, a spatially constrained deposition using a pyramidal template fabricated in silicon by anisotropic etching was proposed for the formation of well-defined opals (Matsuo et al., 2003).

Another problem of opal and inverse-opal fabrication is that self-assembly often results in the formation of poly-crystalline domains. This is caused by a concentration of the colloidal suspension during evaporation, which facilitates multi-site nucleation (Biben et al., 1993). Opals with a poly-crystalline structure are not useful for photonic applications where single crystal order is required. Films of opals sedimented on substrates with pins dry-etched on the surface resulted in monocrystal opals (Mizeikis et al., 2001). However, the pins only imposed the order on three to five adjacent mono layers depending on the size of the microspheres.

The sedimentation process used for forming the opals can also affect their quality. Slow gravitational sedimentation and formation of opals at low evaporation rates are not practical since they are lengthy process and require efficient suppression of mechanical vibrations. Furthermore, the possibility of stacking faults is not negligible when these techniques are used. This is because the high colloidal concentration can lead to multi-site nucleation (Hoogenboom et al., 2002). Sedimentation of opals in 2 and 3D constrained micro volumes could prospectively solve the problems mentioned above because constrained micro volumes impose the order onto the structure.

Transport of electrons and energy in photo-excited molecular systems is in focus of basic and applied research due to prospects to unveil functioning of photosynthesis (Ben-Shem et al., 2003). Different linear and non-linear (Kamada et al., 2007; Hirakawa et al., 2008) optical microscopy methods are used to measure the energy transfer in micro-cavities (Fujiwara et al., 2005) and living cells (Sasaki et al., 1991; Hotta et al., 2006). The principles of photo-excitation and interaction of materials at atomic and molecular level in artificially engineered three-dimensional (3D) structures, such as photonic crystals (PhCs) (John, 1987; Yablonovitch, 1987), high-Q, and random and distributed feed back (DFB) cavities (Wright et al., 2004), are currently under active investigation due to prospects of miniaturization of opto-electronic devices, solar batteries, and creation of alternative light sources.

The 3D confinement of photo-excited materials and control of their (de-)excitation by external cavity or via spatial localization of light–matter interaction has prospects to advance photonics into new multi-disciplinary fields. Better understanding of these processes might lead toward miniaturization of opto-electronic devices.

Here, we demonstrate the formation of well-ordered opals by sedimentation in wet-etched micro-pits in silicon by fast centrifugation (Nishijima, 2009), which the single crystalline fcc opal photonic crystals with large area and short time can be obtained. The resulting three-dimensional structures were characterized by confocal fluorescence and optical diffraction and compared with opal-structures obtained on flat substrates. Lasing of dye-soaked opals showed a well-ordered multi-mode spectrum, and a correlation between the structural quality and lasing spectrum was established. Rhodamine 6G (Rh6G) and sulforhodamine 101 (SRh) were used for the energy transfer studies inside silica opals. These dyes are suitable for a fluorescent and resonance energy transfer system; they have a high fluorescence quantum yield (~1%), similar chemical structures (similar ground state and excited-state dipole momenta), and a high overlap of energy donor (Rh6G) fluorescence and acceptor (SRh) absorption spectra. The lasing properties of these dyes inside opal photonic structures show potential application in fluidic micro-laser field.

2. Experimental

2.1. Fabrication and Characterization of Opal Structures

Single crystal opals were fabricated using an Si template-assisted method (Matsuo et al., 2003). Opal structures were made by centrifuged sedimentation of SiO_2 spheres (diameter 300 nm) at 5,000 rpm for 10 min on a cover glass substrates ~1-mm-thick (Nishijima et al., 2007). After centrifuge, the opal structures were dried and annealed at 500°C/5 h for drying and sintering into a mechanically stable film. Structures were lifted from the pits by adhesive peeling off. Opal films were made by centrifuge sedimentation onto flat glass under the same conditions. The structural quality was inspected by scanning electron microscopy (SEM). The surface of the structure was a (111) face plane of the face-centered-cubic (fcc) structure.

The template had pyramidal pits anisotropically etched along the <111> direction on a (100) surface of Si, covered by 1 μm layer of oxide. The pattern of pits was realized by lithography using a positive photo resist OFPR5000 (Tokyo Ohka Kogyo Co. Ltd.). Following exposure and development of the photoresist, the pattern was dry-etched into the oxide layer using a gas mixture of SF_6 and C_4F_8. The patterned oxide was then used as a mask for wet etching of the silicon using a KOH aqueous solution (33 wt%, with 1 wt% isopropanol). Etching accompanied by sonication was carried out for 4 h, and pyramidal pits were obtained.

Inversion of both the films and pyramids was performed using a sol–gel process. The polystyrene opals were immersed into a tetraethoxysilane (TEOS) sol, which was prepared from TEOS, ethanol, and HCl (at 1:1:0.1 ratio in volume fraction) mixture. The residual polystyrene was removed from the structures by washing in ethyl acetate.

The structural quality was inspected by scanning electron microscopy (SEM). The surface of the structure was a (111) face plane of the face-centered-cubic (fcc) structure.

2.2. Lasing in Opal Structures

The laser dyes were rhodamine 6G (Rh6G, from Wako Pure chemical industries, Ltd.) and sulforhodamine 101 (SRh, from Exciton, Inc.). The molar extinction coefficient of the dyes at the laser excitation wavelength of 532 nm was 99,000 and 23,800 $M^{-1} cm^{-1}$ for Rh6G and SRh, respectively.

The emission spectra of dye-soaked opal films were measured by micro-spectroscopic setup consisting of microscope (IX-71, Olympus) equipped with an objective lens of 40× magnification and numerical aperture of NA = 0.55, and spectrometer (SpectraPro, Acton) with liquid nitrogen-cooled CCD camera. The excitation source was a Nd:YAG laser emitting 7 ns duration pulses at 532 nm (Indi, Spectra Physics) with the repetition rate of 10 Hz. Low repetition rate was allowed to avoid thermal accumulation.

3. Results and Discussion

Figure 1 shows a scanning electron microscopy (SEM) image of a pit on an Si template, a silica opal formed from 300-nm-diameter microspheres, and an inverse-opal obtained via sol–gel inversion from polystyrene opals. Thermal sintering of opals inside the template caused partial adhesion of the microspheres to the template. After peeling the pyramids off, the adhesion caused some defects to the surface. When sol–gel inversion was implemented to make silica pyramids, the side-wall quality was even better (**Figure 1C**. When the same sedimentation conditions were used on cover glass substrates, poly-crystalline opal films with a thickness of up to 100 μm were formed with the typical single crystalline domains having a cross-section of 10–30 μm (as judged by a SEM observation from a single surface layer).

3.1. Optical Characterization of Opals
3.1.1. Confocal Fluorescence Imaging

Since SEM observation does not provide information on the internal 3D structural quality, we applied confocal optical imaging and

diffraction to obtain information of the 3D structure as described below. The relation between the structural 3D ordering and the lasing from the same regions was then investigated.

The internal structure of the opals was carried out by confocal imaging of fluorescence from a dye solution, which was pipette-dropped onto the opal-structures and allowed to permeate them. The soaked structures were then placed on the stage of an Olympus microscope equipped with an FV-300 scanning system, and imaged. a cw-HeNe laser at 543 nm wavelength was used for illumination (principle scheme of characterization setup is shown in **Figure 2**). The beam was scanned across the sample using galvano-mirrors. Focusing was carried out with a 100× magnification oil-immersion objective lens of numerical aperture NA = 1.30. Fluorescence of the rhodamine B dye in an ethylene glycol solution (1.0×10^{-5} M) in the opal structure was detected by a photo-multiplier tube through a confocal pinhole of a 60 μm diameter. The optical setup is shown in **Figure 2A**. The confocal fluorescence imaging was limited to structures comprising microspheres with a diameter of 1 μm or larger. This is because of the optical spatial resolution $\sim\lambda$ of the confocal microscope. Also, the imaging was limited to the small sized pyramids, because of the working distance of the objective lens and the inhibition of scattering from particles. In the past, it has been shown that optical diffraction can be used to assess the 3D internal order of opals (Kanai et al., 2005; Santamaria et al., 2005; Lopez, 2006). Therefore, we used this technique to characterize structures made of smaller particles and larger pyramids.

The optical diffraction pattern represents a Fourier transform of the crystal structure (**Figure 2B**). Lopez et al. demonstrated that the optical conditions necessary to obtain the diffraction patterns are given by Santamaria et al. (2002) and Lopez (2006):

$$\frac{a}{\lambda} \geq m\frac{\sqrt{2}}{n}[1 + \frac{(2s-1)}{3}]^{\frac{1}{2}}, \qquad (1)$$

where a is the lattice constant (424 nm in our case), s and m are two integers, which represent the set of diffraction spots and the diffraction order, respectively. The lowest energy light quanta that satisfy this condition are $\frac{a}{\lambda} = \frac{1.633}{n}(m = s = 1)$ and the next possible diffraction pattern is defined by $\frac{a}{\lambda} = \frac{2.828}{n}(m = 1, s = 2)$.

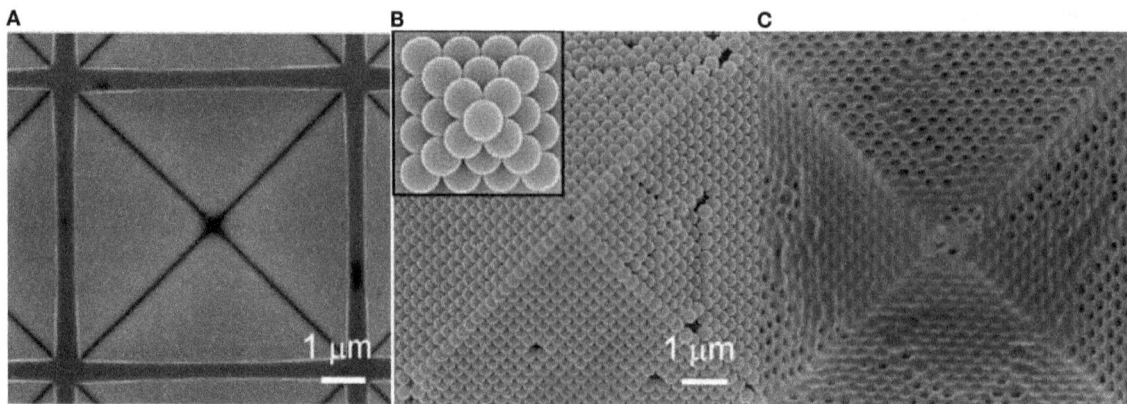

FIGURE 1 | SEM images of a pyramidal pit etched-out in silicon (A), a templated silica opal pyramid (B), and an inverse silica-gel pyramid (an initial opal was templated by polystyrene microspheres) (C). The closest packing surface (111)-plane for fcc crystal structure appears on the walls of the pyramid.

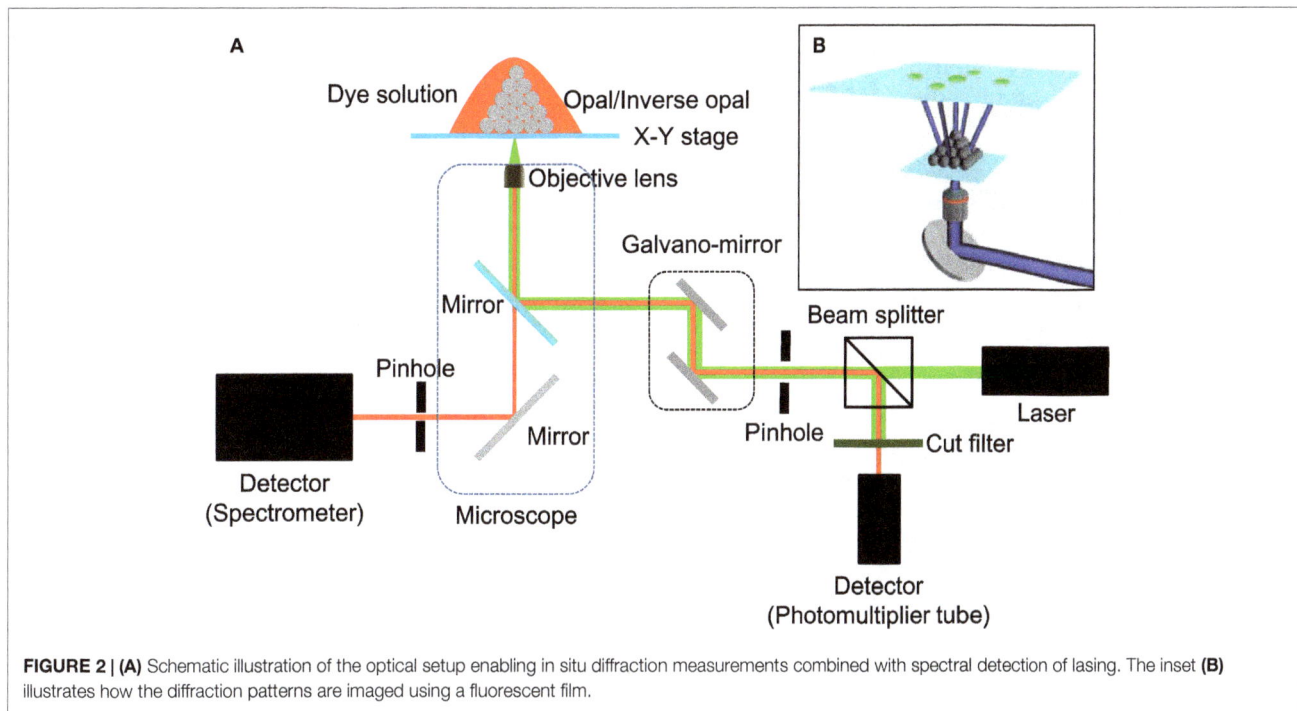

FIGURE 2 | (A) Schematic illustration of the optical setup enabling in situ diffraction measurements combined with spectral detection of lasing. The inset **(B)** illustrates how the diffraction patterns are imaged using a fluorescent film.

In our case, $\frac{a}{\lambda} = 1.18$ and only the lowest energy diffraction pattern can appear. The optical diffraction, or the Fourier transform of three-dimensional crystal structures, was compared with fast Fourier transform (FFT) images of two-dimensional SEM data of the opal surfaces.

In order to test the quality of our structures using diffraction, we used the setup shown in **Figure 2**. The pulses of a femtosecond laser at 360 nm wavelength (82 MHz, 150 fs, the second harmonics of a Tsunami laser, Spectra Physics) were focused on to the bottom side of the samples using a $4\times$ NA = 0.13 objective lens. The diffracted beams were projected on to a coumarine 6-doped thin polystyrene film, and the fluorescence image was collected by a CCD camera.

Laser emission from the structures was also measured. In order to obtain lasing, the samples were soaked in the solution of rhodamine B and ethylene glycol used for confocal imaging. These structures were then pumped with a frequency-doubled Nd:YAG nanosecond laser operating at a wavelength of 532 nm, a repetition frequency of 10 Hz, and with a pulse duration of ~7 ns. The laser beam was coupled into an inverted microscope and focused on the sample using either an objective lenses of $40\times$ magnification and a numerical aperture NA = 0.55 or an objective with $10\times$ magnification and a numerical aperture NA = 0.25. The sample was mounted on a translation stage attached to the microscope and oriented with the bases of the pyramids [or the (111) crystallographic planes of the opal structure] normal to the optical axis of the objective lens. Emission from the rhodamine B ethylene glycol solution (5.0×10^{-4} M) was detected by imaging the surface of the sample (base plane of the pyramids) via the same objective lens used for focusing the pump beam. The collected radiation was analyzed using a spectrometer with 1200 groves/mm grating (resolution <0.1 nm) with a liquid nitrogen-cooled CCD detector (**Figure 2A**).

In order to study the connection between the 3D structural quality of the opals and their lasing properties, optical diffraction and lasing were measured on the same areas using the setup shown in **Figure 2**. First, the optical diffraction was measured on the chosen region/structure. Then, the structure was soaked in dye, and the lasing spectrum was measured. Prior to the optical measurements, the same areas were inspected by SEM. There was an offset of about $10\,\mu$m between the areas measured by SEM and the areas upon which the optical measurements were performed. This offset occurred because the SEM measurements and the optical characterization were carried out with different setups.

Figure 3 shows the fluorescence from an opal structure. The similarity in the refractive indices of silica ($n = 1.46$) and ethylene glycol (1.42) reduced light scattering and facilitated reliable imaging of several layers.

Confocal imaging of the structures showed that the odd and even layers have the same alternating order (**Figure 3**), indicating that the particles formed an fcc structure with the pyramid base being a (100) plane of the fcc structure.

3.1.2. Optical Diffraction

Figure 4 shows the optical diffraction patterns from pyramids of different sizes. The diffraction patterns from larger structures show more scattering. Nevertheless, the fcc structure can be recognized in the diffraction patterns. The presented analysis shows that optical diffraction can be implemented *in situ* (as opposed to *ex situ* SEM analysis) and can be used to assess the structural quality of regions, which were subsequently used for lasing. SEM images of several regions in $10\,\mu$m-thick opal films, along with their Fourier transform and the measured diffraction patterns are shown in **Figure 5**. The Fourier transform of the SEM images (column 2) reveal that hexagonal diffraction pattern from a close packing surfaces along <111> direction of opal films are expected.

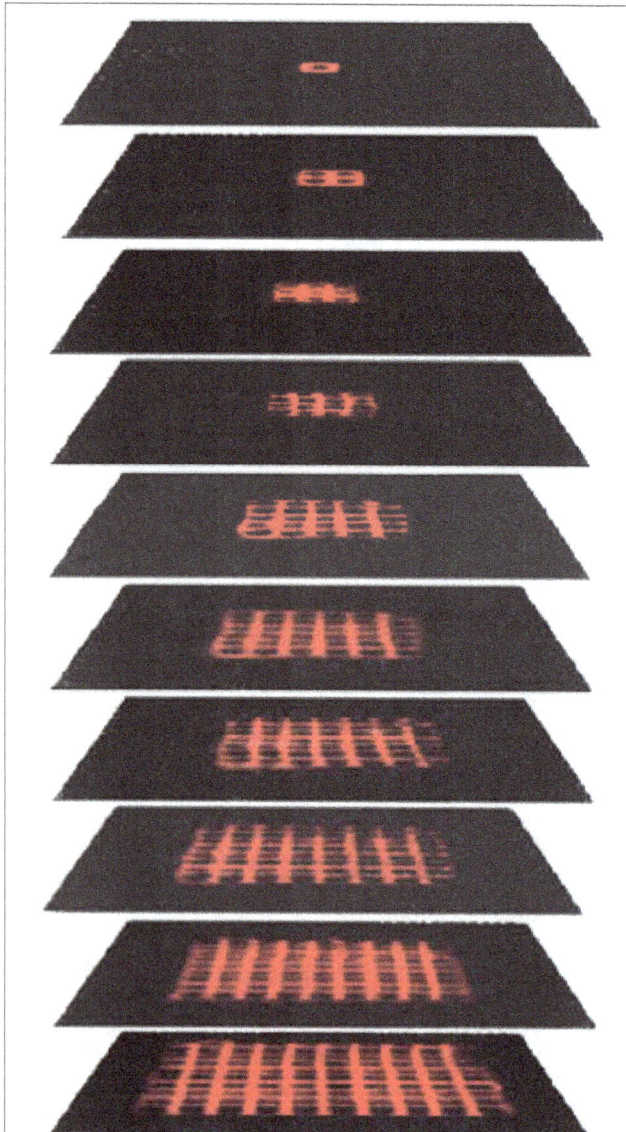

FIGURE 3 | Confocal fluorescent images of a rhodamine B solution in silica opal from the first to the tenth layer. The size of silica spheres was 1 μm; the side length of the pyramid base was 10 μm. A square lattice is observed in each of the planes, which correspond to the (100)-face for a fcc crystal structure.

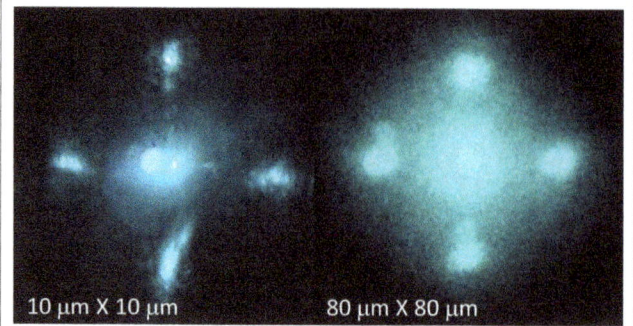

FIGURE 4 | Optical diffraction of pyramidal opals of different sizes. The diameter of the silica microspheres was 300 nm; illumination wavelength 360 nm. Laser beam was directed along to the <100> direction to the fcc crystal lattice (perpendicular to the base of a pyramid).

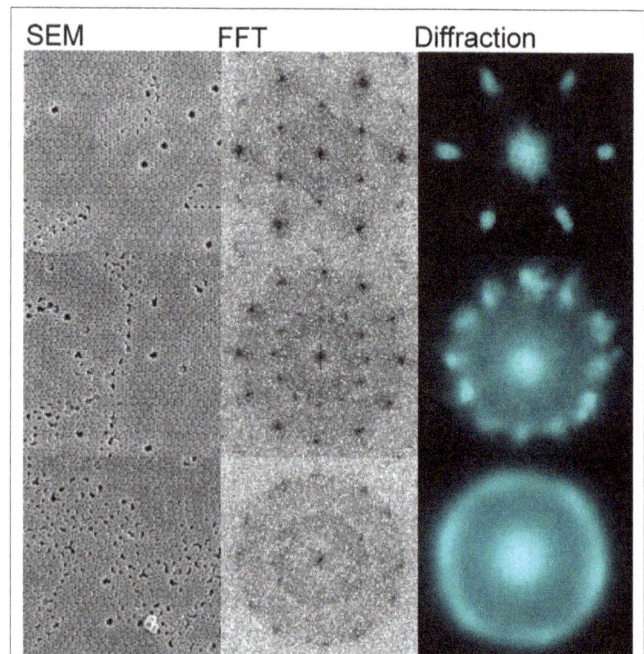

FIGURE 5 | SEM images and their 2D-FFT together with the optical diffraction (Figure 2B) from three typical regions in the opal films. The crystalline order of the regions decreases from top to bottom of the figure. The optical diffraction and SEM were not taken from exactly the same areas and can be compared only qualitatively. The diameter of the silica microspheres was 300 nm and the illumination wavelength was 360 nm. The structure was illuminated along the <111> direction of the opal films (perpendicular to the film).

The most ordered regions with just a few point defects (vacancies) show orderly diffraction patterns. The diffraction from less ordered regions near the grain boundaries shows two superimposed and $\pi/6$-rotated patterns. The $\pi/6$ rotational misalignment is the maximum possible between the two neighboring domains in a <111> plane. Finally, the regions with a large number of grains and smaller than 10 periods in cross-section showed a typical disordered amorphous pattern of Fourier image. These 2D-FFT patterns are consistent with the optical diffraction images (column 3), which reveal a 3D order of the samples.

3.2. Lasing Pyramidal Opals

Figure 6 shows the lasing emission spectrum of rhodamine B dye solution in the poly-crystalline opal film. The lasing threshold

was determined from the power dependence of the emission when the spectrum became spectrally structured. At pump powers above the lasing threshold, the mode intensity increased exponentially with excitation while the spectral intensity outside of the modes decreased. The spectra shown in **Figure 6** were taken at approximately 10% above the lasing threshold.

When the excitation area was larger than $\varphi > 150$ μm in diameter (**Figure 6A**), the spectra were disordered similar to that previously reported from random structures (Frolov et al., 1999; Shkunov et al., 2001). The optical diffraction patterns from such

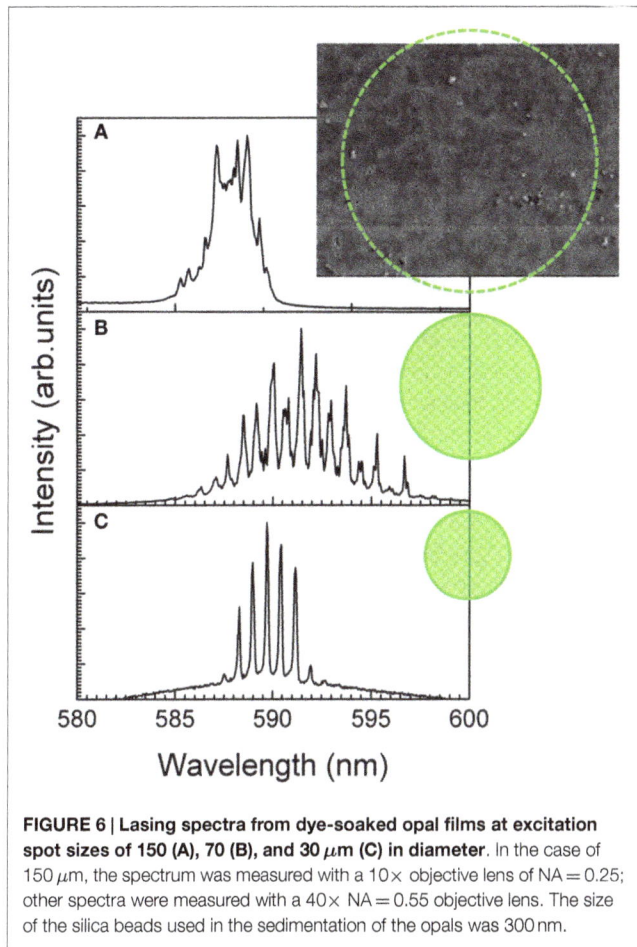

FIGURE 6 | Lasing spectra from dye-soaked opal films at excitation spot sizes of 150 (A), 70 (B), and 30 μm (C) in diameter. In the case of 150 μm, the spectrum was measured with a 10× objective lens of NA = 0.25; other spectra were measured with a 40× NA = 0.55 objective lens. The size of the silica beads used in the sedimentation of the opals was 300 nm.

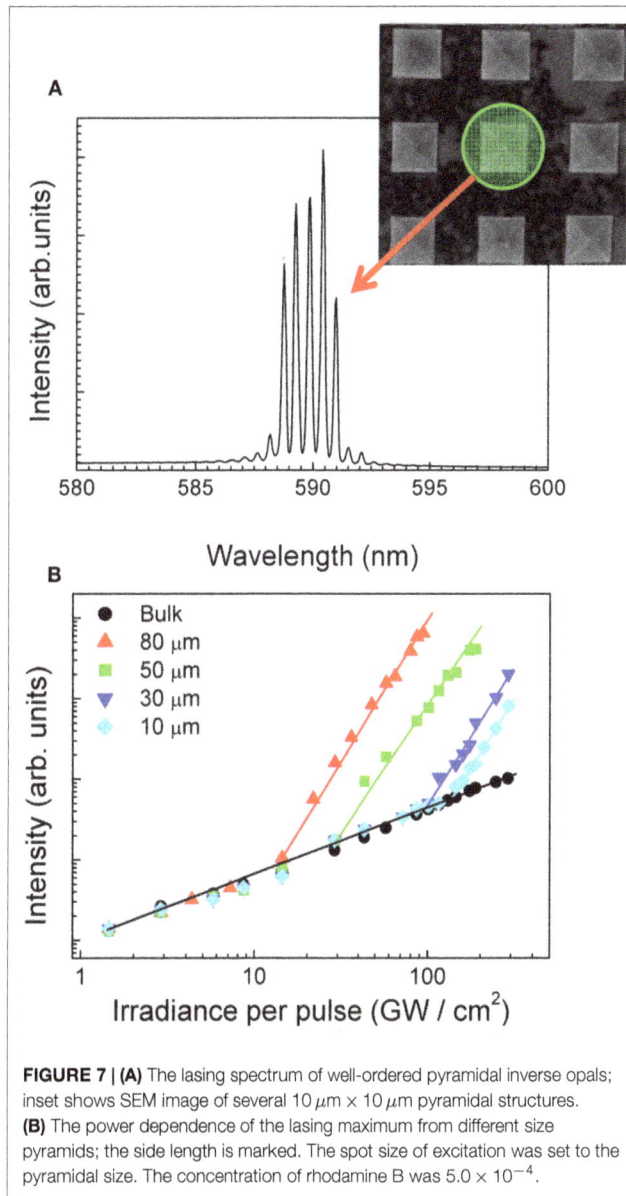

FIGURE 7 | (A) The lasing spectrum of well-ordered pyramidal inverse opals; inset shows SEM image of several 10 μm × 10 μm pyramidal structures. **(B)** The power dependence of the lasing maximum from different size pyramids; the side length is marked. The spot size of excitation was set to the pyramidal size. The concentration of rhodamine B was 5.0×10^{-4}.

regions were amorphous as well (**Figure 5**). In the case of 70 μm areas, more regular spectra were obtained, although there was still some sub-structure and small peaks. The typical optical diffraction from such regions showed a distinct few domain structure (as shown in the middle pattern in **Figure 5**). Well-ordered spectra with regular sharp and periodic peaks were obtained when the spot size was 70 μm and approximately corresponded to a single crystalline domain as observed by the optical diffraction. The lasing peaks were separated by 0.77 nm interval. Similar well-ordered lasing was observed from the inverse opals (**Figure 7A**).

The inverse-opal structures could prospectively be used for gain amplification since the volume fraction of the immersed dye is approximately three times larger than that in direct-opals.

Figure 7A shows the lasing spectrum of a dye-soaked pyramidal single crystal inverse-opal. We measured 10 different regions on the film (as shown in **Figure 6C**) and 10 different pyramids, which showed single crystalline patterns of optical diffraction. The lasing modes of all structures and regions tested had the same spectral positions to within ± 0.3 nm for the opal films and ±0.15 nm for the pyramidal single crystal. This corroborates that the single pyramidal crystals have better wavelength stability of the lasing modes. This type of lasing is not defined by the photonic band gap because of the low refractive index contrast between silica (1.46) and ethylene glycol (1.42) (Kitamura et al., 2007; Nishijima, 2009; Nishijima et al., 2009a). Rather, it is determined

by the distributed feedback (DFB) in the dye as demonstrated recently (Nishijima et al., 2008). The lasing threshold and the onset of lasing were determined from the power dependence of the lasing spectrum shown in **Figure 7B**.

Further studies are necessary to model and experimentally verify the directionality of the lasing modes. In this study, the directionality of excitation and laser emission were integrated due to comparatively tight focusing. The influence of a microsphere size distribution on the lasing spectrum predicted recently (Gottardo et al., 2008) should also be investigated.

The emission spectra of dye-soaked opal films were measured by micro-spectroscopic setup consisting of microscope (IX-71, Olympus) equipped with an objective lens of 40× magnification and numerical aperture of NA = 0.55, and spectrometer (SpectraPro, Acton) with liquid nitrogen-cooled CCD camera. The excitation source was a Nd:YAG laser emitting 7 ns duration pulses at 532 nm (Indi, Spectra Physics) with the repetition rate of 10 Hz. Low repetition rate allowed to avoid thermal accumulation.

Figure 8 shows normalized absorption and fluorescence spectra of the dyes Rh6G and SRh. The marked dashed region shows an overlap of absorption-emission bands of the pair in a 500–650 nm wavelength region. The dyes permeated silica opal film, which was templated using a 300 nm diameter silica microspheres. The concentration of the acceptor dye (SRh) was 1.0 mM, and the donor dye concentration was varied (0, 0.8, 1.4, 1.8, 2.0 mM). **Figure 9** shows the absorption spectra of the each dye solution measured in a 100 μm optical path.

3.3. Förster Energy Transfer

Interaction between two lasing dyes in opal structures is discussed next. The Förster theory of dipole–dipole energy transfer (Forster, 1947, 1948, 1959) defines the efficiency of a radiative excitation state energy transfer in a donor–acceptor system, which is dependent on the separation r and is expressed as (Birks, 1970):

$$k_{ET}(r) = \frac{(9000 \ln 10)\kappa^2}{128\pi^5 n^4 N_A r^6 \tau_D} \int_0^\infty \frac{F_D(\tilde{\nu})\varepsilon_A(\tilde{\nu})}{\tilde{\nu}^4} d\tilde{\nu}$$
$$= \frac{1}{\tau_D}\left(\frac{R_0}{r}\right)^6 \tag{2}$$

where F_D is the fluorescence spectral profile of the donor species, ε_A is the molar extinction coefficient of the acceptor molecule; R_0 is the critical transfer distance, which depends on the donor oscillator strength and spectral overlap, κ is the orientation factor ($\kappa^2 = 2/3$ in the case of a random directional distribution), N_A is the Avogadro number, $\tilde{\nu}$ is the wave number, T_D is the excited-state lifetime of the energy donor (Rh6G in our study), n is the refractive index (ethylene glycol ~1.42).

Equation (2) indicates that a spectrally overlapping absorption–emission of the donor–acceptor pairs is prospective for the efficient energy transfer via a Förster mechanism. Studies of energy transfer in combinations of pyrene–perylene, two different coumarine dyes, and fluorescein-ethidium bromide, have been carried out with applications in biological imaging and bio-analytical science (Takakusa et al., 2003; Fujiwara et al., 2005; de Almeida et al., 2007; Stevens et al., 2009). Overlap of the fluorescence and absorption was estimated from our experiments as 6.9×10^{-15} mol^{-1} cm^6 (**Figure 8**), and the critical Förster radius was calculated as 3.1 nm.

Figure 8 shows normalized absorption and fluorescence spectra of the dyes Rh6G and SRh. The marked dashed region shows an overlap of absorption-emission bands of the pair in a 500–650 nm wavelength region. The dyes permeated silica opal film which was templated using a 300 nm diameter silica microspheres. The concentration of the acceptor dye (SRh) was 1.0 mM, and the donor dye concentration was varied (0, 0.8, 1.4, 1.8, 2.0 mM). Absorption spectra of the each dye solution measured in a 100 μm optical path in shown in **Figure 9**.

Lasing experiments of dye-soaked opal films from the defect-free regions with cross-section of ~20 μm were chosen for excitation-emission measurements. **Figure 10** shows the lasing spectra of dye solutions in the opal film. The Rh6G solution (without SRh) spectra showed lasing with a well-ordered modal structure. The lasing mechanism is due to a gain distributed feedback (DFB) since the refractive index contrast between the silica and dye solution was negligible, and there were no formation of photonic stop gap (Nishijima et al., 2009a). With the addition of SRh (into Rh6G), the lasing of Rh6G changed into amplified spontaneous emission (ASE) while lasing of SRh at the longer wavelength was obtained. The modal structure of the lasing SRh emission was similar to that observed in pure Rh6G.

Lasing requires a population inversion of excited-state dye molecules within their excitation lifetimes which are ~4 and 5 ns for Rh6G and SRh, respectively. The excitation pulse duration of 7 ns was longer than the luminescence lifetime and limits the population inversion (Magde et al., 1999). Inhibition of the R6G lasing clearly indicates an energy transfer between Rh6G and SRh. The pair of dyes becomes a Förster donor–acceptor pair or, in terms of lasing in a four-level system, the donor absorption (1 → 2 transition) feeds the emission (3 → 4). The decay rates of SRh are: radiative, $k_r = 2.0 \times 10^8$ s^{-1}, and non-radiative,

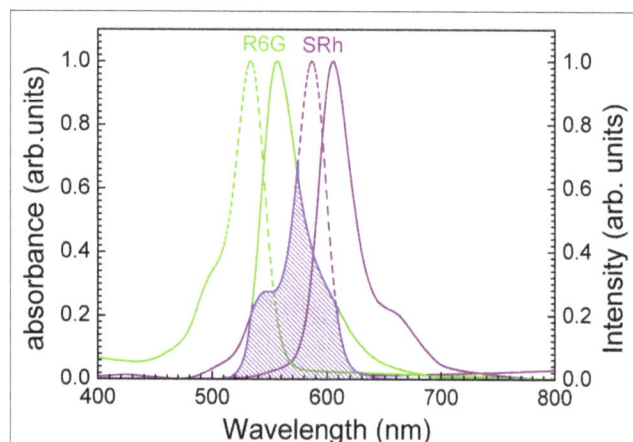

FIGURE 8 | Absorption (dashed lines) and fluorescence (solid lines) spectra of Rh6G and SRh. The donor–acceptor overlap used for calculation of the Förster radius is highlighted.

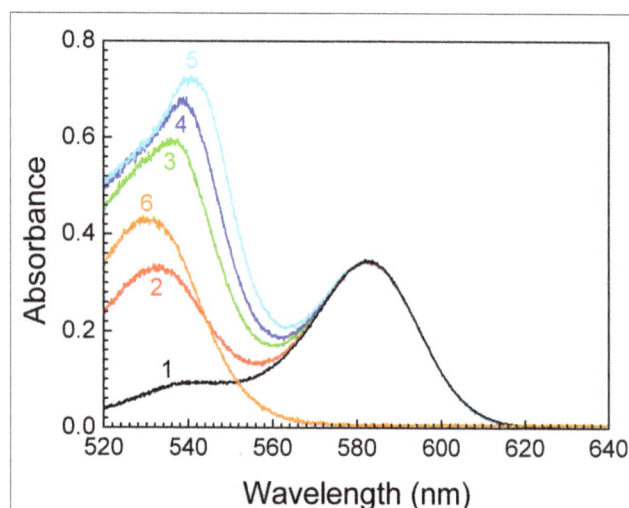

FIGURE 9 | Absorption spectra of dye mixtures: 1 mM of SRh with Rh6G at concentrations 0 (1), 0.8 (2), 1.4 (3), 1.8 (4), 2.0 mM (5) in ethylene glycol; absorption of 1 mM solution of Rh6G is given by curve (6). Sample thickness was 70 μm.

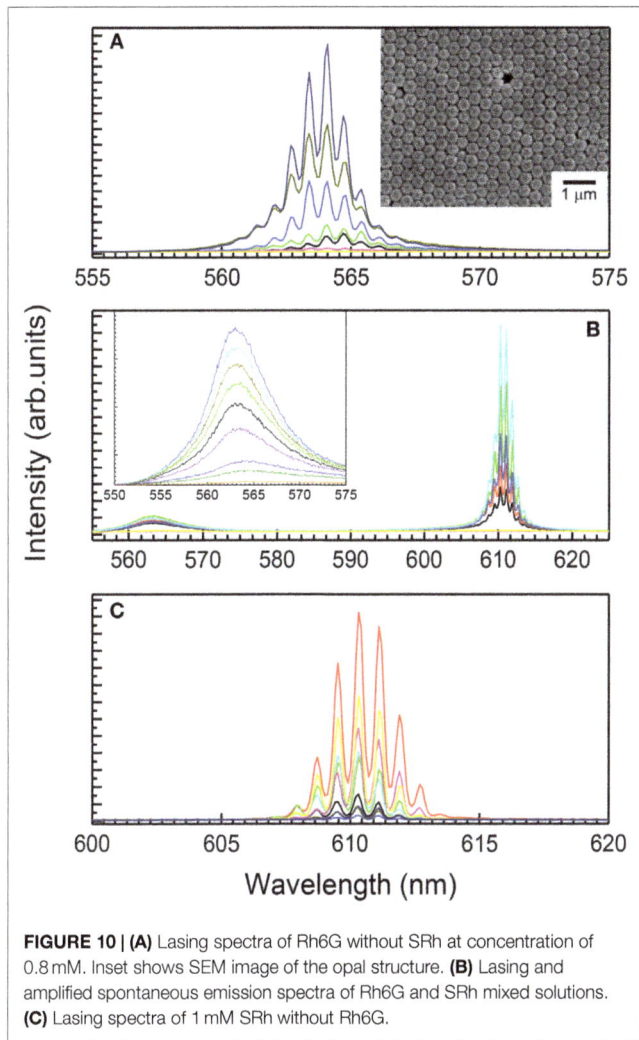

FIGURE 10 | (A) Lasing spectra of Rh6G without SRh at concentration of 0.8 mM. Inset shows SEM image of the opal structure. (B) Lasing and amplified spontaneous emission spectra of Rh6G and SRh mixed solutions. (C) Lasing spectra of 1 mM SRh without Rh6G.

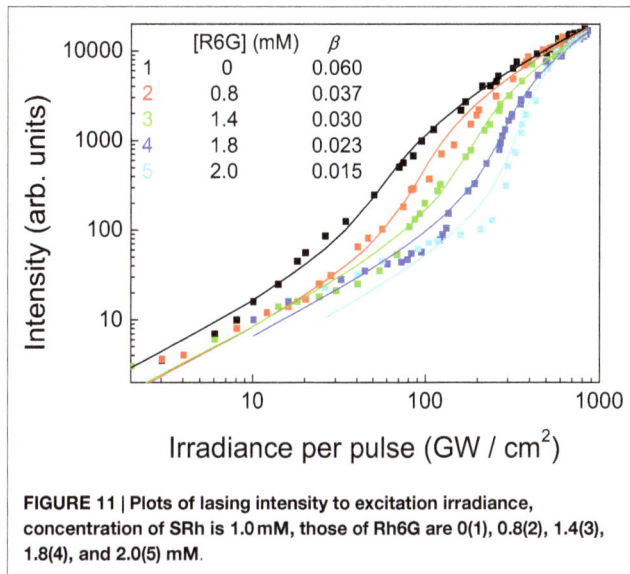

FIGURE 11 | Plots of lasing intensity to excitation irradiance, concentration of SRh is 1.0 mM, those of Rh6G are 0(1), 0.8(2), 1.4(3), 1.8(4), and 2.0(5) mM.

$k_{nr} = 8.6 \times 10^7$ s^{-1}. The energy transfer $2 \to 3$ takes place at the rate $k_{ET}(r) = 2.2 \times 10^{-43} r^{-6}$. Before decaying (within 5 ns), the excited SRh molecules should form a donor–acceptor pair with separation <3 nm.

Figure 11 summarizes data on lasing intensity vs. excitation irradiance. Before lasing, the emission increases linearly on excitation (slope 1 in a log–log presentation). At the lasing threshold, a non-linear increase of intensity was observed. A gain saturation was reached at high irradiance with the onset of linear dependence. Interestingly, the lasing threshold and slope of the non-linear part depended on the concentration of donor dye. The lasing threshold of SRh increased with addition of Rh6G as well as the slope of non-linear part. This demonstrates an inhibition of direct excitation of SRh by a 532 nm excitation laser light with increasing concentration of Rh6G, which is a strong absorber (**Figure 9**). Without Rh6G, the non-linear slope was 1.83 and changes to 2.21 (2), 2.45 (3), 3.34 (4), and 6.30 (5) with increased concentrations of Rh6G. This increase of slopes indicates that effective excitation state formation and population inversion of SRh was achieved due to energy transfer.

The other interesting point of this lasing phenomenon is that the gain-saturated point of lasing is almost the same for each concentration of dye solution. The efficiency of excited-state

energy transfer between Rh6G and SRh is high and is demonstrated in a 3D structure, the opal film. Weak fluorescence of Rh6G still remained, since the energy transfer efficiency is not 100%.

The dependence of a lasing threshold on donor–acceptor concentrations can be quantified via a spontaneous emission coupling ratio β (Horowicz et al., 1992; Takashima et al., 2008). To determine the β value, defined as the fraction of spontaneous emission coupled into a lasing mode with respect to the spontaneous emission into all modes, the input–output power relationship was modeled using a rate equation (Yokoyama and Brorson, 1989; Kippenberg et al., 2006):

$$s = \frac{-1}{2\beta} + \frac{p}{2\Omega} + \frac{\sqrt{(\beta p - \Omega)^2 + 4p\beta^2\Omega}}{2\Omega\beta}, \tag{3}$$

where s is the cavity photon number (proportional to the emission intensity), p is the pump rate (of emitting species), and Ω is the cavity loss rate. This equation was used to fit experimental data with changeable parameters Ω and β. The Ω values were typically 4.0 ± 1 and were not correlated with β.

The best fit β values at different concentration of donor molecules (Rh6G) are plotted in **Figure 12**. With an increasing concentration of Rh6G, β value was decreasing with a well-expressed threshold-like switch-on lasing behavior. In the mixed solution of R6G (energy donor) and SRh (energy acceptor), lasing from Rh6G has never been observed at the used pump powers. This can be explained by an efficient energy transfer via Förster mechanism when, during the life time of the excited donor of approximately few nanoseconds, the energy transfer to the acceptor dye occurred more efficiently and a population inversion of R6G had never occurred. The lowest energy level, the fluorescence band of SRh, was lasing in this system of dye mixture. The lasing threshold increased with increasing concentration of absorbing Rh6G. The non-linear slope of the log–log plot of power dependence can be explained by energy transfer from Rh6G to SRh, which assisted the population inversion of the latter. With an increasing concentration of donor, the probability of energy transfer was also increasing. The small $\beta \sim 10^{-2}$ value represents

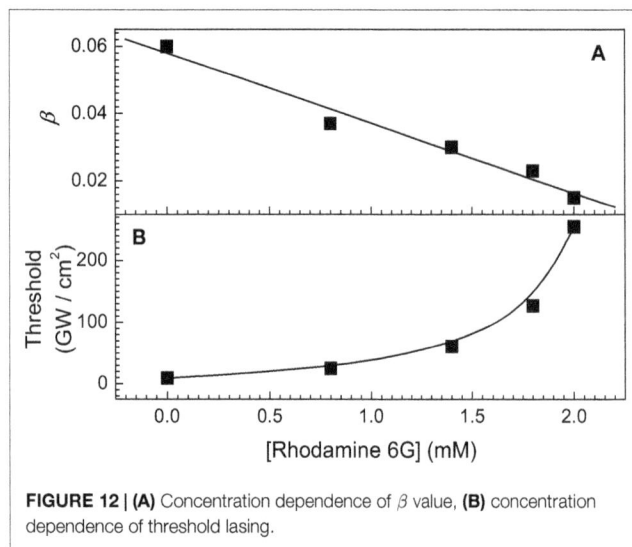

FIGURE 12 | (A) Concentration dependence of β value, **(B)** concentration dependence of threshold lasing.

a strongly non-linear threshold-like switch-on mechanism of lasing, which could find applications in sensor and opto-fluidic applications.

4. Conclusion

We showed that pyramidal templates enable fabrication of single crystalline opal structures. This is because the templates reduce the number of free-surfaces, which always make a destabilizing contribution in structure formation. Hence, by reducing free surfaces (using a template), it is possible to generate well-ordered structures in far-from-equilibrium conditions. Centrifugation helps to facilitate the formation of single crystalline structure in pyramids since destabilizing factors, such as Brownian motion and mechanical vibrations, which plague equilibrium sedimentation, become negligible.

We showed a correlation between the lasing spectrum and the structural quality of opal-structures in both direct and inverse-opal films and pyramidal structures. Single crystalline regions of opal-structures showed well-ordered multi-modal lasing spectra, while poly-crystalline structures showed a less ordered lasing spectra. In both cases, the spectral separation of

modes is defined by the gain-DFB (Nishijima et al., 2009a). Structural 3D characterization of the opals can be carried out using the same setup, which is used to measure photoluminescence and lasing, thereby enabling the correlation between opal structure and its lasing properties to be determined. Establishing the connections between the structural and surface quality of 3D complex structures and their optical functions is required in number of fields in photonics and sensing.

Lasing of a donor–acceptor pair dyes with the presence of strong Förster energy transfer is demonstrated inside silica opals. The Förster energy transfer in a higher refractive index contrast 3D photonic structures with photonic stop gap could further enhance the energy transfer when the donor emission is forbidden by the stop gap and a single-mode lasing is realized (Nishijima et al., 2008). The spectral positions of the lasing modes were stable at the all range of employed laser pumping powers. The lasing mechanism was the gain-DFB. Miniaturized DFB lasers using opal structures as well as micro lasers based on evanescent wave amplification around a colloidal micro-sphere (Datsyuk et al., 2005a,b) are promising directions for miniaturized optically pumped lasers. Such micro-opal patterns can be easily integrated into microfluidic environment (Misawa and Juodkazis, 1999; Juodkazis et al., 2008) providing new functionalities in sensors and laser tweezer applications.

Author Contributions

YN contributed to the concept of the work, designed and carried our experiments, prepared the first draft. SJ contributed to the concept of structural characterization and two dyes lasing, analyzed data, and edited the manuscript.

Acknowledgments

This work is based on the Ph.D. theses of YN. We thank Professor Hiroaki Misawa for continuous support and encouragement during these studies. We are grateful to Professors Shigeki Matsuo, Vitaly Datsyuk, and Sajeev John for discussions. *Funding*: SJ is grateful for the support by ARC Discovery DP130101205 grant.

References

Auer, S., and Frenkel, D. (2001). Prediction of absolute crystal-nucleation rate in hard-sphere colloids. *Nature* 409, 1020–1023. doi:10.1038/35059035

Ben-Shem, A., Frolow, F., and Nelson, N. (2003). Crystal structure of plant photosystem I. *Nature* 426, 630–635. doi:10.1038/nature02200

Biben, T., Hansen, J. P., and Barrat, J. L. (1993). Density profiles of concentrated colloidal suspensions in sedimentation equilibrium. *J. Chem. Phys.* 98, 7330–7344. doi:10.1063/1.464726

Birks, J. B. (1970). *Photophysics of Aromatic Molecules*. London: Wiley-Interscience.

Bolhuis, P. G., Frenkel, D., Mau, S.-C., and Huse, D. A. (1997). Entropy difference between crystal phases. *Nature* 388, 235. doi:10.1038/40779

Bruce, A. D., Wilding, N. B., and Ackland, G. J. (1997). Free energy of crystalline solids: a lattice-switch Monte Carlo method. *Phys. Rev. Lett.* 79, 3002–3005. doi:10.1103/PhysRevLett.79.3002

Choueikani, F., Royer, F., Jamon, D., Siblini, A., Rousseau, J. J., Neveu, S., et al. (2009). Magneto-optical waveguides made of cobalt ferrite nanoparticles embedded in silica/Zirconia organic-inorganic matrix. *Appl. Phys. Lett.* 94, 051113. doi:10.1063/1.3079094

Conti, C., and Fratalocchi, A. (2008). Dynamic light diffusion, three-dimensional Anderson localization and lasing in inverted opals. *Nat. Phys.* 4, 794–798. doi:10.1038/nphys1035

Datsyuk, V. V., Juodkazis, S., and Misawa, H. (2005a). Comparison of the classical and quantum rates of spontaneous light emission in a cavity. *Phys. Rev. A* 72, 025803. doi:10.1103/PhysRevA.72.025803

Datsyuk, V. V., Juodkazis, S., and Misawa, H. (2005b). Properties of a laser based on evanescent-wave amplification. *J. Opt. Soc. Am. B* 22, 1471–1478. doi:10.1364/JOSAB.22.001471

de Almeida, R. F. M., Borst, J., Fedorov, A., Manuel Prieto, Z., and Visser, A. J. W. G. (2007). Complexity of lipid domains and rafts in giant unilamellar vesicles revealed by combining imaging and microscopic and macroscopic time-resolved fluorescence. *Biophys. J.* 93, 539–553. doi:10.1529/biophysj.106.098822

Forster, T. (1947). Experimental and theoretical investigation of intermolecular transfer of electron activation energy. *Z. Naturforsch.* 4a, 321–327.

Forster, T. (1948). Zwischenmolekulare energiewanderung und fluoreszenz. *Ann. Physik.* 437, 55. doi:10.1002/andp.19484370105

Forster, T. (1959). 10th spiers memorial lecture transfer mechanism of electronic excitation. *Discuss. Faraday. Soc.* 27, 7–17. doi:10.1039/df9592700007

Frolov, S., Vardeny, Z., Zakhidov, A., and Baughman, R. (1999). Laser-like emission in opal photonic crystals. *Opt. Commun.* 162, 241–246. doi:10.1016/S0030-4018(99)00089-9

Fujiwara, H., Sasaki, K., and Masuhara, H. (2005). Enhancement of Forster energy transfer withina microspherical cavity. *Chemphyschem* 6, 2410–2416. doi:10.1002/cphc.200500318

Gottardo, S., Sapienza, R., García, P. D., Blanco, A., Wiersma, D. S., and López, C. (2008). Resonance-driven random lasing. *Nat. Phot.* 2, 383–446. doi:10.1038/nphoton.2008.102

Hirakawa, S., Kawamata, J., Suzuki, Y., Tani, S., Murafuji, T., Kasatani, K., et al. (2008). Two-photon absorption properties of azulenyl compounds having a conjugated ketone backbone. *J. Phys. Chem. A* 112, 5198–5207. doi:10.1021/jp800415b

Hoogenboom, J. P., Derks, D., Vergeer, P., and Blaaderen, A. (2002). Stacking faults in colloidal crystals grown by sedimentation. *J. Chem. Phys.* 117, 11320–11328. doi:10.1063/1.1522397

Horowicz, R. J., Heitmann, H., Kadota, Y., and Yamamoto, Y. (1992). Gaas microcavity quantum-well laser with enhanced coupling of spontaneous emission to the lasing mode. *Appl. Phys. Lett.* 61, 392–394. doi:10.1063/1.107893

Hotta, J., Ujii, H., and Hofkens, J. (2006). The fabrication of a thin, circular polymer film based phase shaper for generating doughnut modes. *Opt. Express* 14, 6273–6278. doi:10.1364/OE.14.006273

Hur, J., and Won, Y.-Y. (2008). Fabrication of high-quality non-close-packed 2d colloid crystals by template-guided Langmuir-Blodgett particle deposition. *Soft Matter* 4, 1261–1269. doi:10.1039/b716218a

John, S. (1987). Strong localization of photons in certain disordered dielectric superlattices. *Phys. Rev. Lett.* 58, 2486–2489. doi:10.1103/PhysRevLett.58.2486

Juodkazis, S., Mizeikis, V., Matsuo, S., Ueno, K., and Misawa, H. (2008). Three-dimensional micro- and nano-structuring of materials by tightly focused laser radiation. *Bull. Chem. Soc. Jpn.* 81, 411–448. doi:10.1246/bcsj.81.411

Kamada, K., Antonov, L., Yamada, S., Ohta, K., Yoshikawa, T., Tahara, K., et al. (2007). Two-photon absorption properties of dehydrobenzo[12]annulenes and hexakis(phenylethynyl)benzenes: effect of edge-linkage. *Chemphyschem* 8, 2671–2677. doi:10.1002/cphc.200700555

Kanai, T., Sawada, T., and Kitamura, K. (2005). Air-pulse-drive fabrication of photonic crystal films of colloids with high spectral quality. *Adv. Funct. Mater.* 15, 25–29. doi:10.1002/adfm.200305160

Kippenberg, T. J., Kalkman, J., Polman, A., and Vahala, K. J. (2006). Demonstration of an erbium-doped microdisk laser on a silicon chip. *Phys. Rev. A* 74, 051803(R). doi:10.1103/PhysRevA.74.051802

Kita, S., Kengo, N., and Baba, T. (2008). Refractive index sensing utilizing a cw photonic crystal nanolaser and its array configuration. *Opt. Express* 16, 8174–8180. doi:10.1364/OE.16.008174

Kitamura, N., Fukumi, K., Nishii, J., and Ohno, N. (2007). Relationship between refractive index and density of synthetic silica glasses. *J. Appl. Phys.* 101, 123533. doi:10.1063/1.2748861

Lavrinenko, A. V., Wohlleben, W., and Leyrer, R. J. (2009). Influence of imperfections on the photonic insulating and guiding properties of finite Si inverted opal crystals. *Opt. Express* 17, 747–760. doi:10.1364/OE.17.000747

Li, J., Jia, B., Zhou, G., Bullen, C., Serbin, J., and Gu, M. (2007). Spectral redistribution in spontaneous emission from quantum-dot-infiltrated 3d woodpile photonic crystals for telecommunications. *Adv. Mater.* 19, 3276–3280. doi:10.1002/adma.200602054

Lopez, C. (2006). Three-dimensional photonic bandgap materials: semiconductors for light. *J. Opt. A* 8, R1–R14. doi:10.1088/1464-4258/8/5/R01

Magde, D., Rojas, G. E., and Seybold, P. G. (1999). Solvent dependence of the fluorescence lifetimes of xanthene dyes. *Photochem. Photobiol.* 70, 737–744. doi:10.1111/j.1751-1097.1999.tb08277.x

Matsubara, H., Yoshimoto, S., Saito, H., Yue, J. L., Tanaka, Y., and Noda, S. (2008). Gan photonic-crystal surface-emitting laser at blue-violet wavelengths. *Science* 319, 445–447. doi:10.1126/science.1150413

Matsuo, S., Fujine, T., Fukuda, K., Juodkazis, S., and Misawa, H. (2003). Formation of free-standing micro-pyramid colloidal crystals grown on silicon substrate. *Appl. Phys. Lett.* 82, 4283–4285. doi:10.1063/1.1583138

Misawa, H., and Juodkazis, S. (1999). Photophysics and photochemistry of a laser manipulated microparticle. *Prog. Polym. Sci.* 24, 665–697. doi:10.1016/S0079-6700(99)00009-X

Mizeikis, V., Juodkazis, S., Marcinkevicius, A., Matsuo, S., and Misawa, H. (2001). Tailoring and characterization of photonic crystals. *J. Photochem. Photobiol. C* 2, 35–69. doi:10.1021/ar700143w

Nishijima, Y. (2009), *Spectroscopic Studies of Opal and Inverse Opal Photonic Crystals.* Ph.D. thesis, Hokkaido University, Sapporo

Nishijima, Y., Ueno, K., Juodkazis, S., Mizeikis, V., Fujiwara, H., and Misawa, H. (2009a). Lasing with well-defined cavity modes in dye-infiltrated silica inverse opals. *Opt. Express* 17, 2976–2983. doi:10.1364/OE.17.002976

Nishijima, Y., Ueno, K., Juodkazis, S., Mizeikis, V., Misawa, H., Maeda, M., et al. (2008). Tunable single-mode photonic lasing from zirconia inverse opal photonic crystals. *Opt. Express* 16, 13676. doi:10.1364/OE.16.013676

Nishijima, Y., Ueno, K., Juodkazis, S., Mizeikis, V., Misawa, H., Tanimura, H., et al. (2007). Inverse silica opal photonic crystals for optical sensing applications. *Opt. Express* 15, 12979–12988. doi:10.1364/OE.15.012979

Pronk, S., and Frenkel, D. (1999). Can stacking faults in hard-sphere crystals anneal out spontaneously? *J. Chem. Phys.* 110, 4589–4592. doi:10.1063/1.478339

Pronk, S., and Frenkel, D. (2003). Large difference in the elastic properties of fcc and hcp hard-sphere crystals. *Phys. Rev. Lett.* 90, 255501. doi:10.1103/PhysRevLett.90.255501

Santamaria, F. G., Galisteo-Lopez, J. F., Braun, P. V., and Lopez, C. (2005). Optical diffraction and high-energy features in three-dimensional photonic crystals. *Phys. Rev. B* 71, 195112. doi:10.1103/PhysRevB.71.195112

Santamaria, F. G., Miyazaki, H. T., Urquia, A., Ibisate, M., Belmonte, M., Shinya, N., et al. (2002). Nanorobotic manipulation of microspheres for on-chip diamond architectures. *Adv. Mater.* 14, 1144–1147. doi:10.1002/1521-4095(20020816)14:16<1144::AID-ADMA1144>3.0.CO;2-I

Sasaki, K., Koshioka, M., and Masuhara, H. (1991). Tree-dimensional space- and time resolved fluorescence spectroscopy. *Appl. Spectroscopy* 45, 1041. doi:10.1366/0003702914336336

Shkunov, M. N., DeLong, M. C., Raikh, M. E., Vardeny, Z. V., Zakhidov, A. A., and Baughman, R. H. (2001). Photonic versus random lasing in opal single crystals. *Synth. Met.* 116, 485–491. doi:10.1016/S0379-6779(00)00420-3

Stevens, N., O'Connor, N., Vishwasrao, H., Samaroo, D., Kandel, E. R., Akins, D. L., et al. (2009). Two color RNA intercalating probe for cell imaging applications. *J. Am. Chem. Soc.* 130, 7182–7183. doi:10.1021/ja8008924

Takakusa, H., Kikuchi, K., Urano, Y., and Tetsuo Nagano, H. K. (2003). A novel design method of ratiometric fluorescent probes based on fluorescence resonance energy transfer switching by spectral overlap integral. *Chem. Eur. J.* 9, 1479–1485. doi:10.1002/chem.200390167

Takashima, H., Fujiwara, H., Takeushi, S., Sasaki, K., and Takahashi, M. (2008). Control of spontaneous emission coupling factor β in fiber-coupled microsphere resonators. *Appl. Phys. Lett.* 92, 071115. doi:10.1063/1.2884329

Tawa, K., Hori, H., Kintaka, K., Kiyosue, K., Tatsu, Y., and Nishii, J. (2008). Optical microscopic observation of fluorescence enhanced by grating-coupled surface plasmon resonance. *Opt. Express* 162, 9781–9789. doi:10.1364/OE.16.009781

Velasco, E., Mederos, L., and Navascus, G. (1998). Phase diagram of colloidal systems. *Langmuir* 14, 5652–5655. doi:10.1021/la980126y

Wright, D., Brasselet, E., Zyss, J., Langer, G., and Kern, W. (2004). Dte-doped organic distributed-feedback lasers with index and surface gratings: the role of pump polarization and molecular orientation. *J. Opt. Soc. Am. B* 21, 944–950. doi:10.1364/JOSAB.21.000944

Wu, J., Day, D., and Gu, M. (2008). A microfluidic refractive index sensor based on an integrated three-dimensional photonic crystal. *Appl. Phys. Lett.* 92, 071108. doi:10.1063/1.2840700

Yablonovitch, E. (1987). Inhibited spontaneous emission in solid-state physics and electronics. *Phys. Rev. Lett.* 58, 2059–2062. doi:10.1103/PhysRevLett.58.2059

Yokoyama, H., and Brorson, S. D. (1989). Rate equation analysis of microcavity lasers. *J. Appl. Phys.* 66, 4801–4805. doi:10.1364/OE.18.009909

Conflict of Interest Statement: The authors declare that the research was conducted in the absence of any commercial or financial relationships that could be construed as a potential conflict of interest.

4

Unsaturated and saturated permeabilities of fiber reinforcement: critics and suggestions

Chung Hae Park and Patricia Krawczak *

Polymers and Composites Technology and Mechanical Engineering Department, Ecole Nationale Supérieure des Mines de Douai, Douai, France

In general, permeability measurement results show a strong scattering according to the measurement method, the type of test fluid and the fluid injection condition, even though permeability is regarded as a unique property of porous medium. In particular, the discrepancy between the unsaturated and saturated permeabilities for the same fabric has been widely reported. In the literature, relative permeability has been adopted to model the unsaturated flow. This approach has some limits in the modeling of double-scale porosity medium. We address this issue of permeability measurement by rigorously examining the mass conservation condition. Finally, we identify that the pressure gradient is non-linear with positive curvature in the unsaturated flow and a misinterpretation of pressure gradient is the main reason for the difference between the saturated and unsaturated permeabilities of the same fiber reinforcement. We propose to use a fixed value of permeability and to modify the mass conservation equation if there are air voids, which are entrapped inside the fiber tow. Finally, we also suggest some guidelines and future perspectives to obtain more consistent permeability measurement results.

Keywords: composites manufacturing, permeability, Darcy's law, air voids, pressure gradient, unsaturated flow

Edited by:
Masaki Hojo,
Kyoto University, Japan

Reviewed by:
Ralf Schledjewski,
Montanuniversität Leoben, Austria
Ryosuke Matsuzaki,
Tokyo University of Science, Japan
Woo Il Lee,
Seoul National University,
South Korea

*Correspondence:
Patricia Krawczak,
Polymers and Composites
Technology and Mechanical
Engineering Department, Mines
Douai, 941 rue Charles Bourseul,
Douai 59508, France
patricia.krawczak@mines-douai.fr

Introduction

The most effective way to obtain polymer composites is to impregnate fiber reinforcement by liquid resin. This principle is employed in many composites manufacturing techniques, for example, liquid composite molding where liquid resin permeates into the dry fabric preplaced in a closed mold. Therefore, the analysis of resin flow in fibrous medium is an important phase in the process optimization. It has been general to use Darcy's law, which is an empirical constitutive relation between flow rate and pressure gradient, regarding the fiber reinforcement as porous medium (Darcy, 1856).

$$\vec{u}_{\mathrm{D}} = \frac{Q}{A} = -\frac{\tilde{K}}{\mu}\nabla P \tag{1}$$

where $\vec{u}_{\mathrm{D}}$ is the volume-averaged fluid velocity, Q is the flow rate, A is the flow cross-section, $\tilde{K}$ is the permeability tensor of fiber reinforcement, μ is the fluid viscosity, and P is the fluid pressure. Hence, permeability is a hydraulic conductivity for the fluid motion through porous medium.

Darcy's law can be integrated into the mass conservation equation for incompressible fluid to obtain the pressure field.

$$\nabla \cdot \vec{u}_{\mathrm{D}} = \nabla \cdot \left(-\frac{\tilde{K}}{\mu}\nabla P \right) = 0 \tag{2}$$

It is noticeable that the pressure gradient becomes linear in the case of rectilinear flow if the permeability and the viscosity are constant.

Generally, permeability is considered to be a unique property of fiber reinforcement and it is independent of fluid property such as viscosity as well as of fluid pressure as long as inertial effect is negligible compared with viscous force. Thus, permeability can be determined from the microstructure of porous medium, such as pore size and connectivity, irrespective of fluid type. Currently, there is no standard method for permeability measurement and each laboratory uses its own experimental set-up and method. It has been reported that the experimental results for a same reinforcement can be different according to measurement method, device, and condition (Dungun and Sastry, 2002; Pillai, 2004; Arbter et al., 2011; Naik et al., 2014; Vernet et al., 2014).

The focus of this paper is to investigate a principal reason for the inconsistency of permeability measurement results, especially the dependence on the measurement method. We also propose some remedies and perspectives for this subject.

Permeability Measurement Methods

Many different experimental set-ups and methods for permeability measurement have been proposed during the last several decades (Arbter et al., 2011; Naik et al., 2014; Vernet et al., 2014). We classify permeability measurement methods into two categories according to the flow condition. The one is the saturated permeability (a.k.a. steady-state permeability) method where fiber reinforcement is impregnated by a test liquid, which is introduced into the mold under a constant flow rate. Once the reinforcement is fully saturated and the flow in the mold reaches a steady state, pressure values are measured at two distant points, usually liquid inlet and outlet. Then, the pressure gradient is obtained from the ratio of the pressure drop to the distance between the two pressure sensors assuming the pressure gradient in the mold is linear.

$$K_{sat} = \frac{Q \cdot \mu}{A} \bigg/ \left(-\frac{dP}{dx} \right) = \frac{Q \cdot \mu \cdot L}{A \cdot \Delta P} \tag{3}$$

where K_{sat} is the saturated permeability. L and ΔP are the distance and the pressure difference between the two pressure sensors, respectively.

The other is the unsaturated permeability (a.k.a. unsteady-state or transient permeability) method where the fiber reinforcement in the mold is progressively impregnated by a test liquid injected under a constant pressure at the inlet. During the mold filling process, the flow front position is recorded as a function of time. In general, pressure is monitored only at the liquid inlet assuming that the pressure at the flow front is the atmospheric pressure. According to Darcy's law, we can obtain the relations between flow front position and time in the case of rectilinear flow as following.

$$\frac{dl_f(t)}{dt} = \frac{K_{unsat}}{(1 - V_f) \cdot \mu} \left(-\frac{dP}{dx} \bigg|_{x=l_f} \right) = \frac{K_{unsat}}{(1 - V_f) \cdot \mu} \left(\frac{P_{in}}{l_f(t)} \right) \tag{4}$$

$$K_{unsat} = \frac{l_f^2(t)}{t} \frac{\mu \cdot (1 - V_f)}{P_{in}} \tag{5}$$

where $l_f(t)$ is the distance between the liquid inlet and the flow front, K_{unsat} is the unsaturated permeability, V_f is the fiber volume fraction and P_{in} is the inlet pressure. Generally, square of flow front position is plotted against time in a graphical form. Then, unsaturated permeability value is evaluated from the slope of the straight line by least square fitting of data set.

Issues and Critics

As stated previously, permeability is regarded as an intrinsic property of porous medium. In the literature, however, we can find a strong scatter of permeability measurement results even for a same fiber reinforcement, which depends on the measurement method (saturated permeability vs. unsaturated permeability), the test conditions (injection pressure or flow rate), and the test fluid (Dungun and Sastry, 2002; Pillai, 2004; Arbter et al., 2011; Naik et al., 2014; Vernet et al., 2014). To address this issue of inconsistency, there have been two benchmark exercises of permeability measurement (Arbter et al., 2011; Vernet et al., 2014). At the first benchmark exercise, the permeabilities of two fabrics were measured by different research teams using their own devices and methods (Arbter et al., 2011). A large dispersion of permeability values were observed even for the same reinforcement. Measurement error from human factors was assumed to be the main source for this great variation. At the second benchmark exercise, 12 research teams employed the same measurement procedure to obtain the permeability values for the same carbon fabric (Vernet et al., 2014). Contrary to the first benchmark exercise, a small scatter was observed and the influence from human factors was not significant. Hence, it was proven that the permeability measurement was reproducible if the same procedure and condition were used. However, a large dispersion was observed if different parameters such as injection pressure and test fluid were employed even for the same set-up and procedure.

We can summarize some representative observations from permeability measurement results in the literature as following.

- The ratio of unsaturated to saturated permeabilities varies from 1/4 to 4 (Dungun and Sastry, 2002; Pillai, 2004). In many cases, however, the unsaturated permeability is smaller than the saturated permeability (Dungun and Sastry, 2002; Pillai, 2004; Arbter et al., 2011; Naik et al., 2014; Vernet et al., 2014).

- The test fluid viscosity has been found to have insignificant influence on saturated permeability measurements if it is as low as that of thermoset resin used in the RTM process.

- Many technical sources causing measurement errors and corresponding remedies have been identified: mold deflection (wrong estimation of fiber volume fraction), fluid viscosity variation (strong dependence on temperature), number of fabrics (nesting effect), race-tracking effect, variation of injection pressure or flow rate (especially at the early stage of mold filling), deviation of fabric direction from the principal flow direction in the mold (in the case of rectilinear injection), uncertainty in the flow front tracking in the case of visual observation, etc.

- The role of capillary pressure in the unsaturated permeability method is controversial. Some researchers corrected the unsaturated permeability considering the surface tension effect at the flow front (Ahn et al., 1991; Amico and Lekakou, 2001). On the contrary, some researchers argue that the influence on the capillary pressure is insignificant because the capillary pressure estimated from the characteristic pore size is much smaller than the inlet pressure (Masoodi et al., 2012).

In a subsequent section, we focus on the difference between the unsaturated and saturated permeabilities for the same fabric by investigating the flow physics in the porous medium.

Relative Permeability and Degree of Saturation

Many researchers have proposed a hypothesis that air void formation during the unsaturated flow is a major source for the discrepancy between saturated and unsaturated permeabilities. The microstructure of textile reinforcement is highly heterogeneous showing dual scale porosity distribution, viz., micropore, which is a tiny space between fiber filaments inside the fiber tow and macropore, which is an open gap between fiber tows. Hence, the liquid flow advancement during the mold filling process is non-uniform and air is entrapped at the flow front. Subsequently, air voids remaining inside the fiber reinforcement create partially impregnated zone and the overall hydraulic conductivity may be altered. This phenomenon has been well described in the soil mechanics where Darcy's law had been first proposed. In the experiment where Darcy's law was derived, the volumetric rate of water flowing through a vertical column of soil with a known height was measured to determine the hydraulic conductivity of soil. Therefore, the experimental condition of the original Darcy's experiment corresponds to the saturated permeability measurement method. In the progressive permeation of liquid into porous medium such as sand, however, we can also observe the generation of partially impregnated zone, which is known as "fingering" phenomenon. If there are partially impregnated zones in the soil, the flow rate is decreased for a given height of soil column. To consider this change of overall hydraulic conductivity, relative permeability is frequently adopted which is a ratio of the flow rate in a fully saturated soil to that in partially saturated soil (Van Genuchten, 1980). Thus, relative permeability, k_r can also be considered as a ratio of unsaturated to saturated permeabilities.

$$k_r = \frac{K_{unsat}}{K_{sat}} \tag{6}$$

Generally, relative permeability is defined by degree of saturation, S_w, which is a ratio of liquid volume, V_l to pore volume, V_p in the porous medium.

$$S_w = \frac{V_l}{V_p} \tag{7}$$

As relative permeability varies between the zero and the unity, unsaturated permeability is always smaller than saturated permeability.

The concept of relative permeability has also been adopted to model the unsaturated resin flow for composites manufacturing (De Parseval et al., 1997; Breard et al., 2003; Garcia et al., 2010). Some researchers described the pressure profile in the unsaturated

flow by introducing in the governing equation the relative permeability which varies according to degree of saturation (De Parseval et al., 1997; Breard et al., 2003; Garcia et al., 2010).

$$\nabla \cdot \left(\frac{K_{unsat}}{\mu} \nabla P \right) = \nabla \cdot \left(\frac{k_r\,(S_w) \cdot K_{sat}}{\mu} \nabla P \right) = 0 \tag{8}$$

Generally, the degree of saturation is greater at the upstream than at the downstream. Thus, the relative permeability is greater at the upstream than at the downstream (**Figure 1A**). In this modeling result, therefore, the local negative pressure gradient is smaller at the upstream than at the downstream because negative pressure gradient is inversely proportional to permeability from Darcy's law. Subsequently, the pressure profile should be a non-linear curve with negative curvature (**Figure 1A**) (De Parseval et al., 1997; Breard et al., 2003; Garcia et al., 2010).

Non-Linear Pressure Gradient in the Unsaturated Flow

The relative permeability model suggests that the pressure profile in unsaturated flow with air voids is a non-linear curve with negative curvature. Jung et al. (2010) performed mold filling experiments to study the influences from air voids on the unsaturated flow behavior. It has been widely reported that the type and content of air voids generated during the mold filling process depend on the flow rate (Park and Lee, 2011). In their experiments, test liquid was injected under a pre-assigned constant flow rate into the mold containing a unidirectional stitched fabric to generate a uniform void content at the flow front. This test was repeated for different values of constant flow rate and the pressure profile was observed during the mold filling process. At the high flow rate cases where intra-bundle voids were generated, the pressure profiles were always non-linear as represented in **Figure 1B**, which were deviated from the prediction of classical Darcy's law. Moreover, it should be noted that the curvature of non-linear pressure profile in the experiment was positive whereas it has been always negative in the modeling results using relative permeability approach in the literature. At the low flow rate injection case where inter-bundle voids were formed also resulted in a non-linear pressure profile with positive curvature. Meanwhile, a linear pressure gradient was observed in the case of the intermediate flow rate where neither inter-bundle void nor intra-bundle void was generated. This result can also be confirmed by numerical simulation of transient Stokes–Brinkman equation (**Figure 1C**) (Jung et al., 2010). Near the flow front where fiber tows are partially impregnated, the pressure gradient is non-linear showing positive curvature whereas it is linear at the vicinity of injection gate where the preform is fully saturated (Jung et al., 2010).

These results imply a very important clue to understand the difference between the unsaturated and saturated permeabilities. The basic assumption to evaluate both the saturated and unsaturated permeabilities by Darcy's law is that the pressure gradient is linear (see also Eqs 3 and 4).

$$-\frac{dP}{dx} = \frac{\Delta P}{L} \tag{9}$$

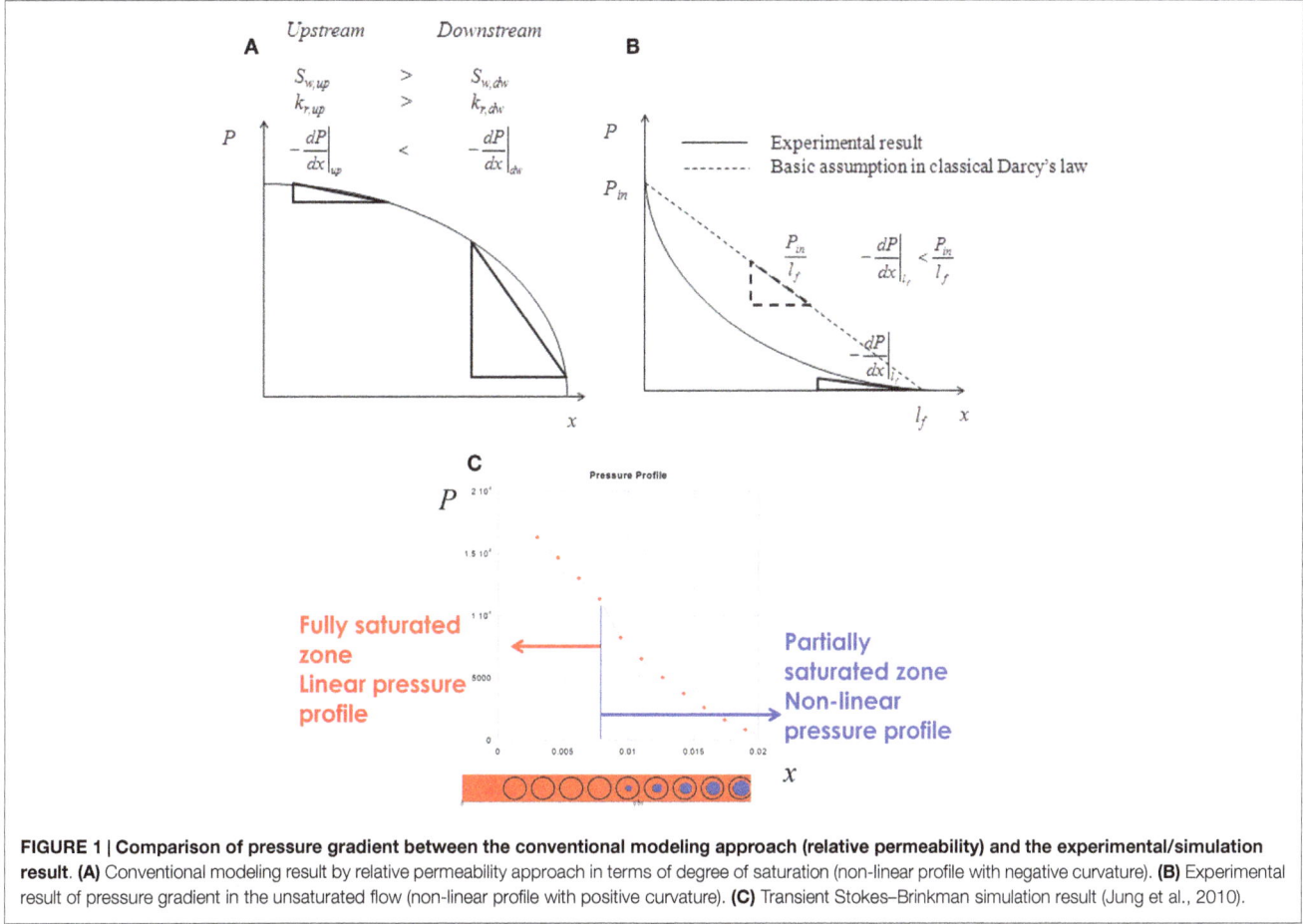

FIGURE 1 | Comparison of pressure gradient between the conventional modeling approach (relative permeability) and the experimental/simulation result. (A) Conventional modeling result by relative permeability approach in terms of degree of saturation (non-linear profile with negative curvature). (B) Experimental result of pressure gradient in the unsaturated flow (non-linear profile with positive curvature). (C) Transient Stokes–Brinkman simulation result (Jung et al., 2010).

As can be seen in the experimental results, however, the pressure gradient is no more linear and cannot be obtained by dividing the pressure difference by the distance between the inlet and outlet (as represented by Eq. 9) if there are air voids.

Especially, it should be kept in mind that the unsaturated permeability is determined by monitoring the flow front advancement with time. During the unsaturated flow experiment, it is the negative pressure gradient at the flow front that decides the flow front advancing velocity. In the case of unsaturated flow, the actual negative pressure gradient at the flow front is smaller than the prediction from the assumption of linear pressure gradient (Figure 1B).

$$-\frac{dP}{dx}\bigg|_{l_f} < \frac{P_{in}}{l_f} \qquad (10)$$

Subsequently, the negative pressure gradient is overestimated and the permeability value is underestimated if Eqs 4 and 5 are used to evaluate unsaturated permeability values.

Mass Conservation Equation Considering Mass Sink Effect

Pillai (2004) proposed to reformulate the mass conservation equation using the same permeability value regardless of flow type. In the case of saturated flow, the inlet liquid flow (Q_{in}) and the outlet liquid flow (Q_{out}) in the control volume are identical and the governing equation represented by Eq. 2 still holds good

(Figure 2A). If fiber tows behind the global flow front are not fully saturated, however, some portion of liquid flows from the macropore into the fiber tow as represented by leakage flow (Q_T) in Figure 2B. Hence, the flow rate is not constant in the flow domain. For example, Darcy's velocity ($\vec{u}_D$) is greater at the upstream than at the downstream where some portion of liquid from the upstream has already been leaked into partially saturated fiber tows. Eventually, the mass continuity equation should be modified by considering this leakage flow, which is represented as a mass sink term.

$$Q_{in} - Q_{out} = Q_T \qquad (11)$$

$$-\nabla \cdot \vec{u}_D = -\nabla \cdot \left(-\frac{\tilde{K}}{\mu}\nabla P\right) = \dot{q} \qquad (12)$$

where $\dot{q}$ is the mass sink term which is the volumetric flow rate entering the fiber tow per unit control volume. We can see that the governing equation for pressure field is Poisson equation with a sourced term in the right hand side. Subsequently, the pressure gradient becomes non-linear in the case of rectilinear flow if this source term is non-zero. If there is no void inside the tow, this mass sink term is zero and the governing equation for pressure becomes Laplace equation where a linear pressure gradient is obtained in the case of rectilinear flow. It should be noted that a fixed value for permeability can be used to model both the saturated and unsaturated flows.

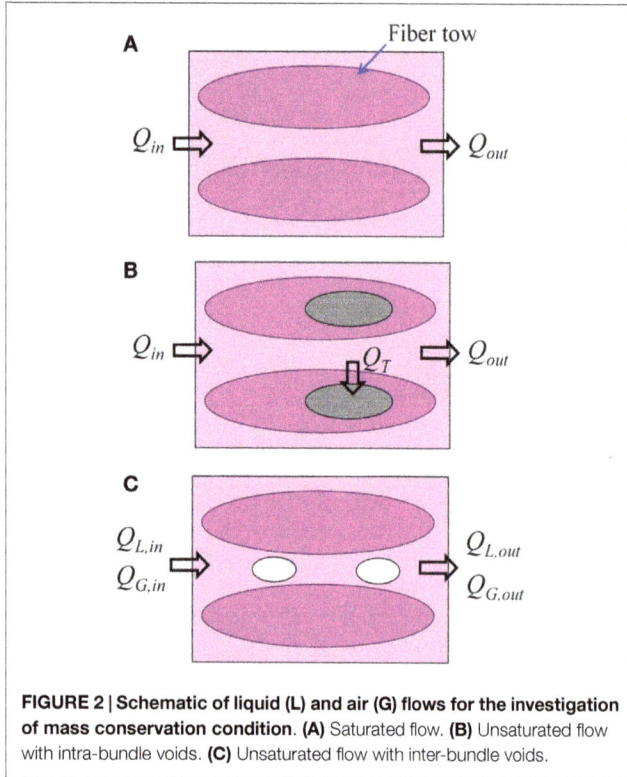

FIGURE 2 | Schematic of liquid (L) and air (G) flows for the investigation of mass conservation condition. (A) Saturated flow. **(B)** Unsaturated flow with intra-bundle voids. **(C)** Unsaturated flow with inter-bundle voids.

This new governing equation can be converted into a dimensionless form in the case of rectilinear flow.

$$\frac{d^2P^*}{dx^{*2}} = \frac{\mu \cdot l_f^2}{K_x \cdot P_{in}}\dot{q} \tag{13}$$

where $P*(= P/P_{in})$ and $x*(= x/l_f)$ are dimensionless pressure and position, respectively. We can see that the non-linearity of pressure profile increases as the flow front position increases and the inlet pressure decreases. Hence, there are significant influences from the mold size and the liquid injection condition upon unsaturated permeability measurement results. The liquid type also affects the unsaturated permeability measurement because the mass sink term is proportional to the test liquid viscosity. Moreover, wetting properties such as surface tension and contact angle play an important role to determine the mass sink effect because the capillary wicking is dominant in the flow inside the fiber tow.

As previously mentioned, the relative permeability model leads to negative curvature of pressure profile whereas positive curvature is obtained in the experiment. This contradictory issue can be addressed by considering the mass sink effect in the mass conservation even if a fixed value of permeability is employed. The curvature of pressure profile is proportional to the mass sink term ($\dot{q}$) on the right hand side of Eq. 13. In the case of intra-bundle void formation, the liquid first fills the open gap between the fiber tows and then enters the fiber tows. Thus, the mass sink term is always positive and the curvature of pressure profile should also be positive. It should be reminded that in the soil mechanics, relative permeability has been adopted to predict the flow rate under a given pressure drop in the soil. Therefore, relative permeability

should be interpreted as a measure of overall flow rate for a pre-assigned global pressure drop and it may be misleading to employ relative permeability for the prediction of local pressure gradient under the assumption of constant flow rate or flow velocity across the entire flow distance.

Conclusion and Perspectives

We investigated the discrepancy between the saturated and unsaturated permeabilities for the same fabric in terms of air void formation during the resin flow. The overestimation of negative pressure gradient in the unsaturated permeability measurement has been identified as the main reason for this inconsistency. The limits of relative permeability model were discussed and some contradictions were demonstrated for the unsaturated flow behavior. To address this issue, the mass conservation equation should be modified by considering the mass sink effect. Subsequently, the non-linear pressure profile with positive curvature can be modeled even if a fixed value of permeability is employed regardless of flow condition. By the dimensionless analysis, we can estimate the influences from the liquid type, the injection pressure, the mold size and the void content upon the unsaturated permeability measurement.

To sum up, we can propose some guidelines and perspectives to improve the reliability of permeability characterization as following.

- It is crucial to correctly obtain the pressure gradient at the flow front in the case of transient mold filling experiment. It is recommended to use a number of pressure transducers in the mold, instead of single (or two) pressure sensor(s), in order to obtain the local pressure gradient at different positions and time instants. It should be kept in mind that the pressure gradient is no more linear if there are air voids in the preform.
- Permeability should be evaluated not by the overall negative pressure gradient $\left(\frac{P_{in}}{l_f}\right)$ but by the local negative pressure gradient at the flow front $\left(-\frac{dP}{dx}\Big|_{x=l_f}\right)$ in the case of unsaturated permeability measurement.

$$l_f(t) = \int_0^t \frac{dl_f(\tau)}{d\tau}d\tau = \frac{K_x}{(1-V_f)\cdot\mu}\left(\int_0^t -\frac{dP}{dx}\Big|_{x=l_f(\tau)}d\tau\right) \tag{14}$$

- Air void migration in the macropore is an important parameter to influence the unsaturated flow behavior. If air voids move along the resin flow, the mass conservation equation for incompressible fluid is no more valid. The mass conservation should be revisited considering both the liquid and air flows (**Figure 2C**).

$$Q_{L,in} - Q_{L,out} \neq 0 \tag{15}$$

$$\left(Q_{L,in} - Q_{L,out}\right) + \left(Q_{G,in} - Q_{G,out}\right) = 0 \tag{16}$$

Furthermore, some air bubbles move faster than the resin flow whereas some bubbles are stationary being blocked between fiber tows. Hence, the air void distribution is not uniform and air voids are usually concentrated at the downstream. This non-uniform air void distribution can make a significant influence on the local pressure gradient. Even though the unsaturated permeability is lower than the saturated permeability in many cases, there are some results reported in the literature where the saturated permeability is greater than the unsaturated permeability (Dungun and Sastry, 2002; Pillai, 2004). To address this issue, the migration velocity and distribution of air voids during the mold filling process should be considered in two-phase flow modeling.

References

Ahn, K. J., Seferis, J. C., and Berg, J. C. (1991). Simultaneous measurements of permeability and capillary pressure of thermosetting matrices in woven fabric reinforcements. *Polym. Compos.* 12, 146–152. doi:10.1002/pc.750120303

Amico, S., and Lekakou, C. (2001). An experimental study of the permeability and capillary pressure in resin-transfer moulding. *Compos. Sci. Technol.* 61, 1945–1959. doi:10.1016/S0266-3538(01)00104-X

Arbter, R., Beraud, J. M., Binetruy, C., Bizet, L., Bréard, J., Comas-Cardona, S., et al. (2011). Experimental determination of the permeability of textiles: a benchmark exercise. *Compos. A* 42, 1157–1168. doi:10.1016/j.compositesa.2011.04.021

Breard, J., Henzel, Y., Trochu, F., and Gauvin, R. (2003). Analysis of dynamic flows through porous media. Part I: comparison between saturated and unsaturated flows in fibrous reinforcement. *Polym. Compos.* 24, 391–408. doi:10.1002/pc.10039

Darcy, H. (1856). *Les fontaines publiques de la Ville de Dijon.* Paris: Dalmont.

De Parseval, Y., Pillai, K. M., and Advani, S. G. (1997). A simple model for the variation of permeability due to partial saturation in dual scale porous media. *Transp. Porous Med.* 27, 243–264. doi:10.1023/A:1006544107324

Dungun, F. D., and Sastry, A. M. (2002). Saturated and unsaturated polymer flows: microphenomena and modeling. *J. Compos. Mater.* 36, 1581–1603. doi:10.1177/0021998302036013179

Garcia, J. A., Gascon, L., Chinesta, F., Ruiz, E., and Trochu, F. (2010). An efficient solver of the saturation equation in liquid composite molding. *Int. J. Mater. Form.* 3, 1295–1302. doi:10.1007/s12289-010-0681-8

Jung, J. W., Hwang, I. J., Park, C. H., Um, M. K., and Lee, W. I. (2010). "Simulation and experimental study on unsaturated flow behavior and void minimization in liquid composite molding process," in *The 7th Asian-Australian Conference on Composite Materials*, Taipei.

Masoodi, R., Pillai, K. M., Grahl, N., and Tan, H. (2012). Numerical simulation of LCM mold-filling during the manufacture of natural fiber composites. *J. Reinf. Plast. Compos.* 31, 363–378. doi:10.1177/0731684412438629

Naik, N. K., Sirisha, M., and Inani, A. (2014). Permeability characterization of polymer matrix composites by RTM/VARTM. *Prog. Aerosp. Sci.* 65, 22–40. doi:10.1016/j.paerosci.2013.09.002

Park, C. H., and Lee, W. I. (2011). Modeling void formation and unsaturated flow in liquid composite molding processes: a survey and review. *J. Reinf. Plast. Compos.* 30, 957–977. doi:10.1177/0731684411411338

Pillai, K. (2004). Modeling the unsaturated flow in liquid composite molding processes: a review and some thoughts. *J. Compos. Mater.* 38, 2097–2118. doi:10.1177/0021998304045585

Van Genuchten, M. T. (1980). A closed-form equation for predicting the hydraulic conductivity of unsaturated soils. *Soil Sci. Am. J.* 44, 892–898. doi:10.2136/sssaj1980.03615995004400050002x

Vernet, N., Ruiz, E., Advani, S., Alms, J. B., Aubert, M., Barburski, M., et al. (2014). Experimental determination of the permeability of engineering textiles: benchmark II. *Compos. A* 61, 172–184. doi:10.1016/j.compositesa.2014.02.010

Conflict of Interest Statement: The authors declare that the research was conducted in the absence of any commercial or financial relationships that could be construed as a potential conflict of interest.

5

Plant biopolymer–geopolymer: organic diagenesis and kerogen formation

Neal S. Gupta*

Department of Science and Technology, Bryant University, Smithfield, RI, USA

Keywords: kerogen, biopolymer degradation, neogenesis, selective preservation, decay, molecular paleontology, organic geochemistry

Edited by:
José Alejandro Heredia-Guerrero,
Fondazione Istituto Italiano di
Tecnologia, Italy

Reviewed by:
Jonathan Watson,
Imperial College London, UK

*Correspondence:
Neal S. Gupta
nealsgupta@cedr-cdl-lab.com

Sedimentary organic matter is formed by diagenesis (reactions in sediments up to 60°C, Tegelaar et al., 1989) and catagenesis (those >100°C induced by thermal cracking, Tissot and Welte, 1984) of biological material introduced during deposition from primary producers. Over 90% of this sedimentary organic matter is a non-hydrolyzable (i.e., immune to acid–base hydrolysis) macropolymer called kerogen (de Leeuw and Largeau, 1993) that does not dissolve in organic solvents and produces petroleum upon catagenesis (Tissot and Welte, 1984). The composition and type of kerogen is heavily dependent on the nature of the biological input (de Leeuw et al., 2006), the environment of deposition (Goth et al., 1998), and the preservation pathway (diagenesis, Briggs, 1999). Both diagenesis and catagenesis can be simulated in the laboratory using P-t apparatus (Stankiewicz et al., 2000) and environmental decay experiments (Briggs, 1999; Gupta et al., 2009).

Kerogen formation is generally attributed to neogenesis (Tissot and Welte, 1984), in which sedimentary organic matter is formed by random intermolecular polymerization and polycondensation of biological residues (e.g., amino acids, sugars, and lipids) including melanoidins or the Selective Preservation of resistant biosynthesized macromolecules that undergo limited chemical change during diagenesis (i.e., they remain morphologically and chemically recognizable as organic remains in the sedimentary rock; Goth et al., 1998). Selective preservation which has gained widespread acceptance as a counter thesis to neogenesis since the mid-1980s posits that aliphatics in fossil organic matter are derived from highly aliphatic and resistant (insoluble and non-hydrolyzable) biopolymers in living organisms, such as algaenan (present in algae; Goth et al., 1998), cutan (present in plant leaves; Mösle et al., 1998), and suberan (present in suberinized vascular tissue, de Leeuw and Largeau, 1993). These survive decay more readily than labile biopolymers such as polysaccharides, proteins, and nucleic acids (Tegelaar et al., 1989). A thorough geochemical analysis of organic fossil matter using high-resolution spectroscopy in conjunction with mass spectrometry [such as in Goth et al. (1998)] and laboratory analytical chemistry methods (de Leeuw and Largeau, 1993) establishes the diagenetic pathway degree of biological preservation (Briggs, 1999) and insights into geochemical transformation of biomolecules to geomolecules (Gupta et al., 2009). Organic sulfurization (Kok et al., 2000) and oxidative reticulation of unsaturated cross-linkages in reacting molecules (Riboulleau et al., 2001) offer mechanistic insights through analysis and case study of modern (extant) and fossil material (discrete) or those disseminated in sediment as kerogen. In light of this, this synthesis opinion article focuses on complimentary and conflicting arguments from plant fossil analysis, their widespread relevance, the biochemical description of contributors, insights from controlled laboratory experiments (decay), and simulated autoclave experiments in pressure–temperature regimes from 270 to 350°C (Stankiewicz et al., 2000).

Morphology and chemical structure of fossil leaves from the Ardèche diatomite (Late Miocene, from southeast France) were detected using pyrolysis–gas chromatography–mass spectrometry, tetramethylammoniumhydroxide (TMAH)-assisted pyrolysis, micro-FTIR, and solid-state ^{13}C NMR spectroscopy to improve understanding of these questions (Gupta et al., 2007a).

These revealed the presence of benzene derivatives, lignin, pristenes, fatty acids, and an aliphatic geopolymer (insoluble in organic solvents, hence recalcitrant) similar in composition to kerogen. Modern leaves failed to reveal the presence of cutan (the aliphatic biopolymer that has been posited to be the direct source of aliphatics in leaf fossils), thereby precluding selective preservation of cutan as the source (Gupta et al., 2007a). TMAH-assisted pyrolysis (an online hydrolysis method) supported this argument by releasing incorporated fatty acyl moieties ranging in C number from C_8 to C_{32} from the macromolecule. The C_{10} to C_{32} acid units are characteristic of the free fatty acid from epicuticular waxes in modern leaves and also phospholipid fatty acid (PLFA) fraction of cell membranes (Gupta et al., 2007a) indicating their direct chemical incorporation in the fossil leaf geopolymer. Such lipid cross-linking is seen in fossil dinoflagellates as well (Versteegh et al., 2004). Cutan, the resistant non-hydrolyzable aliphatic biopolymer present in modern leaves (Mösle et al., 1998), was first reported in the cuticle of *Agave americana* and has generally been considered ubiquitous in leaf cuticles along with the biopolyester cutin. Because leaves and cuticles in the fossil record almost always have an aliphatic composition (Mösle et al., 1998), cutan has been argued as the source biopolymer, especially for the aliphatic component $>C_{20}$. However, Gupta et al. (2006a) analyzed modern leaves using chemical degradation techniques to test for the presence of cutan in multiple taxa concluding its absence in 16 out of 19 taxa analyzed (including gymnosperms and angiosperms). Cutin that forms the protective biopolymer-polyester in modern leaf cuticles (with C_{16} and C_{18} hydroxy fatty acids as abundant components) also acts as a possible source in fossils (Mösle et al., 1998; Gupta et al., 2007a). Other studies that offer similar examples of geopolymer formation in fossils include insects (Briggs, 1999; Stankiewicz et al., 2000), graptolites (Gupta et al., 2006b), kerogen (showing oxidative cross-linking with less selective preservation; Riboulleau et al., 2001), sporopollenin (Watson et al., 2012), and fossil algae (Versteegh et al., 2004).

Degradation of polysaccharides by microbes is dictated by the presence of polysaccharide utilization loci (PUL) in genomes of microbes that dictates which sugars can be utilized by the microbial consortia for energy. Hence, studies have been conducted to understand the relative rate of decay of plant biopolymers (Gupta et al., 2009 and references therein), e.g., by analysis of modern (decaying) and fossil *Metasequoia* leaves. These revealed the presence of structural polyester cutin, guaiacyl lignin units (typical of gymnosperms), and polysaccharides. Analysis of environmentally decayed leaves revealed that the lignin units and cellulose were degraded more relative to cutin, suggesting that cutin is likely more stable than lignin and cellulose in this environmental setting. This is supported by electron microscopy of changes in the

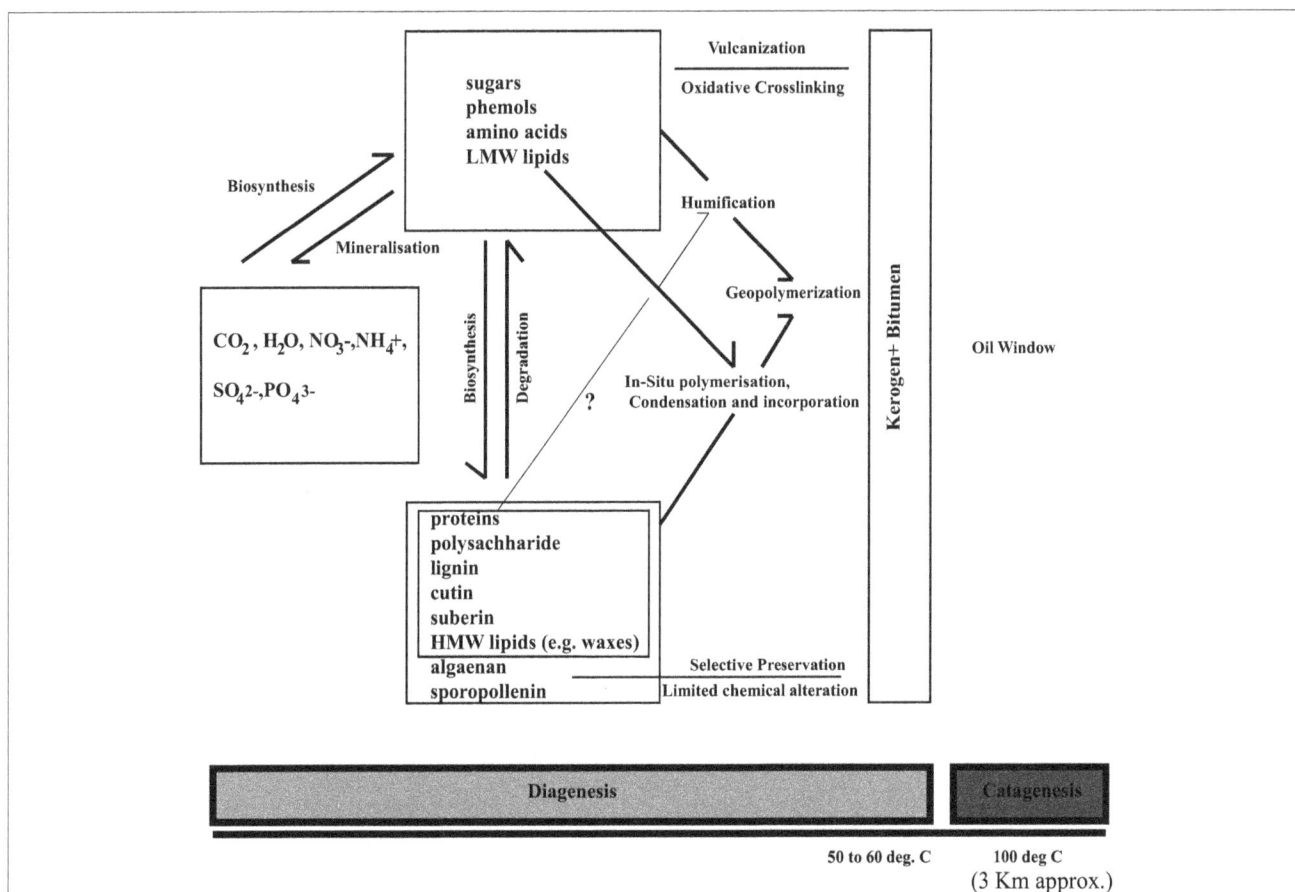

FIGURE 1 | A concept diagram proposing and integrating the pathways of organic matter preservation in tandem with selective preservation (that requires presence and limited change of precursor recalcitrant biopolymer).

cellular structure and cuticle of the modern, decayed, and fossil *Metasequoia* (Gupta et al., 2009).

Experimental heating of cutin is known to generate an aliphatic polymer with carbon chain length $<C_{20}$ (Gupta et al., 2007b), demonstrating that the *n*-alkyl component $>C_{20}$ is a product of incorporation of long-chain plant waxes (indicated by the odd over even predominance of the $>C_{27}$ *n*-alkanes) when cutan is absent in the modern leaf. The resistant nature of cutin compared with lignin and polysaccharides explains the ubiquitous presence of an *n*-alkyl component ($<C_{20}$) in fossil leaves even when polysaccharides are absent and lignin has decayed (Gupta et al., 2009). Experimental maturation of cutan, cutin, lipid waxes, lignin, and polysaccharides in plant leaves (and model compounds – 270 to 350°C, 700 bars) generated a resistant non-hydrolyzable aliphatic macromolecule similar to that detected in plant fossils (Gupta et al., 2007b); Stankiewicz et al. (2000) used similar methods for insects. By comparing the products derived from maturation of different pre-treated plant tissues, these experiments demonstrated that lipids are incorporated into the macromolecular material at 350°C (Gupta et al., 2007b; **Figure 1** shows a new integrated concept diagram). Thus, they indicate that labile organic compounds are a potential source of the aliphatic component of fossil organic matter and kerogen in the absence of a resistant aliphatic precursor in the living organism.

The mechanism leading to the formation of aliphatic components in petroleum from algal sources in marine kerogens is posited to derive from algaenan present in outer cell walls of algae

(Goth et al., 1998). Experiments on the non-algaenan-producing alga *Chlamydomonas reinhardtii* at 260 and 350°C revealed a macromolecule with significant aliphatic component (Gupta et al., 2014) similar to that in the modern leaf experiments, also similar to high H content kerogens; derived directly from saturated and unsaturated C_{16} and C_{18} fatty acids here. The presence of amides, nitriles, and oximes in the heated algae is due to the reaction of lipids with the N-containing protein molecules as also seen in heated insects (Gupta et al., 2014). *Scenedesmus quadricauda* at 350°C (an algaenan-containing green alga – as a control) demonstrated survival of algaenan at these temperatures. Analysis of the solvent-insoluble residue of heated cyanobacterium (*Oscillatoria* sp.) and *Rhodopseudomonas palustris* (purple non-sulfur-containing bacteria) similarly produced a macromolecule with high aliphatic content in these studies. Algaenan thus does not give rise to a ubiquitous aliphatic composition in marine sedimentary organic matter (Goth et al., 1998) as a preserved biopolymer. These renew interests and caveats in organic diagenesis and selective preservation of biopolymers (such as algaenan and cutan) and their effect on long-term carbon sequestration (de Leeuw, 2007), where occurrence of recalcitrant biomolecules is limited in modern taxa.

Acknowledgments

Prof. Derek Briggs (FRS), Prof. Chris Reid (Bryant University), and a Frontiers reviewer are thanked for initial review.

References

Briggs, D. E. G. (1999). Molecular taphonomy of animal and plant cuticles: selective preservation and diagenesis. *Philos. Trans. R. Soc. Lond. B* 354, 7–16. doi:10.1098/rstb.1999.0356

de Leeuw, J. W. (2007). On the origin of sedimentary aliphatic macromolecules: a comment on recent publications by Gupta et al. *Org. Geochem.* 38, 1585–1587. doi:10.1016/j.orggeochem.2007.05.010

de Leeuw, J. W., and Largeau, C. (1993). "A review of macromolecular organic compounds that comprise living organisms and their role in kerogen, coal and petroleum formation," in *Org. Geochem.: Prin. and App*, eds M. H. Engel and S. A. Macko (New York: Plenum Press), 23–62.

de Leeuw, J. W., Versteegh, G. J. M., and van Bergen, P. F. (2006). Biomacromolecules of plants and algae and their fossil analogues. *Plant Ecol.* 189, 209–233. doi:10.1007/s11258-005-9027-x

Goth, K., de Leeuw, J. W., Puttman, W., and Tegelaar, E. W. (1998). Origin of messel oil shale kerogen. *Nature* 336, 759–761. doi:10.1038/336759a0

Gupta, N. S., Briggs, D. E. G., Collinson, M. E., Evershed, R. P., Michels, R., Jack, K. S., et al. (2007a). Evidence for the *in situ* polymerisation of labile aliphatic organic compounds during the preservation of fossil leaves: implications for organic matter preservation. *Org. Geochem.* 38, 499–522. doi:10.1016/j.orggeochem.2006.06.011

Gupta, N. S., Michels, R., Briggs, D. E. G., Collinson, M. E., Evershed, R. P., and Pancost, R. D. (2007b). Experimental evidence for formation of geomacromolecules from plant leaf lipids. *Org. Geochem.* 38, 28–36. doi:10.1016/j.orggeochem.2006.09.014

Gupta, N. S., Collinson, M. E., Briggs, D. E. G., Evershed, R. P., and Pancost, R. D. (2006a). Re-investigation of the occurrence of cutan in plants: implications for the leaf fossil record. *Paleobiology* 32, 432–449. doi:10.1666/05038.1

Gupta, N. S., Briggs, D. E. G., and Pancost, R. D. (2006b). Molecular taphonomy of graptolites. *J. Geol. Soc. Lond.* 163, 897–900. doi:10.1144/0016-7649 2006-070

Gupta, N. S., Steele, A., Fogel, M., Griffin, P., Adams, M., Summons, R. E., et al. (2014). Experimental formation of geomacromolecules from microbial lipids. *Org. Geochem.* 67, 35–40. doi:10.1016/j.orggeochem.2013.11.006

Gupta, N. S., Yang, H., Leng, Q., Briggs, D. E. G., Cody, G. D., and Summons, R. E. (2009). Diagenesis of plant biopolymers: decay and macromolecular preservation of *Metasequoia*. *Org. Geochem.* 40, 802–809. doi:10.1016/j.orggeochem.2009.04.004

Kok, M. D., Schouten, S., and Sinninghe Damsté, J. S. (2000). Formation of insoluble, nonhydrolyzable, sulfur-rich macromolecules via incorporation of inorganic sulfur species into algal carbohydrates. *Geochim. Cosmochim. Acta* 64, 2689–2699. doi:10.1016/S0016-7037(00)00382-3

Mösle, B., Collinson, M. E., Finch, P., Stankiewicz, B. A., Scott, A. C., and Wilson, R. (1998). Factors influencing the preservation of plant cuticles: a comparison of morphology and chemical comparison of modern and fossil examples. *Org. Geochem.* 29, 1369–1380. doi:10.1016/S0146-6380(98)00080-1

Riboulleau, A., Derenne, S., Largeau, C., and Baudin, F. (2001). Origin of contrasting features and preservation pathways in kerogens from the Kashpir oil shales (Upper Jurassic, Russian Platform). *Org. Geochem.* 32, 647–665. doi:10.1016/S0146-6380(01)00017-1

Stankiewicz, B. A., Briggs, D. E. G., Michels, R., Collinson, M. E., and Evershed, R. P. (2000). Alternative origin of aliphatic polymer in kerogen. *Geology* 28, 559–562. doi:10.1130/0091-7613(2000)28<559:AOOAPI>2.0.CO;2

Tegelaar, E. W., de Leeuw, J. W., Derenne, S., and Largeau, C. (1989). A reappraisal of kerogen formation. *Geochim. Cosmochim. Acta* 53, 3103–3106. doi:10.1016/0016-7037(89)90191-9

Tissot, B., and Welte, D. H. (1984). *Petrol. Form. and Occurr*, 2nd Edn. Berlin: Springer-Verlag.

Versteegh, G. M. J., Blokker, P., Wood, G. D., Collinson, M. E., Damsté, J. S. S., and de Leeuw, J. W. (2004). An example of oxidative polymerization of unsaturated fatty acids as a preservation pathway for dinoflagellate organic matter. *Org. Geochem.* 35, 1129–1139. doi:10.1016/j.orggeochem.2004.06.012

Watson, J. S., Fraser, W. T., and Sephton, M. A. (2012). Formation of a polyalkyl macro-molecule from the hydrolysable component within sporopollenin during heating/pyrolysis experiments with *Lycopodium* spores. *J. Anal. App. Pyr.* 95, 138–144. doi:10.1016/j.jaap.2012.01.019

Conflict of Interest Statement: The author declares that the research was conducted in the absence of any commercial or financial relationships that could be construed as a potential conflict of interest.

6

Molecular dynamics study on the distributed plasticity of penta-twinned silver nanowires

Sangryun Lee and Seunghwa Ryu*

Department of Mechanical Engineering, Korea Advanced Institute of Science and Technology, Daejeon, South Korea

Edited by:
Federico Bosia,
University of Torino, Italy

Reviewed by:
Pratyush Tiwary,
ETH Zürich, Switzerland
Shan Jiang,
University of Missouri, USA
Gang Zhang,
Agency for Science, Technology and
Research (A*STAR), Singapore

*Correspondence:
Seunghwa Ryu,
Department of Mechanical
Engineering, Korea Advanced Institute
of Science and Technology,
KAIST Building N7-4, Room 7107,
291 Daehak-ro, Yuseong-gu,
Daejeon 305-701, South Korea
ryush@kaist.ac.kr

The distributed plasticity of pentatwinned silver nanowires has been revealed in recent computational and experimental studies. However, the molecular dynamics (MD) simulations have not considered the imperfections seen in experiments, such as irregular surface undulations, the high aspect ratio of nanowires, and the stiffness of loading devices. In this work, we report the effect of such inherent imperfections on the distributed plasticity of penta-twinned silver nanowires in MD simulations. We find that the distributed plasticity occurs for nanowires having undulations that are less than 5% of the nanowire diameter. The elastic stress field induced by a stacking fault promotes the nucleation of successive stacking fault decahedrons (SFDs) at long distance, making it hard for necking to occur. By comparing the tensile simulation using the steered molecular dynamics method with the tensile simulation with periodic boundary condition (PBC), we show that a sufficiently long nanowire must be used in the constant strain rate simulations with PBC, because the plastic displacement burst caused by the SFD formation induces compressive stress, promoting the removal of other SFDs. Our finding can serve as a guidance for the MD simulation of crystalline materials with large plastic deformation and in the design of mechanically reliable devices based on silver nanowires.

Keywords: molecular dynamics, dislocation, stacking fault decahedrons, distributed plasticity

Introduction

Silver nanowires have been widely used as building blocks in various electronic systems including flexible antennas (Song et al., 2014), highly stretchable strain sensors (Liu and Choi, 2009; Amjadi et al., 2014), transparent conductive films, and flexible conductors (Xu and Zhu, 2012) due to their high conductivity and ease of synthesis (Zhang et al., 2005). Most of those applications are expected to sustain cyclic mechanical loading and be used as components in wearable and flexible devices. For example, silver nanowire–elastomer composite based piezoresistivity sensors experience complex cyclic mechanical loading during their operation (Amjadi et al., 2014) and the nanowire-based touch screen has also been tested under large bending and stretching (Xu and Zhu, 2012). The conductivity of these devices is provided by a percolation network made of silver nanowires and their failure highly depends on the integrity of the network. Therefore, for the reliable design of flexible devices, it is essential to understand the mechanical integrity of the silver nanowire under high strain beyond the plastic limit as well as its recovery under unloading.

In addition, there is a fundamental interest in understanding the plasticity of silver nanowires, for two reasons. First, silver nanowires have a unique penta-twinned structure when chemically synthesized (Zhang et al., 2005; Liu and Choi, 2009; Xu and Zhu, 2012; Amjadi et al., 2014; Song et al., 2014). Nanostructures incorporating nanotwins have been extensively studied with the goal of simultaneously improving their strength and ductility. Second, it is known that the plasticity of a sub-100 nm metal specimen is driven by dislocation nucleation, and thus is very different from bulk counterparts, in which plastic deformation is governed by the motion of existing dislocations. Penta-twinned nanowires can serve as a simple test bed for studying both the role of twin boundaries (TBs) (Zhu and Gao, 2012) and the size effect, together. For example, a few novel properties of penta-twinned nanowires have been revealed including enhance elastic modulus (Zhu et al., 2012), enhanced yield strength (Wu et al., 2006), special multiple conjoint fivefold twins (Jiang et al., 2013a,b), and thermal effects originated from the size effect and the TBs (Wu et al., 2011).

Numerous computational and experimental studies have been devoted to examining the plasticity of penta-twinned nanowires (Cao and Wei, 2006; Wu et al., 2011; Filleter et al., 2012; Sun et al., 2013; Zheng et al., 2014). It was found that the yield of the nanowire initiates with dislocation nucleation from the surface, and plastic deformation is accommodated by a chain of stacking fault decahedrons (SFD) or stacking fault hats (SFH). Multiple distributed plastic zones have been observed for sub-100 nm nanowires, and their mechanism has been explained by the activation and deactivation of SFD chains at stress concentrations due to a surface undulation. Recently, a recoverable plasticity has also been reported by two independent studies (Bernal et al., 2014; Qin et al., 2015) suggesting the crucial role of TBs in blocking the propagation of stacking faults and promoting the recovery under unloading.

However, molecular dynamics (MD) simulation studies have not considered the irregular surface undulation, the high aspect ratio of nanowires, and the stiffness of loading devices that are inherent in experiments. While stress concentrations associated with the irregular surface undulation have been suggested as the origin of distributed plastic zones, there has been no MD simulation that tests the assumptions. The large plastic displacement burst induced by the formation of the SFD chain may significantly affect the stress–strain relationship if nanowires with small aspect ratio are tested in MD simulations. Also, the constant strain rate simulation with a periodic boundary condition (PBC) is equivalent to the use of an infinite stiffness loading device, which may result in a deformation mechanism that is different from the experiments.

In this work, we employ MD simulations to study the effect of the aforementioned factors in the deformation behaviors. We found that the distributed plasticity occurs for nanowires with undulations that are less than 5% of the nanowire diameter. The elastic stress field induced by a stacking fault promotes the delocalization of SFDs, making it hard for necking to occur. For higher diameter undulation, the stress concentration is too high to accommodate a separate plastic zone at a different stress-concentration position. By comparing the tensile simulation using the steered molecular dynamics (SMD) method with the tensile simulation with PBC, we show that a sufficiently long nanowire must be used in the constant strain rate simulations with PBC,

because the plastic displacement burst caused by the SFD formation induces compressive stress, which promotes the removal of other SFHs or SFDs.

Materials and Methods

We employed LAMMPS (Plimpton, 1995) and MD++ (MD++; http://micro.stanford.edu/MDpp) MD packages to study the mechanical behavior of penta-twinned silver nanowires. The interactions between atoms were described by embedded-atom-method (EAM) potential. We tested two versions of EAM potentials developed by Sheng et al. (2011) and Williams et al. (2006). The diameter of the nanowires was around 20 nm, and two different aspect ratios (6 and 12) were considered. The penta-twinned nanowire has a <110> direction along the growth axis and five single face-centered-cubic (FCC) crystals intersect at the {111} TBs (**Figure 1**). To mimic the surface defects and diameter variation observed in experiments (Filleter et al., 2012; Bernal et al., 2014), the nanowire has surface undulations along the axial direction (Agrawal et al., 2009) as shown in **Figure 1B**. At each cross section of the undulation site, we considered an imaginary circle of identical size but centered at a shifted position, and atoms above the circle are removed. The magnitude of shift is a sinusoidal function of the axial position with maximum being 5% of nanowire diameter. We did not observe significant re-arrangements on the undulation sites because there is no sharp surface step. The undulation sites are equally spaced in all nanowires considered in this study. The total number of atoms is about two million for the nanowire with the aspect ratio 6 and four million for the aspect ratio 12. Before the tension, the nanowires are equilibrated at 300 K by utilizing NPT ensemble for 600 ps and NVT ensemble for 600 ps until long period vibration modes decay. Time step and damping parameters are chosen as 1 fs and 0.1, respectively. The

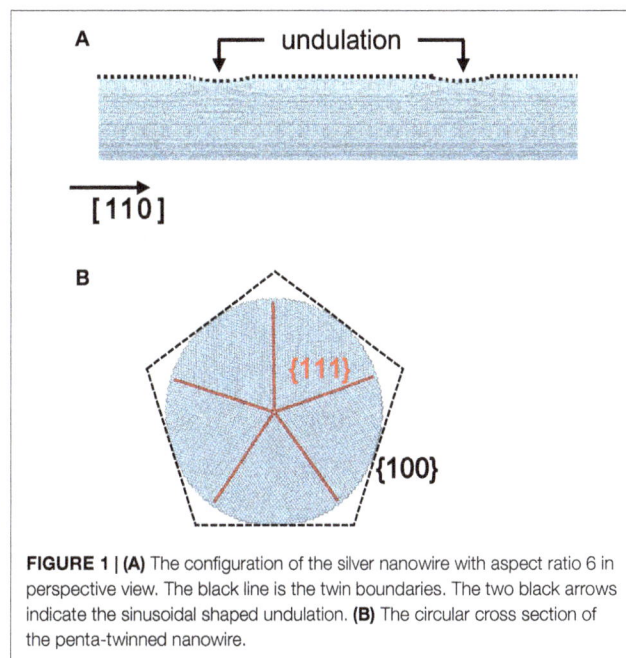

FIGURE 1 | (A) The configuration of the silver nanowire with aspect ratio 6 in perspective view. The black line is the twin boundaries. The two black arrows indicate the sinusoidal shaped undulation. **(B)** The circular cross section of the penta-twinned nanowire.

average length of nanowire during the NPT equilibration is used as the reference length for computing engineering strain.

The tensile loading was performed with two methods. One is the conventional tensile simulation where the simulation box is expanded with a constant rate under PBC. The other is the SMD method where one end of the nanowire is pulled by an imaginary spring while the other end is fixed. For both simulations, we used the time step of 1 fs. The tensile loading simulation of the conventional method was performed at 300 K under canonical ensemble (NVT) with an engineering strain rate of $0.1\,\mathrm{ns}^{-1}$ until fracture occurred. We increase the size of the simulation cell in the loading direction every 10 ps with a fixed strain increment (0.001) and run NVT simulations for 10 ps until it reaches equilibrium. We then measured the average stress, and repeated the elongation and NVT simulation sequence multiple times. The virial stress formula was used to calculate the atomic stress of the nanowires (Pendás, 2002; Zimmerman et al., 2004).

For the SMD method, one end of nanowire is fixed and the other end of nanowire is pulled by a virtual spring. On the onset of plasticity, a large plastic strain burst occurs that cannot be captured by the conventional PBC method. Due to the compliance of the virtual spring, the SMD method is suitable for capturing the large strain while it cannot apply a perfect displacement control. When performing SMD method, we chose the spring constant to be comparable to the stiffness of the 120 nm-long nanowires, and the pulling speed of the spring was chosen to achieve the engineering strain rate of $0.1\,\mathrm{ns}^{-1}$. During SMD simulation, the temperature was kept at 300 K using identical thermostats. To analyze the dislocation structure in the plastic zone, we visualized the dislocation lines with different colors according to their category using OVITO (Alexander, 2010) and Crystal Analysis (Alexander and Karsten, 2010) packages.

Results

First, we compare the deformation behavior of two nanowires with and without diameter undulation (~5% of diameter), as described by the Sheng potential. For the pristine nanowire, the plastic zone is initiated at a random point and extends for a long distance by forming SFD chains, as depicted in **Figure 2A**. In contrast, the deformation of the nanowire with the undulation is accommodated by two distinct plastic zones, activated around the stress concentrated regions, which is similar to the distributed plasticity observed in experiments (Filleter et al., 2012; Bernal et al., 2014) (**Figure 2B**).

Interestingly, for the nanowire with undulations, the plastic zone extends for a large distance before necking occurs, as it does in the pristine nanowire. We found that the toughness moduli, evaluated by integrating the stress–strain curve from zero to fracture strain, are almost identical for the two cases (**Figure 2D**), which indicates that the nanowire accommodates a large plastic deformation despite the presence of stress concentration. Due to this tendency to develop a non-localized plastic zone, the penta-twinned nanowire is tolerant to small surface undulations and can absorb a large amount of energy until failure. This may explain why highly stretchable sensors or touch screens based on the nanowire percolation network can sustain significant cyclic loadings (Xu and Zhu, 2012; Amjadi et al., 2014; Narayanan et al., 2014; Yang et al., 2014; Wang et al., 2015). To analyze the dislocation structures, we drew a double Thompson tetrahedron to visualize the Burgers vector of dislocation of the FCC structure (**Figure 2C**). The most predominant dislocation is the Shockley partial with the Burgers vector $1/6{<}112{>}a$ (the lines painted green color in the Figure) which represents the leading and trailing part of the stacking fault planes in the SFD chains. The next dominant partial dislocation with Burgers vector $1/9{<}222{>}a$ (black

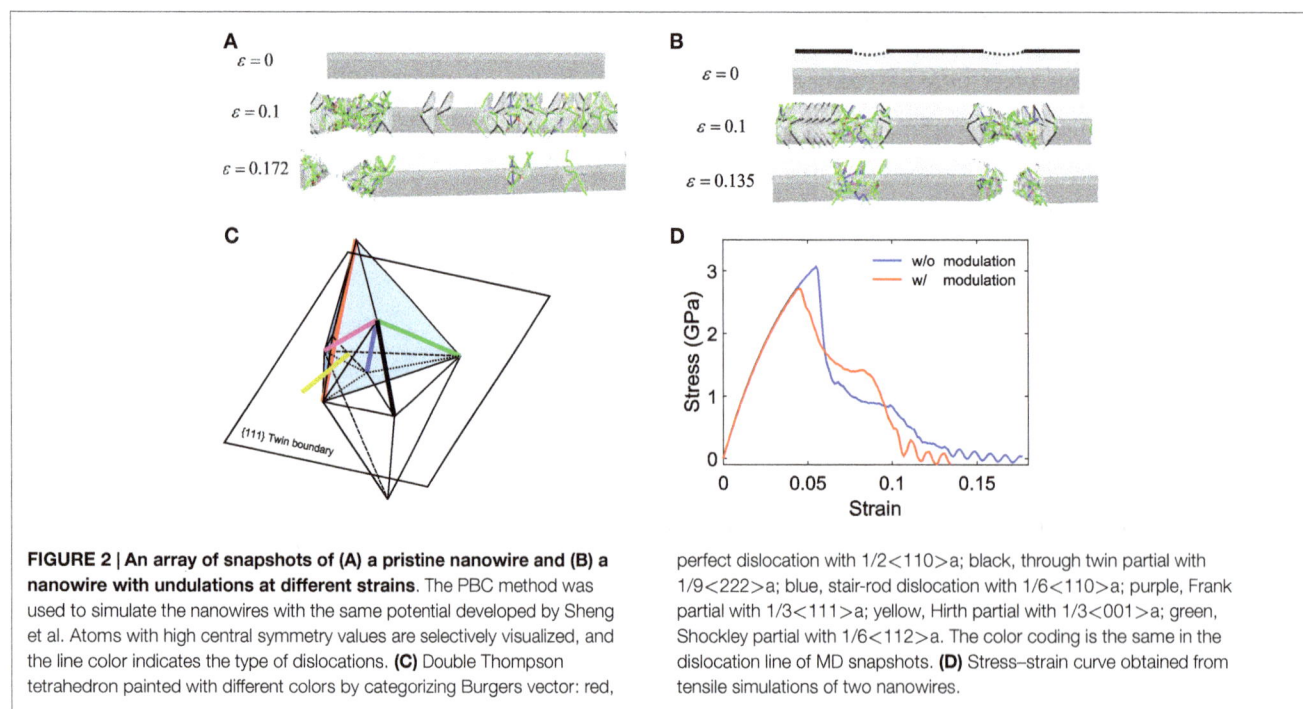

FIGURE 2 | An array of snapshots of (A) a pristine nanowire and (B) a nanowire with undulations at different strains. The PBC method was used to simulate the nanowires with the same potential developed by Sheng et al. Atoms with high central symmetry values are selectively visualized, and the line color indicates the type of dislocations. **(C)** Double Thompson tetrahedron painted with different colors by categorizing Burgers vector: red,

perfect dislocation with $1/2{<}110{>}a$; black, through twin partial with $1/9{<}222{>}a$; blue, stair-rod dislocation with $1/6{<}110{>}a$; purple, Frank partial with $1/3{<}111{>}a$; yellow, Hirth partial with $1/3{<}001{>}a$; green, Shockley partial with $1/6{<}112{>}a$. The color coding is the same in the dislocation line of MD snapshots. **(D)** Stress–strain curve obtained from tensile simulations of two nanowires.

color in figure) is formed at the intersection between the stacking fault and the TB (Bernal et al., 2014).

To probe whether the observed behavior was independent of the empirical potential choice as well as the small aspect ratio, we compared the tensile simulations of nanowires having four undulation sites with larger aspect ratio (12), using both Sheng and Mishin potentials (**Figures 3A–C**). The different results from the potentials are originated from different choices of fitting properties (Williams et al., 2006; Sheng et al., 2011). Still, regardless of the potential type, we observed the activation of a distinct plastic zone around the undulation sites, and found that some of SFD chains remained until the fracture took place, although the amount of plastic deformation was different between the two potential models.

Nevertheless, the MD simulations indicated that the distributed plasticity is an intrinsic property of the penta-twinned nanowire, independent of length, potentials, and the number of surface undulations. Yet, we note that the abrupt reduction of stress following the SFD chain formation has not been observed in tensile tests in experiments (Filleter et al., 2012; Bernal et al., 2014). This may be attributed to the remarkably high strain rate or the much smaller diameter used in simulation. Further research is necessary to clarify this issue.

To understand the origin of the SFD chain formation, we calculated the energy barrier for the formation of a stacking fault plane in the presence of another stacking fault by using the modified nudged elastic band (NEB) methods (**Figure 4A**) (Ryu et al., 2011). Due to high computational cost, we considered nanowires with smaller diameters (6, 8, 10, and 12 nm). For the diameter of 12 nm, the minimum occurs at a distance of around 3.5 nm (3.31 nm for Sheng potential and 3.61 nm for Mishin potential), as shown in **Figure 4C**. Also, the energy barrier is significantly smaller than the initial stacking fault nucleation energy barrier in the pristine nanowire at the same stress. This explains why, with smaller diameter, the propagation of SFH chains occurs away from the stress-concentration region. We found that the most probable successive nucleation distance δ between the first SFH and second SFH is linearly proportional to the diameter of the nanowire D (**Figure 4B**), with the ratio $\delta \sim 0.28D$. This linear relationship implies that the observed energy minimum may be caused by an elastic stress field induced by the first SFH.

We illustrated the elastic stress field in the nanowire when a single stacking fault presents at 0 K, as shown in **Figures 4D,E**. The Von-Mises stress of atoms at the surface maximizes around 3 nm, which is close to the most probable nucleation distance δ. Von-Mises stress is calculated from six components of virial stress of each atom, with atomic volume of silver in bulk (10.3 cc/mol) (Wang et al., 1997). We also visualized the Von-Mises stress field of a single crystal nanowire under the same strain (0.04) at 0 K. As shown is **Figures 4D,E**, the stress field induced from a stacking fault in a single crystal nanowire does not have stress concentration, which clearly shows the role of the TBs in the distributed

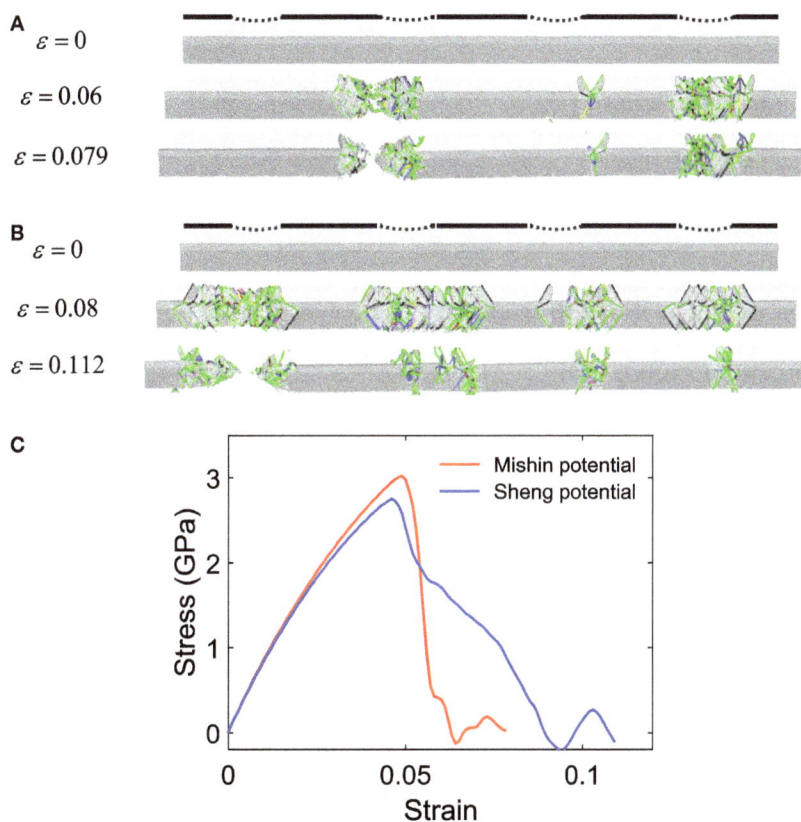

FIGURE 3 | An array of snapshots of AgNWs with different potentials, (A) Mishin and (B) Sheng with the aspect ratio of 12. **(C)** The stress–strain curves of AgNWs which have the same aspect ratio (12) and different EAM potential. The PBC method was applied to obtain the snapshots and stress–strain curves.

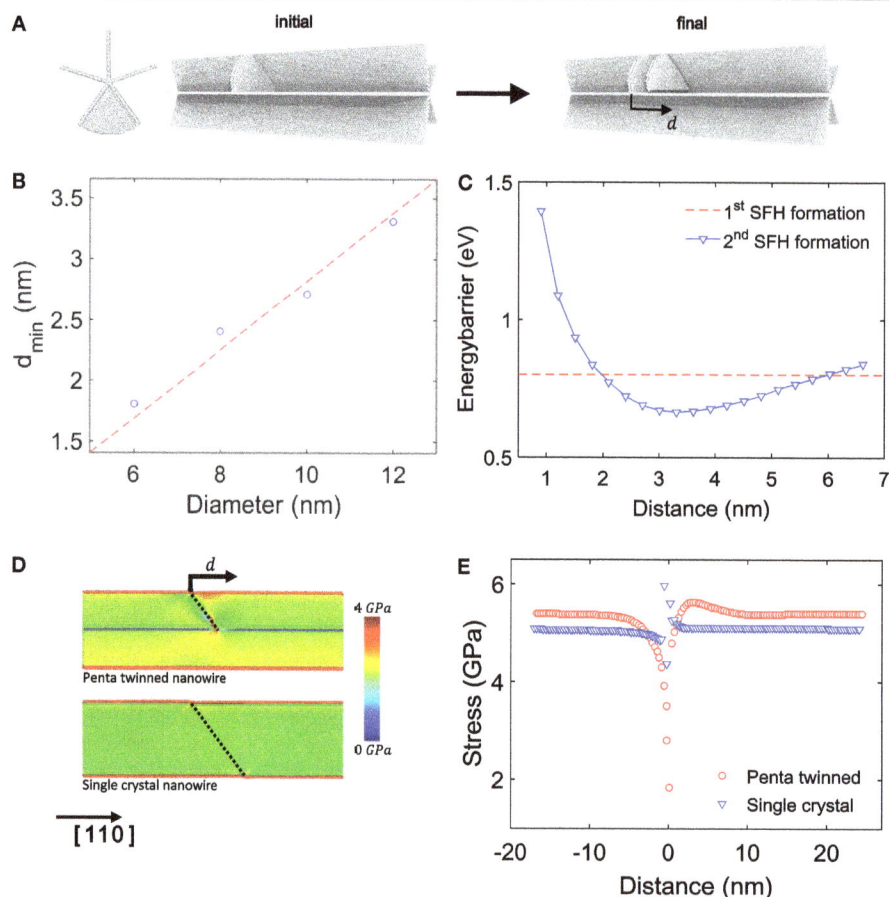

FIGURE 4 | (A) The initial and the final configurations for energy barrier calculations. (B) The distance from the first SFH which makes the optimal value as a function of the diameter of AgNWs. The aspect ratios are the same as 3 and the potential is Sheng. (C) Energy barrier with respect to the distance from the first SFH with Sheng potentials. The diameter of the nanowires is 12 nm and

the aspect ratio is 3. (D) The Von-Mises stress fields in the presence of a stacking fault on the (111) plane for penta-twinned and single crystal nanowire having identical diameter and aspect ratio. Mishin EAM potential is used. (E) The Von-Mises stress of atoms which are located at the surface of two types of AgNWs as a function of the distance.

plasticity of the penta-twinned nanowire. The mismatch between the energy barrier calculation and surface Von-Mises stress plot occurs because the nucleation barrier depends on the integral of the stress field within the dislocation loop at the transition state, not solely on the surface stress. Still, we can conclude that the elastic stress field induced by the stacking fault plane in a penta-twinned nanowire reduces the nucleation barrier of another stacking fault plane, promoting the formation of a long SFH chain, much like a catalytic reaction.

Next, we studied the effect of nanowire length on the deformation behavior. Although the nanowire tested in these experiments was about a few micrometers long, most MD simulations consider a nanowire a few tens of nanometers scale, because of the computational cost (Cao and Wei, 2006; Agrawal et al., 2009; Zheng et al., 2014) with PBC. Nevertheless, that is not enough to fully mimic in simulations the actual experimental conditions. Especially for penta-twinned nanowires, the displacement from a large plastic strain burst is not negligible compared to the nanowire length. For conventional tensile simulations with strain control using PBC, the plastic strain burst in the nanowire induces large stress reduction that effectively causes a significant drop in elastic

stress. To overcome the artifacts from the plastic strain burst, we performed the tensile simulation using the SMD method.

We noticed a few differences in deformation behaviors between the two tensile testing methods. First, the extent of the SFH chain after the fracture is much larger in the case of the SMD simulation (**Figures 5A,B**). The large plastic strain burst from SFH chain formation is accommodated well in the SMD method by length extension. However, the strain burst occurs within a fixed nanowire length in the PBC method and gives rise to internal compressive stress within the nanowire, promoting the removal of SFHs.

The stress–strain curves manifest the differences between the two methods (**Figure 5C**). A much sharper stress drop is observed after the initial SFH chain nucleation in PBC due to the reduction of elastic tensile stress caused by the plastic strain burst. In contrast, **Figure 5D** shows that the nanowire length increases abruptly on the verge of plastic deformation under tensile loading by the SMD method. When the nanowires are stretched to the failure strain, the size of the plastic zone remains smaller for the PBC method because a large portion of SFH are removed by the internal compressive stress.

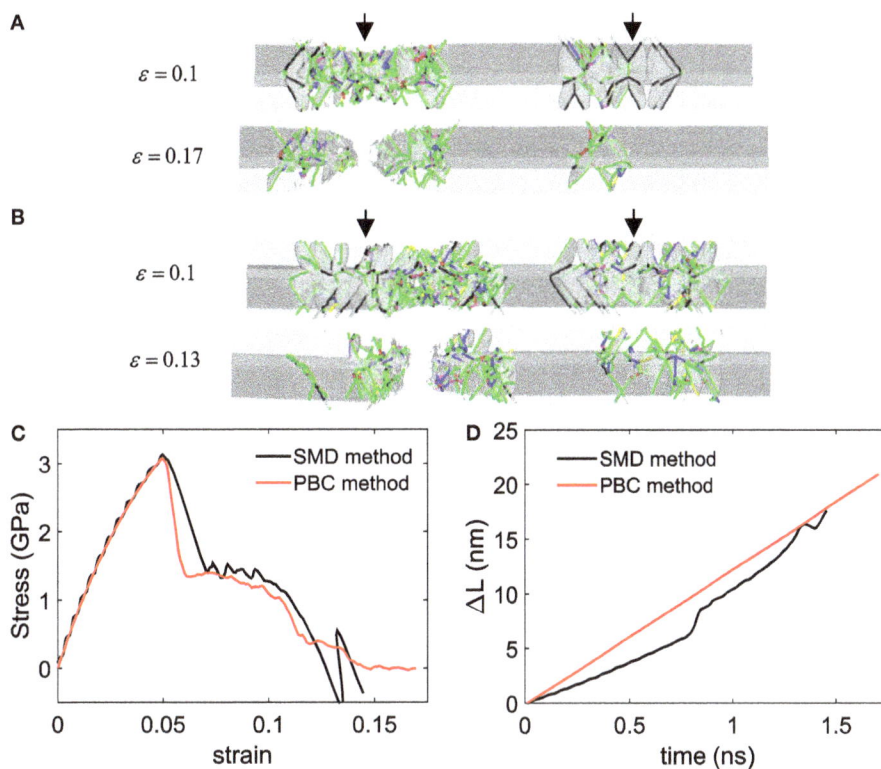

FIGURE 5 | (A) Snapshots of AgNW when the method of simulation is affine deformation by deforming simulation box (PBC method). The black arrows indicate the locations of the undulation sites **(B)** Snapshots obtained from the SMD method. The potential is the same as Mishin and the configurations of the nanowires are also the same ($D = 20\,nm$, AR = 6). **(C)** Stress–strain curve with different methods to simulate uni-axial tensile test, and **(D)** the length increment as a function of time.

Conclusion

In summary, we studied the distributed plasticity of penta-twinned silver nanowires by performing large scale MD simulations. We observed distributed plasticity when the diameter undulation was $<5\%$ for two different empirical potentials. A multiple plastic zone formation was explained by dislocation nucleation from local stress concentrations in the undulated region, which leads to SFH chain propagation. The catalytic formation of the SFH chain is caused by the reduction of the nucleation energy barrier due to the concentrated elastic stress field.

The effects of nanowire length and the loading device stiffness were investigated by comparing two tensile testing methods.

We found that the plastic strain burst by the SFH chain induces a moderate compressive stress in the simulation cell, causing excessive removal of SFHs. Such an artifact of simulation can be overcome by employing the SMD method, in which the nanowire can accommodate large plastic strain burst by length extension.

Acknowledgments

This work is supported by Basic Science Research Program (2013R1A1A010091) and Global Frontier Program (No. 2013M3A6B1078881) through the National Research Foundation of Korea (NRF) funded by the Ministry of Science, ICT and Future Planning.

References

Agrawal, R., Peng, B., and Espinosa, H. D. (2009). Experimental-computational investigation of ZnO nanowires strength and fracture. *Nano Lett.* 9, 4177–4183. doi:10.1021/nl9023885

Alexander, S. (2010). Visualization and analysis of atomistic simulation data with OVITO-the open visualization tool. *Model. Simul. Mater. Sci. Eng.* 18, 015012. doi:10.1088/0965-0393/18/1/015012

Alexander, S., and Karsten, A. (2010). Extracting dislocations and non-dislocation crystal defects from atomistic simulation data. *Model. Simul. Mater. Sci. Eng.* 18, 085001. doi:10.1088/0965-0393/18/8/085001

Amjadi, M., Pichitpajongkit, A., Lee, S., Ryu, S., and Park, I. (2014). Highly stretchable and sensitive strain sensor based on silver nanowire-elastomer nanocomposite. *ACS Nano* 8, 5154–5163. doi:10.1021/nn501204t

Bernal, R. A., Aghaei, A., Lee, S., Ryu, S., Sohn, K., Huang, J., et al. (2014). Intrinsic Bauschinger effect and recoverable plasticity in pentatwinned silver nanowires tested in tension. *Nano Lett.* 15, 139–146. doi:10.1021/nl503237t

Cao, A., and Wei, Y. (2006). Atomistic simulations of the mechanical behavior of fivefold twinned nanowires. *Phys. Rev. B* 74, 214108. doi:10.1103/PhysRevB.74.214108

Filleter, T., Ryu, S., Kang, K., Yin, J., Bernal, R. A., Sohn, K., et al. (2012). Nucleation-controlled distributed plasticity in penta-twinned silver nanowires. *Small* 8, 2986–2993. doi:10.1002/smll.201200522

Jiang, S., Chen, Z., Zhang, H., Zheng, Y., and Li, H. (2013a). Impact-induced bending response of single crystal and five-fold twinned nanowires. *Int. J. Multiscale Comput. Eng.* 11, 1–16. doi:10.1615/IntJMultCompEng.2012003171

Jiang, S., Shen, Y., Zheng, Y., and Chen, Z. (2013b). Formation of quasi-icosahedral structures with multi-conjoint fivefold deformation twins in fivefold twinned metallic nanowires. *Appl. Phys. Lett.* 103, 041909. doi:10.1063/1.4816666

Liu, C. X., and Choi, J. W. (2009). An embedded PDMS nanocomposite strain sensor toward biomedical applications. (1557-170X (Print)).

Narayanan, S., Hajzus, J. R., Treacy, C. E., Bockstaller, M. R., and Porter, L. M. (2014). Polymer embedded silver-nanowire network structures – a platform for the facile fabrication of flexible transparent conductors. *ECS J. Solid State Sci. Technol.* 3, 363–369. doi:10.1149/2.0131411jss

Pendás, A. M. (2002). Stress, virial, and pressure in the theory of atoms in molecules. *J. Chem. Phys.* 117, 965–979. doi:10.1063/1.1484385

Plimpton, S. (1995). Fast parallel algorithms for short-range molecular dynamics. *J. Comput. Phys.* 117, 1–19. doi:10.1006/jcph.1995.1039

Qin, Q., Yin, S., Cheng, G., Li, X., Chang, T. H., Richter, G., et al. (2015). Recoverable plasticity in penta-twinned metallic nanowires governed by dislocation nucleation and retraction. *Nat. Commun.* 6, 5983. doi:10.1038/ncomms6983

Ryu, S., Kang, K., and Cai, W. (2011). Entropic effect on the rate of dislocation nucleation. *Proc. Natl. Acad. Sci. U.S.A.* 108, 5174–5178. doi:10.1073/pnas.1017171108

Sheng, H. W., Kramer, M. J., Cadien, A., Fujita, T., and Chen, M. W. (2011). Highly optimized embedded-atom-method potentials for fourteen fcc metals. *Phys. Rev. B* 83, 134118. doi:10.1103/PhysRevB.83.134118

Song, L., Myers, A. C., Adams, J. J., and Zhu, Y. (2014). Stretchable and reversibly deformable radio frequency antennas based on silver nanowires. *ACS Appl. Mater. Interfaces* 6, 4248–4253. doi:10.1021/am405972e

Sun, M., Cao, R., Xiao, F., and Deng, C. (2013). Surface and interface controlled yielding and plasticity in fivefold twinned Ag nanowires. *Comput. Mater. Sci.* 79, 289–295. doi:10.1016/j.commatsci.2013.06.021

Wang, D., Lee, J., Holland, K., Bibby, T., Beaudoin, S., and Cale, T. (1997). Von Mises stress in chemical-mechanical polishing processes. *J. Electrochem. Soc.* 144, 1121–1127. doi:10.1149/1.1837542

Wang, J., Jiu, J., Araki, T., Nogi, M., Sugahara, T., Nagao, S., et al. (2015). Silver nanowire electrodes: conductivity improvement without post-treatment and application in capacitive pressure sensors. *Nano Micro Lett.* 7, 51–58. doi:10.1007/s40820-014-0018-0

Williams, P. L., Mishin, Y., and Hamilton, J. C. (2006). An embedded-atom potential for the Cu-Ag system. *Model. Simul. Mater. Sci. Eng.* 14, 817. doi:10.1088/0953-8984/21/8/084213

Wu, B., Heidelberg, A., Boland, J. J., Sader, J. E., Sun, X., and Li, Y. (2006). Microstructure-hardened silver nanowires. *Nano Lett.* 6, 468–472. doi:10.1021/nl052427f

Wu, J. Y., Nagao, S., He, J. Y., and Zhang, Z. L. (2011). Role of five-fold twin boundary on the enhanced mechanical properties of fcc Fe nanowires. *Nano Lett.* 11, 5264–5273. doi:10.1021/nl202714n

Xu, F., and Zhu, Y. (2012). Highly conductive and stretchable silver nanowire conductors. *Adv. Mater. Weinheim* 24, 5117–5122. doi:10.1002/adma.201201886

Yang, M., Chon, M. W., Kim, J. H., Lee, S. H., Jo, J., Yeo, J., et al. (2014). Mechanical and environmental durability of roll-to-roll printed silver nanoparticle film using a rapid laser annealing process for flexible electronics. *Microelectron. Reliab.* 54, 2871–2880. doi:10.1016/j.microrel.2014.07.004

Zhang, S. H., Jiang, Z. Y., Xie, Z. X., Xu, X., Huang, R. B., and Zheng, L. S. (2005). Growth of silver nanowires from solutions: a cyclic penta-twinned-crystal growth mechanism. *J. Phys. Chem. B* 109, 9416–9421. doi:10.1021/jp0441036

Zheng, Y. G., Zhao, Y. T., Ye, H. F., Zhang, H. W., and Fu, Y. F. (2014). Atomistic simulation of torsional vibration and plastic deformation of five-fold twinned copper nanowires. *Int. J. Comput. Methods* 11(Suppl.01), 1344010. doi:10.1142/S0219876213440106

Zhu, T., and Gao, H. (2012). Plastic deformation mechanism in nanotwinned metals: an insight from molecular dynamics and mechanistic modeling. *Scr. Mater.* 66, 843–848. doi:10.1016/j.scriptamat.2012.01.031

Zhu, Y., Qin, Q., Xu, F., Fan, F., Ding, Y., Zhang, T., et al. (2012). Size effects on elasticity, yielding, and fracture of silver nanowires: in situ experiments. *Phys. Rev. B* 85, 045443. doi:10.1103/PhysRevB.85.045443

Zimmerman, J. A., WebbIII, E. B., Hoyt, J. J., Jones, R. E., Klein, P. A., and Bammann, D. J. (2004). Calculation of stress in atomistic simulation. *Model. Simul. Mater. Sci. Eng.* 12, S319. doi:10.1088/0965-0393/12/4/S03

Conflict of Interest Statement: The authors declare that the research was conducted in the absence of any commercial or financial relationship that could be construed as a potential conflict of interest.

7

Coating of carbon nanotube fibers: variation of tensile properties, failure behavior, and adhesion strength

Edith Mäder[1,2]*, Jianwen Liu[1], Janett Hiller[1], Weibang Lu[3], Qingwen Li[3], Serge Zhandarov[4] and Tsu-Wei Chou[5]

[1] Leibniz-Institut für Polymerforschung Dresden e. V., Dresden, Germany, [2] Institut für Werkstoffwissenschaft, Technische Universität Dresden, Dresden, Germany, [3] Advanced Materials Division, Suzhou Institute of Nano-Tech and Nano-Bionics, Suzhou, China, [4] "V. A. Bely" Metal-Polymer Research Institute, National Academy of Sciences of Belarus, Gomel, Belarus, [5] Center for Composite Materials, University of Delaware, Newark, DE, USA

Edited by:
Alexander Bismarck,
University of Vienna, Austria

Reviewed by:
Chung Hae Park,
Ecole Nationale Supérieure des Mines de Douai, France
Claudia Merlini,
Federal University of Santa Catarina, Brazil
Tomi Herceg,
Imperial College London, UK

*Correspondence:
Edith Mäder,
Department of Composite Materials,
Leibniz-Institut für Polymerforschung
Dresden e.V., Hohe Strasse 6,
Dresden D-01069, Germany
emaeder@ipfdd.de

An experimental study of the tensile properties of CNT fibers and their interphasial behavior in epoxy matrices is reported. One of the most promising applications of CNT fibers is their use as reinforcement in multifunctional composites. For this purpose, an increase of the tensile strength of the CNT fibers in unidirectional composites as well as strong interfacial adhesion strength is desirable. However, the mechanical performance of the CNT fiber composites manufactured so far is comparable to that of commercial fiber composites. The interfacial properties of CNT fiber/polymer composites have rarely been investigated and provided CNT fiber/epoxy interfacial shear strength (IFSS) of 14.4 MPa studied by the microbond test. In order to improve the mechanical performance of the CNT fibers, an epoxy compatible coating with nano-dispersed aqueous-based polymeric film formers and low viscous epoxy resin, respectively, was applied. For impregnation of high homogeneity, low molecular weight epoxy film formers and polyurethane film formers were used. The aqueous-based epoxy film formers were not crosslinked and able to inter-diffuse with the matrix resin after impregnation. Due to good wetting of the individual CNT fibers by the film formers, the degree of activation of the fibers was improved, leading to increased tensile strength and Young's modulus. Cyclic tensile loading and simultaneous determination of electric resistance enabled to characterize the fiber's durability in terms of elastic recovery and hysteresis. The pull-out tests and SEM study reveal different interfacial failure mechanisms in CNT fiber/epoxy systems for untreated and film former treated fibers, on the one hand, and epoxy resin treated ones, on the other hand. The epoxy resin penetrated between the CNT bundles in the reference or film former coated fiber, forming a relatively thick CNT/epoxy composite layer and thus shifting the fracture zone within the fiber. In contrast to this, shear sliding along the interface between the matrix and the outer fiber layer impregnated with the resin was observed for epoxy resin-coated fibers. These fibers have been successfully pulled out of the matrix droplets and shown that the average local interfacial shear stress value was 63 MPa (with apparent IFSS values 33–60 MPa). The interfacial frictional stress between the fiber and the matrix was rather high (9.5 MPa), which can be attributed to the complex structure of the interface and the fiber twisting.

Keywords: carbon nanotube fibers, coatings, single fiber tensile test, interphase, single fiber pull-out test, fracture surface

Introduction

There exist reports of very high CNT fiber strength which make us assume them as very exciting materials. On the other hand, there is the necessity to further develop new and innovative processing methods in order to gain improved mechanical, electrical, chemical, thermal, and optical properties (Wu et al., 2012a). These multifunctions will make CNT fibers an attractive choice as reinforcement in composites, particularly in the development of electrically conductive structures. Therefore, investigations are concerned with the improvement of both fiber tensile strength and adhesion strength toward polymeric matrices. Besides, it is well known that the ideal internal structure of high-performance fibers for optimum tensile properties consists of perfectly aligned, infinitely long axial fibrils (Chae and Kumar, 2008). **Figure 1** shows a few original images of fibrillated fracture surfaces of very strong organic fibers, such as poly-p-phenylbenzobisoxazol (PBO) fibers, aramid fibers (AR) together with CNT fibers. The tensile strengths of PBO and AR are 6 GPa and 3.2 GPa, respectively, whereas the tensile strength of CNT fibers strongly depends on the processing method. Aerogel-spun fibers spun directly from the chemical vapor deposition synthesis region of a furnace are produced for electrical purposes and possess moderate strength of only 0.19 GPa (Wu et al., 2012a). Continuous CNT fibers spun from CNT carpet, consisting of mainly double- and triple-walled tubes, achieve average tensile strength of 1.2 ± 0.3 GPa (Zu et al., 2012). Chemical vapor produced CNT continuous fibers made of single and dual wall CNTs could improve the tensile strength to 3–5 GPa by post-process stretching and increase of alignment (Wu et al., 2012b). Besides the processing conditions, the objective of this work is to activate the CNTs within the fiber cross section in order to improve the stress transfer ability between the single CNTs or CNT bundles within the fiber and in the interphase, i.e., between the sheath CNTs and the composite matrix. The experimental work focuses on coating the as-received fibers with different aqueous-based film formers and low viscous epoxy resin. Both fiber properties and adhesion strength determinations were performed in order to reveal the variations of tensile, failure, and adhesion features. Furthermore, the resistance changes were investigated simultaneously with the stress–strain behavior and the durability after cyclic loading.

Materials and Methods

Materials

Untreated dry-spun CNT fibers were provided by Advanced Materials Division, Suzhou Institute of Nano-Tech and Nano-Bionics, Suzhou, PR China. **Table 1** shows the characteristics of the CNT fibers used. They were produced at different twisting rates and were differently densified. A 1-cm-wide CNT film was pulled out from the CNT array pre-grown on a silicon substrate, at a constant take-up speed of 20 cm/min. The diameter of the fiber is mainly determined by the twisting rate of the fiber and densification. With higher twisting rate, the diameter gets usually smaller. Also the surface twist angle is determined by the twisting rate. The higher the twisting rate, the higher is the fiber twist angle. The diameter and length of the CNTs are usually the same in all the fibers. The CNT fiber diameters given in **Table 1** are average values determined by digital microscopy and from SEM images.

Three aqueous-based epoxy resin emulsions of different glass transition temperatures (T_g) and epoxy equivalent weights (g/eq), one aqueous-based polyurethane film former, and one low viscous epoxy resin were selected for a dip coating of the CNT fibers in order to achieve coatings of different hardnesses and different impregnations. The concentration of the aqueous epoxy dispersions and polyurethane dispersion, respectively, was kept constant at 2.75 wt%. The low viscous epoxy resin was applied as recommended by the manufacturer. The characteristics of the coating agents as well as the manufacturers are given in detail in **Table 2**.

Methods
Preparation of Coated CNT Fibers

Each CNT fiber piece of about 30 cm length was dipped in the aqueous coating dispersion (for RIM R135 resin, the liquid stoichiometric resin/hardener mixture) with the help of a pair of plastic tweezers and kept in the aqueous emulsion for 1 min. A heat treatment of the fibers was done in an oven for 2 h at 130°C.

Single Fiber Tensile Tests and Characterization of Sensing Abilities

Using the Favigraph semiautomatic fiber tensile tester (Textechno, Germany) equipped with a 1 N force cell, the mechanical tensile

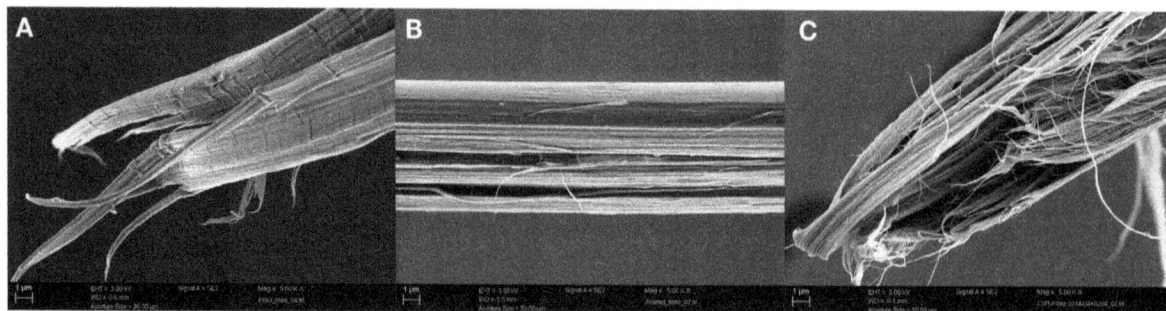

FIGURE 1 | Fracture surfaces of PBO (A), AR (B), and CNT (C) fibers.

stress/strain behavior was tested in accordance with EN ISO 5079 with a test velocity of 0.1 mm/min at a gage length of 2 mm by using special clamp combination of soft and hard elastomers.

The electrical resistance of the coated CNT fibers was measured simultaneously with a Keithley 2001 electrometer having a measurement range from 10^{-6} to $10^9 \, \Omega$ (Keithley Instruments GmbH, Germany) at 23°C and 50% relative humidity (RH). The mean values of at least 10 measurements for every fiber were calculated. To investigate the piezoresistive effect, the electrical resistance was recorded as the specimen underwent either uniaxial tensile or cyclic loading using the semiautomatic equipment described above. The specimen was fixed between two clamps which were coated with conductive silver paste (Acheson Silver DAG 1415M) and served as electrodes. The cyclic loading and unloading tension tests were conducted at a gage length of 2 mm, a cross-head velocity of 0.2 mm/min, and a 1% stepwise increase of the strain amplitude from 1 to 5%. The cyclic loading tests were performed as either load increasing tests with the same specimen to decrease the data scattering or by using different specimens. The deformations were divided into an elastic part, which is recovered when the stress is removed, and a plastic or permanent part. Quantitatively, we used the elastic recovery as elastic extension divided by total extension, which may be plotted against stress or strain. Instead of studying dimensional recovery, one may study and define work recovery (resilience) in a similar manner. It should be noted that hysteresis = (1− work recovery) gives the proportion of the total work that is dissipated as heat (Morton and Hearle, 2008). Simultaneous resistance, strain, and load measurements were integrated with the time scale in a customized data acquisition package TestPoint 2.0.

Scanning Electron Microscopy

The fiber surface morphologies, fracture surfaces after tensile test, and fracture surfaces after pull-out test were studied using the scanning electron microscope (FE–SEM Ultra 55, Carl Zeiss SMT AG, Germany). The samples for SEM observation were coated by ~5-nm-thick platinum layer.

TABLE 1 | Characteristics of CNT fibers.

Designation	Twisting rate/ densification	Nanotube length/ nanotube diameter	Fiber diameter
A	1200 min^{-1}/ethylene	250 μm/4–8 nm	8.5 ± 0.6 μm
B	1000 min^{-1}/no densification	250 μm/4–8 nm	13.0 ± 1.0 μm
C	1446 min^{-1}/no densification	250 μm/4–8 nm	11.2 ± 1.1 μm

Single Fiber Pull-Out Tests

The single CNT fibers were embedded under PC control in matrix resin mixture RIM R135/RIM H137 (100/30 weight parts) to embedded lengths (l_e) ranging from 40 to 100 μm and cured for 60 min at 85°C. Then the specimens were taken from the embedding device and cured for 6 h at 80°C in an oven. After curing, the specimens were stored in a desiccator (silica gel) before pull-out test.

Different interphases in the vicinity of the epoxy matrix/CNT fiber interface were investigated using the single fiber pull-out test. The self-made sample preparation equipment and pull-out apparatus allowed us to record force–displacement curves under quasi-static conditions with the loading rate of 0.01 μm/s at 23°C and 50% RH. The free fiber lengths were kept as short as possible (<50 μm). The diameter of each pulled-out fiber was measured in several cross-sections along the fiber using an optical microscope, and its mean value, d_f, was used in further calculations.

From the experimental data, the following interfacial strength parameters were determined: the apparent interfacial shear strength (IFSS), $\tau_{app} = F_{max}/(\pi d_f l_e)$, where F_{max} is the maximum force recorded during the test; the local IFSS, τ_d, at which interfacial debonding is assumed to start in stress-based models of the pull-out test (Zhandarov and Mäder, 2005); and the critical energy release rate for debonding, G_{ic}, derived in energy-based models (Nairn, 2000; Zhandarov and Mäder, 2003). In addition, we calculated the interfacial frictional stress, τ_f, between the fiber and the matrix. In our calculations, we used the latest approach which relates τ_d and τ_f values to experimentally measured values of F_{max}, the maximum force, and F_b, the force corresponding to the debonding completion. This approach is very useful when the "kink" in a force–displacement curve associated with the debonding onset is hardly discernible; it was described in details elsewhere (Zhandarov and Mäder, 2014). We also used a similar method in the energy-based approach to calculate G_{ic} from F_{max} and F_b.

Atomic Force Microscopy

An atomic force microscope (Digital Instruments D3100, USA) was used as a surface imaging tool. The topography images of samples were studied in tapping mode, while phase shifts, i.e., changes in the phase angle of vibration with respect to the phase angle of the freely oscillating cantilever, recorded simultaneously with height changes, are present as a phase image.

Roughness parameters derived from ASME B46 were calculated: R_a, arithmetic average of the absolute values of the surface

TABLE 2 | Characteristics of aqueous emulsions and epoxy resin used for surface modification of CNT fibers.

Designation	Glass transition temperature, T_g (°C)	Epoxy equiv. weight (g/eq)	Av. particle size (nm)
Filco 348 (Coim, Italy) pure epoxy resin in non-ionic aqueous emulsion	−12	270	150–400
Neoxil 8294 (DSM, Switzerland) flexible epoxy resin (epoxy-ester) in non-ionic aqueous emulsion	20	1300–1700	300–800
Filco 394 (Coim, Italy) modified epoxy resin in non-ionic aqueous emulsion	30	1800	300–800
Hydrosize U5-01 (Michelman, USA) waterborne polyurethane dispersion with an anionic/non-ionic emulsifier	−19		500
RIM R135/RIM H137 (Momentive, USA) very low-viscosity laminating resin system, weight ratio of 100:30	89	166–185 amine value of hardener 400–600 mg KOH/g	

height deviations measured from the mean plane within the box cursor ($5\,\mu m \times 5\,\mu m$); R_{max}, maximum vertical distance between the highest and the lowest data points within the cursor box ($5\,\mu m \times 5\,\mu m$). At least three images of one sample were used for calculation.

Results

Single Fiber Tensile Test on Coated CNT Fibers

Table 3 summarizes results of the single fiber tensile tests for the reference samples in comparison to coated ones. The three reference fibers are characterized by different tensile strengths (σ_{max}) and corresponding strains (ε_{max}) as well as Young's moduli (E) which is due to the different treatment of the CNT fibers within the processing. Fibers A and B achieved much greater strengths and Young's moduli than fiber C. The reason for this difference is due to somewhat overtwisted fiber C. The twisting has both positive and negative effects on the fiber strength. As the positive effect, the twisting would densify the fiber and enhance the load transfer efficiency between CNTs in the fiber. The negative effect is that upon twisting, the CNTs initially aligned in the fiber direction are forced to be inclined to the fiber direction, which degrades the fiber strength. There should be an optimal twisting angle for the fiber strength, which depends upon several factors such as CNT length and diameter, fiber diameter, and others. Therefore, it is not unexpected that fiber C is less strong than fibers A and B.

TABLE 3 | CNT fiber tensile test results as a function of coatings.

Designation	σ_{max} (MPa)	ε_{max} (%)	E (GPa)
A reference	1241 ± 261	4.6 ± 0.4	20.7 ± 6.8
A, Filco 348	1375 ± 187	3.8 ± 0.4	39.2 ± 10.6
A, Filco 394	972 ± 160	3.8 ± 0.6	27.9 ± 8.8
A, Neoxil 8294	1240 ± 246	3.8 ± 0.4	32.8 ± 8.0
B reference	1073 ± 162	4.4 ± 0.4	23.4 ± 5.7
B, Filco 348	1336 ± 119	4.0 ± 0.1	38.6 ± 5.7
B, Filco 394	1455 ± 173	3.4 ± 0.2	53.1 ± 7.7
B, Neoxil 8294	1214 ± 134	3.9 ± 0.2	34.7 ± 7.8
C reference	714 ± 26	5.6 ± 0.3	17.6 ± 4.1
C, PU-05-01	610 ± 8.6	6.1 ± 0.7	13.4 ± 2.3
C, Filco 348	700 ± 48	5.2 ± 0.8	23.7 ± 6.2
C, RIM 135	826 ± 80	3.3 ± 0.4	34.3 ± 7.6

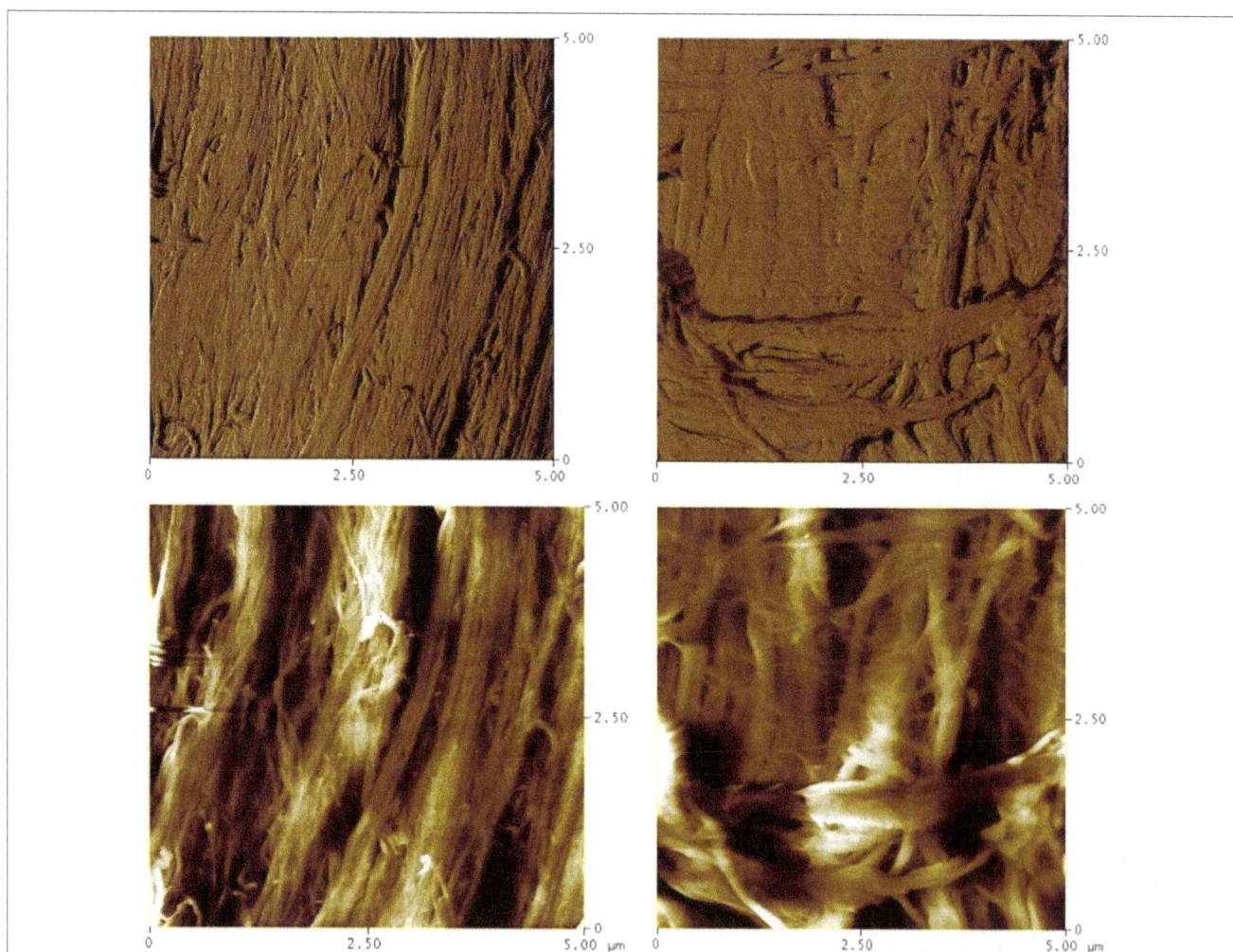

FIGURE 2 | Selected AFM phase and height images (top/bottom), respectively, of reference fibers (left) and Filco 394 treated ones (right).

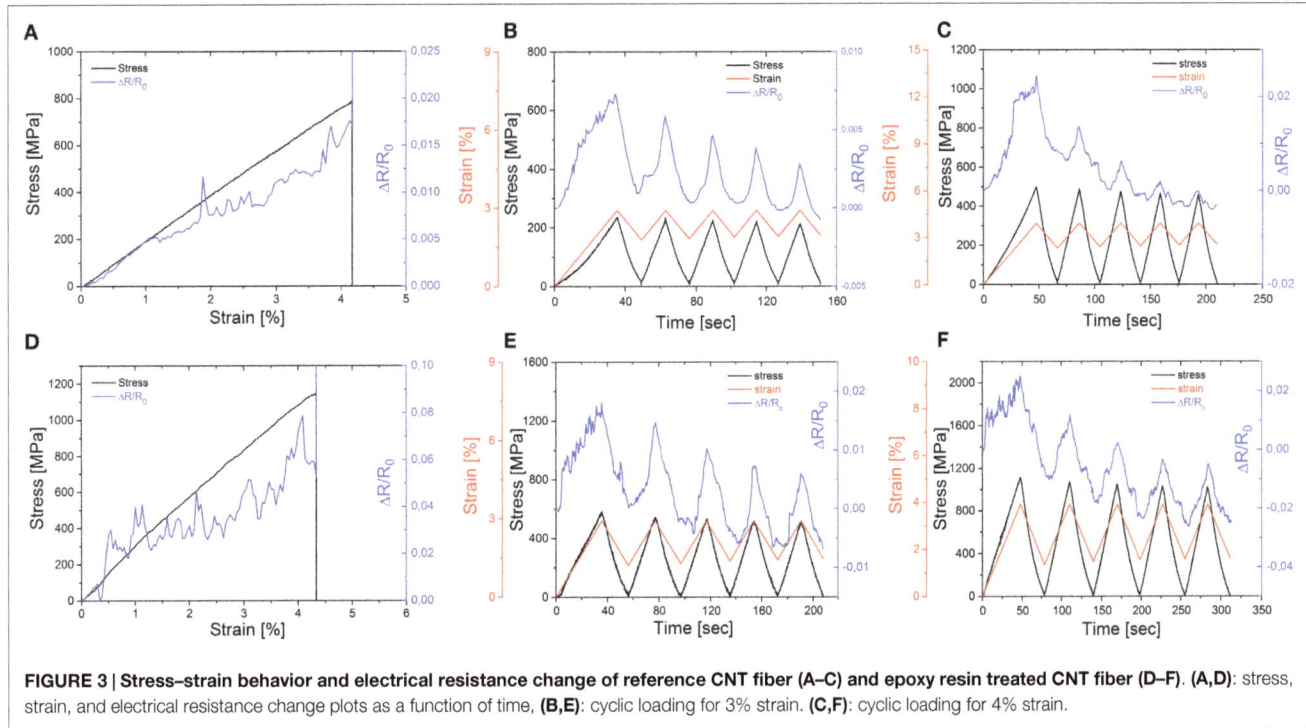

FIGURE 3 | Stress–strain behavior and electrical resistance change of reference CNT fiber (A–C) and epoxy resin treated CNT fiber (D–F). (A,D): stress, strain, and electrical resistance change plots as a function of time, (B,E): cyclic loading for 3% strain. (C,F): cyclic loading for 4% strain.

In dependence of their characteristics, the coatings cause different changes in the tensile properties. Among the three different aqueous-based epoxy film formers, Filco 348 has the lowest T_g and the lowest average particle size. It performs best, which can be explained by improved impregnation and thus increased stress transfer ability between the CNTs or CNT bundles within the CNT fiber due to preferable mechanical interlocking. **Figure 2** shows selected AFM height and phase images of untreated reference and Filco 348 treated CNT fibers which confirm decrease of alignment after coating and significant increase of surface roughness (R_a from 20 to 47 nm, R_{max} from 263 to 534 nm) which can contribute to mechanical interlocking. The better the impregnation, the more CNTs can be activated and participate in the stress transfer. It is noted that the AFM phase images (**Figure 2**, top line) do not indicate material inhomogeneity. Furthermore, the stiffness, determined by Young's modulus, increases due to the impregnation. The other epoxy film formers also show moderate increase of tensile strength and Young's modulus. In contrast, the rather "weak" polyurethane film former leads to a drop in tensile strength and modulus because of the weak interphase created by impregnation between CNTs or CNT bundles. A considerable increase in strength and modulus was detected for CNT fibers with an epoxy resin coating. This is inferred by curing of the reactive system in addition to mechanical interlocking. On the other hand, the brittleness could increase.

Stress–Strain Behavior and Electrical Resistance Change of Coated CNT Fibers

Besides the stress–strain behavior, the electrical resistance change of the CNT fibers was determined. **Figure 3** shows exemplarily the stress–strain and resistance–strain curves of a reference CNT fiber in comparison with an epoxy treated fiber. The epoxy

treated fibers (**Figures 3D–F**) are characterized by much higher stresses at comparable strains related to the reference fibers, as was discussed in Section "Single Fiber Tensile Test on Coated CNT Fibers." The resistance of the reference fiber is 530 Ω, whereas the resistance of the epoxy resin treated fiber increased to 703 Ω due to the polymeric coating. It is interesting to note that the resistance change follows the stress–strain behavior at low strains (**Figures 3A,D**). With increasing strain the resistances deviate to lower resistance changes, which might be due to the increased friction and compression of CNT bundles, resulting in closer contact with increasing strain. Some scatter is visible due to different alignments of CNTs and occurrence of CNT breaks increased with increasing strain. Cyclic loadings at 3% strain achieve a rather slight gradual decay of the resistance change which can be explained again by increasing compression of the CNTs or CNT bundles in the yarn. At 4% strain (near the failure of the CNT fiber), the resistance change of the reference fiber dropped to zero resistance change, whereas the resistance of the epoxy resin treated fiber is characterized by a negative resistance change, i.e., due to further compression and improved cohesion with epoxy treatment, the conductivity of the CNT fiber is improved.

Elastic Recovery and Hysteresis for Coated CNT Fibers

Typical records of stress–strain curves upon cyclic loading of untreated CNT fiber and epoxy resin treated CNT fiber are shown in **Figure 4**. After data evaluation, the elastic recovery and the hysteresis as functions of strain and stress, respectively, are displayed in **Figure 5**. The figures are compared for one specimen and different specimens show a bit greater scatter (<10%), but sufficient reproducibility. Highest elastic recoveries combined with lowest hysteresis values were determined for epoxy resin treated fibers,

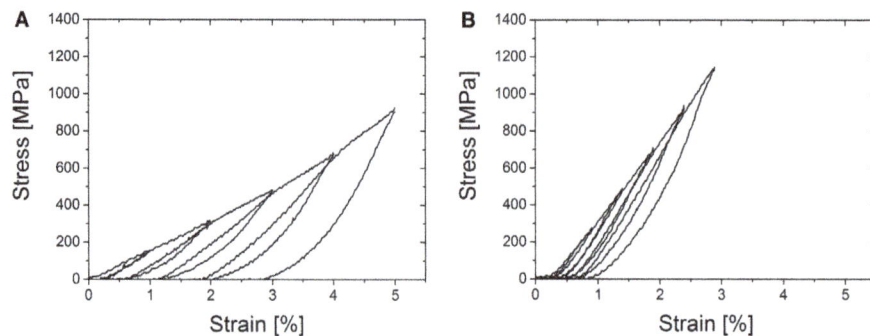

FIGURE 4 | Typical records of stress–strain curves after cyclic loading of untreated (A) and epoxy resin treated (B) CNT fiber.

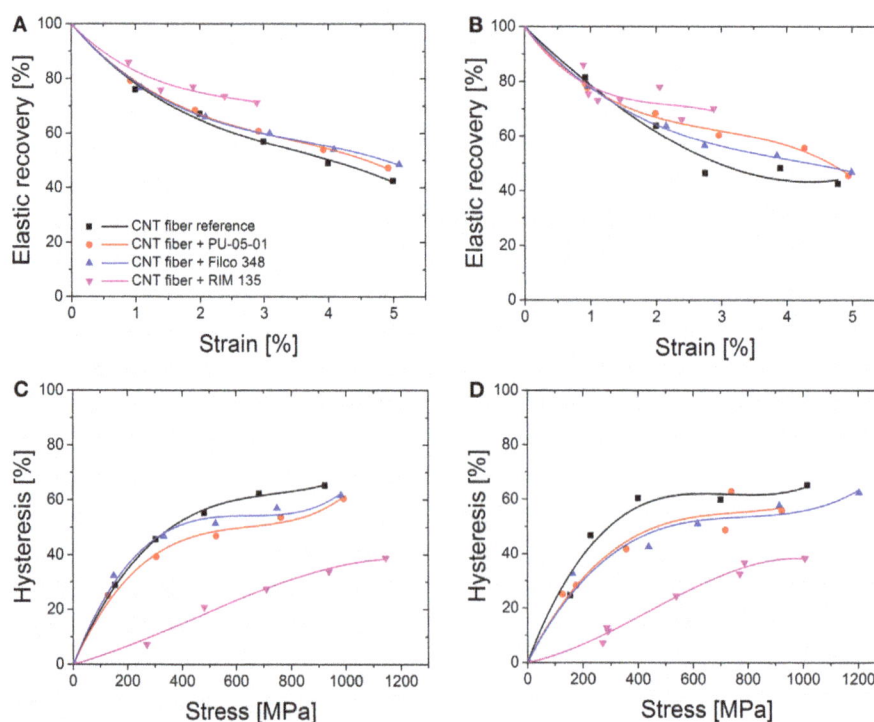

FIGURE 5 | Elastic recovery (A,B) and hysteresis (C,D) plots as a function of strain and stress, respectively, for differently surface treated CNT fibers. (A,C): load increasing tests performed on one specimen; (B,D): different specimens.

while the other treated fibers were very close to the reference samples. This evidences that the epoxy resin treated CNT fiber is more uniform than other fibers; due to resin infiltration between CNT bundles and its subsequent crosslinking, the bundles are bonded with each other and their displacements under loading are more coherent. Consequently, internal friction within this fiber is lower than that in the others, which, in turn, results in lower hysteresis and higher elastic recovery under cyclic loading.

Interfacial Adhesion Strength of CNT Fibers with Epoxy Matrix

The comparison of SEM images of fracture surfaces after the pull-out test for untreated CNT fibers and those coated with RIM 135 resin demonstrated substantial difference in the failure

mechanisms. **Figure 6A** shows the fracture surfaces for uncoated fibers. It can be seen that the epoxy resin had penetrated into the CNT fiber forming a relatively thick (~1 µm) CNT/epoxy composite layer, and the fracture surface located, in part, between this layer and the "core" of the fiber which remained without resin infiltration (radial "steps" in the photographs). Similar behavior was reported by Zu et al. (2012) for continuous CNT fibers in DER 353 epoxy resin. This localization of the fracture zone as well as probably too large embedded lengths (about 70 µm) resulted in fiber breakage before debonding completion, which can easily be seen in **Figure 6A**. It should be noted that pull-out tests of CNT fibers treated with aqueous-based film formers embedded in the same epoxy matrix (results not shown here) also failed in the same mechanism, such as stepwise fracture.

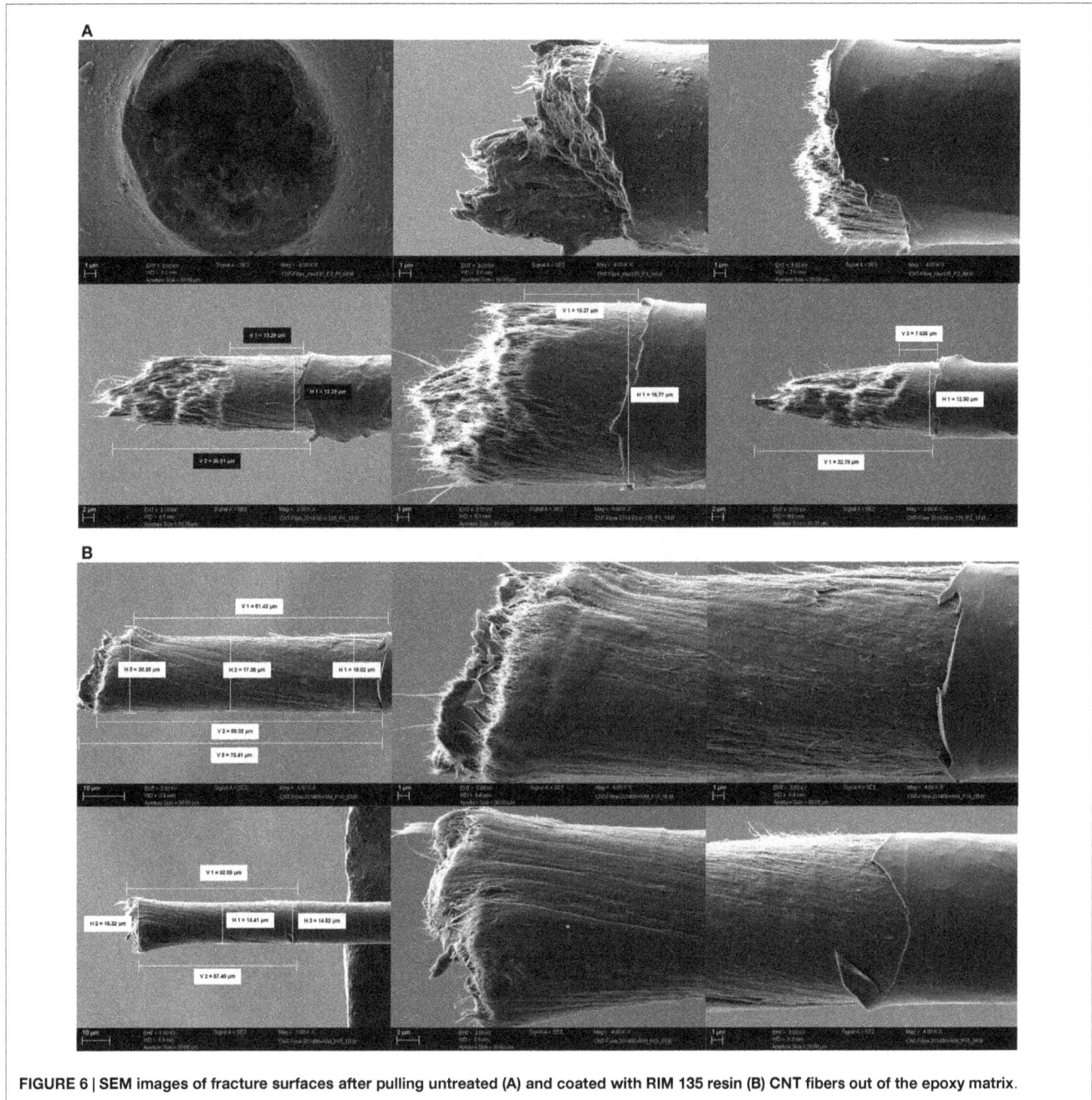

FIGURE 6 | SEM images of fracture surfaces after pulling untreated (A) and coated with RIM 135 resin (B) CNT fibers out of the epoxy matrix.

In contrast to this, most CNT fibers coated with RIM 135 resin have been successfully pulled out of the epoxy matrix droplet. SEM images of fracture surfaces after interfacial debonding (**Figure 6B**) clearly show that shear sliding occurred in this case along the interface between the matrix and the outer fiber layer impregnated with the resin. For coated fibers, resin penetration into a continuous CNT fiber took place at the stage of coating (dipping the fiber in the resin–hardener mixture and subsequent thermal treatment). The conditions of this thermal treatment (relatively short time but the temperature higher than recommended by the manufacturer for this resin) gave rise to the formation of a rather thin surface layer consisting of CNT bundles in partly cured resin with a Tg of 89°C compared to a Tg of 96°C for the cured matrix.

Thus, the interface between it and the bulk matrix in the droplet, formed during the fiber embedment in the matrix and subsequent resin curing, appeared to be the "weak point" of the specimen. The fracture surface basically replicates the structure of the continuous CNT fiber surface, though some particular CNT micro-bundles separated from the fiber can be seen on it. In all probability, the interfacial failure included a cohesive part through the breakage of the weakest outer CNT fascicules.

A typical force–displacement curve for this system is shown in **Figure 7**. Both force values, F_{max} and F_b, required for the calculation of interfacial strength parameters according to our new approach, are easily measurable from this curve, while the "kink" position in the ascending part of the curve, on which

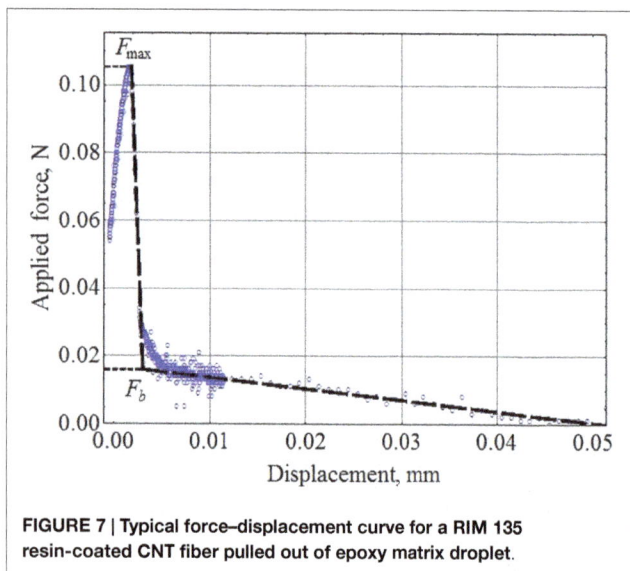

FIGURE 7 | Typical force–displacement curve for a RIM 135 resin-coated CNT fiber pulled out of epoxy matrix droplet.

the traditional approach is based, cannot be determined reliably. We calculated interfacial strength parameters for 10 successfully tested specimens and determined their mean values and SD: the apparent IFSS $\tau_{app} = 45.5 \pm 9.4$ MPa; the local IFSS $\tau_d = 62.7 \pm 10.5$ MPa; the critical energy release rate for interfacial debonding $G_{ic} = 17.0 \pm 6.2$ J/m^2; and the frictional interfacial stress, $\tau_f = 9.5 \pm 3.0$ MPa. The obtained τ_{app} value is much greater than that presented by Zu et al. (2012) (14.4 MPa) but is typical for carbon fibers/epoxy composites [e.g., 30–75 MPa for T300 fibers in the mixture of DER 331 and DER 732 epoxies (Nakamura et al., 2012)]. The values of the local IFSS and the critical energy release rate are also typical for carbon fiber/epoxy systems. However, we should mention that the interfacial frictional stress between the CNT fiber and the epoxy matrix (9.5 MPa) is rather high; in all probability, this can be attributed to the complex structure of the fiber surface, which is formed by a large number of individual CNTs which make finite angles with the fiber axis due to fiber twisting during its spinning from a CNT array.

Conclusion

Three reference CNT fibers, characterized by different twisting rates and densities due to fiber processing, revealed average tensile strength and Young's modulus variations up to 73 and 33%, respectively. The aqueous-based film former coatings lead to considerable changes of strength and modulus values compared to the untreated reference fibers.

Depending on the chemistry of the aqueous-based film former coatings, particle size distribution, and the glass transition

temperature, the film formers differently affected the mechanical properties of the CNT fibers. A very weak polyurethane dispersion (PU-05-01) decreased both fiber's tensile strength and modulus to 85 and 76%, respectively, because of the weak interphases created by impregnation between CNTs or CNT bundles. In contrast to this, especially the epoxy film former Filco 348 (lowest T_g and average particle size) could achieve average strength and modulus improvements of 24 and 65%, respectively. Besides the improved impregnation and thus increased stress transfer efficiency, a mechanical interlocking and misalignment together with an increased surface roughness was revealed by AFM tapping mode images.

An increase of strength (15%) and modulus (95%) related to the reference fibers was found with an epoxy resin coating caused by curing of the reactive system in addition to mechanical interlocking. It should be noted that the electrical resistance increased from 530 to 703 Ω due to the resin coating. Upon tensile loading, the electrical resistance deviated to lower values with increasing strain caused by the closer contact of the CNTs. Cyclic loadings achieved a gradual decay of the resistance change depending on the strain due to increasing compression of the CNTs. Again, the epoxy resin treated fibers lead to lowest resistances due to improved cohesion between the CNTs. The epoxy treated fibers exhibited the greatest elastic recovery and the lowest hysteresis during the cyclic loading.

Interfacial adhesion strength determinations of CNT fibers with epoxy matrix showed different failure mechanisms after pull-out tests for epoxy resin treated fibers compared to untreated and film former coated ones. SEM fracture surfaces revealed about 1-μm-thick CNT/epoxy composite layer for uncoated fibers which shifted the fracture zone inside the fiber and resulted in stepwise fiber fracture before debonding completion.

In contrast, epoxy resin-coated fibers were successfully pulled out of the epoxy resin matrix after interfacial debonding, and clear shear sliding along the interface between the matrix and the outer CNT layer of the fiber impregnated with the resin was observed. Due to the curing conditions of the resin coating, the interphase became the "weak point" of the single fiber model composite.

The determination of the interfacial strength parameters according to our new approach revealed apparent and local IFSS values of 45 and 63 MPa, respectively. Compared with literature data, these values are typical for carbon fiber/epoxy composites. However, the evaluated interfacial frictional stress between the CNT fiber and the epoxy resin matrix (9.5 MPa) is rather high and can be attributed to the complex structure of the fiber surface, which is formed by a large number of individual CNTs having a distinct density and a finite angle with the fiber axis due to the manufacturing process.

References

Chae, H. G., and Kumar, S. (2008). Making strong fibers. *Science* 319, 908–909. doi:10.1126/science.1153911

Morton, W. E., and Hearle, J. W. S. (2008). *Physical Properties of Textile Fibres*, 4th Edn. Boca Raton, FL: Woodhead Publishing Limited and CRC Press LLC.

Nairn, J. A. (2000). Analytical fracture mechanics analysis of the pull-out test including the effects of friction and thermal stresses. *Adv. Compos. Lett.* 9, 373–383.

Nakamura, R., Netravali, A. N., and Hosur, M. V. (2012). Effect of halloysite nanotube incorporation in epoxy resin and carbon fibre ethylene/ammonia plasma treatment on their interfacial property. *J. Adhes. Sci. Technol.* 26, 1295–1312. doi:10.1163/156856111X593612

Wu, A. S., Chou, T. W., Gillespie, G. W., Lashmore, D., and Rioux, J. (2012a). Electromechanical response and failure behavior of aerogel-spun carbon nanotube fibres under tensile loading. *J. Mater. Chem.* 22, 6792–6798. doi:10.1039/c2jm15869h

Wu, A. S., Nie, X., Hudspeth, M. C., Chen, W. W., Chou, T. W., Lashmore, D. S., et al. (2012b). Strain rate-dependent tensile properties and dynamic electromechanical response of carbon nanotube fibres. *Carbon N. Y.* 50, 3876–3881. doi:10.1016/j.carbon.2012.04.031

Zhandarov, S., and Mäder, E. (2003). Indirect estimation of fibre/polymer bond strength and interfacial friction from maximum load values recorded in the microbond and pull-out tests. Part II: critical energy release rate. *J. Adhes. Sci. Technol.* 17, 967–980. doi:10.1163/156856103322112879

Zhandarov, S., and Mäder, E. (2005). Characterization of fibre/matrix interface strength: applicability of different tests, approaches and parameters. *Compos. Sci. Technol.* 65, 149–160. doi:10.1016/j.compscitech.2004.07.003

Zhandarov, S., and Mäder, E. (2014). An alternative method of determining the local interfacial shear strength from force-displacement curves in the pull-out and microbond tests. *Int. J. Adhes. Adhes.* 55, 37–42. doi:10.1016/j.ijadhadh.2014.07.006

Zu, M., Li, Q., Zhu, Y., Dey, M., Wang, G., Lu, W., et al. (2012). The effective interfacial shear strength of carbon nanotube fibers in an epoxy matrix characterized by a microdroplet test. *Carbon N. Y.* 50, 1271–1279. doi:10.1016/j.carbon.2011.10.047

Conflict of Interest Statement: The authors declare that the research was conducted in the absence of any commercial or financial relationships that could be construed as a potential conflict of interest.

8

Group IV direct band gap photonics: methods, challenges, and opportunities

Richard Geiger, Thomas Zabel and Hans Sigg*

Laboratory for Micro- and Nanotechnology, Paul Scherrer Institut, Villigen, Switzerland

The concept of direct band gap group IV materials may offer a paradigm change for Si-photonics concerning the monolithic implementation of light emitters: the idea is to integrate fully compatible group IV materials with equally favorable optical properties as the chemically incompatible group III–V-based systems. The concept involves either mechanically applied strain on Ge or alloying of Ge with Sn, which permits to drastically improve the radiative efficiency of Ge. The favorable optical properties result from a modified band structure transformed from an indirect to a direct one. The first demonstration of such a direct band gap laser has recently been accomplished in GeSn. This demonstration proves the capability of this new concept, which may permit a qualitative as well as a quantitative expansion of Si-photonics in not only traditional but also new areas of applications. This review aims to discuss the challenges along this path in terms of fabrication, characterization, and fundamental understanding, and will elaborate on evoking opportunities of this new class of group IV-based laser materials.

Keywords: Si photonics, germanium, strain, GeSn, direct band gap, laser

Edited by:
Koji Yamada,
National Institute of Advanced
Industrial Science and Technology,
Japan

Reviewed by:
Shinichi Saito,
University of Southampton, UK
Krishna C. Saraswat,
Stanford University, USA

*Correspondence:
Hans Sigg,
Laboratory for Micro- and
Nanotechnology, Paul Scherrer
Institut, Villigen PSI, CH 5232,
Switzerland
hans.sigg@psi.ch

Introduction

The Si-based optical platform is rapidly changing the landscape of photonics by offering powerful solutions, for example, for data links (Miller, 2010) and sensing (Passaro et al., 2012) to name only two out of many. This development has taken place in spite of the fact that Si itself is a poor emitter of light. This is without a doubt due to the fact that Si technology as used in very large-scale integration (VLSI) and complementary metal-oxide-semiconductor (CMOS) technology is extremely mature and advanced. This fact seemingly compensates for the shortfalls in concepts for Si to generate light.

Nowadays, group III–V materials are implemented to integrate active light sources onto the Si platform by using involved coupling schemes and/or heterogeneous integration (Fang et al., 2013). However, because these materials are chemically intolerant to Si, their integration bears a lot of burdens, which raises the fabrication costs. Strongly preferred are materials that are compatible to Si, tolerated by the technology (preferentially CMOS), and capable of producing light similar in efficiency to traditional group III–V semiconductor systems.

In direct band gap systems, light generation is based on radiative recombination of electrons and holes, both with practically the same momentum as schematically shown in **Figure 1A**. In unstrained, i.e., "regular" bulk Ge, however, the excited electrons will preferentially occupy the lower conduction band energy states of the L-valley. In Ge, the momentum of the electrons does, thus, not match those of the holes, which occupy the degenerated heavy- and light-mass valence bands at the Γ-point (c.f. **Figure 1B**). The appearing momentum mismatch requires a phonon for the recombination. But note that except the position of the indirect L-valley, which is in Ge 140 meV below the Γ valley minimum, the band alignments near the Γ-point in system A (say InGaAs, one of the most prominent group III–V systems used for lasing) and B are very similar. To achieve the favorable direct recombination

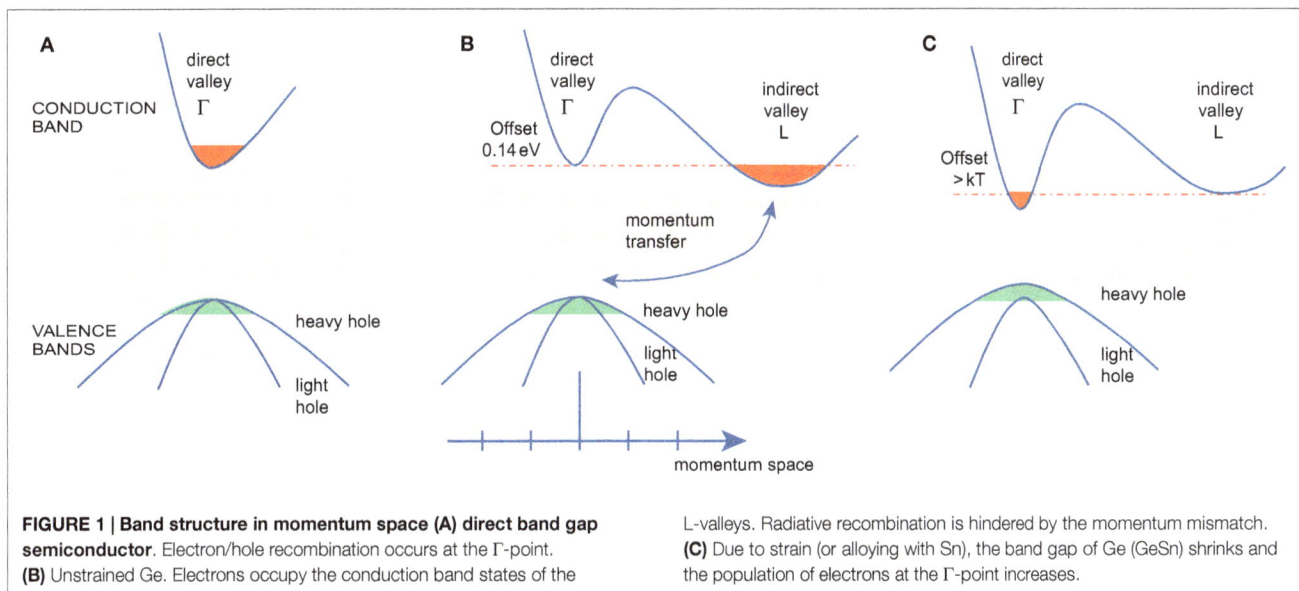

FIGURE 1 | Band structure in momentum space (A) direct band gap semiconductor. Electron/hole recombination occurs at the Γ-point. (B) Unstrained Ge. Electrons occupy the conduction band states of the L-valleys. Radiative recombination is hindered by the momentum mismatch. (C) Due to strain (or alloying with Sn), the band gap of Ge (GeSn) shrinks and the population of electrons at the Γ-point increases.

condition also for Ge, we need to find a way to inject electrons into that conduction band valley with its energetic minimum at the Γ-point. This is realized in the most straightforward fashion when all unwanted electron levels are energetically shifted above the Γ-states, which is equivalent to transfer the system from a fundamentally indirect to a fundamentally direct one.

We will discuss the two methods that make this conversion possible. One involves the application of tensile strain, while the second approach relies on alloying Ge with Sn. Thus, the obtained band alignments are depicted in **Figure 1C**. With either one of the methods, the Γ-valley can be reduced below the indirect one at L enabling efficient carrier injection into the Γ-valley. Moreover, the VB degeneracy is lifted depending on the strain state and its loading, biaxial or uniaxial, c.f. Section "Modeling" for more details.

In **Figure 2**, we show the state-of-the-art of the strain and alloying approach toward the realization of a direct band gap group IV material. For our discussion, we selected those approaches that are potentially compatible with CMOS fabrication and are suited for optical applications. Very thin membranes (Sánchez-Pérez et al., 2011), nanowires clamped in bulky mechanical strain apparatus (Greil et al., 2012), Ge bulk layers on III–V substrates (Huo et al., 2011), etc., are not considered here because they are unpractical for integration on Si. Not considered either is light emission from Si-based quantum wells and defects; for a recent review, see Saito et al. (2014). In our compilation, **Figure 2**, we benchmark the two strain loadings (uniaxial and biaxial) and Sn alloy composition against the achieved relative band offset, $\Delta E/E_0$, where an offset ΔE of 100% is equal to $E_0 \sim 140$ meV for the case of unstrained Ge. An offset parameter of 0 meV (0%) corresponds, thus, to Γ- and L-valleys having their band edges at the same energy.

The black arrow on the left hand side of the second line in **Figure 2** marks the case of highly n-doped Ge (Liu et al., 2010), where a maximum of 0.25% biaxial strain is accomplished. This value of 0.25% is the one typically obtained from direct epitaxy of Ge on Si. It arises due to the difference between the thermal expansion coefficients of Si and Ge (Michel et al., 2010). High n-doping

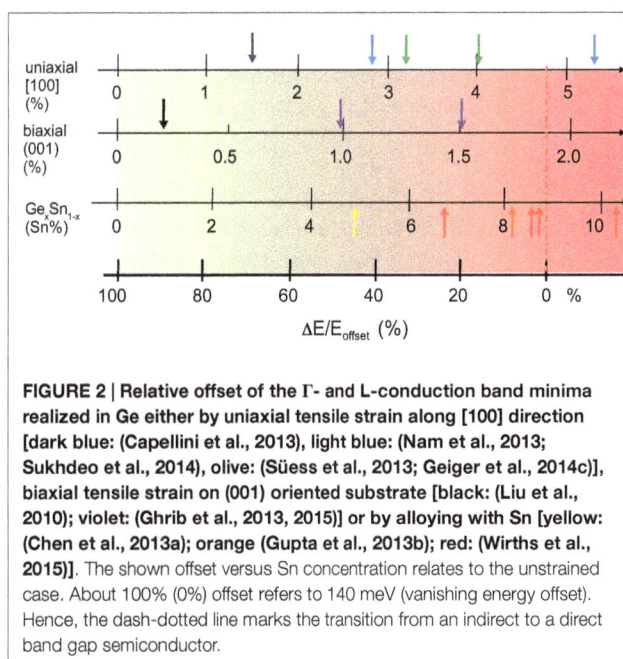

FIGURE 2 | Relative offset of the Γ- and L-conduction band minima realized in Ge either by uniaxial tensile strain along [100] direction [dark blue: (Capellini et al., 2013), light blue: (Nam et al., 2013; Sukhdeo et al., 2014), olive: (Süess et al., 2013; Geiger et al., 2014c)], biaxial tensile strain on (001) oriented substrate [black: (Liu et al., 2010); violet: (Ghrib et al., 2013, 2015)] or by alloying with Sn [yellow: (Chen et al., 2013a); orange (Gupta et al., 2013b); red: (Wirths et al., 2015)]. The shown offset versus Sn concentration relates to the unstrained case. About 100% (0%) offset refers to 140 meV (vanishing energy offset). Hence, the dash-dotted line marks the transition from an indirect to a direct band gap semiconductor.

is introduced to fill the parasitic indirect states (Xiaochen et al., 2010). Such doping does not transform the material into a direct gap system, but it appeared that under optical excitation and electrical pumping the light emission shows an intensity threshold as well as linewidth narrowing (Liu et al., 2010; Camacho-Aguilera et al., 2012). These results became widely known as the optically and electrically pumped Ge-laser. However, since the time when these announcements were made in 2010 and 2012, only one other demonstration of these effects has been reported so far (Koerner et al., 2015). This very recent and only result concerns a Ge diode structure with an unstrained active region doped at 3×10^{19} cm^{-3}. The obtained emission spectra are similar to the one from the original work from the MIT group. However, as we will show below, these spectra significantly differ in several aspects – the intensity, the linewidth, and the Fabry–Perot (FP) multi-mode behavior – from those obtained with the here discussed direct band gap lasers. Moreover, as will be discussed in Section "Lifetime, Gain, and Loss," these Ge-lasing observations are contradicted by gain experiments (Carroll et al., 2012) as well as by theoretical analysis performed by several groups (Liu et al., 2007; Chow, 2012; Dutt et al., 2012; Peschka et al., 2015) when the lasing current density threshold is calculated using realistic non-radiative lifetimes (Geiger et al., 2014a). As the experimental foundation for understanding this peculiar threshold behavior of low strained (and in one case even unstrained) Ge is ambiguous, we will focus here on reports concerning direct band gap group IV systems, which includes the first unmistakable proof of interband lasing in a group IV system. Without doubt, with the advent of direct band gap systems showing unambiguous lasing, an excellent opportunity is created, which will help to unravel in the very near future above raised questions regarding the lasing in highly n-doped Ge.

Coming back to **Figure 2**: the arrows colored in violet depict 1.0 and 1.5% biaxially tensilely strained structures that have been achieved via deposition of Si–Nitride (SiN) stressor layers (Ghrib et al., 2013, 2015). This strain is equivalent to a band offset of ~70 and 30 meV, which corresponds to 50 and 30% of the unstrained band offset value, respectively. So far, the highest strain values are obtained in suspended microbridges under uniaxial loading as is shown on the top stroke. There, the ~0.25% biaxial prestrain is enhanced and transformed into uniaxial strain. The arrows in olive (Süess et al., 2013; Geiger et al., 2014c) and blue (Nam et al., 2013; Sukhdeo et al., 2014) mark recent achievements from the two leading groups. The latest result (Sukhdeo et al., 2014) indicates that the bridge technology can indeed provide direct band gap strained Ge. SiN stressor layers on suspended microbridges or FP cavities deliver far less strain and offset reductions (Capellini et al., 2013, 2014). As shown by the red arrows on the third stroke, alloying Ge with Sn also provides optical group IV material with a fundamental direct band gap. The transition from fundamental indirect to direct occurs at a Sn concentration of ~9% for relaxed GeSn. Depending on the strain loading, i.e., tensile or compressive, the crossover shifts to a higher or lower Sn concentration. Hence, a 20-nm thick GeSn layer with 8% Sn sandwiched between Ge claddings and processed into microdisks is not as close to the direct transition as a relaxed layer with 6% Sn because of the −1% biaxial compressive loading (Chen et al., 2013a). The GeSn alloy above the crossover in **Figure 2** exhibits 0.7% in-plane strain at a Sn concentration of 13%. This system shows lasing at low temperature (Wirths et al., 2015). We will present this recent result and will, thereby, clarify the characteristic of the experimental observation of lasing.

The availability of direct band gap group IV semiconductors as compiled in **Figure 2**, together with the rise of promising results, in particular, the demonstration of lasing in the GeSn system, has motivated the writing of this review. It is meant to present the current understanding evoked from the research undertaken at many places worldwide. Although some of the following descriptions are exemplified for only one of the systems (strain or Sn alloying), we will argue that the physics of this two direct gap systems can be understood by analogy. By merely emphasizing the

similarities of the physics and the characterization methods used for investigations, we hope to provide a comprehensive overview that will support and interest many scientists to enter this highly relevant field of research.

In Section "Direct Band Gap Group IV Materials," the band structure in Ge is given in dependence of strain. We then summarize the fabrication steps for strain engineering and Sn alloying. In Section "Characterization Methods: Optical Properties," several optical characterization methods are introduced, such as pump and probe spectroscopy developed for this very purpose at the infrared beamline of the Swiss light source (SLS). Gain and loss studies performed on Ge layers as well as carrier lifetime measurements are shown in Section "Lifetime, Gain, and Loss." These results impact the discussion on lasing in n-doped Ge, which is briefly repeated to exemplify the capability of these experimental methods. The analysis of temperature-dependent photoluminescence (PL) is found to deliver a quantitative measure for the directness of GeSn layers, shown in Section "Photoluminescence – Direct Band Gap," and narrow emission spectra together with an intensity versus excitation-threshold represent the first observation of lasing in a direct band gap group IV system, shown in Section "Optically Pumped Laser." Investigation challenges, such as the quantitative analysis of the Auger recombination and the carrier transport, are appointed in Section "Challenges" together with other fundamental device-related issues, such as cavity design, band gap renormalization, and thermal budgets for alloys. We speculate about the opportunities for Si photonics offered by an efficient monolithically integrated laser source in Section "Opportunities," and furthermore discuss the prospect of a Ge and/or GeSn electro-optical data processing platform. We conclude in Section "Conclusion and Outlook" and give a short outlook.

Direct Band Gap Group IV Materials

Modeling
Band Structure

The effect of tensile strain on Ge's band edges shown in **Figure 3** illustrates the path of the transitions' energies going from an indirect to a direct band gap system. The energies for interband- (solid lines) and intervalence-band transitions (broken lines) between the respective conduction- and valence-band edges are calculated via deformation potential theory as implemented in the nextnano® modeling software (Birner et al., 2007). Due to the fact that the Γ-valley energy reduces faster than the one of the L-valley, Ge transforms into a direct band gap semiconductor at ~4.7% uniaxial strain along [100] when the direct transition (black line) decreases below the energy of the indirect recombination (green line). For Ge under biaxial tensile strain or GeSn alloys, the band edges behave similarly with an indirect-to-direct band gap crossover at ~1.6–2.0% strain (El Kurdi et al., 2010; Virgilio et al., 2013; Wen and Bellotti, 2015) or and at a Sn-content of ~9% (Low et al., 2012; Gupta et al., 2013b; Wirths et al., 2015) for a fully relaxed layer.

In the valence band, strain lifts the degeneracy of light hole and heavy hole bands and introduces a mixing such that this distinction becomes meaningless, especially under high strain. For low strain, VB1 and VB2 in **Figure 3** are mostly "heavy hole"- and

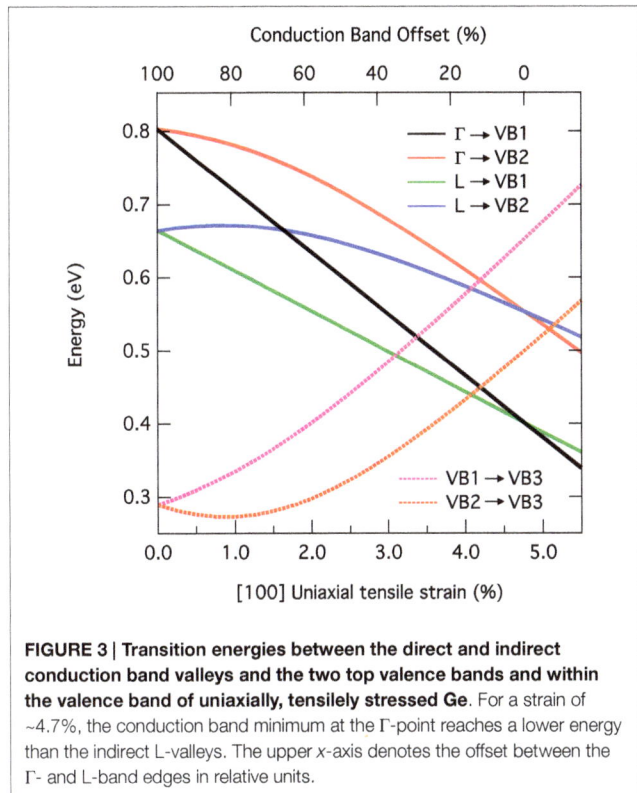

FIGURE 3 | Transition energies between the direct and indirect conduction band valleys and the two top valence bands and within the valence band of uniaxially, tensilely stressed Ge. For a strain of ~4.7%, the conduction band minimum at the Γ-point reaches a lower energy than the indirect L-valleys. The upper *x*-axis denotes the offset between the Γ- and L-band edges in relative units.

"light hole"-like, respectively. VB3 refers to the split-off band. The energetic order of the heavy and light hole bands is reverted when moving from the uniaxial to the biaxial case.

Most of the theoretical work concerning Ge light emission utilizes $k \cdot p$ theory including 6 bands (Aldaghri et al., 2012; Chang and Cheng, 2013; Virgilio et al., 2013), 8 bands (Zhu et al., 2010; Wirths et al., 2013b), or 30 bands (El Kurdi et al., 2010). The latter is not restricted to the Brillouin-zone center but describes the full energy dispersion. In other works, the empirical pseudopotential method (Dutt et al., 2013; Wen and Bellotti, 2015), density functional theory (Tahini et al., 2012), and the tight-binding model (Dutt et al., 2012) are employed. The agreement between the models is generally found to be satisfactory.

Gain

In **Figure 4**, we show gain calculations for uniaxially stressed Ge in dependence of *n*-type doping and conduction band offset (c.f. the scale of the upper *x*-axis in **Figure 3**). The band structure was computed with an 8-band $k \cdot p$ approach (Birner et al., 2007). The gain was calculated via Fermi's golden rule, assuming cylindrical symmetry for the valence bands to simplify the calculation of the joint density of states (JDOS), c.f. Virgilio et al. (2013). More details of the calculation can be found in Süess et al. (2013), supplementary information. The peak gain at room temperature (RT) is plotted after subtraction of the loss following the experimentally determined electron- and hole-absorption cross-sections from Carroll et al. (2012) (Süess et al., 2013). The black, broken line indicates when transparency is reached. As an example, for a system at the crossover to a direct band gap system, transparency

is nearly reached at a low injection of 1×10^{18} cm^{-3} as depicted in **Figure 4A**. Applying a moderate doping of 1×10^{19} cm^{-3} results in a gain of >500 cm^{-1}, which is sufficient to overcome typical resonator losses. When the doping level is increased to 2×10^{19} cm^{-3}, the gain approximately triples to ~1500 cm^{-1} according to our model. This suggests that *n*-doping is a very effective method to promote high gain for correspondingly low excitation in direct band gap systems. This is due to the fact that as long as the offset is not much larger than kT (Sukhdeo et al., 2014), electrons will nevertheless spread into the L-band from where they cannot contribute to gain.

Transparency can also be achieved for an undoped and still indirect, strained Ge system under higher excitation. We obtain transparency at an injection of 1×10^{19} cm^{-3} for a system with a remaining offset of 25% (35 meV) and an *n*-doping level below 1×10^{18} cm^{-3} (see **Figure 4B**). When the direct band gap is reached (0% offset), the net gain amounts to 1000 cm^{-1}, which can be increased up to 2000 cm^{-1} at a doping of 2×10^{19} cm^{-3}. In contrast to the indirect band gap Ge, a reduction in temperature helps to increase the gain as soon as the Γ valley constitutes the lowest conduction band energy due to condensation of the carriers into the direct gap states. For example, an intrinsic direct band gap Ge system with 25 meV band offset exhibits a net gain of the order of 4500 cm^{-1} at a temperature of 20 K and an injection of 1×10^{19} cm^{-3} compared to 1700 cm^{-1} at RT.

When comparing gain predictions in literature, we experience larger differences than between predictions of energy levels and their relative positions. The reason for this stems from the uncertainty in the loss. For weakly strained and relaxed Ge, experimental values are available as discussed in Section "Lifetime, Gain, and Loss." Hence, the overall agreement of the predictions is largely coherent. For example, calculations consistently predict gain for Ge with a large offset (80%) only for the case of very high doping of >5×10^{19} cm^{-3}. For strained and alloyed systems, however, the interband energies approach the one of the intervalence band transitions. The energies may even cross, as shown in **Figure 3**. Hence, loss processes related to these transitions will become critical. Furthermore, the gain as predicted by a Green's functional approach (Wen and Bellotti, 2015) tend to be smaller than the commonly used joint density of state formalism as applied for **Figure 4**.

Fabrication
Microbridges

Strain engineering is nowadays a standard tool in microelectronics to improve device performance, where the lattice mismatch between Si and Ge is used to generate strain via epitaxy. However, the pseudomorphic deposition of Ge on Si leads to compressive strain, which deteriorates the light emission efficiency and is, furthermore, limited to small layer thicknesses. Therefore, the main method used to introduce strain is the application of external stressor layers, such as silicon nitride (SiN), which is compatible with CMOS processing. Some work following this approach includes the deposition of stressors on the back side of Ge membranes (Nam et al., 2011, 2012), on micropillars (Velha et al., 2013), or on selectively grown Ge (Oda et al., 2013).

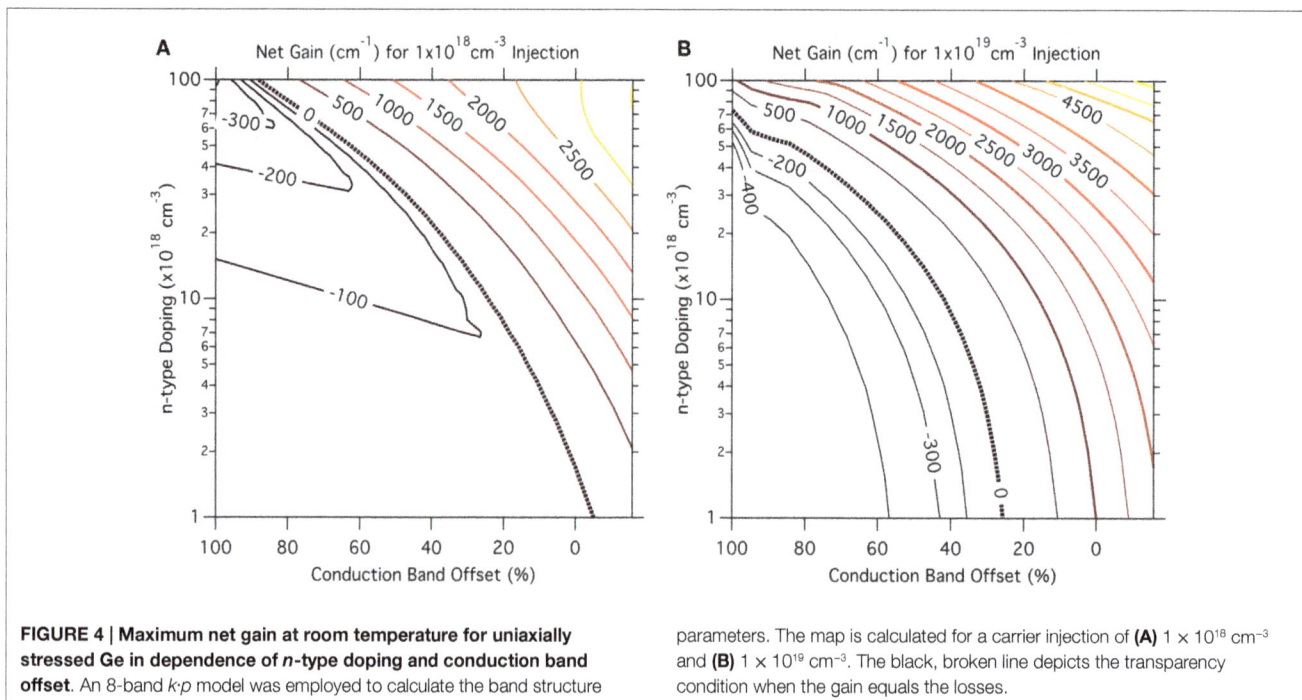

FIGURE 4 | Maximum net gain at room temperature for uniaxially stressed Ge in dependence of *n*-type doping and conduction band offset. An 8-band *k·p* model was employed to calculate the band structure parameters. The map is calculated for a carrier injection of **(A)** 1×10^{18} cm^{-3} and **(B)** 1×10^{19} cm^{-3}. The black, broken line depicts the transparency condition when the gain equals the losses.

An advantage of using external stressor layers is the simplicity to combine the strain transfer with standard cavity structures like FP waveguides (Capellini et al., 2014). However, the achieved strain is so far limited to a predominantly uniaxial strain of 1.5%. In other efforts, SiN layers were deposited on Ge microdisks, resulting in a biaxial strain of 1.0% (Ghrib et al., 2013) and 1.5% (Ghrib et al., 2015). However, these stressor layer approaches suffer from a large strain inhomogeneity across the Ge layer, and elaborated all-around stressor techniques using wafer transfer and bonding. These results are included in **Figure 2**.

Following a different route, it was shown that high levels of tensile strain can be locally induced without the use of any external stressor layers (Minamisawa et al., 2012; Süess et al., 2013). In the approach by Süess et al., the starting substrate is the commonly used tensilely strained Ge layer with a biaxial strain of ~0.2%. Subsequently, the layer is patterned into a microbridge with a narrow central cross-section (the "constriction") and larger outer cross-sections (the "pads") as shown in **Figures 5A,B**. As last processing step, the structure is underetched by selectively removing the underlying buried oxide with hydrofluoric acid, c.f. **Figure 5C**. Releasing the structure leads to a relaxation of the strain in the pads, which in turn increases the strain in the constriction. Due to Hooke's law and force balance, strain accumulated in the constriction will depend on the ratio of pad and constriction widths as well as the ratio between their lengths (Minamisawa et al., 2012; Süess et al., 2013). Hence, following this principle, any strain can be generated in the constriction by solely varying the geometrical parameters independent of the actual dimensions of the structure. In contrast to external stressors where the achievable strain is limited by the efficiency of strain transfer, this strain enhancement is only limited by the material strength. **Figure 5A** shows enhancement factors of more than 20× realized

from 0.15% biaxial strain in Ge on Si (blue squares) or Ge on silicon-on-insulator (SOI) (green circles and red triangles) for bridges with varying geometrical dimensions. The agreement of the experimental values with the ones predicted by finite element modeling (red triangles) is excellent. For Ge on SOI, the highest strain achieved in 6 μm × 2 μm constrictions is 3.1%. When starting from 200 nm thick germanium-on-insulator substrates (GOI), which feature a significantly reduced dislocation density (Akatsu et al., 2006; Hartmann et al., 2010), a strain of 5.7% was observed in a 5.0 μm × 0.2 μm constriction (Sukhdeo et al., 2014). According to **Figures 2** and **3**, such a strain is by far sufficient to transform Ge into a direct band gap material showing the prospect of the strain-enhancement technique given a starting material with high-crystal quality.

GeSn Alloying

The epitaxial growth of GeSn alloys poses several challenges, such as a large lattice mismatch between α-Sn and Si (17%) or Ge (15%), and a low solid-solubility of <1%. Therefore, the fabrication of high quality and smooth epilayers was a demanding task for many years and the development of new growth processes to deposit GeSn under non-equilibrium conditions at low temperatures was required. Whereas the first attempts to grow GeSn alloys were based on molecular beam epitaxy (MBE) in 1980s and 1990s (Pukite et al., 1989; Harwit et al., 1990; Wegscheider et al., 1990; Fitzgerald et al., 1991; He and Atwater, 1997), device-grade GeSn epilayers could be synthesized since the early 2000s when the first chemical vapor deposition (CVD) processes were developed (Bauer et al., 2003).

Nowadays, several groups established growth processes for GeSn utilizing either MBE (Bratland et al., 2003; Chen et al., 2011a; Bhargava et al., 2013; Oehme et al., 2013) or CVD

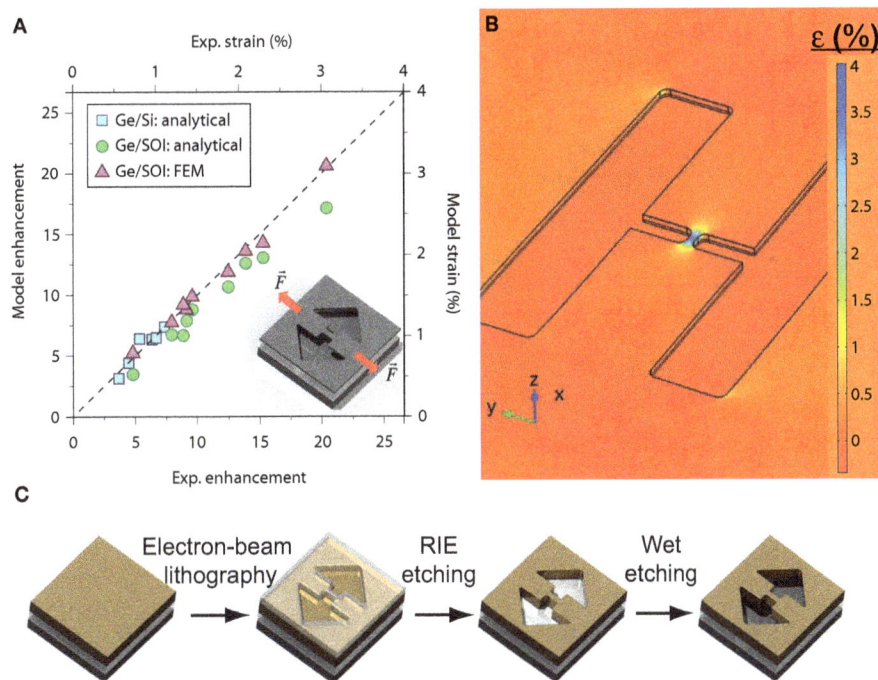

FIGURE 5 | Suspended microbridges from thermally pre-strained Ge.
(A) Experimental and modeled strain for Ge microbridges fabricated on Si and SOI. The analytical strain-enhancement model is given in Süess et al. (2013). The enhancement of 22× corresponds to 3.1% uniaxial strain (Süess et al., 2013).

(B) Strain profile of a suspended bridge structure as obtained by finite element modeling (FEM). Due to the relaxation in the pads, the strain in the central constriction is enhanced. **(C)** Process flow for the fabrication of suspended microbridges.

(Vincent et al., 2011; Chen et al., 2013b; Wirths et al., 2013a; Xu et al., 2013; Du et al., 2014) for a variety of applications, e.g., photodiodes, photodetectors, or MOSFETs. Here, due to the reduced lattice mismatch compared to Si, Ge is preferred as virtual substrate (VS) in order to ensure layers of high monocrystalline quality. Regarding the epitaxial growth of direct band gap GeSn alloys, nearly strain relaxed or even tensilely strained layers are highly desired, since for compressively strained GeSn layers, i.e., GeSn coherently grown on Ge VS, higher Sn contents are necessary for the indirect to direct transition (Gupta et al., 2013b). Owing to an advantageous relaxation mechanism for GeSn layers on Ge VS, dislocations seem to mostly protrude into the Ge VS rather than into the GeSn layer, which is beneficial for optical properties as the density of non-radiative recombination centers is reduced (Takeuchi et al., 2006; Senaratne et al., 2014; Wirths et al., 2015). Although relaxation takes place, a certain level of compressive biaxial strain (typically between −0.6 to −0.8%) remains nevertheless, which, as already said in connection with **Figure 2**, shifts the indirect-to-direct band gap crossover to higher Sn concentrations with respect to fully relaxed GeSn. Therefore, several approaches are being followed to reduce the compressive strain, such as growth on lattice-matched InGaAs VS (Chen et al., 2011a), which is not acceptable within a CMOS processing line, or deposition of ever thicker layers to enforce further strain relaxation (Senaratne et al., 2014; Wirths et al., 2015). Gupta et al. (2013b) introduced a robust etching approach enabling to selectively dry etch the Ge VS underneath

the epitaxial GeSn layers. The authors envision their method to enable the fabrication of direct band gap GeSn micro disks.

Figures 6A–C show transmission electron microscopy images of a GeSn layer with 13% Sn grown via reduced pressure CVD on a Ge VS (Wirths et al., 2015). The advantageous relaxation mechanism mentioned above can be seen here with dislocation-loops (blue arrows) emitted into the Ge VS. Despite the high-Sn content, the thickness of the GeSn layer could be increased up to 560 nm without deteriorating the high-crystalline quality. Owing to the large thickness, a relaxation of 60% could be achieved such that only a mild compressive strain of −0.6% was present. As will be shown in Section "Photoluminescence – Direct Band Gap," this epilayer was proven to be a direct band gap group IV semiconductor that provides net gain and, hence, shows lasing under optical pumping.

We conclude this section on the fabrication of GeSn alloys by summarizing the list of beneficial assets GeSn epitaxy brings to the current Si technology facilitating future developments and integration. Apart from the prospect to fabricate a fundamental direct band gap group IV material, GeSn alloys are attractive because of (i) low-temperature deposition on Si(001) compatible with existing CMOS processes; (ii) strain relaxation with reasonably low threading dislocation density; (iii) available option for selective growth on silicon, which is attractive for photonic integration; (iv) GeSn/SiGeSn heterojunction layers to generate carrier confinement in quantum wells; (v) and therefore, tunability of the lattice constant offering opportunities to combine the alloys with the strained membrane method.

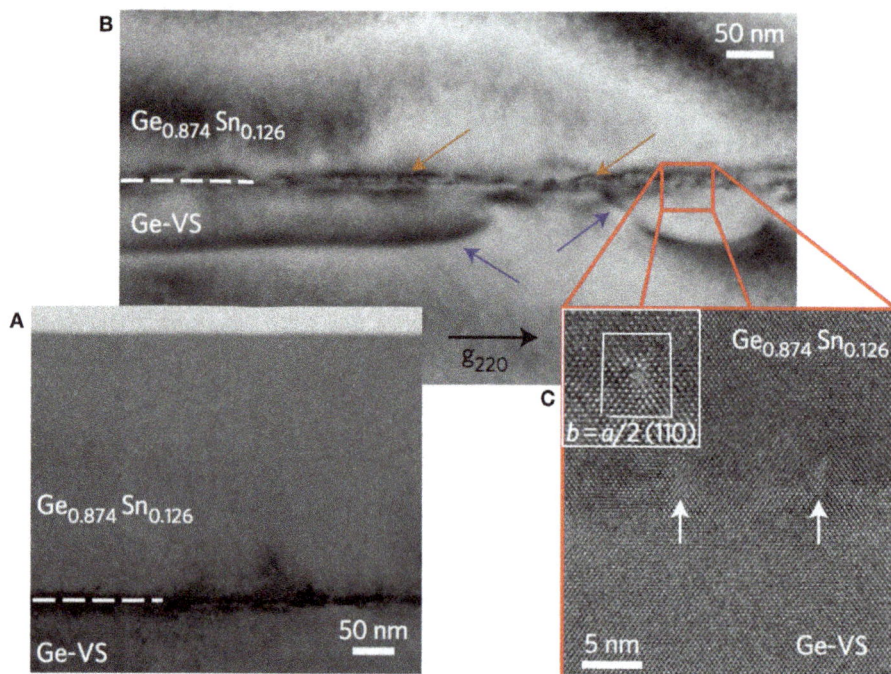

FIGURE 6 | Cross-sectional transmission electron microscopy (TEM) image of a Ge$_{0.87}$Sn$_{0.13}$ alloy. (A) Expanded view showing the high-crystalline quality of the GeSn epilayer. The defects are located near the interface to the Ge virtual substrate. **(B)** Dislocation-loops (blue arrows) emitted below the GeSn/Ge-interface (orange arrows) penetrating only into the Ge virtual substrate. **(C)** High-resolution TEM image of the interface used for Burgers vector calculations. Lomer dislocations with $b = a/2[110]$ are observed (Wirths et al., 2015).

Characterization Methods: Optical Properties

Lifetime, Gain, and Loss

When describing a material with regards to its suitability as an efficient laser source, key properties that decide upon adequacy are the gain and loss, i.e., the material's ability to amplify light, as well as the non-radiative lifetime, which determines the internal quantum efficiency as well as the achievable steady-state carrier density. These characteristics can be extracted in a direct way using broadband, time-resolved pump–probe transmission, and reflection spectroscopy. Possible ways for performing such experiments could be via tunable lasers or supercontinuum sources. However, particularly synchrotron-based infrared pump–probe spectroscopy has been shown to offer advantageous conditions for measuring the carrier density, their lifetime as well as gain and loss due to its extended bandwidth and suitable pulse lengths (Carroll et al., 2012; Geiger et al., 2014a,b). At the infrared beamline of the SLS, 100 ps long pulses of infrared light are supplied from the synchrotron and serve as broadband probe pulses, whereas the excess charge carriers are optically excited by a 100 ps Nd:YAG laser at 1064 nm (Carroll et al., 2011). The delay time between pump and probe pulses can be varied electronically, which offers the possibility to follow the dynamics of a system over a long time period by probing at different times after excitation. In the following, we review some of the pump–probe measurements performed at the SLS and give the most important results that challenge the

interpretation of the Ge-lasing observations. We renarrate this discussion at the end of this chapter.

Figure 7A shows the mid-infrared reflection of a Ge layer grown on Si plotted as the ratio of pumped (R_P) and unpumped (R_U) reflection signal. The different colors depict different optical excitation strengths between 1 and 160 MW cm^{-2}. All of the spectra were taken for a pump–probe delay time of 250 ps. The distinct minimum observed in the spectra is attributed to the carriers' plasma frequency. For an increasing excitation power, the minimum shifts to higher energy and becomes at the same time more pronounced. As the plasma frequency shifts in first order proportional to the square root of the total amount of charge carriers in the system, such reflection measurements facilitate a convenient method for the quantitative determination of the carrier density. Thus, the extracted carrier concentration in dependence of the optical pump power for delay times of 0 and 250 ps is shown in the inset of **Figure 7A**. Moreover, by analyzing the carrier density at a fixed pump power for varying delay times, the reflection spectra can be used to extract the carrier decay times. In the case shown here, the carrier density drops to ~4×10^{19} cm^{-3} within 250 ps for all generated carrier concentrations larger than 4×10^{19} cm^{-3}. This behavior indicates an increasingly faster decay time at high-carrier concentrations, which is attributed to Auger recombination (Carroll et al., 2012).

While the analysis of mid-infrared reflection spectra enables to directly access charge carrier concentration and decay time, the latter can also be extracted from near-infrared transmission

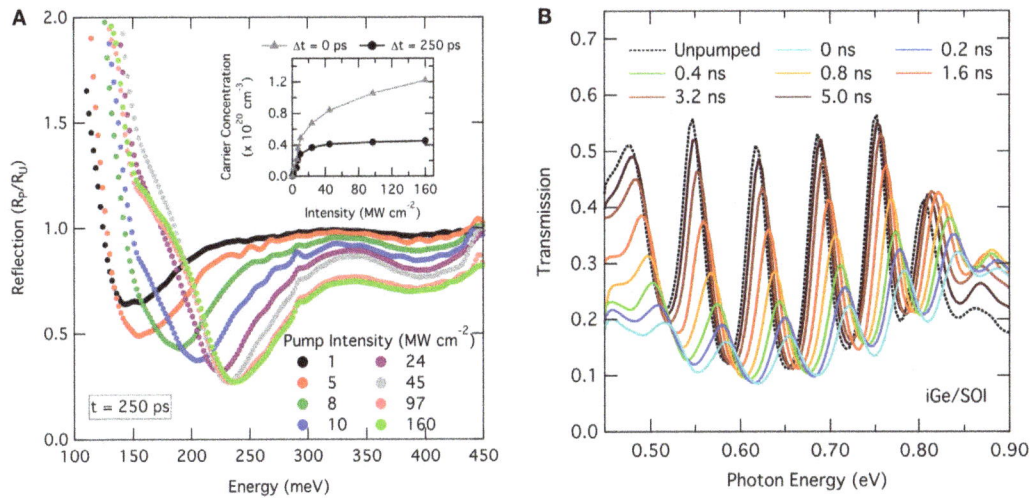

FIGURE 7 | Time-resolved infrared reflection and transmission spectroscopy: (A) mid-infrared reflection spectra of Ge on Si expressed as the ratio of pumped (R_P) and unpumped reflection (R_U) for varying excitation power at a pump–probe delay of 250 ps. The resonance in the spectra is attributed to the carrier plasma frequency, which enables to extract the total amount of charge carriers. The inset shows the carrier concentration for 0 and 250 ps delay time in dependence of the excitation power. **(B)** Normal-incidence pump–probe transmission spectra for Ge on SOI for varying delay times. Strong Fabry–Perot oscillations are observed from the thin film interference. Analyzing the peak-shifts facilitates the extraction of the decay time.

measurements. In **Figure 7B**, normal-incidence transmission spectra of intrinsic Ge are plotted, while the delay time between pump and probe is varied. As SOI is used as substrate, distinct FP oscillations are observed due to standing wave interferences between the Ge/air and Si/SiO$_2$ interfaces. For short delay times, the transmission is significantly reduced due to absorption. Above the direct band gap of ~0.8 eV, there is an increase compared to the unpumped transmission due to gain or bleaching. By following the shifts of the minima or maxima, the dynamics of the refractive index is obtained, which enables the extraction of the carriers' decay time. Compared to the decay time analysis from the mid-infrared reflection, the sensitivity to detect small carrier densities is higher in such a measurement because the refractive index – and, hence, the oscillation extrema – follows the carrier densities linearly, which enables to follow the decay processes within an extended time window.

In **Figure 8**, the time-dependent FP peak shifts are shown for differently prepared Ge layers. The shifts were normalized to unity at $t = 0$ ns and the decay fitted to an exponential curve (Geiger et al., 2014a). The defective Ge/Si interface was identified as the main non-radiative loss channel, as (i) Ge selectively grown via ultrahigh vacuum CVD (selGe in **Figure 8**) and a full epilayer grown via low-energy plasma-enhanced CVD (iGe in **Figure 8**) feature the same surface recombination velocity (SRV) – i.e., the carrier lifetime normalized to the layer thickness – of ~800 m s^{-1}, (ii) a built-in field introduced by modulation doping (nGe/iGe) increases the lifetime compared to iGe by keeping electrons away from the interface, and (iii) the longest lifetime was observed for an overgrown GOI wafer, where the defective Ge/Si interface is removed (SRV = 490 m s^{-1}). These results demonstrate the importance of engineering the material- and, in the case of Ge on Si, especially the interface quality to obtain a high-internal quantum

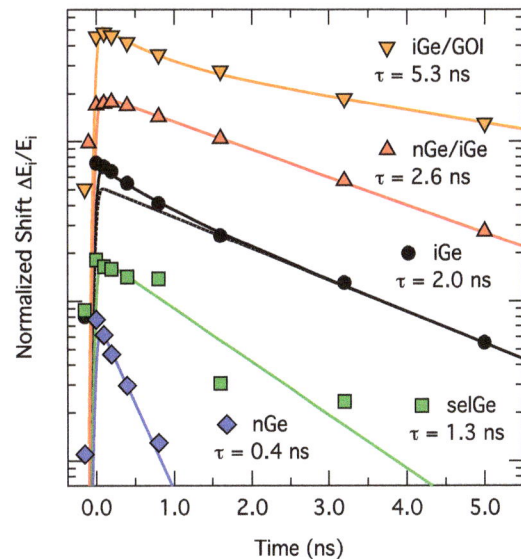

FIGURE 8 | Normalized peak shifts taken from normal-incidence transmission spectra (as, e.g., in Figure 6B) for a series of differently prepared Ge layers: iGe, nGe, selGe refers to intrinsic, *n*-doped, and selectively grown Ge, respectively. GOI refers to Ge on insulator. The non-radiative lifetime is obtained through an exponential fit to the data (Geiger et al., 2014a).

efficiency and, thus, a low-threshold laser. Furthermore, similar pump–probe transmission studies on strained microbridges showed that neither strain, at least up to ~2%, nor processing affects the lifetime (Geiger et al., 2014b), indicating that a high-crystal quality can be maintained using the microbridge strain-enhancement technology.

For the analysis of gain and loss, the transmission spectra should be recorded under the Brewster angle such that the obstructing FP resonances do not occur, c.f. **Figure 9A** for the case of unstrained Ge on Si for different pump–probe delay times. $\Delta t = 0$ ps refers to the maximum overlap between pump and probe and, hence, to the highest carrier density. The thick lines in blue and red show modeled transmission spectra for the unpumped and pumped case. Under excitation, a strong absorption occurs with a linear dependence on energy. At the direct band gap, the absorption gets reduced due to gain, but the gain is too small to generate a negative absorption and, hence, light amplification. This situation holds true for all other delay times, i.e., carrier concentrations, as well.

In **Figure 9B**, the situation for $\Delta t = 0$ ps is shown again in terms of the absorption coefficient with the modeled functions being plotted separately for (i) direct gap absorption before pumping (blue, thick line), (ii) direct gap absorption under excitation (red, thick line), and (iii) the featureless pump-induced absorption decreasing linearly in energy (red, thin line). Even though a gain of ~850 cm^{-1} is observed as displayed by a negative absorption, the loss from that spectrally distributed absorption at the same energy is >6000 cm^{-1} making light amplification impossible. To show the contrast to an established laser material featuring a direct band gap, the same absorption properties are plotted in **Figure 9C** for the case of InGaAs. Here, the pump-induced losses are independent on energy and amount to ~1000 cm^{-1}, which is compensated by a direct gap gain of ~1700 cm^{-1} such that a net gain of 700 cm^{-1} is revealed. We should mention here that the theoretical analysis of Carroll et al. (2012) has been questioned (Dutt et al., 2012) concerning the strength of the gain (red line, **Figure 9B**) but not the experiments, which clearly show that the loss is by far larger than the gain.

From the preceding analysis, it is clear that a solid understanding and consideration of the loss processes is required for an accurate description of gain in Ge. For illustration, the absorption cross-sections for three Ge samples (Ge#1: $N_d = 0$, $\varepsilon_{xx} = 0$; Ge#2: $N_d = 2.5 \times 10^{19}$ cm^{-3}, $\varepsilon_{xx} = 0$; Ge#3: $N_d = 0$, $\varepsilon_{xx} = 0.25\%$) are plotted in **Figure 10** in dependence of the total carrier density $N_T = N_d + N_P$, where N_d refers to the doping concentration and N_P to the pump-induced carrier density. As a comparison, the cross-sections of three InGaAs layers (InGaAs#1: $N_d = 0$, InGaAs#2: $N_d = 5.3 \times 10^{18}$ cm^{-3}; InGaAs#3: $N_d = 2.1 \times 10^{19}$ cm^{-3}) are plotted as well. Therefore, **Figure 10** reveals that the absorption scales predominantly with N_P indicating that the absorption cross-section from holes σ_h is much larger than the cross-section for electrons σ_e. Indeed, describing the absorption via a linearly dependent cross-section as $\alpha = \sigma_e N_e + \sigma_h N_h$ (where subscripts e and h refer to electrons and holes, respectively) offers a good representation of the experimental data with $\sigma_h/\sigma_e > 10$. The absorption cross-section for holes is significantly larger than for electrons, because in addition to the non-momentum conserving intraband or Drude-type free carrier absorption, the holes can undergo vertical intervalence band transitions (Newman and Tyler, 1957), which are hereby identified as the main loss channel in Ge. A similar conclusion concerning the cross-section ratio can be deduced from the InGaAs data shown in **Figure 10** in agreement with common knowledge for direct band gap lasing materials (Adams et al., 1980; Childs et al., 1986). Furthermore, the absolute values of the hole cross-section are in a similar range but larger for InGaAs than for Ge, c.f. **Figure 10**. However, the absorption is much higher in Ge due to a much larger total carrier density needed to achieve a gain like in the InGaAs sample.

Finally, we would like to relate the above presented data of gain and loss as well as of lifetimes in Ge layers on Si to the observation

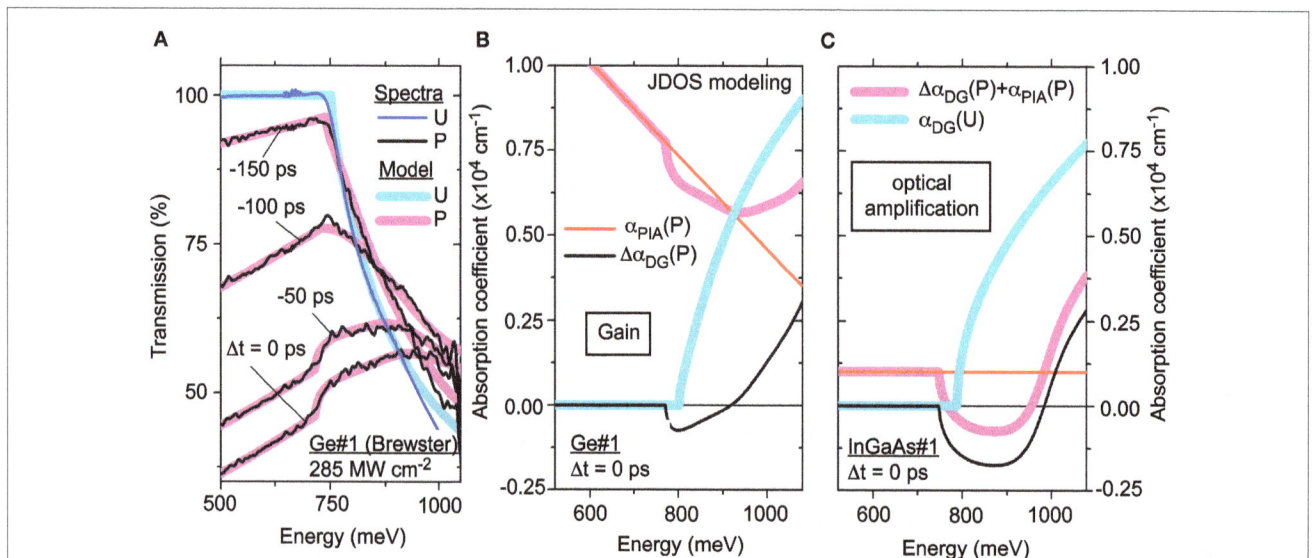

FIGURE 9 | Analysis of pump and probe measurements.
(A) Transmission spectra for Ge on Si measured under Brewster angle for different delay times. **(B)** Modeled direct gap absorption unpumped (blue, thick line) and pumped (red, thick line) as well as pump-induced

absorption (red, thin line) obtained from the spectra in **(A)** at 0 time delay.
(C) Similar extraction of direct gap gain and losses for undoped InGaAs showing light amplification, as the gain surpasses the pump-induced losses (Carroll et al., 2012).

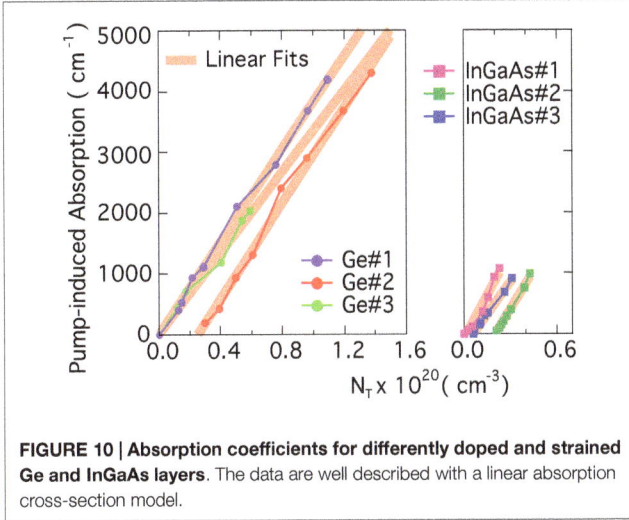

FIGURE 10 | Absorption coefficients for differently doped and strained Ge and InGaAs layers. The data are well described with a linear absorption cross-section model.

of lasing in highly n-doped and weakly strained Ge (Liu et al., 2010; Camacho-Aguilera et al., 2012), which was recently repeated at the University of Stuttgart with an unstrained, highly n-doped light emitting diode (LED) (Koerner et al., 2015). First, as is shown above by the gain and loss experiment (Carroll et al., 2011), the loss in Ge layers strongly exceeds the gain at all the investigated carrier densities up to 10^{20} cm^{-3} and in all investigated cases, i.e., Ge with and without weak strain and/or n-doping. Hence, Carroll's results are apparently in conflict with the observation of lasing (Liu et al., 2010) in similar but not identical material. Second, the non-radiative lifetimes, which have been determined for such Ge layers only recently by Geiger et al. (2014a), c.f **Figure 8**, shine a new light on previous and recent gain and threshold current density calculations (Dutt et al., 2012; Peschka et al., 2015). Using the obtained carrier lifetime of the order of 1–2 ns for threshold current estimates, the calculated threshold of the order of 100 kA cm^{-2} – obtained by assuming a lifetime of 100 ns – needs to be rescaled by a factor of 50–100. Surely, such a current density is above the material's limit and also exceeds the observed threshold values by ~2 orders of magnitude. Therefore, not only the gain/loss experiments but also the theory (when fed with properly valued parameters) shows that more research is needed to understand the MIT results. The recent paper by the Stuttgart group (Koerner et al., 2015) may give the directions for further thinking: a "lasing" threshold was reached only shortly before their devices failed, hinting at a carrier breakthrough. The heat pulse related to the breakthrough may have caused the peaked emission signal.

Photoluminescence – Direct Band Gap

Photoluminescence spectroscopy offers a convenient tool for probing the changes of the electronic band structure induced via strain or Sn alloying. As shown in **Figure 11A**, the reduced offset between Γ- and L-valleys manifests in an increased emission intensity of the PL signal (Süess et al., 2013). A similar effect has also been observed by Chen et al. (2011a).

A quantitative analysis of the relative alignment between Γ- and L can be obtained from the temperature dependence of the

PL, as has been demonstrated recently in Wirths et al. (2015). In **Figure 11B**, the temperature-dependent PL intensity for a set of samples with Sn content from 8 to 13% is shown. The data have been normalized to unity at 300 K. For the sample with the lowest Sn content, a rapid drop in intensity on lowering the temperature is observed, whereas for the three other samples a steady increase in intensity can be seen with the intensity increase being dependent on the Sn concentration. Qualitatively, the increase from sample to sample can be explained by the reduced conduction band offset with increasing Sn, whereas cooling down leads to a condensation of the carriers into the lowest energy states such that the direct gap emission either vanishes for indirect band gap materials as sample Ge$_{0.92}$Sn$_{0.08}$ because the Γ valley is not populated anymore or increases strongly when the electrons condense at the minimum of the Γ valley. To quantify the band offsets for the set of GeSn samples, the emission efficiency is calculated via a similar JDOS model as the one used for calculating the gain in **Figure 4**. Therein, the offset ΔE between Γ- and L-valleys as well as the injected charge carrier density N_i represent the free fitting parameters. Furthermore, the temperature dependence of the lifetime is assumed to be identical for all samples and follows the Shockley–Read–Hall (SRH) recombination characteristics (Shockley and Read, 1952; Schubert, 2006) that describes non-radiative recombination via trap states. Using this model, an excellent agreement with the experimental data is obtained (Wirths et al., 2015). The GeSn sample with 13% Sn content is, hence, identified as a true direct band gap group IV semiconductor with its Γ-valley being 25 meV below the indirect L-valleys. From the second fit parameter, a RT carrier density N_i of 4×10^{17} cm^{-3} is deduced consistent with a carrier lifetime of 0.35 ns, which corresponds to a SRV of 570 m s^{-1}. This value is close to the ones reported for elemental Ge on Si (Geiger et al., 2014a), c.f. **Figure 8**, which is an indication of the high-crystalline quality of the investigated GeSn epilayers.

The temperature dependence of the non-radiative carrier lifetime is obtained as:

$$\tau = (1/\tau_0 + 1/\tau_{SRH} + 1/\tau_{Auger})^{-1} \qquad (1)$$

where τ_0 describes the lifetime at low temperature, and τ_{SRH} describes the decay due to the capture of charge carriers by mid-gap states, i.e., $\tau_{SRH} = A \times (1 + \cosh(E_T/kT))$. τ_{Auger} describes the Auger recombination time, which can be neglected here due to the low-carrier densities. Furthermore, E_T is the difference between the trap level energy and the intrinsic Fermi-level, k is the Boltzmann constant, and A is to normalize τ to 0.35 ns at 300 K as obtained from the temperature-dependent PL. For $\Delta E = 19$ meV and $\tau_0 = 2.1$ ns, a good agreement between the extracted lifetimes and the lifetime model is obtained (Wirths et al., 2015). For temperatures >50 K, there is a drastic decrease in carrier lifetime from ~2 ns to 350 ps for higher temperatures. As the temperature dependence of this process is well described via the SRH model, the lifetime decay is attributed to the capture of carriers via mid-gap states originating from defects (Wirths et al., 2015). These defects could potentially be related to defects located at the GeSn/Ge-interface (Geiger et al., 2013; Wirths et al., 2015), but further studies are needed to unambiguously identify the origin of this deterioration and, subsequently, improve the material quality.

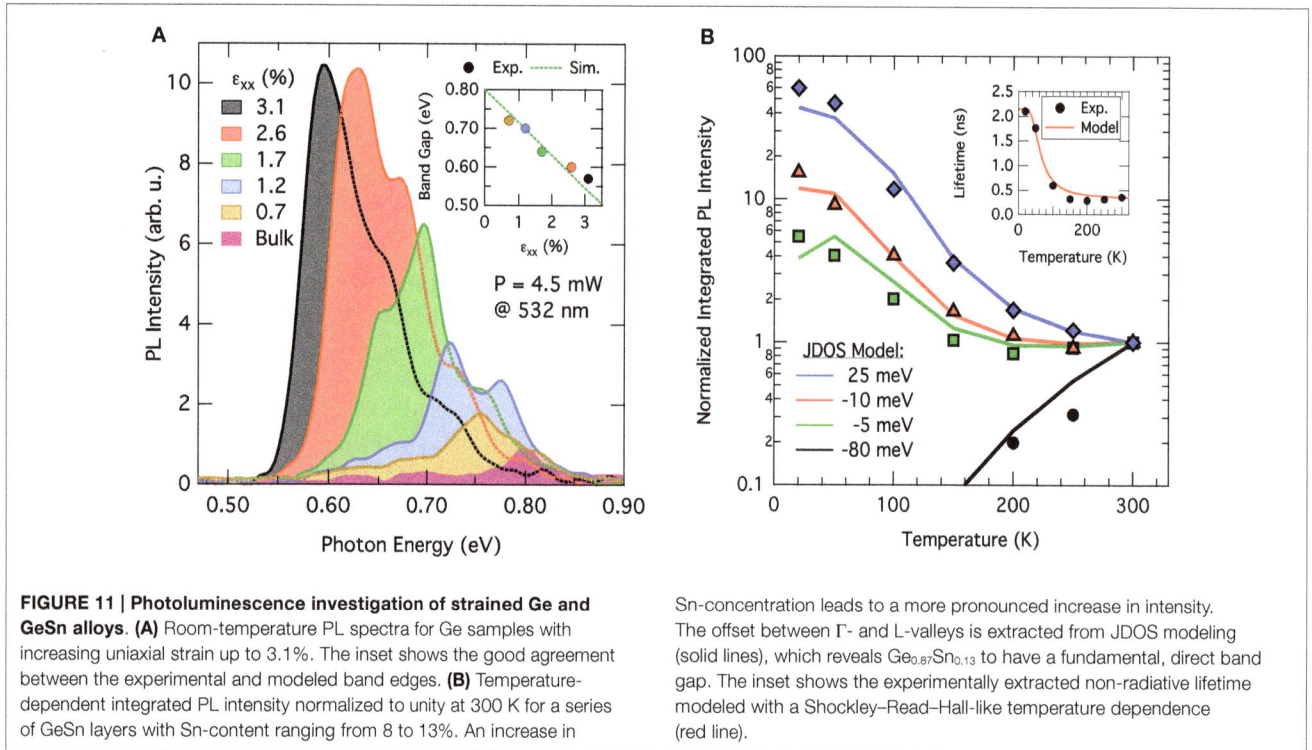

FIGURE 11 | Photoluminescence investigation of strained Ge and GeSn alloys. (A) Room-temperature PL spectra for Ge samples with increasing uniaxial strain up to 3.1%. The inset shows the good agreement between the experimental and modeled band edges. **(B)** Temperature-dependent integrated PL intensity normalized to unity at 300 K for a series of GeSn layers with Sn-content ranging from 8 to 13%. An increase in

Sn-concentration leads to a more pronounced increase in intensity. The offset between Γ- and L-valleys is extracted from JDOS modeling (solid lines), which reveals $Ge_{0.87}Sn_{0.13}$ to have a fundamental, direct band gap. The inset shows the experimentally extracted non-radiative lifetime modeled with a Shockley–Read–Hall-like temperature dependence (red line).

Optically Pumped Laser

According to the modeling results shown in Section "Modeling," a direct band gap Ge-based system should feature a net gain and, hence, enable light amplification at low excitation. In the previous analysis of low-temperature PL on GeSn alloys in Section "Photoluminescence – Direct Band Gap," a fundamental direct band gap could be identified for a Sn content of 13% in a strain-relaxed layer with 0.7% compressive biaxial strain. To show lasing, a 560-nm epilayer of such $Ge_{0.87}Sn_{0.13}$ material was grown providing an overlap of 60% for the fundamental transverse electric mode in a 5 μm wide FP cavity (Wirths et al., 2015). For this layer, modal gain could be observed at 20 K via the variable-stripe-length (VSL) method under pulsed optical excitation at 1064 nm with a differential gain of ≈ 0.4 cm kW^{-1} and a threshold excitation density of ≈ 325 kW cm^{-2} (c.f. **Figure 12**). Above threshold, the gain increases linearly with excitation and can easily pass 100 cm^{-1}. The stripe length-dependent PL analysis is a widely applied technique to measure net modal gain, but it does not allow to resolve the gain and loss as by pump and probe spectroscopy. More evidentially, a gain statement, such as provided by **Figure 12**, becomes respected only after showing lasing.

Indeed, when pumping a FP cavity over its full length, a strongly enhanced emission and narrowing of the line spectra is observed as soon as the modal gain surpasses the cavity losses. This behavior is shown in **Figure 13** where the edge-emission spectra from a 1-mm long FP cavity at 20 K are plotted for varying optical excitation powers. The curves are offset for clarity. The threshold obtained from the lasing experiments matches well with the one obtained from the VSL method. For an excitation density of 1 MW cm^{-2}, lasing could be observed up to 90 K. This

temperature is equivalent to the temperature range where the lifetime was found to drop substantially from ~2 ns to 350 ps. Hence, it is tempting to attribute the limitation of lasing to temperatures <100 K to the carrier capture by defect-induced mid-gap states, as has appealed from the analysis shown in **Figure 11B**, inset. However, carrier transfer to the L-valleys and carrier out diffusion into the Ge may be a determining factor, as well.

Despite the breakthrough of presenting for the first time a direct band gap group IV material that is lasing under optical pumping, there still remain open questions. For example, with an excitation power of 325 kW cm^{-2}, a non-radiative lifetime of 2 ns, as shown in **Figure 11**, and a typical absorbance of 1×10^4 cm^{-1} at 1064 nm, a steady-state carrier density of ~3.5×10^{19} cm^{-3} is estimated. With this number, the gain at low temperature from our model is found to be >5000 cm^{-1}. And, interpolating from **Figure 4**, at excitation density of 0.6×10^{18} cm^{-3} we would expect for a system with positive offset of about 15%, a material gain of ~300 cm^{-1} at RT. We assign this large discrepancy from what is observed at low temperature and what a RT calculation predicts to resonant intervalence band absorption. As mentioned above, due to the lack of experimental data for direct gap Ge or GeSn, the energy dependence of the loss as measured by pump–probe experiments for Ge (Carroll et al., 2012) has been used for **Figure 4**. Its proper inclusion possibly adds significant contribution to the loss (Wen and Bellotti, 2015).

In order to improve such gain calculations, which critically depend on the knowledge of the band structure, mappings of the entire valence, and conduction band in reciprocal space would certainly be highly valuable. This could be possible via angle-resolved

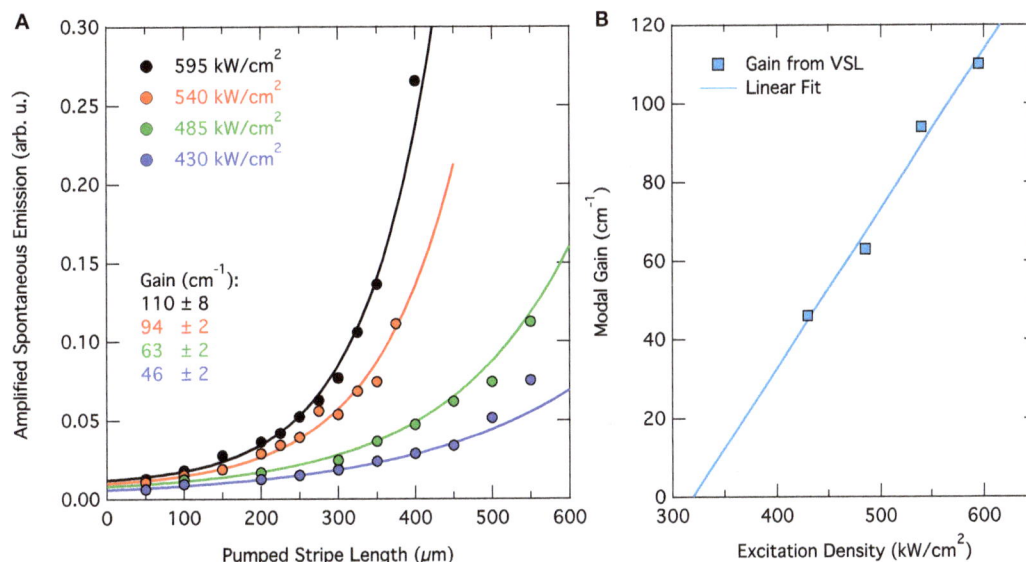

FIGURE 12 | Gain extraction via the variable-stripe-length method (VSL). (A) Edge-emitted intensity of $Ge_{0.87}Sn_{0.13}$ at 20 K in dependence of the pumped waveguide length for varying excitation densities. The modal gain is extracted by an exponential fit to the data. **(B)** Differential gain extracted from the spectra in **(A)** indicating a linear dependence on excitation density with a threshold at 325 kW cm^{-2}.

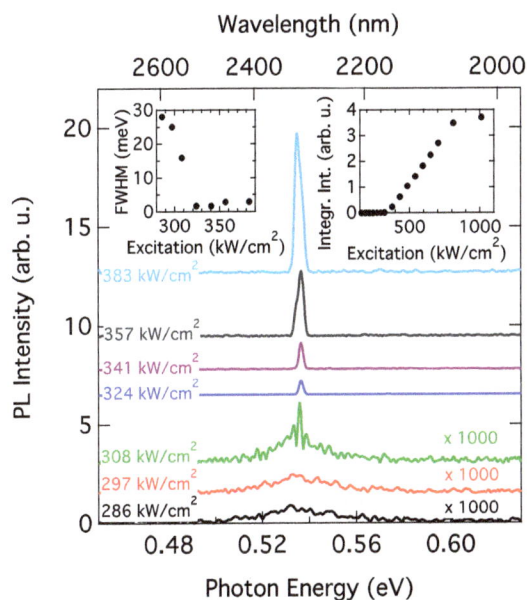

FIGURE 13 | Lasing emission spectra measured from the facet of a 5-μm wide and 1 mm long FP waveguide cavity under optical pumping at 20 K. A clear threshold behavior can be observed in the spectra with respect to output intensity and linewidth, c.f. inset to the right and left hand side, respectively.

photoelectron spectroscopy (ARPES) at high energy (Gray et al., 2011) or the soft x-ray regime (Strocov et al., 2014). Other promising experimental techniques not covered because of lack of space include lifetime measurements via time-resolved PL measurements (He and Atwater, 1997; Nam et al., 2014; Saito et al., 2014).

In fact, similar lifetimes for Ge as obtained from synchrotron measurements (Geiger et al., 2014a) were found (Nam et al., 2014).

Challenges

In the previous section, we reviewed experiments and results related to the dependence of the optical properties on strain and alloying of Ge with Sn. Furthermore, we summarized investigations concerning the first lasing of a direct band gap group IV semiconductor and expounded on the temperature dependence of the PL as a powerful tool to determine the directness of a group IV material. We illustrated optical methods based on pump and probe spectroscopy using synchrotron light to determine the carrier lifetime, gain, and loss under optical pumping related to the injected carrier density.

Future experiments along these lines on both, the strained Ge system and GeSn alloys at various strain and Sn concentration, respectively, will allow to establish the fundamentals of lasing in direct band gap group IV systems. The impact of doping on gain, loss, and carrier lifetime should also be addressed in dependence of the directness of the respective system to verify the picture elucidated by **Figure 4** of Section "Gain."

As an example, intervalence band absorption, Auger recombination, and the electrical injection are some of the many fundamental aspects of group IV direct band gap lasing pending to be understood and quantified.

Intervalence Band Absorption

One of the most essential parameters determining the efficiency of a laser is associated to the parasitic absorption due to the injected holes (Adams et al., 1980; Childs et al., 1986). As shown experimentally by Carroll et al. (2012) for Ge, this absorption depends linearly on the excitation and inclines with decreasing energy. This can be understood from the Drude dependence of the free carrier absorption modified by dipole allowed intervalence band transitions. Because the emission wavelength increases when approaching the direct band gap configuration, and the initially degenerate heavy and light hole bands split due to strain, the parasitic absorption will strongly increase in direct band gap systems and may, thus, obstruct the efficiency of lasing (Wen and Bellotti, 2015). Applying the above introduced optical characterization methods should allow to investigate these effects in detail, which, together with evolving theoretical results, will enable to complete our understanding.

Auger Recombination

On the material level, the performance of optical devices depends strongly on the charge carrier recombination lifetime similarly as described by rate Eq. 1. Here, both the radiative and the Auger recombination lifetime depend on the carrier density n: $1/\tau_{rad} = B \times n$, $1/\tau_{Auger} = C \times n^2$. The quadratic carrier density dependence implicates that Auger recombination becomes a dominant loss mechanism at high-charge carrier densities, which can be the order of $>10^{19}$ cm^{-3} for typical laser devices.

Extensive theoretical work based on perturbation theory has shown that, despite its indirect band gap, the band structure of both, Si and Ge, is favorable for direct Auger recombination (Huldt, 1974; Lochmann, 1978) with Auger recombination coefficients of the order of 10^{-30} cm^{6}s^{-1}. This is comparable to direct band gap materials like GaAs or GaN used for optical devices in the visible part of the spectrum but still much smaller than for low band gap materials like InAs (Metzger et al., 2001).

Significantly, less is known about direct band gap group IV Auger recombination. For direct band gap $Ge_{0.9}Sn_{0.1}/Ge_{0.75}Si_{0.1}Sn_{0.15}$ multi-quantum-well structures, Sun et al. recently showed theoretically that the RT Auger recombination lifetime is of the order of 50 ns compared to a radiative lifetime of 10 ns (Sun et al., 2010). Other work on GeSn (Dutt et al., 2013) and n-doped or tensile strained Ge (Liu et al., 2007; Jain et al., 2012) only refers to unstrained bulk Ge recombination coefficients to include in their gain models.

We believe, however, that this is unjustified considering that there is an exponential dependence of the Auger lifetime on the band gap and effective masses (Beattie and Landsberg, 1959; Huldt, 1971; Adams et al., 1980), both being strongly altered in direct band gap Ge. Using the simple exponential dependence derived by Beattie and Landsberg (1959) to scale the experimentally determined Auger recombination coefficient of $C_{Ge} = 10^{-30}$ cm^{6}s^{-1} (Carroll et al., 2012) via the effective masses and band gaps of direct band gap Ge, an Auger coefficient of the order of 10^{-26}–10^{-27} cm^{6}s^{-1} is obtained. Despite the overly strong simplicity of this comparison, it shows that Auger coefficients will most probably increase and need to be addressed and investigated in the future.

Carrier Injection

From an electronic point of view, Ge is one of the most interesting materials as it offers both, high electron and hole mobilities

(μ_e = 3900 cm^2/Vs and μ_h = 1900 cm^2/Vs, respectively) (Golikova et al., 1962; Jacoboni et al., 1981). For optical devices, this is appealing because it results in diffusion lengths of several 100 μm, more than sufficient for, e.g., typical detector absorber sizes and laser cavity lengths. However, at the same time, Ge suffers from its low band gap, which causes large leakage currents in Ge pn junctions (Metzger et al., 2001; Satta et al., 2006), thus, requiring extensive work on surface passivation due to the lack of a native oxide.

The already beneficial mobility properties can be further improved by employing tensile strain as has been shown for both, uniaxially and biaxially strained Ge (Schetzina and McKelvey, 1969; Chu et al., 2009; Chen et al., 2011b). Here, we would like to highlight that all these studies have been performed on indirect band gap Ge where the increase in electron mobility is mediated by a reduction of the effective mass in the L valley. Even without strain, the Γ-valley already offers an ~8 times smaller effective mass and correspondingly higher electron mobility. Similarly, an increase in hole mobility is expected due to the lifting of the valence band degeneracy (Beattie and Landsberg, 1959; Fischetti and Laux, 1996). This is advantageous as a high mobility strongly reduces the resistivity of the device allowing an efficient injection and extraction of charge carriers.

GeSn emerged as a material of interest in electronics only recently; therefore, less transport data are available. However, theoretical studies predict very large electron mobilities as well as hole mobilities of the order of 4500 cm^2/Vs for direct band gap GeSn (Sau and Cohen, 2007). The first reported experimental mobility study has been done on low Sn content (<6%) indirect band gap GeSn layers yielding a Hall mobility of the order of ~200–300 cm^2/Vs (Nakatsuka et al., 2010). Slightly better results have been obtained thereafter investigating p-MOSFETs hole channel mobility (Gupta et al., 2013a; Wang et al., 2013).

In summary, we see that a vast amount of knowledge concerning the mobility exists leaving a good base for further studies. Moreover, many electrical devices and the corresponding fabrication techniques, e.g., passivation, contacting, or annealing, have been conceived allowing for a fast implementation in optical devices.

However, besides the tremendous changes in the carrier mobility, there are additional effects coming into play with electrical injection of charge carriers from indirect to direct band gap Ge. Exemplarily in **Figure 14**, such an injection scheme in form of a pin diode is discussed for the case of tensile strained Ge bridges where the strain profile is shown in **Figure 5B**.

Far from the strained constriction, electrons can be injected into the L-valleys of the conduction band as in standard Ge diodes. However, close to the center the strain profile alters the band structure with L- and γ-valley starting to cross, which allows for intervalley scattering (Boucaud et al., 2013) from a high- into a low-effective mass valley with a higher mobility, a process inverse to the Gunn effect (Gunn, 1963). This may support current extraction and injection in optical devices. However, an actual impact still needs to be proven.

Cavity Design

For the usage of direct band gap materials in lasing structures, high-quality factor optical resonators are necessary confining the

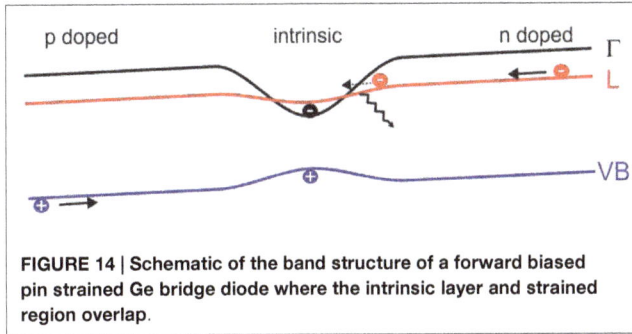

FIGURE 14 | Schematic of the band structure of a forward biased pin strained Ge bridge diode where the intrinsic layer and strained region overlap.

light in the gain region and, thus, allowing for stimulated emission. Helpful in this regard is the high-refractive index contrast of the Ge–air and GeSn–air interfaces (Kasper et al., 2013), which should allow for good confinement properties. The design of suitable cavities seems to be straightforward for GeSn lasers where the wafer-scale direct band gap on Si or Ge already facilitated the implementation of well-known laser cavities, such as FP cavities (Wirths et al., 2015) or microdisks (Cho et al., 2011). At the same time, optical microdisk cavities with quality factors of ~1400 have been demonstrated in tensile strained Ge using external SiN stressors (Ghrib et al., 2013) as well as waveguide cavities (Capellini et al., 2014).

The situation is much more complex for uniaxially strained Ge bridges where patterning of the bridge inevitably relaxes the strain and, hence, prohibits a fundamental direct band gap. This excludes many popular cavity designs, in particular, microdisks, photonic crystals, and FP cavities. Hence, distributed feedback structures, which do not rely on patterning of the strained region, are currently under investigation (Marin et al., 2015).

Band Gap Renormalization and Material Stability

Relevant for both here discussed direct band gap systems is a quantitative analysis of the band gap renormalization of the involved Γ and L valleys in dependence of their respective carrier population. So far, experiments suggest that the renormalization corrections are comparable for the two valleys. Hence, the offset between the Γ and L states would not depend on the injection density, which is essential for a stable injection.

Moreover, material specific investigation concerns, for example, the thermal stability of GeSn and SiGeSn metastable alloys with regards to Sn diffusion and segregation where extensive segregation can result in changes of the emission wavelength and/or emission efficiency. Recently, investigations have been pursued to examine the temperature budget a GeSn or SiGeSn device would be able to withstand, e.g., by *in situ* studies (Fournier-Lupien et al., 2014) or annealing experiments (Wirths et al., 2014). First, *in situ* results indicate phase separation of a 12% Sn containing ternary SiGeSn and binary GeSn alloys at ~420 and 460°C, respectively, which is surprising considering that the higher mixing entropy usually results in a higher thermal stability of ternary alloys (Fournier-Lupien et al., 2014). Annealing experiments revealed distinct Sn diffusion at 300°C for GeSn with approximately the same composition (Wirths et al., 2014).

Finding the intrinsic stress limits of Ge is another item of interest in this context.

By saying this, we conclude our listing of fundamental and materials-related challenges. This list may be incomplete. However, it confirms that the research and development of a laser source from a group IV material will involve many disciplines from fundamental to device physics and from wave optics to material and transport properties. To progress fast, a collaborative effort is demanded.

Opportunities

Photonics

Direct band gap group IV laser systems may permit a qualitative as well as a quantitative expansion of Si-photonics (group IV photonics) into traditional but also new areas of applications. However, it is requested that such lasers can be operated energy efficiently, under ambient conditions and can be fully integrated with current Si technology. An answer to whether this is possible cannot be given yet as the research is at an early stage. We can only speculate about the specifications of such a laser and, thus, have to guess which of the applications would profit most from a successful implementation of group IV lasers. Hence, for the following discussion, let us assume that this all-group-IV laser does indeed exist and it operates (i) under electrical injection, (ii) at RT or above, and (iii) with reasonable power conversion efficiency. What could we do with such a device, where is the highest impact, and what is the platform of choice?

A high return is merely achievable when this laser device will be combined with the current Si photonics by using the same platform. Most advanced are photonic elements fabricated on SOI, except for applications in the visible part of the spectrum – not covered here – where SiN-based structures are often used. SOI for photonics typically consists of Si layers with a thickness of ~200–250 nm and a several micrometer thick buried oxide to avoid leakage of the propagating modes into the Si substrate. For strain engineering, the compatibility with SOI has already been shown (Süess et al., 2013), c.f. Section "Microbridges," and, as mentioned above, bridges with even higher mechanical strength are fabricated from GOI using wafer transfer (Sukhdeo et al., 2014). In fact, wafer-scale fabrication of GOI using the SmartCut® process has already been established several years ago for electronics (Augendre et al., 2009). GOI for photonic applications, where thicker layers and a thicker BOX are required, has been presented recently by Reboud et al. (2015). A photonic platform based on GOI, in comparison to SOI, has the advantage that all photonic elements, such as waveguides, bends, and the resonant structures, can be reduced in size because of the larger refractive index contrast. This allows for the potential fabrication of more dense optical circuits and, hence, for an easier integration with electronics. Furthermore, Ge provides coverage of the longer wavelengths toward 10 μm and more. Moreover, by using processes that are selective for either Ge or Si, the GOI platform may provide additional fabrication opportunities. The high quality (Si)GeSn presented by Wirths et al. (2013b) has been deposited on a Ge VS on Si(001) indicating that the growth on SOI and certainly GOI is possible, as well.

With a laser device implemented on the currently used SOI or similarly on GOI, various new applications will emerge. Before speculating, we may picture the many already existing photonic elements. To select a few: Low-loss (<1 dB cm^{-1}) single mode waveguides in various designs, tapers to adiabatically match the waveguide modes to fibers, and low-loss grating couplers (<1 dB) in 1D or 2D providing polarization splitting. Other standard elements include directional couplers, Mach Zehnder interferometers, ring resonators, and light modulators based on free carrier injections or quantum confined Stark effect. Detectors based on slightly tensile strained (0.2%) Ge provide more than 10 GHz speed for high-data rate transmission. Other device elements include add-drop filters, buffers, and switches, which can be integrated with fluidic channels for (bio-) sensing. The short wavelength infrared at which the here discussed group IV direct band gap lasers emit (see **Figure 3**) is certainly a clear asset for sensing applications (Soref, 2010; Nedeljkovic et al., 2013).

Silicon-on-insulator is (and GOI could become) a very convenient platform to realize high-performing photonic crystal structures enabling unique photonic circuits, such as compact high-Q cavities, which can operate stably at single and dual-wavelength, and as wavelength division multiplexer as desired for optical signal processing. This listing can be extended almost indefinitely, naming, e.g., switching and steering of optical signals, slow light, pulse compression, customized reflectors, and filters. Together with the expectation that such photonic circuits will be very cost effective, compact, reliable, and efficient, a monolithically integrated laser source will certainly bring new functionality, in particular when optics can be merged with electronics.

CMOS Integration

The combination of optics with CMOS electronics to realize an on-chip data distribution network (Heck and Bowers, 2014) is – without any doubt – one of the most advanced and challenging applications for direct band gap group IV lasers. The requirements are so complex (Miller, 2009) that before the start of such a development, many fundamental questions have to be answered, such as the efficiency issues among other challenges, which have been addresses in the previous section. However, once these hurdles are taken, we expect to arise a highly competitive and attractive platform solution for future data processing applications. In fact, the extension of CMOS by integration of Ge and (Si)GeSn may not just resolve the demands for a monolithic laser gain medium, but, as discussed widely elsewhere (Kao et al., 2014), (Si)GeSn would already advance the performance of the electronic circuits. This appealing double benefit, together with the potential compatibility to CMOS of such an all-group-IV solution, bears an essential advantage in comparison to other emerging technologies, such as spin- and/or valley-based electronics, which rely in part on non-conform chemical elements and non-CMOS fabrication processes.

Hence, we expect that as soon as the fundamental lessons of direct band gap lasing are learnt and a gain medium well-qualified for injection pumping at RT is defined, research and development of a new opto-electronic platform will quickly advance. Experts in CMOS technology, group IV epitaxy, laser physics, and Si electronics and photonics will cooperate and define the routes to opto-electronics for fast and energy-efficient data processing.

Conclusion and Outlook

We reviewed the methods for achieving a direct band gap in group IV semiconductors in the most promising material system for the prospect of a Si compatible laser, namely, Ge modified either via tensile strain or by alloying with Sn. We expanded on the methods to characterize these systems and gave examples on their optical properties. The recent advances in numerous approaches to achieve a direct band gap have finally concluded in the first demonstration of lasing in a direct band gap GeSn alloy (Wirths et al., 2015).

With this demonstration, we are at the beginning of an exciting journey in the field of silicon photonics. As shown in great detail, the many optical characterization tools at hand allow us to address a large amount of fundamental questions, including band gap renormalization, various recombination processes, and doping level-dependent lasing performance, but also material- and technology-related issues, such as high Q-factor cavity design, diffusive carrier transport, stress, and thermal diffusion limits.

We hope that with outlining these challenges, we can motivate a vast amount of new researchers from various backgrounds in optics, material science, and device physics to join this interesting research field. We believe that combined efforts will converge in a reasonable time to a demonstration of a practical laser source being electrically pumped, highly efficient, and fully integrated on an electro-optical CMOS platform. This building block will finally pave the way for true monolithic on-chip integration of photonics and CMOS electronics for new sensors in the long wavelength infrared, and will eventually enable to build an on-chip or off-chip electro-optical data distribution network for high-performance computing.

Acknowledgments

We would like to acknowledge the many scientific collaborators we were fortunate to work with over the last few years. They supported us in building up a strong portfolio in the investigation and understanding of lasing in group IV systems, and the fabrication of direct band gap group IV materials. In particular, we thank our previous group members Gustav Schiefler, Martin J. Süess, and Renato Minamisawa for their contributions, which led to this appealing strain concept, and the group of Dan Buca (FZ Jülich), who contacted us for investigating their high quality material and thus gave us the opportunity to learn also about GeSn alloys. The tremendous progress achieved in a short time is a shining example of our good collaboration. We also thank Jérôme Faist and Ralph Spolenak (ETHZ) for their whole-hearted support to this subject and their many essential contributions. Finally, we acknowledge the Swiss Science Foundation (SNF) for supporting part of the here reviewed studies over several years.

References

Adams, A. R., Asada, M., Suematsu, Y., and Arai, S. (1980). The temperature dependence of the efficiency and threshold current of In 1- xGa xAs yP 1- y lasers related to intervalence band absorption. *Jpn. J. Appl. Phys.* 19, L621–L624. doi:10.1143/JJAP.19.L621

Akatsu, T., Deguet, C., Sanchez, L., Allibert, F., Rouchon, D., Signamarcheix, T., et al. (2006). Germanium-on-insulator (GeOI) substrates – a novel engineered substrate for future high performance devices. *Mater. Sci. Semicond. Process.* 9, 444–448. doi:10.1016/j.mssp.2006.08.077

Aldaghri, O., Ikonić, Z., and Kelsall, R. W. (2012). Optimum strain configurations for carrier injection in near infrared Ge lasers. *J. Appl. Phys.* 111, 053106. doi:10.1063/1.3691790

Augendre, E., Sanchez, L., Benaissa, L., Signamarcheix, T., Hartmann, J.-M., Le Royer, C., et al. (2009). Challenges and progress in germanium-on-insulator materials and device development towards ULSI integration. *ECS Trans.* 25, 351–362. doi:10.1149/1.3203972

Bauer, M. R., Tolle, J., Bungay, C., Chizmeshya, A. V. G., Smith, D. J., Menéndez, J., et al. (2003). Tunable band structure in diamond-cubic tin-germanium alloys grown on silicon substrates. *Solid State Commun.* 127, 355–359. doi:10.1016/S0038-1098(03)00446-0

Beattie, A. R., and Landsberg, P. T. (1959). Auger effect in semiconductors. *Proc. R. Soc. A Math. Phys. Eng. Sci.* 249, 16–29. doi:10.1098/rspa.1959.0003

Bhargava, N., Coppinger, M., Gupta, J. P., Wielunski, L., and Kolodzey, J. (2013). Lattice constant and substitutional composition of GeSn alloys grown by molecular beam epitaxy. *Appl. Phys. Lett.* 103, 041908. doi:10.1063/1.4816660

Birner, S., Zibold, T., Andlauer, T., Kubis, T., Sabathil, M., Trellakis, A., et al. (2007). Nextnano: general purpose 3-D simulations. *IEEE Trans. Electron Devices* 54, 2137–2142. doi:10.1109/TED.2007.902871

Boucaud, P., Kurdi El, M., Ghrib, A., Prost, M., de Kersauson, M., Sauvage, S., et al. (2013). Recent advances in germanium emission. *Photon. Res.* 1, 102. doi:10.1364/PRJ.1.000102

Bratland, K., Foo, Y., Soares, J., Spila, T., Desjardins, P., and Greene, J. (2003). Mechanism for epitaxial breakdown during low-temperature Ge(001) molecular beam epitaxy. *Phys. Rev. B* 67, 125322. doi:10.1103/PhysRevB.67.125322

Camacho-Aguilera, R. E., Cai, Y., Patel, N., Bessette, J. T., Romagnoli, M., Kimerling, L. C., et al. (2012). An electrically pumped germanium laser. *Opt. Express* 20, 11316. doi:10.1364/OE.20.011316

Capellini, G., Kozlowski, G., Yamamoto, Y., Lisker, M., Wenger, C., Niu, G., et al. (2013). Strain analysis in SiN/Ge microstructures obtained via Si-complementary metal oxide semiconductor compatible approach. *J. Appl. Phys.* 113, 013513. doi:10.1063/1.4772781

Capellini, G., Reich, C., Guha, S., Yamamoto, Y., Lisker, M., Virgilio, M., et al. (2014). Tensile Ge microstructures for lasing fabricated by means of a silicon complementary metal-oxide-semiconductor process. *Opt. Express* 22, 399–410. doi:10.1364/OE.22.000399

Carroll, L., Friedli, P., Lerch, P., Schneider, J., Treyer, D., Hunziker, S., et al. (2011). Ultra-broadband infrared pump-probe spectroscopy using synchrotron radiation and a tuneable pump. *Rev. Sci. Instrum.* 82, 063101–063101–9. doi:10.1063/1.3592332

Carroll, L., Friedli, P., Neuenschwander, S., Sigg, H., Cecchi, S., Isa, F., et al. (2012). Direct-gap gain and optical absorption in germanium correlated to the density of photoexcited carriers, doping, and strain. *Phys. Rev. Lett.* 109, 057402. doi:10.1103/PhysRevLett.109.057402

Chang, G.-E., and Cheng, H. H. (2013). Optical gain of germanium infrared lasers on different crystal orientations. *J. Phys. D Appl. Phys.* 46, 065103. doi:10.1088/0022-3727/46/6/065103

Chen, R., Gupta, S., Huang, Y.-C., Huo, Y., Rudy, C. W., Sanchez, E., et al. (2013a). Demonstration of a Ge/GeSn/Ge quantum-well microdisk resonator on silicon: enabling high-quality Ge(Sn) materials for micro- and nanophotonics. *Nano Lett.* 14, 37–43. doi:10.1021/nl402815v

Chen, R., Huang, Y.-C., Gupta, S., Lin, A. C., Sanchez, E., Kim, Y., et al. (2013b). Material characterization of high Sn-content, compressively-strained GeSn epitaxial films after rapid thermal processing. *J. Cryst. Growth* 365, 29–34. doi:10.1016/j.jcrysgro.2012.12.014

Chen, R., Lin, H., Huo, Y., Hitzman, C., Kamins, T. I., and Harris, J. S. (2011a). Increased photoluminescence of strain-reduced, high-Sn composition Ge1−xSnx alloys grown by molecular beam epitaxy. *Appl. Phys. Lett.* 99, 181125. doi:10.1063/1.3658632

Chen, Y. T., Lan, H. S., Hsu, W., Fu, Y. C., Lin, J. Y., and Liu, C. W. (2011b). Strain response of high mobility germanium n-channel metal-oxide-semiconductor field-effect transistors on (001) substrates. *Appl. Phys. Lett.* 99, 022106. doi:10.1063/1.3604417

Childs, G. N., Brand, S., and Abram, R. A. (1986). Intervalence band absorption in semiconductor laser materials. *Semicond. Sci. Technol.* 1, 116–120. doi:10.1088/0268-1242/1/2/004

Cho, S., Chen, R., Koo, S., Shambat, G., Lin, H., Park, N., et al. (2011). Fabrication and analysis of epitaxially grown Ge1-xSnx microdisk resonator with 20-nm free-spectral range. *IEEE Photon. Technol. Lett.* 23, 1535–1537. doi:10.1109/LPT.2011.2163929

Chow, W. W. (2012). Model for direct-transition gain in a Ge-on-Si laser. *Appl. Phys. Lett.* 100, 191113. doi:10.1063/1.4714540

Chu, M., Sun, Y., Aghoram, U., and Thompson, S. E. (2009). Strain: a solution for higher carrier mobility in nanoscale MOSFETs. *Annu. Rev. Mater. Res.* 39, 203–229. doi:10.1146/annurev-matsci-082908-145312

Du, W., Ghetmiri, S. A., Conley, B. R., Mosleh, A., Nazzal, A., Soref, R. A., et al. (2014). Competition of optical transitions between direct and indirect bandgaps in Ge1−xSnx. *Appl. Phys. Lett.* 105, 051104. doi:10.1063/1.4892302

Dutt, B., Lin, H., Sukhdeo, D. S., Vulovic, B. M., Gupta, S., Nam, D., et al. (2013). Theoretical analysis of GeSn alloys as a gain medium for a Si-compatible laser. *IEEE J. Sel. Top. Quantum Electron.* 19, 1502706–1502706. doi:10.1109/JSTQE.2013.2241397

Dutt, B., Sukhdeo, D. S., Nam, D., Vulovic, B. M., Ze, Y., and Saraswat, K. C. (2012). Roadmap to an efficient germanium-on-silicon laser: strain vs. n-type doping. *IEEE Photon. J.* 4, 2002–2009. doi:10.1109/JPHOT.2012.2221692

El Kurdi, M., Fishman, G., Sauvage, S., and Boucaud, P. (2010). Band structure and optical gain of tensile-strained germanium based on a 30 band k·p formalism. *J. Appl. Phys.* 107, 013710. doi:10.1063/1.3279307

Fang, Z., Chen, Q. Y., and Zhao, C. Z. (2013). A review of recent progress in lasers on silicon. *Opt. Laser Technol.* 46, 103–110. doi:10.1088/0034-4885/76/3/034501

Fischetti, M. V., and Laux, S. E. (1996). Band structure, deformation potentials, and carrier mobility in strained Si, Ge, and SiGe alloys. *J. Appl. Phys.* 80, 2234–2252. doi:10.1063/1.363052

Fitzgerald, E. A., Xie, Y. H., Green, M. L., Brasen, D., Kortan, A. R., Michel, J., et al. (1991). Totally relaxed GexSi1−x layers with low threading dislocation densities grown on Si substrates. *Appl. Phys. Lett.* 59, 811. doi:10.1063/1.105351

Fournier-Lupien, J. H., Chagnon, D., Levesque, P., AlMutairi, A. A., Wirths, S., Pippel, E., et al. (2014). In situ studies of germanium-tin and silicon-germanium-tin thermal stability. *ECS Trans.* 64, 903–911. doi:10.1149/06406.0903ecst

Geiger, R., Frigerio, J., Süess, M. J., Christina, D., Isella, G., Spolenak, R., et al. (2014a). Excess carrier lifetimes in Ge layers on Si. *Appl. Phys. Lett.* 104, 062106–062106. doi:10.1063/1.4865237

Geiger, R., Süess, M. J., Bonzon, C., Frigerio, J., Christina, D., Isella, G., et al. (2014b). *Carrier Lifetimes in Uniaxially Strained Ge Micro Bridges.* IEEE, 227–228.

Geiger, R., Süess, M. J., Bonzon, C., Spolenak, R., Faist, J., and Sigg, H. (2014c). *Strained Ge Microbridges to Obtain a Direct Bandgap Laser.* IEEE, 7–8.

Geiger, R., Frigerio, J., Süess, M. J., Minamisawa, R. A., Chrastina, D., Isella, G., et al. (2013). *Excess Carrier Lifetimes in Ge Layers on Si.* IEEE, 103–104.

Ghrib, A., Kurdi El, M., de Kersauson, M., Prost, M., Sauvage, S., Checoury, X., et al. (2013). Tensile-strained germanium microdisks. *Appl. Phys. Lett.* 102, 221112. doi:10.1364/OE.23.006722

Ghrib, A., Kurdi El, M., Prost, M., Sauvage, S., Checoury, X., Beaudoin, G., et al. (2015). All-around SiN stressor for high and homogeneous tensile strain in germanium microdisk cavities. *Adv. Opt. Mater.* 3, 353–358. doi:10.1002/adom.201400369

Golikova, O. A., Moizhes, B. Y., and Stilbans, L. S. (1962). Hole mobility of germanium as a function of concentration and temperature. *Sov. Phys. Solid State* 3, 2259–2265.

Gray, A. X., Papp, C., Ueda, S., Balke, B., Yamashita, Y., Plucinski, L., et al. (2011). Probing bulk electronic structure with hard X-ray angle-resolved photoemission. *Nat. Mater.* 10, 759–764. doi:10.1038/nmat3089

Greil, J., Lugstein, A., Zeiner, C., Strasser, G., and Bertagnolli, E. (2012). Tuning the electro-optical properties of germanium nanowires by tensile strain. *Nano Lett.* 12, 6230–6234. doi:10.1021/nl303288g

Gunn, J. B. (1963). Microwave oscillations of current in III–V semiconductors. *Solid State Commun.* 1, 88–91. doi:10.1016/0038-1098(63)90041-3

Gupta, S., Chen, R., Vincent, B., Lin, D., Magyari-Kope, B., Caymax, M., et al. (2013a). (Invited) GeSn channel n and p MOSFETs. *ECS Trans.* 50, 937–941. doi:10.1149/05009.0937ecst

Gupta, S., Magyari-Köpe, B., Nishi, Y., and Saraswat, K. C. (2013b). Achieving direct band gap in germanium through integration of Sn alloying and external strain. J. Appl. Phys. 113, 073707. doi:10.1063/1.4792649

Hartmann, J. M., Sanchez, L., Van Den Daele, W., Abbadie, A., Baud, L., Truche, R., et al. (2010). Fabrication, structural and electrical properties of compressively strained Ge-on-insulator substrates. Semicond. Sci. Technol. 25, 075010. doi:10.1088/0268-1242/25/7/075010

Harwit, A., Pukite, P. R., Angilello, J., and Iyer, S. S. (1990). Properties of diamond structure SnGe films grown by molecular beam epitaxy. Thin Solid Films 184, 395–401. doi:10.1016/0040-6090(90)90437-I

He, G., and Atwater, H. A. (1997). Interband transitions in SnxGe1−xAlloys. Phys. Rev. Lett. 79, 1937–1940. doi:10.1103/PhysRevLett.79.1937

Heck, M. J. R., and Bowers, J. E. (2014). Energy efficient and energy proportional optical interconnects for multi-core processors: driving the need for on-chip sources. IEEE J. Sel. Top. Quantum Electron. 20, 332–343. doi:10.1109/JSTQE.2013.2293271

Huldt, L. (1971). Band-to-band auger recombination in indirect gap semiconductors. Phys. Status Solidi A 8, 173–187. doi:10.1002/pssa.2210080118

Huldt, L. (1974). Auger recombination in germanium. Phys. Status Solidi A 24, 221–229. doi:10.1088/0957-4484/22/43/435401

Huo, Y., Lin, H., Chen, R., Makarova, M., Rong, Y., Li, M., et al. (2011). Strong enhancement of direct transition photoluminescence with highly tensile-strained Ge grown by molecular beam epitaxy. Appl. Phys. Lett. 98, 011111. doi:10.1063/1.3534785

Jacoboni, C., Nava, F., Canali, C., and Ottaviani, G. (1981). Electron drift velocity and diffusivity in germanium. Phys. Rev. B 24, 1014–1026. doi:10.1103/PhysRevB.24.1014

Jain, J. R., Hryciw, A., Baer, T. M., Miller, D. A. B., Brongersma, M. L., and Howe, R. T. (2012). A micromachining-based technology for enhancing germanium light emission via tensile strain. Nat. Photonics 6, 398–405. doi:10.1038/nphoton.2012.111

Kao, K.-H., Verhulst, A. S., Rooyackers, R., Douhard, B., Delmotte, J., Bender, H., et al. (2014). Compressively strained SiGe band-to-band tunneling model calibration based on p-i-n diodes and prospect of strained SiGe tunneling field-effect transistors. J. Appl. Phys. 116, 214506. doi:10.1063/1.4903288

Kasper, E., Kittler, M., Oehme, M., and Arguirov, T. (2013). Germanium tin: silicon photonics toward the mid-infrared [Invited]. Photon. Res. 1, 69. doi:10.1364/PRJ.1.000069

Koerner, R., Oehme, M., Gollhofer, M., Schmid, M., Kostecki, K., Bechler, S., et al. (2015). Electrically pumped lasing from Ge Fabry-Perot resonators on Si. Opt. Express 23, 14815–14822. doi:10.1364/OE.23.014815

Liu, J., Sun, X., Camacho-Aguilera, R., Kimerling, L. C., and Michel, J. (2010). Ge-on-Si laser operating at room temperature. Opt. Lett. 35, 679. doi:10.1364/OL.35.000679

Liu, J., Sun, X., Pan, D., Wang, X., Kimerling, L. C., Koch, T. L., et al. (2007). Tensile-strained, n-type Ge as a gain medium for monolithic laser integration on Si. Opt. Express 15, 11272. doi:10.1364/OE.15.011272

Lochmann, W. (1978). Phonon-assisted auger recombination in indirect gap semiconductors. Phys. Status Solidi A 45, 423–432. doi:10.1002/pssa.2210450208

Low, K. L., Yang, Y., Han, G., Fan, W., and Yeo, Y.-C. (2012). Electronic band structure and effective mass parameters of Ge1-xSnx alloys. J. Appl. Phys. 112, 103715. doi:10.1063/1.4767381

Marin, E., Bonzon, C., Geiger, R., Zabel, T., Sigg, H., and Faist, J. (2015). Proceedings CLEO-Europe 2015. (in print)

Metzger, W. K., Wanlass, M. W., Ellingson, R. J., Ahrenkiel, R. K., and Carapella, J. J. (2001). Auger recombination in low-band-gap n-type InGaAs. Appl. Phys. Lett. 79, 3272–3274. doi:10.1063/1.1418032

Michel, J., Liu, J., and Kimerling, L. C. (2010). High-performance Ge-on-Si photodetectors. Nat. Photonics 4, 527–534. doi:10.1038/nphoton.2010.157

Miller, D. A. B. (2009). Device requirements for optical interconnects to silicon chips. Proc. IEEE 97, 1166–1185. doi:10.1109/JPROC.2009.2014298

Miller, D. A. B. (2010). Optical interconnects to electronic chips. Appl. Opt. 49, F59. doi:10.1364/AO.49.000F59

Minamisawa, R. A., Süess, M. J., Spolenak, R., Faist, J., David, C., Gobrecht, J., et al. (2012). Top-down fabricated silicon nanowires under tensile elastic strain up to 4.5%. Nat Commun 3, 1096. doi:10.1038/ncomms2102

Nakatsuka, O., Tsutsui, N., Shimura, Y., Takeuchi, S., Sakai, A., and Zaima, S. (2010). Mobility behavior of Ge1-xSnx layers grown on silicon-on-insulator substrates. Jpn. J. Appl. Phys. 49, 04DA10. doi:10.1143/JJAP.49.04DA10

Nam, D., Kang, J.-H., Brongersma, M. L., and Saraswat, K. C. (2014). Observation of improved minority carrier lifetimes in high-quality Ge-on-insulator using time-resolved photoluminescence. Opt. Lett. 39, 6205. doi:10.1364/OL.39.006205

Nam, D., Sukhdeo, D., Cheng, S.-L., Roy, A., Chih-Yao Huang, K., Brongersma, M., et al. (2012). Electroluminescence from strained germanium membranes and implications for an efficient Si-compatible laser. Appl. Phys. Lett. 100, 131112. doi:10.1063/1.3699224

Nam, D., Sukhdeo, D., Roy, A., Balram, K., Cheng, S.-L., Huang, K. C.-Y., et al. (2011). Strained germanium thin film membrane on silicon substrate for optoelectronics. Opt. Express 19, 25866–25872. doi:10.1364/OE.19.025866

Nam, D., Sukhdeo, D. S., Kang, J.-H., Petykiewicz, J., Lee, J. H., Jung, W. S., et al. (2013). Strain-induced pseudoheterostructure nanowires confining carriers at room temperature with nanoscale-tunable band profiles. Nano Lett. 13, 3118–3123. doi:10.1021/nl401042n

Nedeljkovic, M., Khokhar, A. Z., Hu, Y., Chen, X., Penades, J. S., Stankovic, S., et al. (2013). Silicon photonic devices and platforms for the mid-infrared. Opt. Mater. Express 3, 1205. doi:10.1364/OME.3.001205

Newman, R., and Tyler, W. (1957). Effect of impurities on free-hole infrared absorption in p-type germanium. Phys. Rev. 105, 885–886. doi:10.1103/PhysRev.105.885

Oda, K., Okumura, T., Tani, K., Saito, S.-I., and Ido, T. (2013). Improvement of photoluminescence from Ge layer with patterned Si3N4 stressors. Thin Solid Films 557, 355–362. doi:10.1016/j.tsf.2013.08.117

Oehme, M., Buca, D., Kostecki, K., Wirths, S., Holländer, B., Kasper, E., et al. (2013). Epitaxial growth of highly compressively strained GeSn alloys up to 12.5% Sn. J. Cryst. Growth 384, 71–76. doi:10.1016/j.jcrysgro.2013.09.018

Passaro, V., Tullio, C., Troia, B., Notte, M., Giannoccaro, G., and Leonardis, F. (2012). Recent advances in integrated photonic sensors. Sensors 12, 15558–15598. doi:10.3390/s121115558

Peschka, D., Thomas, M., Glitzky, A., Nurnberg, R., Gartner, K., Virgilio, M., et al. (2015). Modeling of edge-emitting lasers based on tensile strained germanium microstrips. IEEE Photon. J. 7, 1–15. doi:10.1109/JPHOT.2015.2427093

Pukite, P. R., Harwit, A., and Iyer, S. S. (1989). Molecular beam epitaxy of metastable, diamond structure SnxGe1-x alloys. Appl. Phys. Lett. 54, 2142–2144. doi:10.1063/1.101152

Reboud, V., Widiez, J., Hartmann, J.-M., Dias, G. O., Fowler, D., Chelnokov, A., et al. (2015). Structural and optical properties of 200 mm germanium-on- insulator (GeOI) substrates for silicon photonics applications. Proc. SPIE 9367, Silicon Photonics X 2015. doi:10.1117/12.2079393

Saito, S., Gardes, F. Y., Al-Attili, A. Z., Tani, K., Oda, K., Suwa, Y., et al. (2014). Group IV light sources to enable the convergence of photonics and electronics. Front. Mater. 1:15. doi:10.3389/fmats.2014.00015

Sánchez-Pérez, J. R., Boztug, C., Chen, F., Sudradjat, F. F., Paskiewicz, D. M., Jacobson, R. B., et al. (2011). Direct-bandgap light-emitting germanium in tensilely strained nanomembranes. Proc. Natl. Acad. Sci. U.S.A. 108, 18893–18898. doi:10.1073/pnas.1107968108

Satta, A., Nicholas, G., Simoen, E., Houssa, M., Dimoulas, A., De Jaeger, B., et al. (2006). Impact of germanium surface passivation on the leakage current of shallow planar p-n junctions. Mater. Sci. Semicond. Process. 9, 716–720. doi:10.1016/j.mssp.2006.08.037

Sau, J., and Cohen, M. (2007). Possibility of increased mobility in Ge-Sn alloy system. Phys. Rev. B 75, 045208. doi:10.1103/PhysRevB.75.045208

Schetzina, J. F., and McKelvey, J. P. (1969). Strain dependence of the minority carrier mobility in p-type germanium. Phys. Rev. 181, 1191. doi:10.1103/PhysRev.181.1191

Schubert, E. F. (2006). Light-Emitting Diodes, 2nd Edn. Cambridge University Press.

Senaratne, C. L., Gallagher, J. D., Jiang, L., Aoki, T., Smith, D. J., Menendez, J., et al. (2014). Ge1-ySny (y = 0.01-0.10) alloys on Ge-buffered Si: synthesis, microstructure, and optical properties. J. Appl. Phys. 116, 133509. doi:10.1063/1.4896788

Shockley, W., and Read, W. (1952). Statistics of the recombinations of holes and electrons. Phys. Rev. 87, 835–842. doi:10.1103/PhysRev.87.835

Soref, R. (2010). Mid-infrared photonics in silicon and germanium. Nat. Photonics 4, 495–497. doi:10.1038/nphoton.2010.171

Strocov, V. N., Wang, X., Shi, M., Kobayashi, M., Krempasky, J., Hess, C., et al. (2014). Soft-X-ray ARPES facility at the ADRESS beamline of the SLS: concepts, technical realisation and scientific applications. J. Synchrotron. Radiat. 21, 32–44. doi:10.1107/S1600577513019085

Süess, M. J., Geiger, R., Minamisawa, R. A., Schiefler, G. L., Frigerio, J., Chrastina, D., et al. (2013). Analysis of enhanced light emission from highly strained germanium microbridges. Nat. Photonics 7, 466–472. doi:10.1038/nphoton.2013.67

Sukhdeo, D. S., Nam, D., Kang, J.-H., Brongersma, M. L., and Saraswat, K. C. (2014). Direct bandgap germanium-on-silicon inferred from 5.7% ⟨100⟩ uniaxial tensile strain [Invited]. *Photon. Res.* 2, A8. doi:10.1364/PRJ.2.0000A8

Sun, G., Soref, R. A., and Cheng, H. H. (2010). Design of a Si-based lattice-matched room-temperature GeSn/GeSiSn multi-quantum-well mid-infrared laser diode. *Opt. Express* 18, 19957–19965. doi:10.1364/OE.18.019957

Tahini, H., Chroneos, A., Grimes, R. W., Schwingenschlögl, U., and Dimoulas, A. (2012). Strain-induced changes to the electronic structure of germanium. *J. Phys. Condens. Matter* 24, 195802. doi:10.1088/0953-8984/24/19/195802

Takeuchi, S., Sakai, A., Yamamoto, K., Nakatsuka, O., Ogawa, M., and Zaima, S. (2006). Growth and structure evaluation of strain-relaxed Ge 1− xSn xbuffer layers grown on various types of substrates. *Semicond. Sci. Technol.* 22, S231–S235. doi:10.1088/0268-1242/22/1/S54

Velha, P., Dumas, D. C., Gallacher, K., Millar, R., Myronov, M., Leadley, D. R., et al. (2013). *Strained Germanium Nanostructures on Silicon Emitting at >2.2 μm Wavelength.* IEEE, 142–143.

Vincent, B., Gencarelli, F., Bender, H., Merckling, C., Douhard, B., Petersen, D. H., et al. (2011). Undoped and in-situ B doped GeSn epitaxial growth on Ge by atmospheric pressure-chemical vapor deposition. *Appl. Phys. Lett.* 99, 152103. doi:10.1063/1.3645620

Virgilio, M., Manganelli, C. L., Grosso, G., Schroeder, T., and Capellini, G. (2013). Photoluminescence, recombination rate, and gain spectra in optically excited n-type and tensile strained germanium layers. *J. Appl. Phys.* 114, 243102. doi:10.1063/1.4849855

Wang, L., Su, S., Wang, W., Gong, X., Yang, Y., Guo, P., et al. (2013). Strained germanium-tin (GeSn) p-channel metal-oxide-semiconductor field-effect-transistors (p-MOSFETs) with ammonium sulfide passivation. *Solid State Electron.* 83, 66–70. doi:10.1016/j.sse.2013.01.031

Wegscheider, W., Eberl, K., Menczigar, U., and Abstreiter, G. (1990). Single-crystal Sn/Ge superlattices on Ge substrates: growth and structural properties. *Appl. Phys. Lett.* 57, 875. doi:10.1063/1.104264

Wen, H., and Bellotti, E. (2015). Rigorous theory of the radiative and gain characteristics of silicon and germanium lasing media. *Phys. Rev. B* 91, 035307. doi:10.1103/PhysRevB.91.035307

Wirths, S., Buca, D., Mussler, G., Tiedemann, A. T., Holländer, B., Bernardy, P., et al. (2013a). Reduced pressure CVD growth of Ge and Ge1-xSnx alloys. *ECS J. Solid State Sci. Technol.* 2, N99–N102. doi:10.1149/2.006305jss

Wirths, S., Ikonić, Z., Tiedemann, A. T., Holländer, B., Stoica, T., Mussler, G., et al. (2013b). Tensely strained GeSn alloys as optical gain media. *Appl. Phys. Lett.* 103, 192110. doi:10.1063/1.4829360

Wirths, S., Geiger, R., den Driesch von, N., Mussler, G., Stoica, T., Mantl, S., et al. (2015). Lasing in direct-bandgap GeSn alloy grown on Si. *Nat. Photonics* 9, 88–92. doi:10.1038/nphoton.2014.321

Wirths, S., Stange, D., Pampillón, M.-A., Tiedemann, A. T., Mussler, G., Fox, A., et al. (2014). High-k gate stacks on low bandgap tensile strained Ge and GeSn alloys for field-effect transistors. *ACS Appl. Mater. Interfaces* 7, 62–67. doi:10.1021/am5075248

Xiaochen, S., Jifeng, L., Kimerling, L. C., and Michel, J. (2010). Toward a germanium laser for integrated silicon photonics. *IEEE J. Sel. Top. Quantum Electron.* 16, 124–131. doi:10.1109/JSTQE.2009.2027445

Xu, C., Beeler, R. T., Jiang, L., Grzybowski, G., Chizmeshya, A. V. G., Menéndez, J., et al. (2013). New strategies for Ge-on-Si materials and devices using non-conventional hydride chemistries: the tetragermane case. *Semicond. Sci. Technol.* 28, 105001. doi:10.1088/0268-1242/28/10/105001

Zhu, Y.-H., Xu, Q., Fan, W.-J., and Wang, J.-W. (2010). Theoretical gain of strained GeSn0.02/Ge1−x−y′SixSny′ quantum well laser. *J. Appl. Phys.* 107, 073108. doi:10.1063/1.3329424

Conflict of Interest Statement: The authors declare that the research was conducted in the absence of any commercial or financial relationships that could be construed as a potential conflict of interest.

9

Shape memory polymer nanofibers and their composites: electrospinning, structure, performance, and applications

Fenghua Zhang[1], Zhichun Zhang[1], Tianyang Zhou[1], Yanju Liu[2] and Jinsong Leng[1]*

[1]*Centre for Composite Materials and Structures, Harbin Institute of Technology, Harbin, China,* [2]*Department of Astronautical Science and Mechanics, Harbin Institute of Technology, Harbin, China*

Edited by:
Ilkwon Oh,
Korea Advanced Institute
of Science and Technology,
South Korea

Reviewed by:
Naba Kumar Dutta,
Indian Institute of Technology, India
Hyacinthe Randriamahazaka,
Université Paris Diderot, France

*Correspondence:
Jinsong Leng,
Centre for Composite Materials and
Structures, Harbin Institute of
Technology (HIT), No. 2 YiKuang
Street, PO Box 3011,
Harbin 150080, China
lengjs@hit.edu.cn

Shape memory polymers (SMPs) have been defined as a kind of smart materials under great investigation from academic research to industry applications. Research on SMPs and their composites now incorporates a growing focus on nanofibers which offers new structures in microscopic level and the potential of enhanced performance of SMPs. This paper presents a comprehensive review of the development of shape memory polymer nanofibers and their composites, including the introduction of electrospinning technology, the morphology and structures of nanofibers (non-woven fibers, oriented fibers, core/shell fibers, and functional particles added in the fibers), shape memory performance (thermal and mechanical properties, stimulus responsive behavior, multiple and two-way shape changing performance), as well as their potential applications in the fields of biomedical and tissue engineering.

Keywords: shape memory polymers, nanofibers, electrospinning, structure, applications

Introduction

Since the discovery of shape memory polymers (SMPs) in 1980s, they have the capability to change sizes, shapes, stiffness, or strain when triggered by an external stimulus (heat, electric and magnetic field, water, or light) (Huang et al., 2005; Lendlein et al., 2005; Lan et al., 2009; Liu et al., 2009; Leng et al., 2011; Kumar et al., 2012). SMPs and their composites have captured great attention due to their excellent features, such as multi-shape memory effect, low cost, easy processing, and large deformation (Liu et al., 2007; Xie, 2010; Leng et al., 2011). These unique characteristics guarantee a broad range of applications (Behl and Lendlein, 2007; Liu et al., 2007; Leng et al., 2009, 2011), including actuating, sensing, remote control, smart textiles, robotics and biomedical tissue, as well as aerospace, and so on (Gall et al., 2002; Lendlein and Langer, 2002; Nagahama et al., 2009; Hu et al., 2012; Liu et al., 2014). However, the simple structure limits the development of shape changing materials and practical application. The various fabrication processes result in a diversity of structures of smart polymers. From a structural view point, SMPs can be classified as shape memory blocks, shape memory foams, shape memory fibers, and shape memory films (Leng et al., 2011; Hu et al., 2012). Academic and industrial research on shape memory nanofibers reap huge fruits in the field of fundamental and applied science. Electrospinning method has enjoyed great attention since the 1990s, which is extensively applied to produce nanofibers and their composites. A number of multiple functional materials have been electrospun into nanofibers used in different application fields. SMPs in the form of fibers are generating great interest in structural and functional applications because they have well-developed surface area, porous structure, and are easy adsorption, filtration, and penetration, as well as deformation (Cha et al., 2005; Ji et al., 2006; Kotek, 2008; Zhuo et al., 2008a; Chung et al., 2011).

Based on the above advantages, SMPs and their composites with fibrous structure have been increasingly investigated. In recent years, some researches on shape memory nanofibers fabricated by electrospinning have been reported (Meng et al., 2007; Zhuo et al., 2008b; Zhang et al., 2013a, 2014a). Cha et al. synthesized shape memory polyurethane (SMPU) block copolymers that were electrospun into nanofibers at 2005 (Cha et al., 2005). Hu et al. of the Hong Kong Polytechnic University fabricated the smart polyurethane fibers with shape memory effect (Ji et al., 2006; Zhuo et al., 2008a). Then, the same group further demonstrated that the non-woven SMPU nanofibers (the diameter in a range of 50–700 nm) can be triggered by heat and the shape recovery was 98% after several cycles (Zhuo et al., 2011a). Afterwards, Jing-Nan et al. found that, compared to the bulk SMPU film, SMPU microfibers showed a much quicker shape recovery process when heated in a water bath. The shape recovered from 10 to 90%, the microfiber film only took one-fourth of the time of that needed for the bulk film (Zhang et al., 2011). More recent studies by Lendlein et al. verified that microscaled non-woven fabrics prepared by electrospinning from a chloroform solution of a degradable multiblock copolymer (PDLCL) consisting of crystallizable poly(x-pentadecalactone) hard segments (PPDL) and poly(e-caprolactone) switching segments (PCL) exhibited excellent shape memory performance, with potential application in biomedicine (Matsumoto et al., 2012). An electrospun SMP film with a stable micro/nano-fibrous structure was stable and reversible for at least three cycles of shape memory tests (Chen et al., 2012). In addition, some novel shape memory fibrous composites are being fabricated by adding some functional fillers or synthesized by some special groups can realize the reversible shape changing, two-way and multiple shape memory effect, multiple stimulus actuation, etc. (Gong et al., 2012; Zhang et al., 2013b). Nanotechnology enables SMPs to be utilized in a slice of particular application fields, such as biomedical devices, textiles, nanosensors, and nanoactuators. Most widely recognized potential of SMPs is for artificial blood vessels, drug release, scaffolds, and cell growth template (Mather and Luo, 2011; Mather et al., 2014). New advances in shape memory polymer nanofibers (SMPNs) exhibit a trend of broader applications of small size and flexible materials.

Shape memory fibers in micro- or nano-size are smart, soft, portable, and wearable, promoting the academic and industrial research. This review is mainly seeking to present and discuss the current systematic and comprehensive advancement of electrospun shape memory nanofibers, exhibit ongoing application studies, and provide future perspective. To stimulate further research on smart polymer nanofiber, the content of this work covers the fabrication methods-electrospinning, characterizations, morphologies and structures of electrospun nanofibers, dual- and multi-shape memory effect, thermal–mechanical shape memory cycle, reversible and multifunctional properties, as well as potential applications.

Electrospinning Technology

Nanofibers are a kind of one-dimensional material, due to these special features mentioned above, and have attracted great attention. Current fiber formation processes include melt spinning, dry spinning, wet spinning, gel spinning, emulsion spinning, shear spinning, and electrospinning. Electrospinning technology, with several unique advantages compared with other methods, has been recognized as the most high-efficiency way to electrospin polymers into the continuous and direct nanofibers (Huang et al., 2003). At present, more than 100 polymers have been successfully fabricated into fibers. The advantages of electrospinning, including low cost, easy processing, and the desirable features for resulting nanofibers such as small diameter, various structures, as well as multifunction usages, mean that electrospun nanofibers have potential applications in biomedical engineering (drug release, hemostatic gauze, etc.), tissue engineering (such as artificial vessels, scaffolds, etc.), textiles, nanosensors, filters (Andreas and Joachim, 2007).

The improvement of electrospinning equipment has led to the nanofibers with desirable structures and properties. The traditional needle electrospinning equipment that contains three parts, including a syringe needle, high voltage power supply, as well as a collector, has been used to prepare nanofibers quite extensively (Li and Xia, 2004). According to the experimental results, some theories and models have been developed to investigate the electrospinning process. The quintessential illustration of the electrospinning setup is shown in Figure 2 in Li and Xia, 2004. During the electrospinning process, the polymer solution drop overcomes the surface tension to form the jet when the electrical field is strong enough. Under the electrostatic interaction, the droplet is distorted into a conical shape named the Taylor cone. The jet flow is drawn into a straight line, and then curves to the spiral (Reneker and Chun, 1996). The electrostatic force induces the jet flow to be stretched and divided into ultrathin fibers. At last, the nanofibers are obtained on the collector after the solvent evaporating. On macro-scale, electrospinning, eventually, generates a fibrous network membrane. The applied voltage, distance between tip and collector, solution concentration, and flow rate, as well as viscosity and conductivity of polymers, have a critical influences on electrospinning results (Shin et al., 2001). The various structures and diameters of fibers can be designed by changing the electrospinning conditions. The electrospinning process is usually elaborately controlled and modified to meet the practical requirements.

Morphology and Structure of SMP Nanofibers

It is worth pointing out that shape memory fibers based on SMPs and their composites are fabricated by electrospinning technology, which exhibit different structures and morphologies based on the processing parameters. The affect factors include the changes in the solution concentration, applied electric field, spinneret structure, the distance between spinneret and collector, as well as the collectors (Deitzel et al., 2001; Li and Xia, 2004). In addition to the above mentioned factors, temperature and humidity are also important factors in determining the morphology and diameters of electrospun fibers. Research papers and patents demonstrate that nanofibers obtained through electrospinning with some special morphology and structures can be synthesized if the appropriate electrospinning processing conditions are employed. Hu et al. have discussed the dependence of the morphology of SMPU nanofibers on the applied voltage, solution concentration, and feeding rate. In particular, the solution concentration played an important role in transforming the polymer solution into ultrafine fibers. In general, increased concentration will lead to increased fiber diameters (Meng et al., 2007).

The non-woven fibers are the most simple and traditional structure, the porous morphology can be programmable during the shape changing process (Deitzel et al., 2001; Matsumoto et al., 2012). Figure 2 in Zhuo et al. (2011a) shows the morphology of the typical non-woven nanofibers examined by SEM, the diameters of which vary from 600 to 700 nm. With the use of a similar electrospinning setup, other researchers also get the same structure when the various polymers are electrospun into fibers (Chen et al., 2012; Gong et al., 2012; Zhang et al., 2013b). Non-woven fibrous structure shows an excellent pore network and can be used as filters or catalysts.

For practical applications, such as photonic or electrical products, as well as fiber reinforced composites, a number of polymers are indispensable. These are then made into the well-oriented and highly ordered fibers (Kameoka et al., 2003). Various approaches are used to fabricate the aligned fibers. As an example, some researchers modified the electrospinning collectors into rotating drum or a pair of split electrodes as collectors for forming the aligned fibers (Andreas and Joachim, 2007). The resulting fibers formed by different collectors show varied degrees of orientation. Some oriented fibers exhibiting excellent shape memory effect have been prepared. Ji et al. collected SMPU fibers with oriented morphology (**Figure 1**) (Ji et al., 2006). It was found that the smart fibers exhibited excellent recovery for application of mechanical force, to an extreme extent, providing a possibility to generate the actuators with high-performance and high recovery stress. It is necessary to control the morphology and structure that accelerate the development of smart nanofibers. Such architectures will be useful in improving shape recovery and shape fixity behaviors.

In addition, the specific secondary structure, such as core/shell, has been fabricated by designing the spinneret comprising two coaxial capillaries (Sun et al., 2003; Rana et al., 2013). SMPU fibers with core/shell nanostructure prepared by co-electrospinning

two different concentration solutions have been demonstrated (Zhuo et al., 2011b). As shown in Figure 3 in Zhuo et al. (2011b) transmission electron microscopy (TEM) images show the bead-on-string structure (I) and core-shell structure (II), obtained from the core solution of polycaprolactone-based SMPU and shell solution of pyridine-containing polyurethane. This kind of composite nanofibers with antibacterial activity was proposed to be used as smart materials and antibacterial materials. Historically, electrospinning was limited to the feature of polymer, which means that some kinds of polymers cannot be electrospun into fibers due to their poor spinnability. To this end, our group has recently fabricated the PCL/epoxy composite fibers with core/shell structure by co-electrospinning two polymers that phase separate (Zhang et al., 2014b, 2015). Co-electrospinning serving as a template provides an easy way to make the polymers without spinnability into fibrous structure, broadening the applications in micro- or nano-scale devices.

Besides the fibers with special structures, some multifunctional fibers filled with functional materials, such as carbon nanotubes, carbon blacks, carbon nanofibers, and magnetic particles (Fe_3O_4, Ni nanowires, etc.), have been developed (Leng et al., 2007; Sahoo et al., 2010; Zhang et al., 2014c). Introduction of nanoparticles into polymers for electrospinning to generate well-defined properties has been demonstrated. For example, incorporation of Fe_3O_4 into polycaprolactone (PCL) nanofibers has been reported (Gong et al., 2012). The results indicated that the composite fibers were able to be triggered by alternating magnetic field, which was an easy method for implementing remote control. Furthermore, the Alamar blue assay suggested that the electrospun composite fibers possessed good biocompatibility and could be applied in biomedical fields.

Shape Memory Performance

As seen in the increasing number of various research papers and patents, research on shape memory behaviors of SMPNs has captured worldwide attention (Mather and Luo, 2011; Sauter et al., 2012; Mather et al., 2014). Other properties of SMPNs, which also affect the shape memory effect, have been investigated. The thermal, mechanical, and chemical properties of SMPNs are characterized by differential scanning calorimetry (DSC), thermogravimetric analysis (TGA), infrared camera, dynamic mechanical analysis (DMA), mechanical tension test, Fourier transform infrared spectroscopy (FTIR), and so on (Leng et al., 2011). DMA and mechanical testing are also used to measure the shape memory performance of SMPNs. And the scanning electron microscopy (SEM) and TEM were carried out to observe the morphology and structure of SMPNs. Recently, in order to analyze the mechanisms of shape memory effect based on the shape changing of the internal microstructures of electrospun fibers, the microscopic deformation was conducted.

Leng's group has reported the microscopic shape changes. As shown in **Figure 2**, the microscopic shape memory behavior of electrospun Nafion nanofiber membrane could be programmable (Zhang et al., 2014a). At a microscopic level, the original non-woven fibrous structure was easily deformed to the temporary shape, and then recovered its original structure when the sample

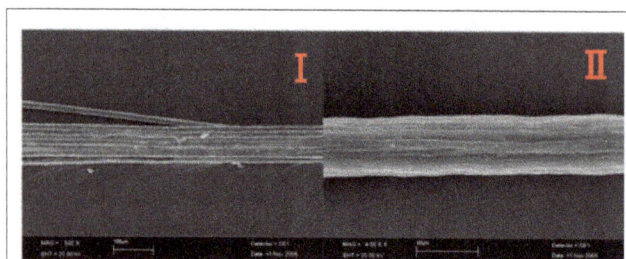

FIGURE 1 | SEM of SMF-3. View of multifilament (500x) (Ji et al., 2006 © IOP Publishing. Reproduced with permission. All rights reserved).

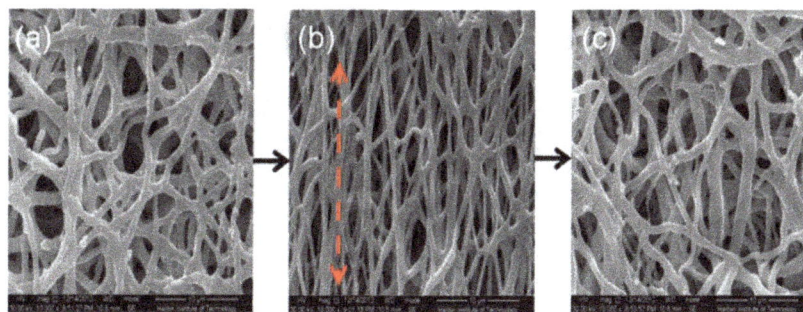

FIGURE 2 | SEM images of Nafion nanofibers, (a) the original fibers, (b) the stretched fibers, and (c) recovered fibers. (With kind permission from Springer Science+Business Media Zhang et al., 2014a).

was reheated. This result implied that the nanofibers' structure was stable after cycles of shape memory test. The same phenomenon was also studied (Zhuo et al., 2011a). Furthermore, Chen et al. have reported a kind of polyurethane nanofibers showing reversible shape memory effect (Zhang et al., 2011). The fibers showed different structures before and after deformation. After the fibrous film was stretched and fixed, the fibers tended to orientate along the strain direction. The orientated fibers were actuated to random form when the macro-film recovered the original shape in each cycle.

Other performances and variations for SMP fibers could be achieved by designing molecular ratios. Bao et al. have demonstrated that they have successfully fabricated a kind of biomedical fibrous scaffolds, made from biodegradable poly(D,L-lactide-co-trimethylene carbonate) fibers with shape memory effect (Bao et al., 2014). The monomer ratio of DLLA: TMC varied from 5:5 to 9:1, which resulted in the glass transition temperature T_g of the fibrous scaffolds to increase from 19.2 to 44.2°C. As shown in **Figure 3**, the shape memory effect of fibrous PLMC scaffolds in the form of spiral and cylindrical bar are successfully proved. The spiral shape was able to recover to the strip shape when dipped in the water bath at 39°C (**Figure 3A**). The macroscopic deformation in the 3-D shapes of "S," "M," and "P" completely recovered quickly to their original shapes with a speed of 12 s at 39°C (**Figure 3B**). The densely packed and partially fused fibrous structure influenced the fibers in contact with the hot surrounding because the response rate is dependent on the heat transfer rate. The rapid shape recovery process of these electrospun fibers were attributed to their high specific surface area. Therefore, SMPNs with fibrous structure were more suitable to be used as the immediate activation or constructing devices due to a long activation time. The PLMC fibers exhibited outstanding shape fixity and shape recovery properties with both shape recovery and shape fixity ratios above 90%.

The typical programing of a thermomechanical cycle test for bulk SMPs can be carried out to investigate the shape memory properties of SMPNs. The dual-shape memory behavior of thermo-induced SMPNs is presented in **Figure 4**, consisting of two steps (shape fixity and shape recovery) and showing the effect of temperature on the strain and the stress (Zhang et al., 2014a). The electrospun nanofiber membrane was heated at 130°C for 10 min,

then cooled down to 40°C. During this process, an external force was applied to the sample inducing the change in strain from 0 to 56%. This was the shape fixing step. When the membrane was exposed at 40°C for a certain time, it could recover its original shape after reheating above the transition temperature (130°C), which was defined as shape recovery process. The electrospun Nafion nanofibers showed perfect dual-shape memory property when deformed and recovered at 130°C. Equations 12 are used to calculate the shape fixity ratio (R_f) and shape recovery ratio (R_r) of SMPNs (Xie, 2010). The R_f and R_r were all above 90% when triggered by heat.

$$R_f\left(X \rightarrow Y\right) = \frac{\varepsilon_y - \varepsilon_x}{\varepsilon_{y,\text{load}} - \varepsilon_x} \times 100\% \qquad (1)$$

$$R_r\left(Y \rightarrow X\right) = \frac{\varepsilon_y - \varepsilon_{x,\text{rec}}}{\varepsilon_y - \varepsilon_x} \times 100\% \qquad (2)$$

where X and Y denoted different shapes, $\varepsilon_{y,\text{load}}$ was the maximum strain under load, ε_y and ε_x were the fixed strains after cooling and load removal, and $\varepsilon_{x,\text{rec}}$ was the strain after recovery.

In order to discuss the repeatability and durability, the shape memory cycles of SMPNs should be tested more than once. For instance, the electrospun non-woven fibers were fabricated from degradable polyesterurethanes named PDLCL containing poly(x-pentadecalactone) hard segments (PPDL) and poly(e-caprolactone) switching segments (PCL), demonstrating excellent shape memory properties. The strain–stress curves of the thermomechanical cycles are shown in Figure 4 in Matsumoto et al. (2012), which describes the shape memory cycles over three times. As it can be seen from this figure, compared to the first cycle, the shape recovery ratio of the second cycle deduces to 89%. However, the shape fixity ratios in all cycles stay constant at 83%. These results suggested that the electrospun PDLCL fibers enjoyed favorable reproducible shape-memory properties. Matsumoto et al., 2012 also describes the four subsequent cycles of shape memory test, the ε_m increases from 25 to 50%. The shape recovery ratio was lower than the aforementioned experiments when ε_m was 25% after the second cycle. The stress showed linear increase until the strain reached the last cycle value, and then stress relaxation was generated, which gave rise to the recovery rate decrease.

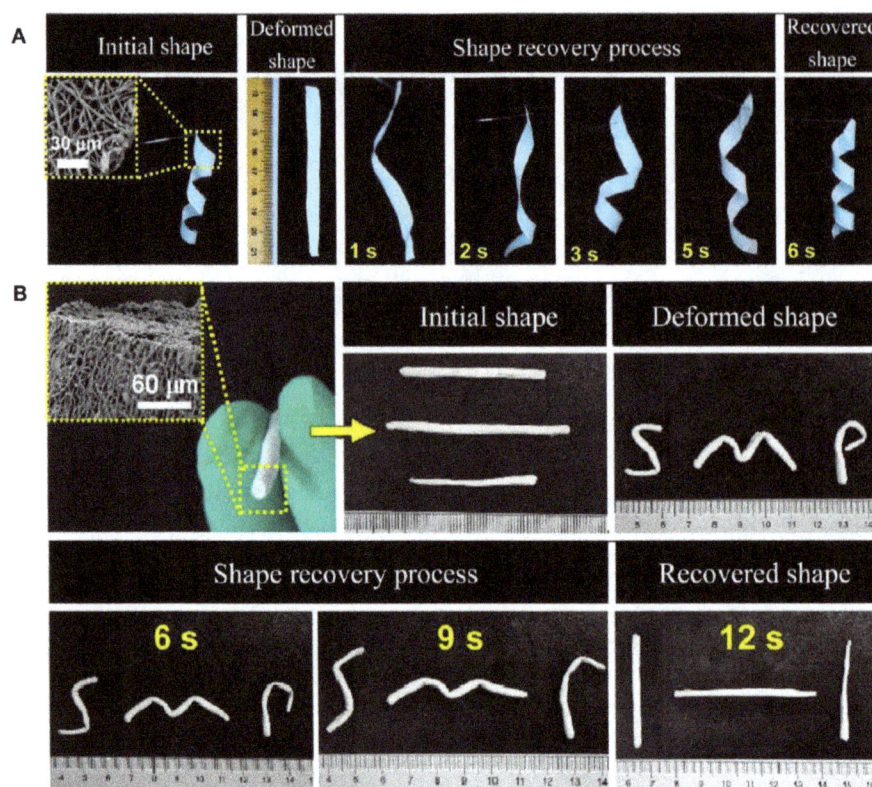

FIGURE 3 | Macroscopic demonstration of the shape memory effect for electrospun fibrous PLMC. (A) Shape recovery process of a spiral in membranous form (2-D); (B) shape recovery process of the cylindrical bars made of electrospun fibers (inset image), in transition from the temporary shape ("SMP") to the permanent straight bars. Reprinted with permission from Bao et al. (2014). Copyright (2014) American Chemical Society.

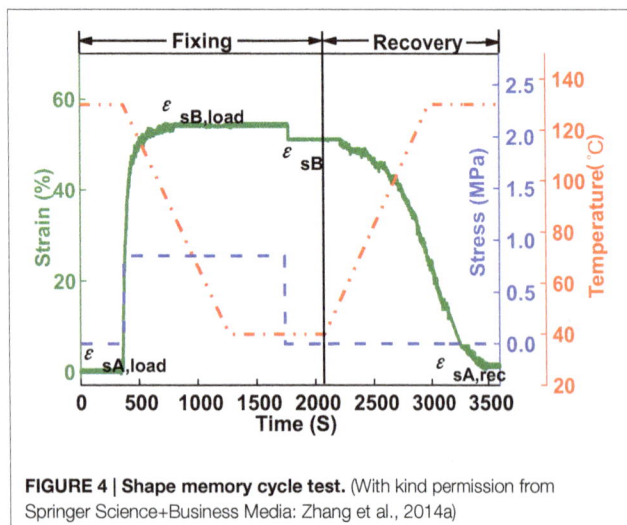

FIGURE 4 | Shape memory cycle test. (With kind permission from Springer Science+Business Media: Zhang et al., 2014a)

Expanding the multifunction of SMPNs, researchers design the special methods to stimulate the shape recovery processes. Zhou et al. found that the electrospun spider silk protein had a crystallization memory property when it was stimulated by a high pressure (Peng et al., 2009). Furthermore, the moisture-responsive Kraft lignin-based materials prepared

by electrospinning blends of Kraft lignin fractions with different physical properties are reported, as shown in **Figure 5**. Moreover, the different material morphologies, including submicrometer fibers, bonded non-wovens, porous films, and smooth films were activated by the differences in thermal mobility between lignin fractions influencing the degree of inter fiber fusion occurring during oxidative thermostabilization of electrospun non-woven fabrics (Dallmeyer et al., 2013). The size of nanopores/micropores in SMPU membranes prepared by electrospinning with two-way shape memory effect was investigated successfully. According to the temperature change, the pore size can be changed from 150 to 440 nm, demonstrating that the SMPN membrane can serve as a smart membrane to selectively separate substances by controlling temperature only (Ahn et al., 2011).

We note that simple shape memory behaviors are limited to the practical applications. It has been reported that the electrospun Nafion fibrous membranes can memorize more than two shapes, which is named multiple shape memory effect (Zhang et al., 2013b). Nafion, consisting of polytetrafluorethylene backbone and perfluoroether sulfonic acid side chains, possesses promising properties that give rise to numerous applications of the materials in a variety of fields. Nafion solution with concentration 5% can be electrospun into nanofibers by addition of a very small amount of poly(ethyleneoxide) (PEO). The resulting fibers, with

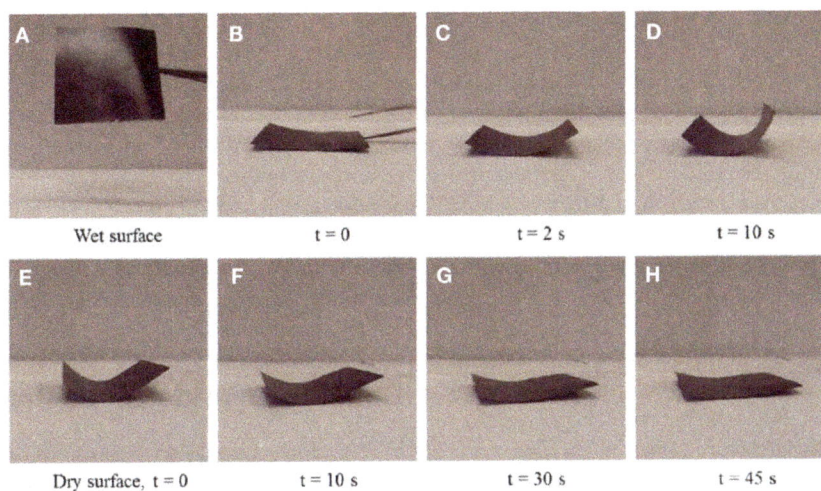

FIGURE 5 | Thermostabilized (5°C/min) 50/50 F4/F1−3 film placed on moist paper (A−D), then moved to dry paper (E−H). Reprinted with permission from Dallmeyer et al. (2013). Copyright (2013) American Chemical Society.

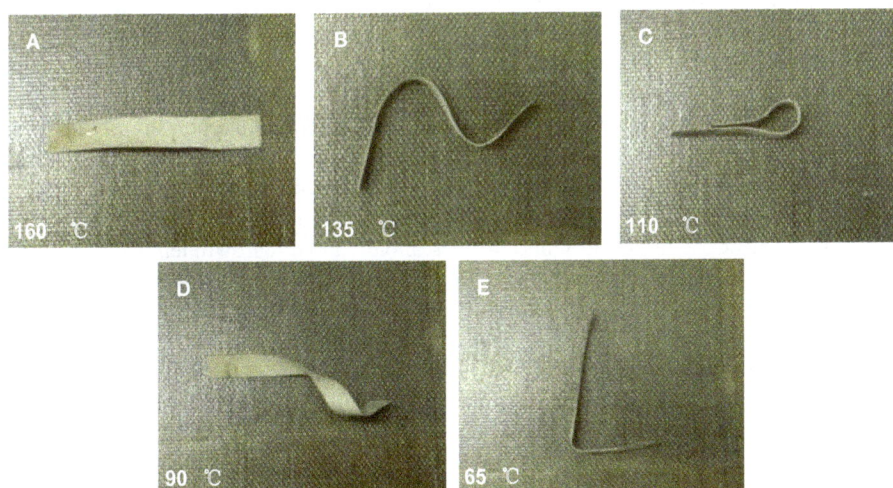

FIGURE 6 | Photos of electrospun Nafion nanofiber membranes with quintuple-shape memory effect. (Zhang et al., 2013b, © IOP Publishing. Reproduced with permission. All rights reserved).

a broad transition temperature range from 60 to 170°C, show uniform and continuous non-woven structure, which is designed to remember five shapes at different transition temperatures, as shown in **Figure 6**. The multiple temporary shapes of electrospun Nafion fibrous membranes can be induced by heat in a single triple-, quadruple-, quintuple-shape memory cycle with a high speed, respectively. After the quintuple-shape memory cycle, the samples still keep the stable nanofiber morphology, providing potential applications in smart textiles, sensors, and actuators, as well as artificial skins.

Further efforts to integrate novel functions of SMPNs into applications will likely prove advantages and an important step forward advantage. For instance, multi-stimulus responsive shape memory fibers may remotely control the cell grow in different shape changing template surfaces. Two-way shape memory effect enables fibrous membranes or textiles to change porous structure under tunable conditions, for example, in intelligent clothes. Moreover, improvement to the eletrospinning processing will permit practical applications of nanofibers in the future.

Applications

SMPs can be fixed into pre-designed shapes and recover their permanent shapes, which can be triggered by heat, moisture, solutions, light, electricity, magnetism, or multi-stimulus. The unique feature provides SMPs' unlimited potential uses. In recent years, SMPNs as a novel kind of smart fibrous materials also have wide ranging applications (Fu et al., 2009; Ko et al., 2013; Chung et al., 2014), such as smart fabrics and biomedical materials (e.g., biomimetic scaffolds, cell growth template, self-healing application, medical devices, antibacterial nanomaterials) (Zhuo et al., 2011b; Schneider et al., 2012; Tseng et al., 2012; Zhang and Li, 2013). In particular, shape memory scaffolds with high porosity and specific surface area that have also found potential for use in tissue engineering in the future. The traditionally used structures are static physical structures and poorly suited to mimicking the complex dynamic behavior of *in vivo* microenvironments. Shape memory nanofibers provide an attractive class of supports for biotechnology and tissue engineering thanks to their large surface areas and small diameters.

An interest in biodegradable and biocompatible SMPs and progress in biotechnology will result in the commercial production of biomedical tissues. For example, the shape memory poly(D,L-lactide-co-trimethylene carbonate) fibers fabricated by electrospinning have potential application as bone scaffolds (Bao et al., 2014). Authors analyzed the proliferation and morphology of osteoblasts cultured onto the electrospun nanofibrous scaffolds as revealed in **Figure 7**. Nanofibrous scaffolds of PLMC with ratio (8:2) (Tg = 36.7°C) and PLMC with ratio (9:1) (Tg = 44.2°C) were used to investigate their properties. The cytocompatibility was calculated by MTT assay, the results of which are presented in **Figure 7A**. The cellular spreading, adhesion, and growth abilities showed similar tendency (**Figure 7B**). **Figure 7C** exhibits the SEM images of morphology of osteoblasts attached and proliferated on the nanofibrous PLMC scaffolds. It can be seen from **Figure 7D**, the cells were stained after 4 and 7 days of culture, in order to monitor the cell morphology of proliferating osteoblasts on the nanofibers. All these results indicated that the electrospun fibrous scaffolds with shape memory effect could be applied in repairing various bone defects, including healing bone crew holes and guided bone regeneration.

Tseng et al. have demonstrated that the electrospun SMPU fiber scaffold is capable of being programed to change macroscopic shape and microscopic architecture during cell culture. This smart fibrous scaffold was adopted to test the hypothesis that a shape-memory-actuated change in scaffold fiber alignment can be used to control the behavior of attached and viable cells, as shown by Tseng et al. (2013). The shape memory triggered changes in scaffold fiber alignment were demonstrated by the attached and viable cells, controlling the cell morphology. Scaffolds, with shape memory effect, that play an important role in the biological fields, are potential to provide *in vitro* platforms for the investigation of cells and tissues.

Furthermore, some improved and useful SMPs and SMPNs have also been developed. Dye-containing polymers that have potential applications in medical fields, such as medical devices, or drug delivery have attracted a host of researchers. To this end, Torbati et al. have systematically investigated a light-emitting shape memory poly(vinyl acetate) (PVAc) web fabricated by electrospining with indocyanine green (Torbati et al., 2014). As shown in Figure 6 in Torbati et al. (2014) the electrospun webs are cut into different shapes, which are able to shrink when immersed in 25°C water or heated at 50°C, respectively. The feature of water-induced shape changes provides

the possible applications in medical devices, including feeding tubes and catheters.

As described in previous sections, SMPNs are remarkably useful and flexible materials for a number of applications. In addition to the fiber reinforced composites, biomedical usages, smart clothes, and smart sensors, SMPNs will expand the usages of smart catalysts, sacrificial templates, and smart electronic devices, as well as smart optical devices in the future. It is expected that research on SMPNs will be deepen and broaden to include interdisciplinary research and technological convergence, generating more smart products to promote the development in smart materials and structures.

Conclusion and Perspective

In this paper, the various dimensions of SMPNs have been systematically discussed, including the fabrication methods, characterization, structures, properties, and potential applications in broad fields in the future. The electrospinning technology as an effective approach to fabricate nanofibers has been used to obtain the different kinds of SMPNs which show excellent thermal and mechanical properties, as well as shape memory effect after several cycles. The diversity of fibers' structures is mainly based on the electrospinning conditions, including needle's structure, collector, applied voltages, and concentration of polymeric solutions. SMPNs are capable of experiencing a large recoverable and revisable deformation upon some certain stimuli, including heat, magnetic field, electricity, moisture, light, etc. SMPNs and their composites are attractive because of their potential applications, especially in biomedical engineering, including drug release, scaffolds, and cell growth templates in the future.

The merits of SMPNs and their composites include fast response, good flexibility, low weight, easy processing, multiple function, and diverse structures. However, compared with the block SMPs, SMPNs show low strength. Moreover, some of SMPs are not able to be electrospun into fibers. Recently, functional particles, such as carbon nanotubes and carbon blacks, Fe_3O_4, have been loaded into the fibers to enhance the mechanical and magnetic properties. In addition, co-electrospinning method, a good choice to form the core/shell fibers, makes some polymers without spinnability into fibers. Furthermore, SMPNs are better candidates for biomedical applications. One of the critical changes is to realize the effective actuation. One possible way to solve this problem is to add some magnetic particles in the fibers to achieve the remote control by alternating magnetic field.

Therefore, based on the previous works, the future direction of SMPN will focus on the following research:

(1) New kinds of SMPN. Prospective SMPN may cover two-way SMPN, thermosetting SMPN, and SMPN with diverse structure, etc.

(2) Multiple stimulus of SMPN. At present, most of SMPN can be triggered by electricity, heat, or moisture, and so on. In order to meet the practical applications, remote stimulus is necessary. And some special fields also require the multi-stimulus which would enable the materials to tune the stimulus under different conditions.

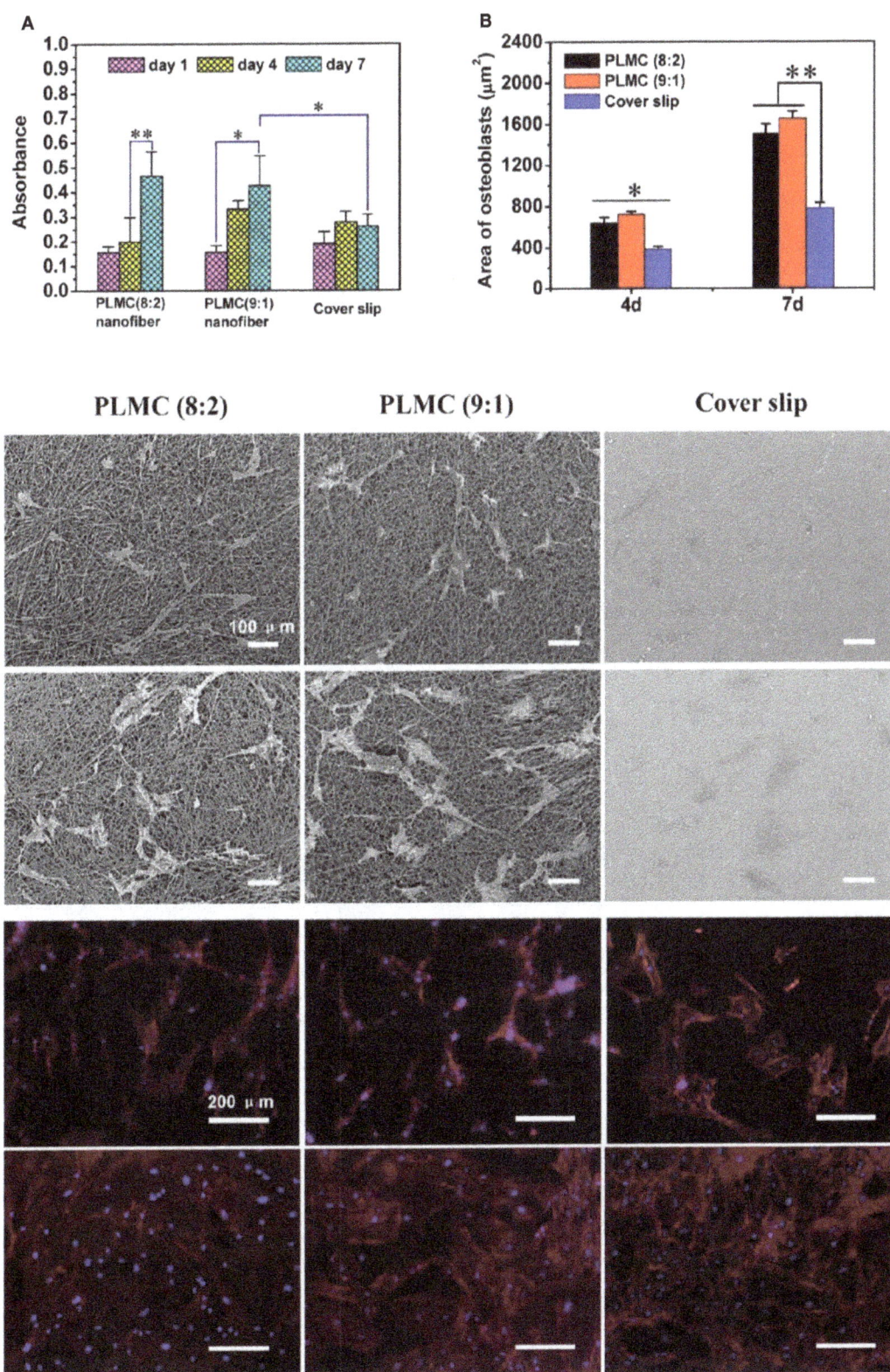

FIGURE 7 | **Proliferation and morphology observation of osteoblasts cultured onto the electrospun nanofibrous scaffolds of PLMC (8:2) and PLMC (9:1), and the coverslip control for up to 7 days.** (A) Histogram of osteoblast proliferation by MTT assay; (B) histogram of area of osteoblasts; (C) SEM micrographs; (D) fluorescence staining of osteoblasts. F-actin filaments (red) and nucleus (blue) were stained by Phalloidin and DAPI, respectively. Reprinted with permission from Bao et al. (2014). Copyright (2014) American Chemical Society.

(3) Multifunctional SMPN and broad applications. Except for shape memory effect, some functional materials can be added into the SMPN, which will possibly induce novel functions and applications in the future.

References

Ahn, J. S., Yu, W. R., Youk, J. H., and Ryu, H. Y. (2011). In situ temperature tunable pores of shape memory polyurethane membranes. *Smart Mater. Struct.* 20, 105024. doi:10.1088/0964-1726/20/10/105024

Andreas, G., and Joachim, H. W. (2007). Electrospinning: a fascinating method for the preparation of ultrathin fibers. *Angew. Chem. Int. Ed. Engl.* 46, 5670–5703. doi:10.1002/anie.200604646

Bao, M., Lou, X., Zhou, Q., Dong, W., Yuan, H., and Zhang, Y. (2014). Electrospun biomimetic fibrous scaffold from shape memory polymer of PDLLA-co-TMC for bone tissue engineering. *ACS Appl. Mater. Inter.* 6, 2611–2621. doi:10.1021/am405101k

Behl, M., and Lendlein, A. (2007). Shape-memory polymers. *Mater. Today* 10, 20–28. doi:10.1016/S1369-7021(07)70047-0

Cha, D. I., Kim, H. Y., Lee, K. H., Jung, Y. C., Cho, J. W., and Chun, B. C. (2005). Electrospun nonwovens of shape-memory polyurethane block copolymers. *J. Appl. Polym. Sci.* 96, 460–465. doi:10.1002/app.21467

Chen, H. L., Cao, X. Y., Zhang, J. N., Zhang, J. J., Ma, Y. M., Shi, G., et al. (2012). Electrospun shape memory film with reversible fibrous structure. *J. Mater. Chem.* 22, 22387–22391. doi:10.1039/C2JM33970F

Chung, S. E., Park, C. H., Yu, W. R., and Kang, T. J. (2011). Thermoresponsive shape memory characteristics of polyurethane electrospun web. *J. Appl. Polym. Sci.* 120, 492–500. doi:10.1002/app.33167

Chung, Y. C., Yang, K., Choi, J. W., and Chun, B. C. (2014). Characterization and application of polyurethane copolymers grafted with photoluminescent dyes. *Color. Technol.* 130, 305–313. doi:10.1007/s12221-014-0008-3

Dallmeyer, I., Chowdhury, S., and Kadla, J. F. (2013). Preparation and characterization of Kraft lignin-based moisture-responsive films with reversible shape-change capability. *Biomacromolecules* 14, 2354–2363. doi:10.1021/bm400465p

Deitzel, J. M., Kleinmeyer, J., Harris, D., and Beck Tan, N. C. (2001). The effect of processing variables on the morphology of electrospun nanofibers and textiles. *Polymer* 42, 261–272. doi:10.1016/S0032-3861(00)00250-0

Fu, G. D., Xu, L. Q., Yao, F., Li, G. L., and Kang, E. T. (2009). Smart nanofibers with a photoresponsive surface for controlled release. *ACS Appl. Mater. Inter.* 1, 2424–2427. doi:10.1021/am900526u

Gall, K., Dunn, M. L., Liu, Y. P., Finch, D., Lake, M., and Munshib, N. A. (2002). Shape memory polymer nanocomposites. *Acta Mater.* 50, 5115–5126. doi:10.1016/S1359-6454(02)00368-3

Gong, T., Li, W., Chen, H., Wang, L., Shao, S., and Zhou, S. (2012). Remotely actuated shape memory effect of electrospun composite nanofibers. *Acta Biomater.* 8, 1248–1259. doi:10.1016/j.actbio.2011.12.006

Hu, J. L., Zhu, Y., Huang, H. H., and Lu, J. (2012). Recent advances in shape-memory polymers: structure, mechanism, functionality, modeling and applications. *Prog. Polym. Sci.* 37, 1720–1763. doi:10.1016/j.progpolymsci.2012.06.001

Huang, W. M., Yang, B., An, L., Li, C., and Chan, Y. S. (2005). Water-driven programmable polyurethane shape memory polymer: demonstration and mechanism. *Appl. Phys. Lett.* 86, 114105. doi:10.1063/1.1880448

Huang, Z. M., Zhang, Y. Z., Kotaki, M., and Ramakrishna, S. (2003). A review on polymer nanofibers by electrospinning and their applications in nanocomposites. *Compos. Sci. Technol.* 63, 2223–2253. doi:10.1016/S0266-3538(03)00178-7

Ji, F. L., Zhu, Y., Hu, J. L., Liu, Y., Yeung, L. Y., and Ye, G. D. (2006). Smart polymer fibers with shape memory effect. *Smart Mater. Struct.* 15, 1547–1554. doi:10.1088/0964-1726/15/6/006

Kameoka, J., Orth, R., Yang, Y., Czaplewski, D., Mathers, R., Coates, G. W., et al. (2003). A scanning tip electrospinning source for deposition of oriented nanofibres. *Nanotechnology* 14, 1124–1129. doi:10.1088/0957-4484/14/10/310

Ko, Y. I., Kim, B. S., Bae, J. S., Kim, Y. A., and Kim, I. S. (2013). Silicone-coated elastomeric polylactide nanofiber filaments: mechanical properties and shape memory behavior. *RSC Adv.* 3, 20091–20098. doi:10.1039/c3ra42361a

Kotek, R. (2008). Recent advances in polymer fibers. *Polym. Rev.* 48, 221–229. doi:10.1080/15583720802020038

Kumar, U. N., Kratz, K., Behl, M., and Lendlein, A. (2012). Shape memory properties of magnetically active triple-shape nanocomposites based on a grafted polymer network with two crystallizable switching segments. *Express Polym. Lett.* 6, 26–40. doi:10.3144/expresspolymlett.2012.4

Lan, X., Liu, Y. J., Lv, H. B., Wang, X. H., Leng, J. S., and Du, S. (2009). Fiber reinforced shape-memory polymer composite and its application in a deployable hinge. *Smart Mater. Struct.* 18, 024002. doi:10.1088/0964-1726/18/2/024002

Lendlein, A., Jiang, H., Junger, O., and Langer, R. (2005). Light-induced shape-memory polymers. *Nature* 434, 879–882. doi:10.1038/nature03496

Lendlein, A., and Langer, R. (2002). Biodegradable, elastic shape-memory polymers for potential biomedical applications. *Science* 296, 1673–1676. doi:10.1126/science.1066102

Leng, J. S., Lan, X., Liu, Y. J., and Du, S. Y. (2011). Shape-memory polymers and their composites: stimulus methods and applications. *Prog. Mater. Sci.* 56, 1077–1135. doi:10.1016/j.pmatsci.2011.03.001

Leng, J. S., Lv, H. B., Liu, Y. J., and Du, S. Y. (2007). Electroactivate shape-memory polymer filled with nanocarbon particles and short carbon fibers. *Appl. Phys. Lett.* 91, 144105. doi:10.1063/1.2790497

Leng, J. S., Lv, H. B., Liu, Y. J., Huang, W. M., and Du, S. Y. (2009). Shape-memory polymers – a class of novel smart materials. *MRS Bull.* 34, 848–855. doi:10.1557/mrs2009.235

Li, D., and Xia, Y. N. (2004). Electrospinning of nanofibers: reinventing the wheel? *Adv. Mater.* 16, 1151–1170. doi:10.1002/adma.200400719

Liu, C., Qin, H., and Mather, P. T. (2007). Review of progress in shape-memory polymers. *J. Mater. Chem.* 17, 1543–1558. doi:10.1039/B615954K

Liu, Y. J., Du, H. Y., Liu, L. W., and Leng, J. S. (2014). Shape memory polymers and their composites in aerospace applications: a review. *Smart Mater. Struct.* 23, 023001. doi:10.1088/0964-1726/23/2/023001

Liu, Y. J., Lv, H. B., Lan, X., Leng, J. S., and Du, S. Y. (2009). Review of electro-active shape-memory polymer composite. *Compos. Sci. Technol.* 69, 2064–2068. doi:10.1016/j.compscitech.2008.08.016

Mather, P. T., and Luo, X. F. (2011). *Shape Memory Elastomer, Useful e.g. to Form Adaptive Seals Such as Heat-Shrinkable Seals that Prevent Water Leaking and Produce Configurable Surgical Tools, Comprises Non-Woven Mat, and Resin Matrix Infiltrated Throughout the Mat.* US2011021097-A1.

Mather, P. T., Torbati, A., and Mather, R. (2014). *Near Infrared Fluorescent Marker for Use in e.g. Medical Device During Surgical Procedure in e.g. Mouse, has Electrospun Fibrous Web Formed from Shape Memory Polymer and Near Infrared Dye with Excitation and Emission Wavelength.* US2014303490-A1.

Matsumoto, H., Ishiguro, T., Konosu, Y., Minagawa, M., Tanioka, A., and Richau, K. (2012). Shape-memory properties of electrospun non-woven fabrics prepared from degradable polyesterurethanes containing poly(x-pentadecalactone) hard segments. *Eur. Polym. J.* 48, 1866–1874. doi:10.1016/j.eurpolymj.2012.07.008

Meng, Q. H., Hu, J. L., Zhu, Y., Lu, J., and Liu, Y. (2007). Morphology, phase separation, thermal and mechanical property differences of shape memory fibres prepared by different spinning methods. *Smart Mater. Struct.* 16, 1192–1197. doi:10.1088/0964-1726/16/4/030

Nagahama, K., Ueda, Y., Ouchi, T., and Ohya, Y. (2009). Biodegradable shape-memory polymers exhibiting sharp thermal transitions and controlled drug release. *Biomacromolecules* 10, 1789–1794. doi:10.1021/bm9002078

Peng, H., Zhou, S., Jiang, J., Guo, T., Zheng, X., and Yu, X. (2009). Pressure-induced crystal memory effect of spider silk proteins. *J. Phys. Chem. B* 113, 4636–4641. doi:10.1021/jp811461b

Rana, S., Kim, S. D., and Cho, J. W. (2013). Conducting core-sheath nanofibers for electroactive shape-memory applications. *Polym. Adv. Technol.* 24, 609–614. doi:10.1002/pat.3118

Reneker, D. H., and Chun, I. (1996). Nanometre diameter fibres of polymer, produced by electrospinning. *Nanotechnology* 7, 216–223. doi:10.1088/0957-4484/7/3/009

Acknowledgments

This work is supported by the National Natural Science Foundation of China (Grant No. 11225211 and No.11272106).

Sahoo, N. G., Rana, S., Cho, J. W., Li, L., and Chan, S. H. (2010). Polymer nanocomposites based on functionalized carbon nanotubes. *Prog. Polym. Sci.* 35, 837–867. doi:10.1016/j.progpolymsci.2010.03.002

Sauter, T., Kratz, K., and Lendlein, A. (2012). Shape-memory properties of electrospun non-wovens prepared from amorphous polyetherurethanes under stress-free and constant strain conditions. *MRS Proc.*, 1403(1), 105–112. doi:10.1557/opl.2012.251

Schneider, T., Kohl, B., Sauter, T., Kratz, K., Lendlein, A., Ertel, W., et al. (2012). Influence of fiber orientation in electrospun polymer scaffolds on viability, adhesion and differentiation of articular chondrocytes. *Clin. Hemorheol. Microcirc.* 52, 325–336. doi:10.3233/CH-2012-1608

Shin, Y. M., Brenner, M. P., and Rutledge, G. C. (2001). Electrospinning: a whipping fluid jet generates submicron polymer fibers. *Appl. Phys. Lett.* 78, 1149–1151. doi:10.1063/1.1345798

Sun, Z. C., Zussman, E., Yarin, A. L., Wendorff, J. H., and Greiner, A. (2003). Compound core-shall polymer nanofibers by co-electrospinning. *Adv. Mater.* 15, 1929–1932. doi:10.1002/adma.200305136

Torbati, A. H., Mather, R. T., Reeder, J. E., and Mather, P. T. (2014). Fabrication of a light-emitting shape memory polymeric web containing indocyanine green. *J. Biomed. Mater. Res. B Appl. Biomater.* 102, 1236–1243. doi:10.1002/jbm.b.33107

Tseng, L. F., Mathe, P. T., and Henderson, J. H. (2012). "A programmable shape-changing scaffold for regenerative medicine," in *Bioengineering Conference (NEBEC), 2012 38th Annual Northeast* (Philadelphia, PA: IEEE), 227–228. doi:10.1109/NEBC.2012.6207046

Tseng, L. F., Mather, P. T., and Henderson, J. H. (2013). Shape-memory-actuated change in scaffold fiber alignment directs stem cell morphology. *Acta Biomater.* 9, 8790–8801. doi:10.1016/j.actbio.2013.06.043

Xie, T. (2010). Tunable polymer multi-shape memory effect. *Nature* 464, 267–270. doi:10.1038/nature08863

Zhang, F. H., Zhang, Z. C., Liu, Y. J., Cheng, W. L., Huang, Y. D., and Lenga, J. (2015). Thermosetting epoxy reinforced shape memory composite microfiber membranes: fabrication, structure and properties. *Compos. Part A Appl. Sci. Manuf.* 76, 54–61. doi:10.1016/j.compositesa.2015.05.004

Zhang, F. H., Zhang, Z. C., Liu, Y. J., and Leng, J. S. (2013a). "Fabrication of shape memory nanofibers by electrospinning method," in *SPIE Proc* (San Diego), 8687.

Zhang, F. H., Zhang, Z. C., Liu, Y. J., Lu, H. B., and Leng, J. S. (2013b). The quintuple-shape memory effect in electrospun nanofiber membranes. *Smart Mater. Struct.* 22, 085020. doi:10.1088/0964-1726/22/8/085020

Zhang, F. H., Zhang, Z. C., Liu, Y. J., and Leng, J. S. (2014a). Shape memory properties of electrospun Nafion nanofibers. *Fibers Polym.* 15, 534–539. doi:10.1007/s12221-014-0534-z

Zhang, F. H., Leng, J. S., Liu, Y. J., and Zhang, Z. C. (2014b). *Thermosetting-Thermoplastic Core-Shell Structured-Shape Memory Composite Fiber is Obtained by Providing Thermosetting Epoxy Resin as Core Layer and Thermoplastic Polycaprolactone as Shell Layer, and Recovering to Original Shape.* CN104032409-A.

Zhang, F. H., Zhang, Z. C., Liu, Y. J., and Leng, J. S. (2014c). Electrospun nanofiber membranes for electrically activated shape memory nanocomposites. *Smart Mater. Struct.* 23, 065020. doi:10.1088/0964-1726/23/6/065020

Zhang, J. N., Ma, Y. M., Zhang, J. J., Dan, X., Yang, Q. L., Guan, J.-G., et al. (2011). Microfiber SMPU film affords quicker shape recovery than the bulk one. *Mater. Lett.* 65, 3639–3642. doi:10.1016/j.matlet.2011.06.083

Zhang, P. F., and Li, G. Q. (2013). Structural relaxation behavior of strain hardened shape memory polymer fibers for self-healing applications. *J. Polym. Sci. Pol. Phys.* 51, 966–977. doi:10.1002/polb.23295

Zhuo, H. T., Hu, J. L., and Chen, S. J. (2008a). Electrospun polyurethane nanofibres having shape memory effect. *Mater. Lett.* 62, 2074–2076. doi:10.1016/j.matlet.2007.11.018

Zhuo, H. T., Hu, J. L., Chen, S. J., and Yeung, L. Y. (2008b). Preparation of polyurethane nanofibers by electrospinning. *J. Appl. Polym. Sci.* 109, 406–411. doi:10.1002/app.28067

Zhuo, H. T., Hu, J. L., and Chen, S. J. (2011a). Study of the thermal properties of shape memory polyurethane nanofibrous nonwoven. *J. Mater. Sci.* 46, 3464–3469. doi:10.1007/s10853-011-5251-z

Zhuo, H. T., Hu, J. L., and Chen, S. J. (2011b). Coaxial electrospun polyurethane core-shell nanofibers for shape memory and antibacterial nanomaterials. *Express Polym. Lett.* 5, 182–187. doi:10.3144/expresspolymlett.2011.16

Conflict of Interest Statement: The authors declare that the research was conducted in the absence of any commercial or financial relationships that could be construed as a potential conflict of interest.

10

Recent progress in the growth and applications of graphene as a smart material: a review

Brahim Aïssa[1,2]*, Nasir K. Memon[1], Adnan Ali[1] and Marwan K. Khraisheh[1]

[1] *Qatar Environment and Energy Research Institute (QEERI), Qatar Foundation, Doha, Qatar,* [2] *Department of Smart Materials and Sensors for Space Missions, MPB Technologies Inc., Montreal, QC, Canada*

Edited by:
Maenghyo Cho,
Seoul National University,
South Korea

Reviewed by:
Jianbo Yin,
Northwestern Polytechnical
University, China
Joo-Hyung Kim,
Inha University, South Korea

*Correspondence:
Brahim Aïssa,
Department of Smart Materials and
Sensors for Space Missions, MPB
Technologies Inc., 151 Hymus
Boulevard, Pointe-Claire,
Montreal, QC H9R1E9, Canada
baissa@qf.org.qa

Innovative breakthroughs in fundamental research and industrial applications of graphene material have made its mass and low-cost production as a necessary step toward its real world applications. This one-atom thick crystal of carbon, gathers a set of unique physico-chemical properties, ranging from its extreme mechanical behavior to its exceptional electrical and thermal conductivities, which are making graphene as a serious alternative to replace many conventional materials for various applications. In this review paper, we highlight the most important experimental results on the synthesis of graphene material, its emerging properties with reference to its smart applications. We discuss the possibility to successfully integrating graphene directly into device, enabling thereby the realization of a wide range of applications, including actuation, photovoltaic, thermoelectricity, shape memory, self-healing, electrorheology, and space missions. The future outlook of graphene is also considered and discussed.

Keywords: gas-phase growth, graphene material, smart applications

Introduction

Graphene material is considered as the first lab-made 2D atomic crystal. Because of their unique physical and chemical properties – such as mechanical stiffness, strength and elasticity, and extremely high electrical and thermal conductivity (Geim and Novoselov, 2007a; Geim, 2009) – graphene is described to be a serious alternative to replace many conventional materials in various applications, and could enable many disruptive innovation and potentially existing markets. For example, the combination of optical transparency, electrical and thermal conductivities, and mechanical elasticity will find application either in flexible electronics and/or transparent coatings, and the list of such combinations is continuously growing (**Figure 1**). Basically, graphene is a single 2D layer of carbon atoms, with a typical thickness of 0.34 nm. It is sp^2 hybridized, where carbon atoms are covalently bonded to three other atoms in a hexagonal lattice structure (Geim, 2009; Layek and Nandi, 2013). Recently, graphene has been extensively investigated, both in terms of fundamental research and R&D applications. Graphene was isolated for the first time by Novoselov et al. (2004) what was worth to them the 2010 Nobel Prize in Physics for their groundbreaking work. Their unprecedented structural and physico-chemical properties (especially its mechanical and electrical behaviors) in addition to its carrier mobility – the highest know to date, at room temperature – makes the research on graphene one of the most important topics in all materials science fields (Basu and Bhattacharyya, 2012). On the other hand, graphene' structure serves as the basic shape of almost all other carbonaceous materials, including fullerene (Muge and Chabal, 2011), single and multi-walled carbon nanotubes (Hassan, 2012), and even graphite, which is simply a multiple layers graphene

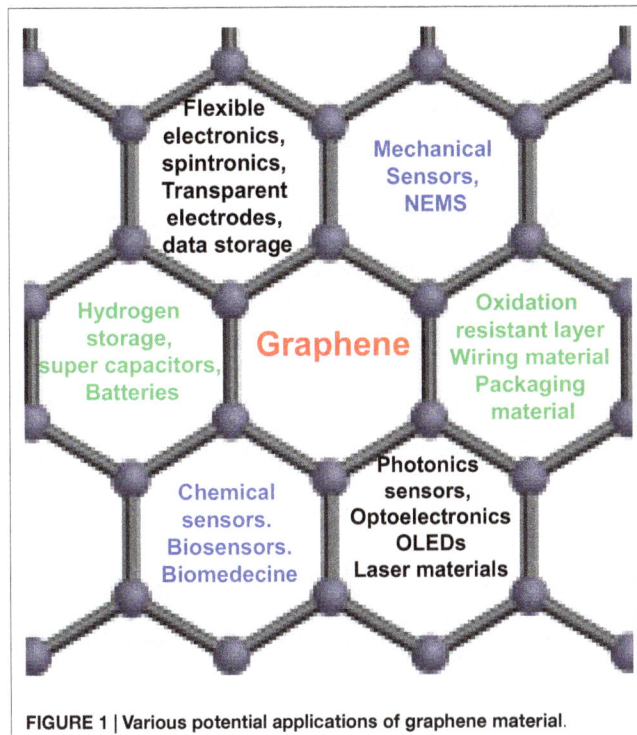

FIGURE 1 | Various potential applications of graphene material.

(Basu and Bhattacharyya, 2012). Literature survey shows that graphene material has numbers of potential applications, including nanoelectronic like-devices, gas sensors, hydrogen storage, and polymer-based nanocomposites (Boukhvalov et al., 2008; Ponnamma et al., 2010; Schwierz, 2010; Casolo et al., 2011), and could serve as an ideal prototype to investigate the properties of many other 2D nanosystems, such as 2D silicon and silicon carbide (2D-SiC), zinc oxide, boron nitride, and germanium (Elias et al., 2009; Bekaroglu et al., 2010; Houssa et al., 2010; Tang and Cao, 2010; Voon et al., 2010; Zhang et al., 2010a).

Experimentally measured properties of graphene have not only exceeded those obtained in any other material but also reached very often its theoretically predicted limits. A typical example is its room-temperature carrier mobility, of 2.5×10^5 cm²/Vs, which is found to be very close to the theoretical limit of 2×10^5 cm²/Vs (Mayorov et al., 2011). Many other representative examples could be found in the relevant literature, just to cite a few: a Young's modulus of 1 TPa very close to that predicted by theory (Liu et al., 2007; Lee et al., 2008; Morozov et al., 2008); a thermal conductivity of 3000 W mK⁻¹ (Balandin, 2011); an optical absorption of 2.3% in the infrared (Nair et al., 2008); its property to be completely impermeable to gases (Bunch et al., 2008); its ability to carry one million time higher densities of electrical current than copper (Moser et al., 2007), and its potential to be chemically functionalized (Elias et al., 2009; Loh et al., 2010; Nair et al., 2010). It is worth noting that the majority of these properties have been experimentally measured for a high-quality graphene samples, deposited on specific substrates, such as hexagonal boron nitride (Dean et al., 2010; Mayorov et al., 2011). However, similar properties have not been observed so far on graphene material prepared using classical techniques, although these conventional processes

are continuously improving (Neto et al., 2009; Sarma et al., 2011). We conclude then that the challenge related to find markets of graphene applications is mainly related to the real progress realized in its mass production with appropriate characteristics.

The number of publications related to the graphene material is continually growing has increased dramatically especially in last years (from about 4000 in 2010 to more than 14,000 in 2014) (Choi et al., 2010). **Figure 2** shows how the number of refereed articles dealing with graphene material has steadily increased since 2004, based on data collected from the Engineering Village web-based information service.

The same tendency is also recorded with patents applications, which have downright doubled within 2 years only [i.e., from 2010 to 2012 with a total of 8416 patents worldwide by February 2013 (Zhang et al., 2013)].

In the recent years, there have been many review works, related either to theoretical and/or experimental studies, discussing the topics of synthesis and application of graphene material. To cite just few recent works, Neto et al., (2009) reviewed the electrical properties of graphene, and then focused on its electronic transport properties (Neto et al., 2009). Other experimental reviews included detailed discussions of synthesis (Zhang et al., 2010a) and Raman spectroscopy processes of transport mechanisms (Ni et al., 2008; Avouris, 2010; Giannazzo et al., 2011), related to electronic applications graphene, including transistors like-devices, bandgap engineering (Loh et al., 2010), and optoelectronic technologies (Bonaccorso et al., 2010; Schwierz, 2010). However, among all the published articles on the matter, only 18 review works have been conducted on the smart applications of the graphene materials (**Figure 3**), and to the best of our knowledge, only one review-article has been published on 2015. In sum, along with the increase in the number of publications in this area comes a need for a comprehensive review article, and the objective of this paper is to address this need. The literature is indeed lacking a comprehensive review of the recent experimental advancements on graphene material and its smart applications. This is the aim of our article. However, due to the huge number of various works that are involved, and often the unavailability of access to many conference proceedings, the emphasis of this paper was on the most accessible refereed journal articles. Obviously, it was not possible and practical to cover all of these articles, especially since a lot of them had already been covered by previous paper review; an attempt was made to select representative articles in each of the relevant categories. This review should be particularly well suited to graduate students who desire an introduction to the study of graphene that will provide them with many references for further reading.

Graphene Scalable Synthesis Perspectives

Initially discovered by micromechanical exfoliation of graphite (Geim and Novoselov, 2007b), graphene has generated widespread interest as a smart material. However, for graphene to make a significant impact within industry, it is important to develop methods for scalable synthesis of high-quality graphene. The current common production methods for graphene include liquid exfoliation, ultrahigh vacuum processes,

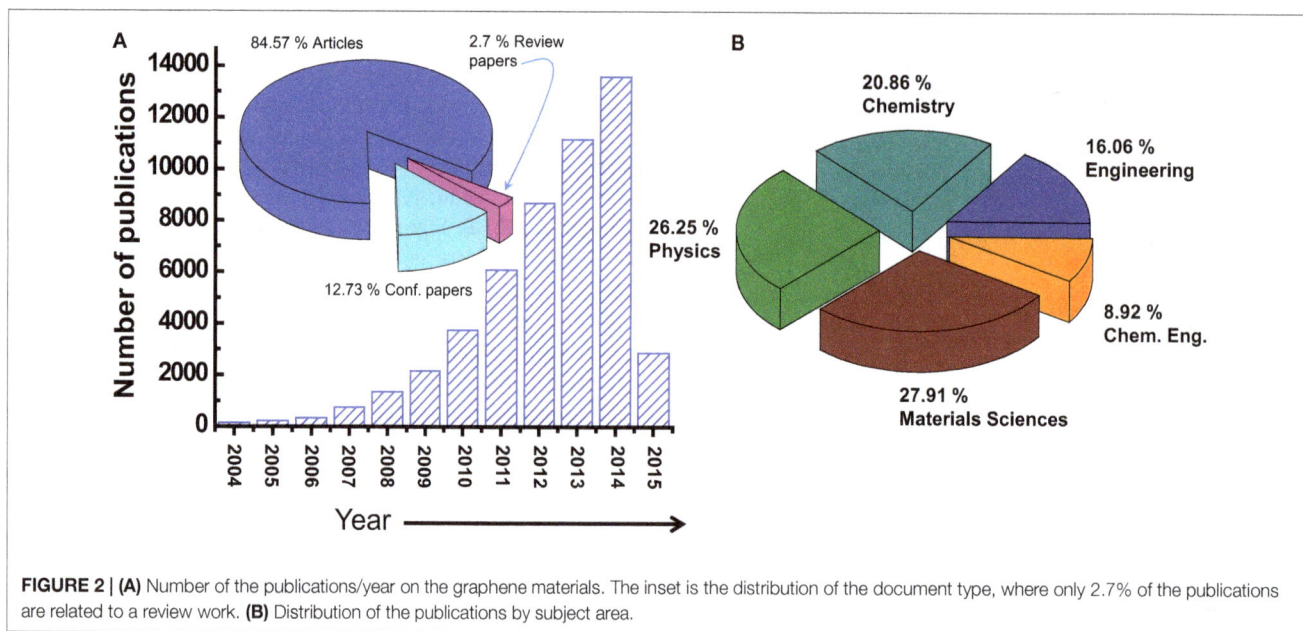

FIGURE 2 | (A) Number of the publications/year on the graphene materials. The inset is the distribution of the document type, where only 2.7% of the publications are related to a review work. **(B)** Distribution of the publications by subject area.

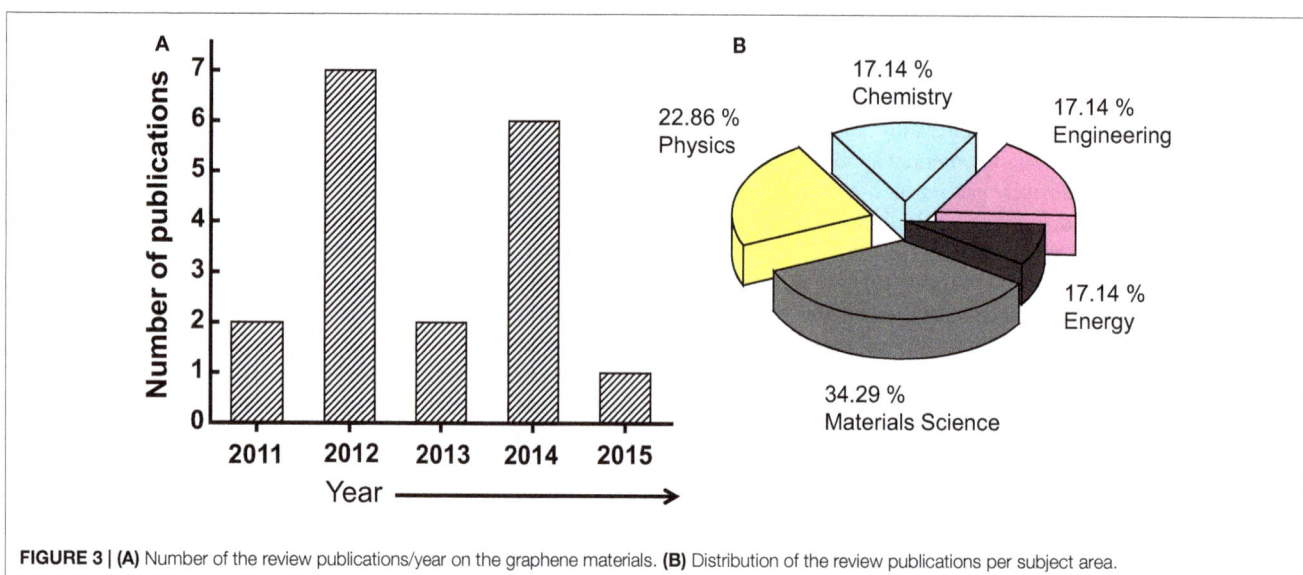

FIGURE 3 | (A) Number of the review publications/year on the graphene materials. **(B)** Distribution of the review publications per subject area.

annealing of silicon carbide (*SiC*), and obviously, chemical vapor deposition (CVD). Other methods, which could be used for scalable graphene synthesis, include plasma enhanced CVD, flame synthesis, and pulsed laser deposition (PLD). These methods will be discussed in this review with a focus on identifying processes that can be translated for commercial production of graphene. This review does not cover methods for the production of graphene oxide.

Micromechanical Exfoliation

Micromechanical exfoliation involves peeling highly ordered pyrolytic graphite (HOPG) using adhesive tape (Novoselov et al., 2005). Since each layer of graphene is bonded to the other layer by van der Waals bonding, it is feasible to cleave HOPG. Normally,

the peeling is performed several times. This process can also be used to produce few layers graphene (FLG). While this is the simplest method for the production of graphene and is commonly used in laboratory experiments, the production method is not scalable for large-scale graphene growth.

Liquid-Phase Exfoliation

Liquid-phase exfoliation (LPE) involves using a solvent to exfoliate graphite by ultrasonication (Hernandez et al., 2008; Lotya et al., 2009). Commonly used solvents include acetic acid, sulfuric acid, and hydrogen peroxide (Singh et al., 2011). The ultrasonication time is typically 60 min with a power of 250–500 W. Green and Hersam (2009) reported the use of sodium cholate as a surfactant for the exfoliation of graphene. Moreover, they were able to separate the sheets by density gradient ultracentrifugation, which enabled the isolation of graphene from FLG. LPE can be used for the production of graphene nanoribbons (GNRs) (Li et al., 2008), where the width is <10 nm. While LPE represents a scalable method for the production of graphene, large scale film growth remains really challenging.

Chemical Vapor Deposition Based Synthesis

Chemical vapor deposition of graphene involves the use of transition metals, where nickel (Ni) (Reina et al., 2008; Chae et al., 2009; Kim et al., 2009; Losurdo et al., 2011) and copper (Cu) (Li et al., 2009; Bae et al., 2010; Guermoune et al., 2011; Suk et al., 2011; Wang et al., 2011) are suitable for large scale production of graphene.

Graphene growth based on CVD has shown exceptional device properties (**Figure 4**) (Bae et al., 2010), with electron mobility of 7350 cm^2 V^{-1}s^{-1} (Novoselov et al., 2005). In addition, large scale production of 30″ graphene films was demonstrated using roll-to-roll CVD (Mattevi et al., 2011). The graphene obtained from this process was of high quality, with a sheet resistance of ~125 Ω/square and 97.4% optical transmittance.

Graphene growth using CVD is fairly straightforward, where a copper or nickel substrate is placed in a reactor at temperatures normally around 1000°C. The initial step in the process is to introduce hydrogen in the reactor. This step is critical to eliminate any oxide layer present in the metal, for the case of Cu this will reduce any native layers of CuO and Cu$_2$O. The hydrogen atmosphere also enables the growth of grain boundaries (Mattevi et al., 2011), which is necessary for the growth of high-quality graphene. Afterwards, a hydrocarbon gas (typically methane) is added to the reactor. The hydrocarbon gas provides the necessary carbon species used in the growth of graphene. The hydrocarbon gas to hydrogen ratio plays an important role in the growth of graphene. If insufficient hydrogen is present, this could result in oxidized metal layers being present, which can lead to a disordered graphene structure. By contrast, excess hydrogen has shown to etch away graphene. On polycrystalline substrates, the graphene flakes tend to have different lattice orientations.

Using CVD, graphene is grown onto transition metals, which enables a low-energy pathway by forming intermediate compounds for the growth of graphene. The first row of transition metals Fe, Co, Ni, and Cu is of great interest due to their low cost and high availability. The difference in the carbon solubility between these metals impacts the growth quality, where Fe has the highest and Cu has the lowest carbon solubility. For this reason, Cu is an ideal metal for growing single layer graphene. When using Ni and Co it is common to get up to 10 layers of graphene. Similarly, on Fe it is common to have FLG.

Most practical applications of graphene require that the underlying surface be insulating. For this reason, graphene must be transferred to an insulating surface, such as SiO$_2$ (Bhaviripudi et al., 2010). Additionally, this transfer is required to measure the optoelectronic properties of the synthesized graphene. The commonly used process to transfer graphene is to first deposit and cure poly (methylmethacrylate) (PMMA) on the metal sheet. Afterwards, etch the Cu metal sheet using iron chloride. This gives a floating sheet of PMMA and graphene, which is rinsed in deionized water. Subsequently, transfer this layer to an insulating surface and use acetone to remove the PMMA layer.

Plasma Enhanced Chemical Vapor Deposition Synthesis

Plasma enhanced CVD (PECVD) is another method used for the synthesis of graphene that is comparable to the thermal CVD process (Zhu et al., 2007; Yuan et al., 2009; Kim et al., 2011; Bo et al., 2013). PECVD is based on a number of plasma sources,

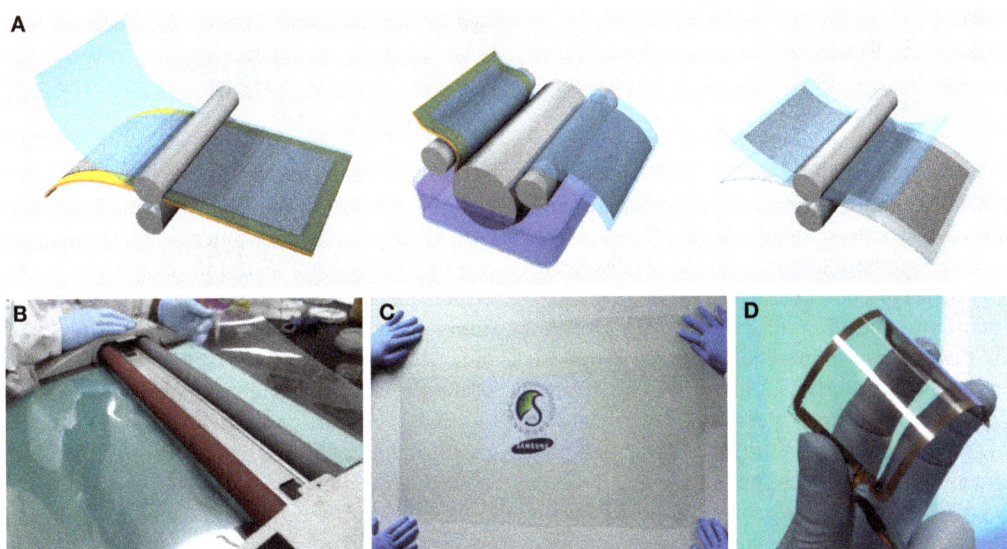

FIGURE 4 | (A) Schematic of the roll-to-roll production of graphene films grown on a copper foil. **(B)** Roll-to-roll transfer of graphene films from a thermal release tape to a PET film at 120°C. **(C)** A transparent large-area graphene film transferred on a 35″ PET sheet. **(D)** An assembled graphene/PET touch panel showing mechanical flexibility. Reproduced with permission of Bae et al. (2010). Copyright 2010, Nature Nanotechnology.

such as microwave (MW) (Malesevic et al., 2008), radio frequency (RF) (Wang et al., 2004), and direct current (dc) arc discharge (Krivchenko et al., 2012) have been utilized in the growth of graphene. Copper and nickel are typically used as the substrate for PECVD graphene growth; however, a number of additional substrates have also been used (Dato et al., 2008; Bo et al., 2013). A particularly exciting technique is a substrate free method based on the decomposition of ethanol in MW-based PECVD reactor (Guermoune et al., 2011). Such methods provide can be used for scalable production of graphene powder. Typical growth conditions of PECVD graphene on a substrate are 5–100% CH_4 in H_2 with a substrate temperature of 500–800°C (Singh et al., 2011; Bo et al., 2013). The power of the plasma is 900 W. Such processes can enable the growth of graphene at lower temperatures and shorter duration (<5 min). However, the quality of the graphene film is typically lower when compared to thermal CVD.

The growth of graphene films at low temperature is important for a number of applications. For instance, fabrication of graphene for high-performance display glass has to be at a temperature lower than 660°C (Ren et al., 1998; Bo et al., 2013). Furthermore, production at lower temperature can provide new manufacturing opportunities in the area of flexible electronics based on plastics (Tung et al., 2009). Critical challenges still need to be overcome, as at lower temperatures the graphene film tends to have a higher disorder.

Flame Synthesis

Flame synthesis is extensively used to produce commercial quantities of nanoparticles. Of the most widely used nanoparticles, carbon black, fumed silica, and titania, flame synthesis is the dominant method in the production of these materials. Volumetric production of the flame synthesis industry is on the order of 100 metric tons per day (Kammler et al., 2001). A key advantage of flames is that it readily provides the high temperature necessary for gas-phase synthesis along with a carbonizing or oxidizing environment.

With respect to graphene, flame synthesis is not as commonly studied when compared to CVD, but it offers several important advantages, such as scalability and cost effectiveness. The most commonly used flame types include premixed, normal diffusion, inverse diffusion, and co-flow (Inoue et al., 2010; Memon et al., 2013). Since early 2000s, a number of researchers have focused on the use of flames for CNT synthesis (Inoue et al., 2010). However, the development of flame synthesis for graphene is still in its early stage. In addition to flame type, other parameters including temperature, species concentration, and velocity impact the growth process. Graphene being a two-dimensional material requires large-scale production across a substrate. Due to the temperature and species gradients that occur in most flames it is difficult to scale the growth of graphene across an entire substrate. Moreover, a reduced environment with carbon rich species, which is necessary for graphene growth, is difficult to achieve in most flames. Nevertheless, flame synthesis has the potential to economically enable the mass production of graphene.

Similar to earlier CNT flame synthesis papers, where the growth of CNTs was observed near the soot region of a premixed flame, carbon particles containing graphene were observed in Bunsen (propane) flame (Ossler et al., 2010). These particles were collected by placing a transmission electron microscopy grid 2 cm above the tip of the burner. The grid was held within the flame for 10–50 ms. The graphene films were several hundreds of nanometers in size.

In an attempt to grow graphene on copper, Li et al. (2011b) investigated the growth of graphene using an ethanol burner. The substrate was placed within the flame at a temperature of 550–700°C and the flame was extinguished using a cap to prevent the oxidation of the copper foil. The growth of an amorphous carbon film was observed on the substrate and XPS confirmed the formation of sp^2, sp^3, and C–O bonded atoms. Graphene was not observed due to the low temperature and the presence of oxygen within the flame. In a different experiment, Li et al. (2011a) were able to synthesis graphene successfully on nickel. The process utilized two different burners (burner 1 and burner 2), with the substrate situated within the interior region of the flame structure itself (**Figure 5**). Burner 1 (alcohol burner) surrounded the substrate for the entire time, where it prevented air oxidation and served as the carbon source. Burner 2 (butane-fueled Bunsen burner) provided the additional heating of the substrate and served as the carbon source for graphene growth. The flame was extinguished using a cap. There are still numerous challenges in using flame synthesis for the growth of graphene, specifically in developing methods that result in higher quality graphene.

Epitaxial Growth on Silicon Carbide Substrate

Yannopoulos et al. (2012) have investigated the thermal decomposition of *SiC* surface, which was providing an epitaxial growth of graphene material (**Figure 6**). They reported a new process using a CO_2 laser as the heating step for a fast and one-step growth process of large uniform graphene film on *SiC*. This method can control the stacking order of epitaxial graphene and is cost-effective since it does not involve any pretreatment step or high-vacuum process. The decomposition operated at low temperature and proceeded in the second time scale, thus providing a means to engineering graphene patterns on *SiC* by focused laser beams.

Pulsed Laser Deposition

The PLD process is definitely considered as one of the most versatile growth approaches. Since the laser energy source is located outside the deposition chamber; the use of either ultrahigh vacuum or ambient gas becomes possible (Krebs et al., 2003). Combined with a stoichiometry transfer between ablated target and substrate where the material is deposited, this flexibility allows depositing theoretically all possible kinds of materials, including polymers or fullerenes (Eason et al., 2006). This technique was first employed by Smith et al. (1965) (Krebs et al., 2003) in 1965 to elaborate semiconductors and dielectric thin films. In 1987, it was then fully developed by Dijkkamp et al. (1987) for the deposition of high-temperature superconductors. Their work allowed to define the main characteristics of PLD, namely, the stoichiometry transfer between target and deposited film (Smith et al., 1965; Dijkkamp et al., 1987; Chrisey and Hubler, 1994; Eason et al., 2006). Since the work of Dijkkamp et al., the

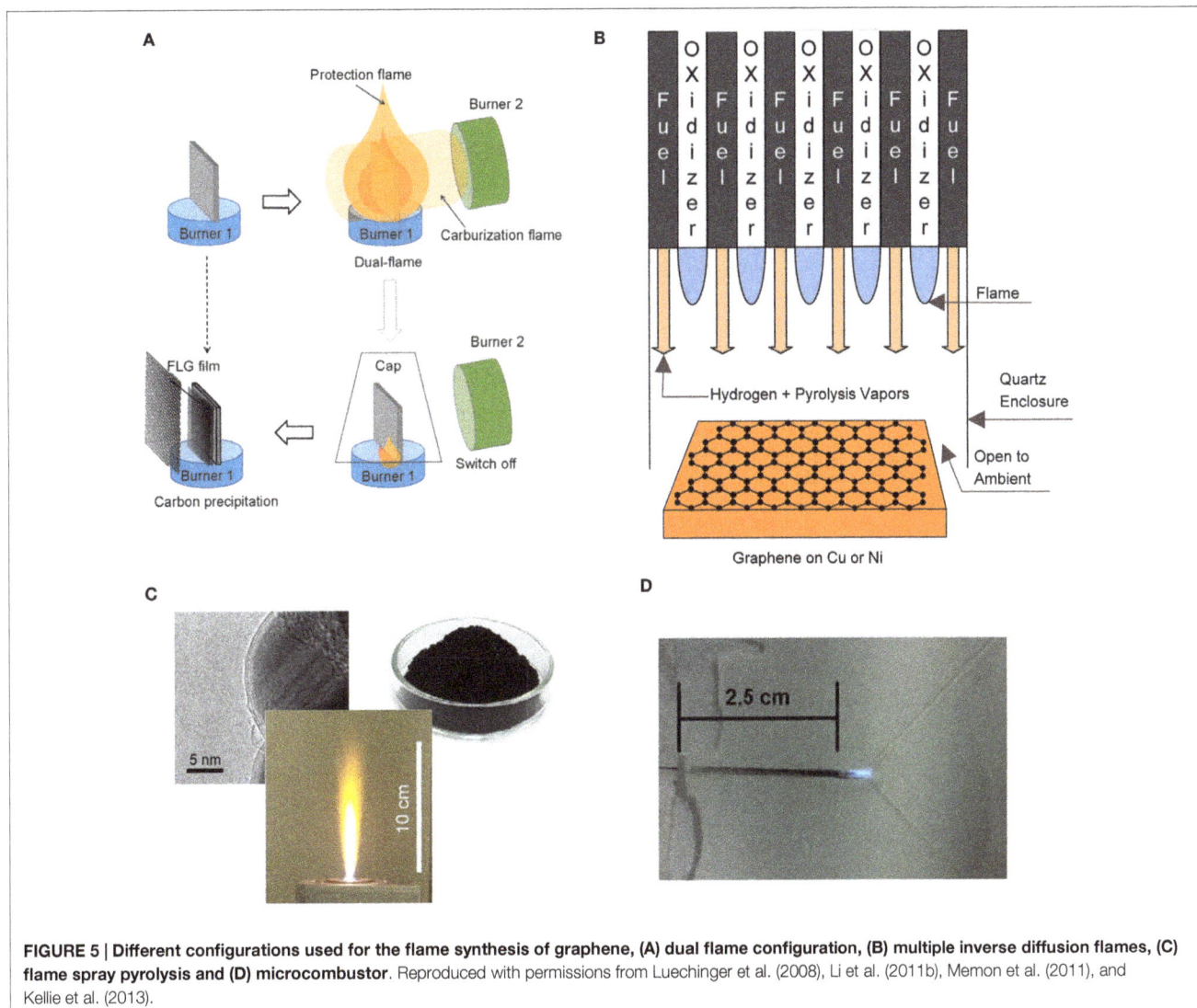

FIGURE 5 | Different configurations used for the flame synthesis of graphene, (A) dual flame configuration, (B) multiple inverse diffusion flames, (C) flame spray pyrolysis and (D) microcombustor. Reproduced with permissions from Luechinger et al. (2008), Li et al. (2011b), Memon et al. (2011), and Kellie et al. (2013).

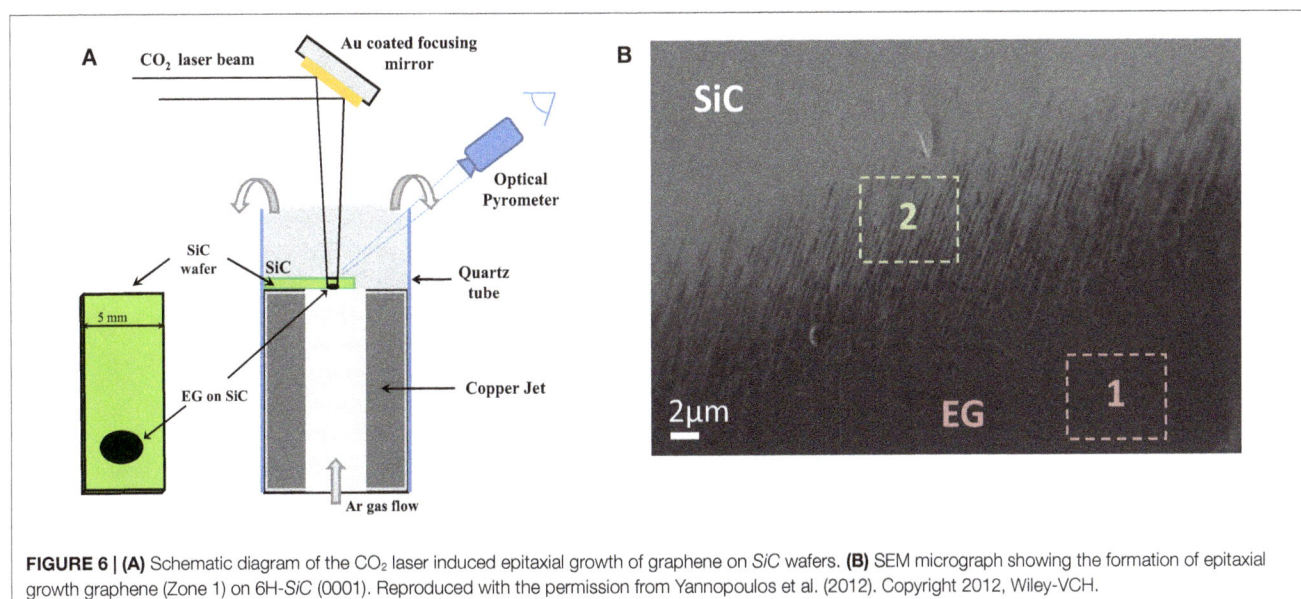

FIGURE 6 | (A) Schematic diagram of the CO_2 laser induced epitaxial growth of graphene on *SiC* wafers. **(B)** SEM micrograph showing the formation of epitaxial growth graphene (Zone 1) on 6H-*SiC* (0001). Reproduced with the permission from Yannopoulos et al. (2012). Copyright 2012, Wiley-VCH.

deposition technique has been extensively used for all kinds of oxides, nitrides, carbides and also for preparing metallic systems and even polymers or fullerenes (Krebs et al., 2003). During PLD, almost all experimental parameters can be adjusted to control the film growth, and ranging from the laser parameters (Fluence, wavelength, pulse-duration, and repetition rate), to the deposition conditions (target-to-substrate distance, temperature, nature of the gas, pressure, etc.).

A representative schematic diagram for PLD (Krebs et al., 2003) is shown in **Figure 7**. Inside the vacuum chamber (ultrahigh vacuum, UHV), targets of elementary or alloy elements are struck at an angle of 45° by a high energy focused pulsed laser beam. The atoms and ions ablated from the target(s) are then deposited directly on the substrate (Krebs et al., 2003). In the majority of the

cases, the substrates are attached with their surfaces parallel to the target(s) surfaces at a distance of 2–10 cm.

To the benefit of the reader, **Figure 8** summarizes the main first-deposited materials since the introduction of PLD in 1987, with respect to the year for deposition and corresponding reference.

As mentioned earlier, in the PLD process, one of the main advantages is the fact that during deposition, the stoichiometry of the deposited material is very close to the target (Krebs et al., 2003). Consequently, it is possible to prepare stoichiometric thin films from a single alloy bulk target (Krebs et al., 2003).

In the context of the graphene growth, and in parallel to the CVD deposition method, physical vapor deposition has also been reported for the growth of graphene (Koh et al., 2010; Zhang and Feng, 2010b). In UHV, PLD chambers graphite is normally used as the target with a transition metal as the substrate (**Figure 9**). A substrate temperature of 1300°C was reported for 1–2 layers of high-quality graphene (**Figure 9**) (Zhang and Feng, 2010b). No carbide formation occurs at the interface of graphene and the metal (Zhang and Feng, 2010b). While numerous metals can be used as a catalyst, nickel appears to be the most promising for low temperature growth. Numerous parameters, such as the cooling rate and laser power, impact the quality of graphene films (Koh et al., 2010).

Laser-Based Chemical Vapor Deposition

A continuous wave (CW) laser is utilized for laser-based CVD in an enclosed chamber (**Figure 10A**) (Park et al., 2011). The precursor gases used include methane and hydrogen, with Ni foil as the substrate. The synthesis mechanism is based on a vapor–liquid–solid that only takes nanoseconds to picoseconds. The spectra of the Raman spectroscopy showing the different graphene layers are illustrated in **Figure 10B**. A key advantage of this process is that it can be used for graphene lithography,

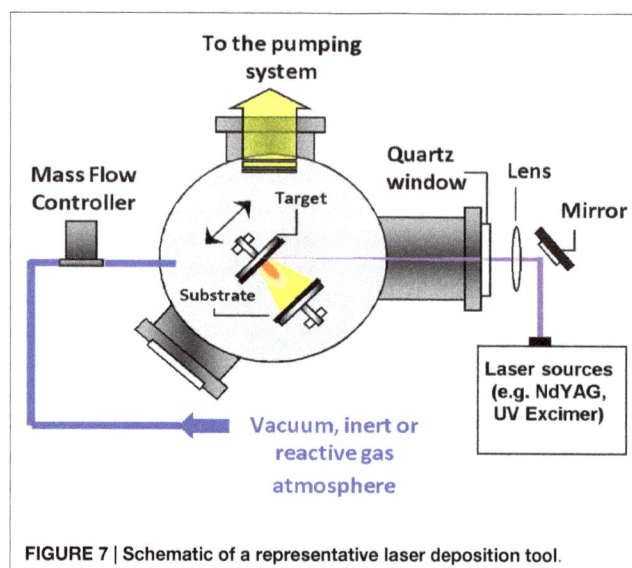

FIGURE 7 | Schematic of a representative laser deposition tool.

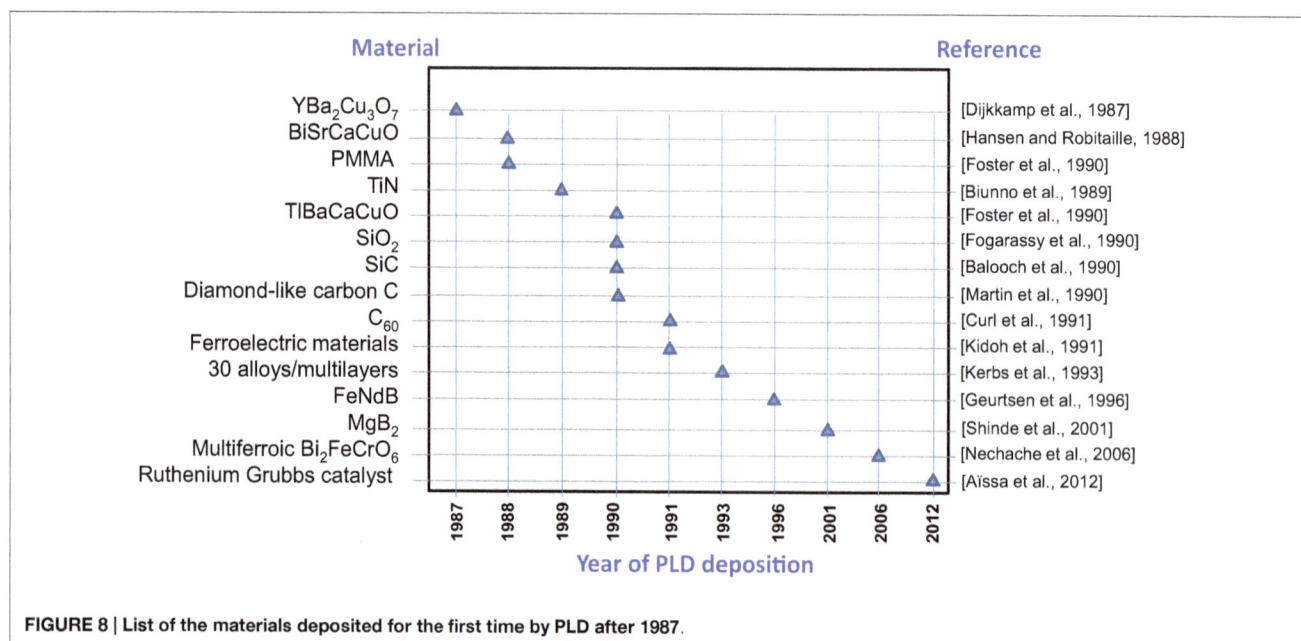

FIGURE 8 | List of the materials deposited for the first time by PLD after 1987.

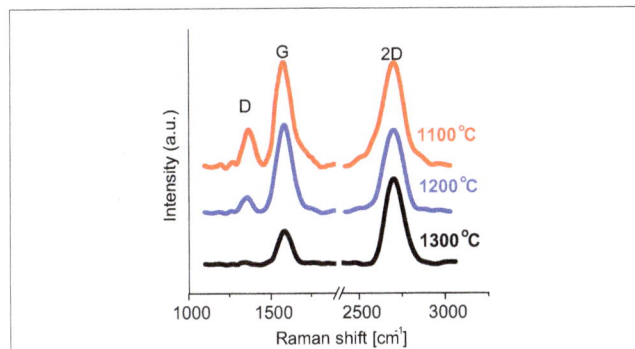

FIGURE 9 | Graphene deposition by means of PLD. Measured Raman spectra with respect to the growth temperature. Reproduced with permission from Zhang and Feng (2010b). Copyright 2010, Elsevier.

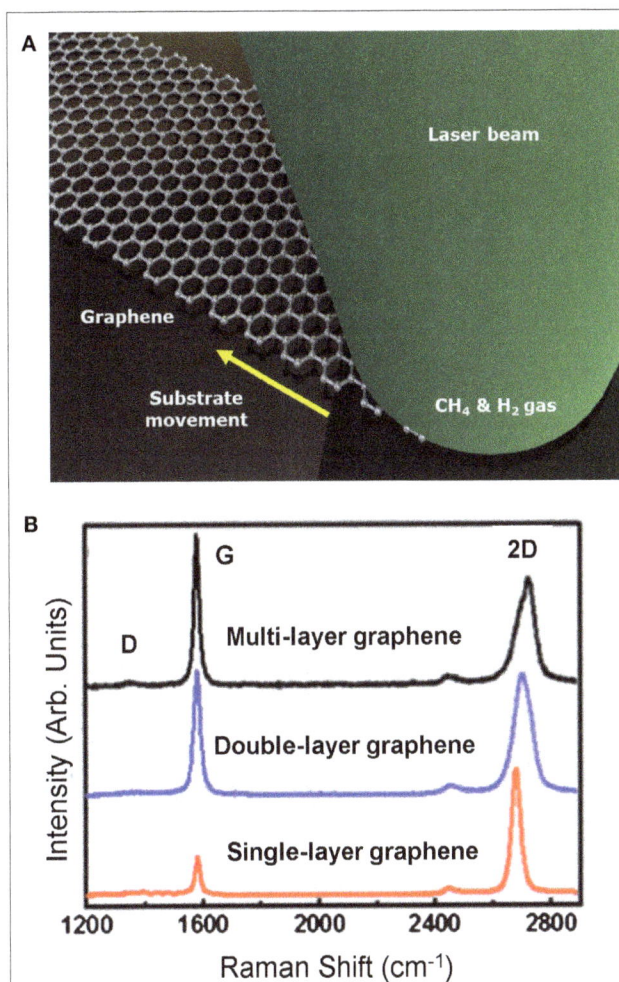

FIGURE 10 | (A) Schematic of the laser induced CVD process, and **(B)** mea-sured Raman as a function of the number of graphene layers. Reprinted with permission from Park et al. (2011). Copyright 2011, American Institute of Physics.

where the laser can be scanned on specific areas of a metal catalyst enabling direct growth.

Laser Growth Directly on Silicon and Quartz Substrates

Sun et al. (2010) produced graphene on Cu and Ni film using PMMA. Due to the existence of the metal films, the graphene films need be transferred to other substrate using polydimethyl-siloxane (PDMS) or PMMA (Kim et al., 2009; Reina et al., 2009; Sun et al., 2010). Silicon wafer is the most important single-crystal substrate used for semiconductor devices and integrated circuits (ICs). Suemitsu et al. (2010) produced epitaxial graphene on Si substrate. In their approach, a *SiC* film of about 100 nm-thick was deposited on the Si wafer before growth, so graphene was grown on the *SiC* surface. Direct growth of graphene on bare Si substrate without any other material is very attractive. Graphene films can form a Schottky junction with Si, which can produce a built-in electric field and realize electron-hole separation, and has been used to fabricate solar cells (Li et al., 2010a; Gunst et al., 2011; Karamitaheri et al., 2011).

Wei and Xu (2012) have demonstrated the growth of FLG directly on a silicon substrate using a laser irradiation. Silicon sub-strates were coated with PMMA, which was then evaporated using a CW laser beam. The laser beam also melts the silicon surface and carbon atoms from PMMA separates from the silicon upon cooling to form FLG. A substrate of 1 cm × 2 cm p-type (111)-oriented Si wafer was used to grow graphene. The silicon wafer was first cleaned and the native oxide layer was removed in buffer hydrofluoric acid (HF) solution to form H-terminated silicon surface. A PMMA layer was coated on the Si surface by spin coating, then covered by a quartz wafer of the same size, and then fixed on a sample stage using two spring clamps. The purpose of using the quartz wafer is to maintain a high enough concentration of carbon after PMMA is evaporated and dissociated by laser irradiation. The growth was conducted in a vacuum chamber using a CW that was directed on the Si surface to melt the surface of the Si wafer.

For the synthesis of graphene on metal, two main growth mechanisms were proposed. On Ni, graphene was produced via carbon dissolution and precipitation (Yu et al., 2008). On Cu, the growth can be explained by surface-catalyzed process, which involves carbon nucleation on the Cu surface, and the growth of graphene with the addition of carbon to the edges (Wei and Xu 2012). However, both of these growth mechanisms cannot explain the graphene growth on Si. Cu or Ni maintains solid in the graphene growth process. They found that if the laser power was below the melting point of silicon, there was no graphene grown on the silicon surface.

Similar work has been achieved by Wei et al. (2013), who grow FLG (2–3 layers) on quartz substrate by using a continuous-wave laser by suing a photoresist S-1805 coated on the quartz wafer (thickness 30 nm).

Applications of Graphene Materials in Functional Devices

Graphene research has skyrocketed since the Nobel Prize winners Andre Geim and Konstantin Novoselov published

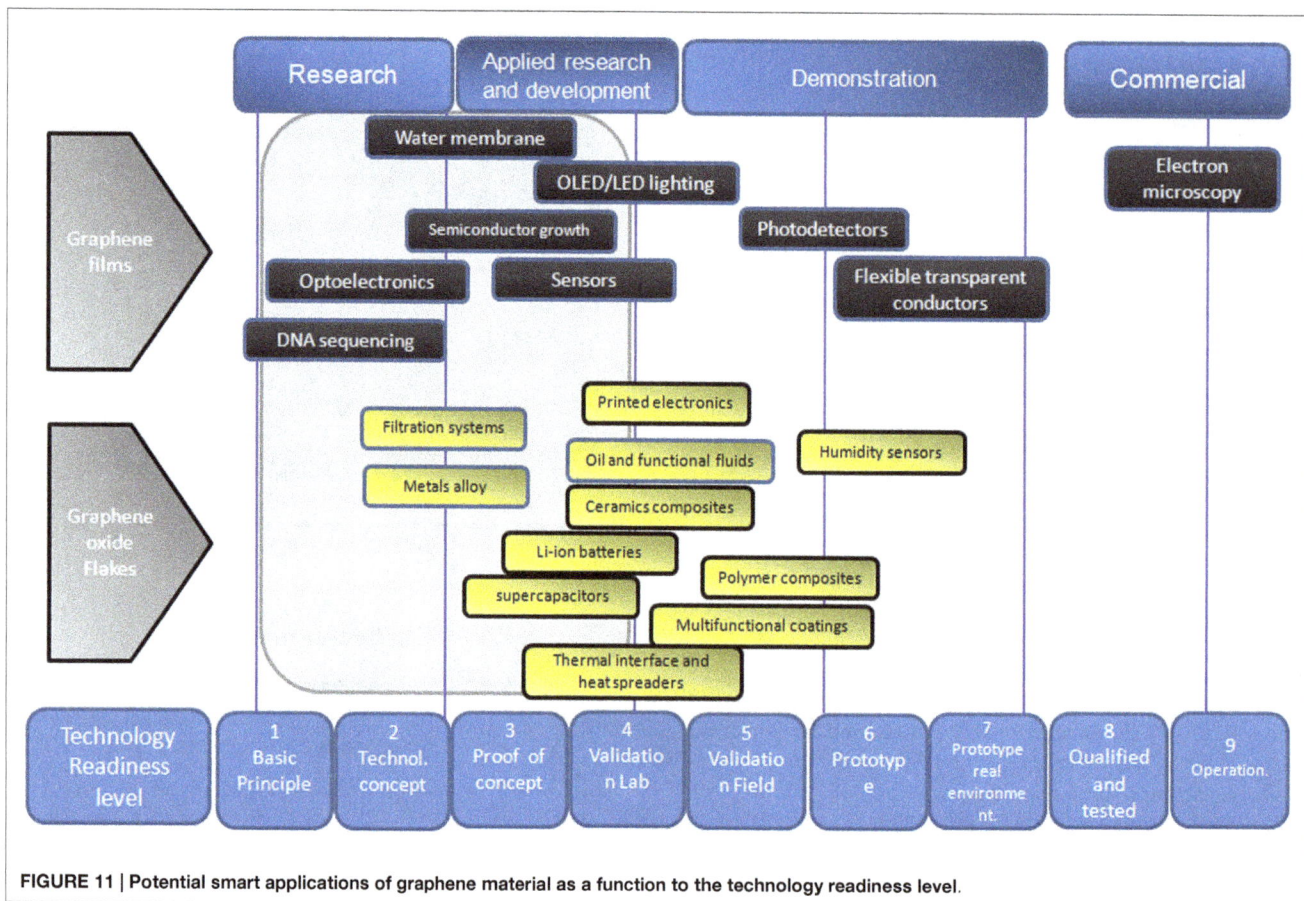

FIGURE 11 | Potential smart applications of graphene material as a function to the technology readiness level.

a number of papers on the discovery of graphene 11 years ago. Since then, there has been a sharp increase in our scientific knowledge of graphene, as shown by the number of publications and patent applications on graphene (**Figure 2**). However, only a handful of graphene-based devices have entered the market, thus we are in an early stage in the commercialization of this material. Further research and development of graphene are critical to help achieve the full potential of this material and the sections below relate to research work of using graphene in smart applications. **Figure 11** shows attempt of classification of some potential smart applications of graphene material as a function to the corresponding technology readiness level.

Photovoltaic Cell

Photovoltaic (PV) cell is the device which converts light to electricity (Chapin et al., 1954; Wang et al., 2008). So far, graphene-based solar cells have been demonstrated in dye-sensitized PV cells (Choe et al., 2010; Jo et al., 2012), organic bulk-heterojunction PV cells (Li et al., 2010b; Yin et al., 2010), hybrid ZnO/poly(3-hexylthiophene) (P3HT) PV cells (Li et al., 2010c), Si based PV cells (Shim et al., 2011), and InGaN p-i-n PV cells (Gomez De Arco, 2010).

The functionalization of graphene material either during synthesis process (*in situ*) or post-treatment has demonstrated not only the possibility to control the properties of the surfaces and

interfaces but also tailoring its work function (Jo et al., 2010; Guo et al., 2011; He et al., 2011; Wan et al., 2011). In an organic PV cell, the difference in work function between the two conductors creates an electrical field in the organic layer (Jo et al., 2012). To date, the power conversion efficiency (PCE) of graphene-electrode organic solar cells (OSCs) has been reported to be in the range of 0.08–2.60% (Wan et al., 2011; Jo et al., 2012), which is indeed much lower than those of conventional OSCs made with ITO electrodes (8.37%) (Stankovich et al., 2006; Chen et al., 2010a; He et al., 2011; Wan et al., 2011; Yu et al., 2012; Saravanakumar et al., 2013; Yokomizo et al., 2013; Sharma et al., 2014). However, the PCE of graphene-electrode OSCs needs to be improved to make graphene film a serious candidate for OSCs (Jo et al., 2012).

Transparent and Flexible Electronics

Currently, most of the research groups in electronics devices fabrication are investigating different routes to fabricate flexible and transparent electronic devices for various types of applications, including smart windows, IC cards, displays, LEDs, solar cells, etc. (Wager et al., 2003; Yu et al., 2011; Jo et al., 2012). In materials, graphene is one of the viable aspirants, which have all the required properties, at the same time, such as optical transparency, mechanical flexibility, and high conductivity. In recent years, different research groups have reported the integration of graphene-based composite electrodes, including graphene/pentacene and graphene/SWCNT, into transparent and flexible

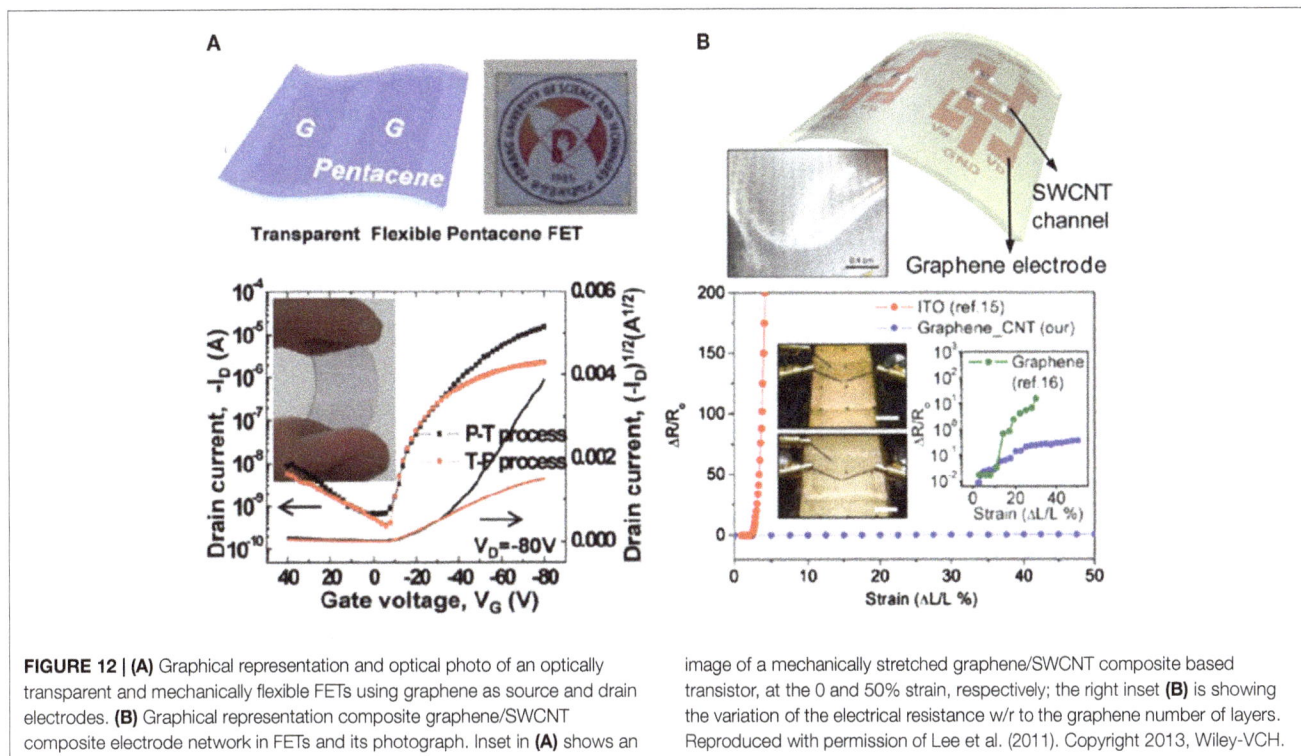

FIGURE 12 | (A) Graphical representation and optical photo of an optically transparent and mechanically flexible FETs using graphene as source and drain electrodes. **(B)** Graphical representation composite graphene/SWCNT composite electrode network in FETs and its photograph. Inset in **(A)** shows an image of a mechanically stretched graphene/SWCNT composite based transistor, at the 0 and 50% strain, respectively; the right inset **(B)** is showing the variation of the electrical resistance w/r to the graphene number of layers. Reproduced with permission of Lee et al. (2011). Copyright 2013, Wiley-VCH.

electronic like-devices, such as transistors, memory devices, and ICs (Wager et al., 2003; Ji et al., 2011; Lee et al., 2011; Yu et al., 2011; Jo et al., 2012).

Pentacene-based organic FETs-devices were elaborated onto a flexible substrate by means of patterning and transfer or transfer and patterning processes (P-T) and (T-P) techniques that are defining the graphene electrodes (**Figure 12A**) (Lee et al., 2011; Jo et al., 2012). As a gate electrodes, a plastic substrate based on PEDOT/PSS was used, and poly-4-vinylphenol (PVP) as cross-linker. The typical transfer response of these organic FET is shown in the **Figure 12A**, where a carrier mobility of 0.01 and 0.12 cm^2/Vs were systematically estimated for the (T-P) and (P-T) processes, respectively (Jo et al., 2012). In addition to the field effect carrier mobility, the overall electronic characteristics of the organic transistors dealing with the (P-T) technique of graphene are superior (Jo et al., 2012). **Figure 12B** is showing the variation of the electrical resistance as a function of the number of graphene layers. However, there is large room for improving these electronic properties, especially the carrier mobility and to make it more feasible to use in large scale applications (Jo et al., 2012).

Smart Applications of Graphene

Thermoelectric Application of Graphene

Thermoelectric materials (TEM) achieve the conversion between thermal and electrical energy and vice versa. This field has regained renewed attention because of the huge potential of TEM to be applied in Peltier coolers and thermoelectric power generators. It is well established that the performance of TEM are determined mainly by its dimensionless figure of merit, namely ZT. To date, the performance research of TEM have mainly focused on inorganic semiconductors, such as PbTe, Bi$_2$Te$_3$, CoSb$_3$, SnSe, and theirs alloys or composites (Li et al., 2010d). The challenge to develop TEM for a crystalline system with high performance is know how to tailor the interconnected thermoelectric physical parameters, including Seebeck coefficient, the electrical conductivity, and thermal conductivity.

The efficiency of a TEM-based device is usually characterized by the following figure of merit:

$$ZT = S^2\sigma T / \kappa,$$

where S is the Seebeck coefficient (or thermopower, µV/K), σ is the electronic conductance (S/m), T is the thermal conductivity including contributions from both phonons and electrons (W/mK), and κ is the absolute temperature (K).

Figure 13 shows a schematic illustration of a thermoelectric module for (a) power generation (Seebeck effect) and (b) active generation (Peltier effect). **Figure 13A** shows an applied temperature difference, which causes charge carriers in the material (electron or holes), to diffuse from the hot side to the cold one, resulting in a current flow through the circuit. **Figure 13B** is schematic of the heat that evolves at the upper junction and is adsorbed at the lower junction when a current is made to flow through the circuit.

Recently, great effort has been made in improving the TEM dimensionless figure of merit (ZT). The difficulty in to simultaneously optimizing them, which causes thermoelectric research to stagnate for a while, until great reductions in

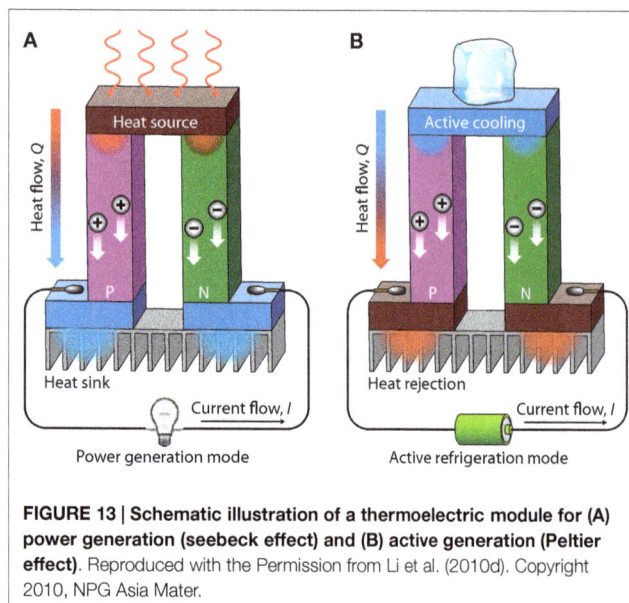

FIGURE 13 | Schematic illustration of a thermoelectric module for (A) power generation (seebeck effect) and (B) active generation (Peltier effect). Reproduced with the Permission from Li et al. (2010d). Copyright 2010, NPG Asia Mater.

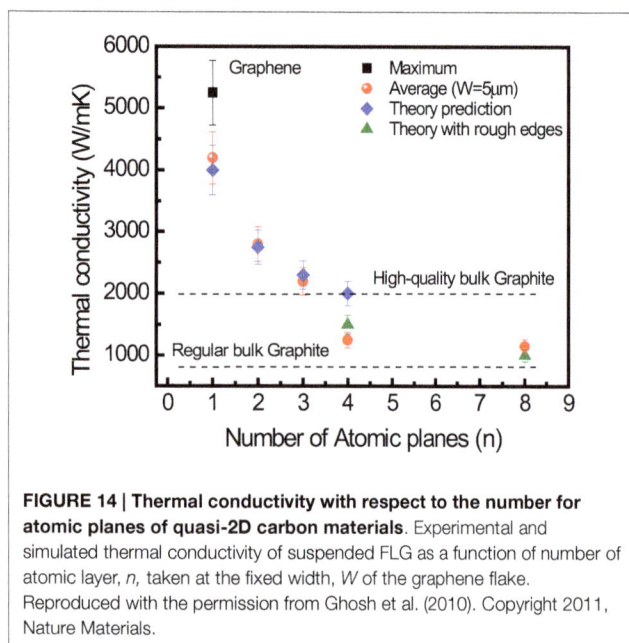

FIGURE 14 | Thermal conductivity with respect to the number for atomic planes of quasi-2D carbon materials. Experimental and simulated thermal conductivity of suspended FLG as a function of number of atomic layer, n, taken at the fixed width, W of the graphene flake. Reproduced with the permission from Ghosh et al. (2010). Copyright 2011, Nature Materials.

thermal conductivity were theoretically and experimentally proven in nanostrtructured materials.

Ghosh et al. (2010) examined the evolution of the thermal properties of FLG with respect to increasing thickness (i.e., the number of graphene layer, n). The results obtained by Raman spectroscopy have shown that, overall, the thermal conductivity decreases with increasing the FLG thickness, approaching that of the bulk graphite limit (**Figure 14**) (Ghosh et al., 2010). The experimental data points in **Figure 14** were all normalized to the same width of graphene sheet of 5 μm (Ghosh et al., 2010). The detailed procedure is described in Ghosh et al. (2010).

The evolution observed from 2D graphene to bulk graphite is explained by the cross-plane coupling of the low-energy phonons and changes in the phonon scattering, since more states are available for scattering owing to the increased number of phonon branches. The thermoelectric power (TEP) is the voltage developed across a sample when a constant temperature gradient is applied. TEP of 80 μV/K was recently measured in graphene at room temperature (300 K) (Zuev et al., 2009). Similar to the quantum Hall effect in electronic transport, quantized TEP has also been observed in graphene at high-magnetic fields (Zuev et al., 2009). The TEP can be tuned in graphene, even to negative values, under the application of a gate bias or chemical potential (Wei et al., 2009). Very large TEP values have been predicted for GNRs, for instance, 4 mV/K for a 1.6 nm wide ribbon (Haskins et al., 2009). In comparison, the highest value experimentally reported so far is 850 μV/K for two-dimensional electron gases in $SrTi_2O_3$ heterostructures (Ohta et al., 2007), while only a few μV/K has been reported for bulk graphite (Tyler et al., 1953). The TEP power of single walled carbon nanotubes (SWNTs) has been theoretically and experimentally shown to be 60 μV/K (Hone et al., 1998), inferior to that of graphene. A giant thermoelectric coefficient of 30 mV/K was reported in metallic electrodes periodically patterned over graphene, deposited on SiO_2 substrate (Chang et al., 2007).

On the other hand, thermoelectric properties of graphene have attracted increased interest as well, since it can convert heat to electricity and vice versa. A high thermopower value of 80 μV/K was reported for graphene (Zhu et al., 2009; Zuev et al., 2009). Various structures of graphene have been examined, including nanoribbons (GNRs) (Haskins et al., 2009; Sevinçli et al., 2009; Ouyang and Hu, 2010; Huang et al., 2011; Mazzamuto et al., 2011), quantum dots (Yan et al., 2012), graphene junctions, and chevron-type structures (Chen et al., 2010b). It is worth noted here that ZT of zigzag GNRs can exceed 3 (Ouyang and Hu, 2010).

Recently, many works have been conducted on the creation of graphene band gap by means of making an array of holes (antidots) into the graphene layer (Pedersen et al., 2008; Ouyang et al., 2011; Petersen et al., 2011; Chang and Nikolic, 2012). This is an inevitable property for the integration of graphene material directly into transistor architecture (Ouyang et al., 2011; Petersen et al., 2011). The gap can be engineered by controlling the lattice geometry (i.e., the antidot size and the hole-to-hole separation) (Shen et al., 2008; Eroms and Weiss, 2009; Bai et al., 2010; Kim et al., 2010; Gunst et al., 2011; Karamitaheri et al., 2011). Further, thermoelectric properties of these structures (2D graphene antidot lattices) have also been investigated where ZT up to 0.3 was found (Li et al., 2010d).

Graphene in Shape Memory Materials

Shape memory polymers (SMPs) are defined as smart materials, with the faculty to respond to an external stimulus, typically heat, and have a wide-range of applications from biomedical devices to space materials (Han and Chun., 2014). Polyurethane (PU) is the most attractive SMP material, commonly referred to as SMP polyurethane (SMPU) (Lee et al., 2014). Based on the specific molecular design, SMPU can form either a crystalline or

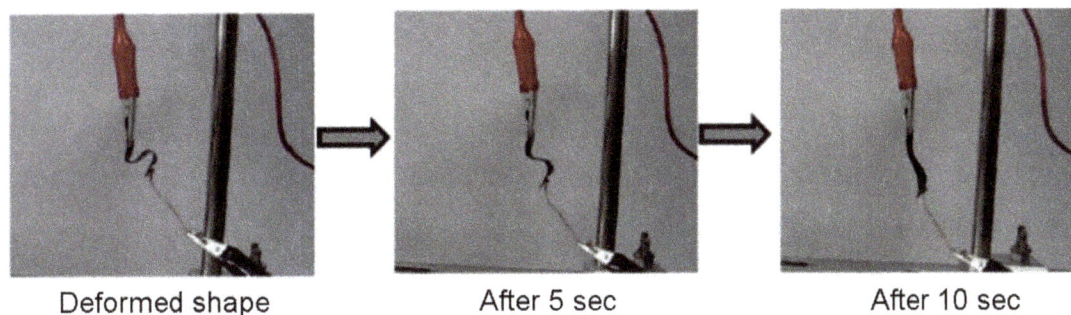

FIGURE 15 | Shape recovery behavior of graphene-cross linked PU composites, for a temporary helix shape (left) to a permanent shape when applying a constant voltage of 50 V. Reproduced with permission from Rana et al. (2013). Copyright 2013, Royal Society of Chemistry.

amorphous structure, with a broad actuation temperature range from −20 to 150°C (Kim et al., 2014). SMP composites based on nanofillers have shown to increase the overall shape memory properties. An exciting research area is the use of carbon-based nanomaterials, particularly graphene, as nanofillers for SMPUs.

Han and Chun (2014) prepared a graphene/PU composite material by functionalizing graphene with diazonium salts carrying phenethyl alcohol. The resulting material showed enhanced shape memory, thermal, and mechanical properties. The resulting material showed shape fixity up to 98 with 94% shape recovery ratio after four cycles. Hysteresis loss can be low as 2%. Rana et al. (2013) prepared a flexible and conductive shape memory composite based on PU and functionalized graphene sheets. The graphene sheets were functionalized using phenyl isocyanate and poly diol. The composite resulted in 97% shape recovery and 95% shape fixity (see **Figure 15**). Modifications to PU can also improve the shape memory performance with graphene. Thakur et al. (2013) have demonstrated that castor oil-modified hyperbranched PU and graphene-oxide (without functionalization) can result in a shape recovery of ~99.5% with shape fixity of ~90%. Park et al. (2014) have demonstrated that PU and allyl isocyanate modified graphene oxide can also be actuated by infrared absorption, with a resulting shape recovery of 90%.

Graphene in Self-Healing Materials

Long-term stability and durability of materials are one of the main challenges faced today for structural and coating applications, especially when using polymeric composites materials. Indeed, there are many parameters affecting the degradation of materials, including environmental conditions of large types (temperature gradient, UV irradiation, oxygen erosion, corrosion, etc.). As indicated by their name, self-healing materials are first conceived and then designed and elaborated to have the potential to heal mainly the mechanical properties of the materials when damaged. Representative applications of self-healing materials are found in composite polymers, various metals, ceramics, and their related composites, which are subjected to a wide variety of healing principles.

Recently, a few studies have demonstrated the potential of graphene as an additive for self-healing materials. Dong et al. (2013) synthesized a composite material of poly(acrylamide) (PAM), poly(acrylic acid) (PAA), and graphene that exhibits shape memory effect and self-healing ability. Graphene content was in the range of 10–30% and the material can be recovered after 20 cycles of cut and self-healing. The self-healing capabilities are illustrated in **Figure 16**, where 10 wt.% graphene sample is cut in the middle and the two pieces were healed together at 37°C for 20 min. The results suggest that the self-healing ability and shape memory effect occur due to a "zipper effect" of PAM–PAA that forms or dissociate the hydrogen-bond network, where such effects are limited without the addition of graphene. Huang et al. (2013) have demonstrated the use of FLG with thermoplastic polyurethane (TPU) as self-healing material initiated using an electric stimulus. When the FLG loading rate is 5 wt.%, the material can be healed to an efficiency higher than 98% in 3 min. Furthermore, upon increasing the FLG loading rate to 8 wt.%, the same healing efficiency of 98% can be achieved in 15 s only. It is expected that graphene efficiently converts electrical energy to thermal energy, which improves the self-healing ability of TPU diffusion and re-entanglement of the TPU chains. More recently, Wang et al. (2013) and Sullivan et al. (1977) have recently reported the fabrication of a composite based on a cross-linked (CL) hydrogen bonding polymer with graphene oxide. This composite material was found to enable a fast self-healing capability, with high efficiency, occurring at room-temperature. More importantly, the healing reaction was produced without the need of any external stimuli, such as electrical bias, light, and heat (Zhang et al., 2014). The introduction of the graphene material in this CL-based polymer was found to be crucial to reduce the needed amount of CL sites that are necessary for the healing reaction. It is worth noting here that a huge amount of CL sites are not only negatively impacting the mechanical properties of the polymer (including flexibility) but also its dynamic characteristics (Zhang et al., 2014). Such nanocomposite materials involving CL-hydrogen bonding polymer with graphene-oxide can be useful in many applications, e.g., protecting barrier for electronic devices, sealing layer for gas systems, and stretchable self-healing conductive wires (Zhang et al., 2014).

Graphene in Photomechanical Actuators

Actuators are materials, which change their shape or dimensions under the application of external stimulus. To date, the well-recognized materials for actuation are piezoelectrics (Nakamura et al., 1989), ferro-electrics (Kuribayashi et al., 1989), shape

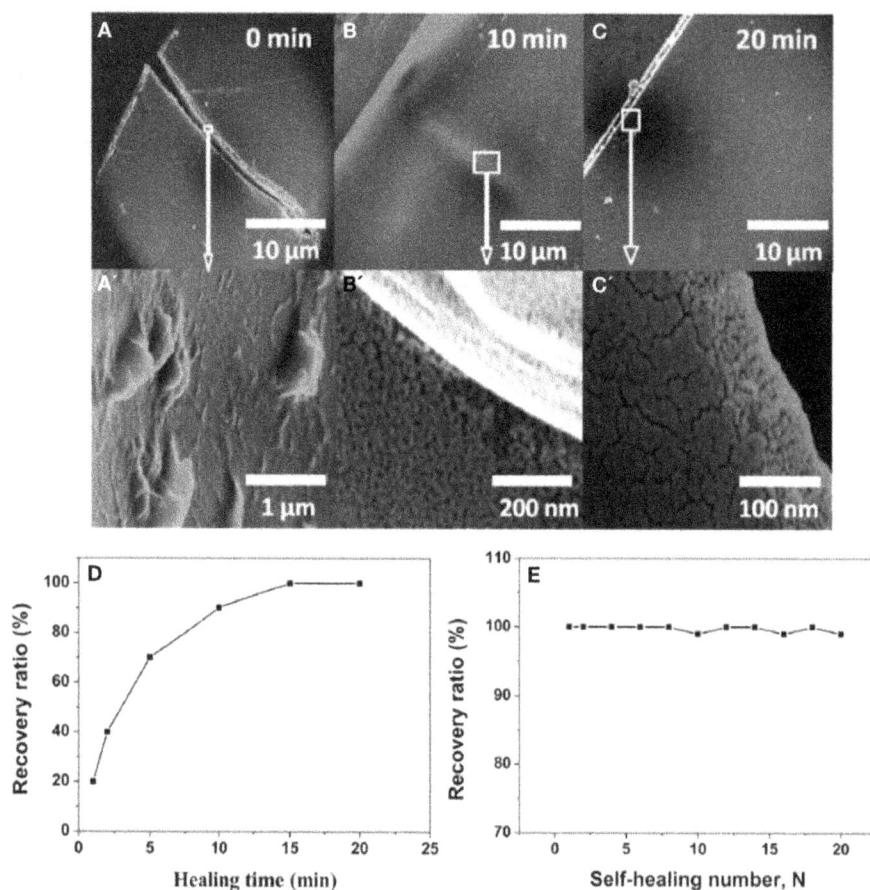

FIGURE 16 | SEM image of G-PAM-PAA strip heating at 37°C in different times, (A) 0 min, (B) 10 min, and (C) 20 min. (A′–C′) are close-up view of the factures occurred in (A–C), respectively. Percentage of the recovery ratio as a function of (D) healing-time and (E) self-healing number (N). Reproduced with permission from Dong et al. (2013). Copyright 2013, Wiley-VCH.

memory alloys (Damjanovic and Newnham, 1992), electrostrictive materials (Damjanovic and Newnham, 1992), and conducting polymers (Smela et al., 1999). Recently, large stresses and strains from low-voltage electromehcanical actuation have exhibited by carbon nanotubes (CNTs) (Baughman et al., 1999) and porous metallic nanoparticles (Ahir and Terentjev, 2005). Beside this, both single walled CNT (SWCNT) and multiwalled CNT (MWCNT)-based composites have been reported to undergo photomechanical actuation (Dreyer et al., 2010).

Several articles have reported about graphene-based composites, mostly in sheets form and have derived from graphite oxide or graphite intercalation compounds (GICs). It has been observed that intrusion of graphite oxide or graphite intercalation compounds derived fillers can enhance electrical conductivities of polymeric matrices (Gómez-Navarro et al., 2008), the Young's moduli (Potts et al., 2011), and could easily be functionalized to tailor to host polymer properties (Sun et al., 2009). Now, it is well-known that by tailoring the number of layers of graphene nanoplateletes (GNP) and GNR, the overall properties of composite material change [such as saturable absorption (Casiraghi et al., 2007) and electric field assisted band gaps (Zhang et al., 2009)].

Loomis et al. (2012) have reported that a simple polymer composite system with photomechanical responses is realized solely by incorporation of a homogeneous dispersion of GNPs within a PDMS elastomer matrix. It has been observed that the actuation responses of GNPs/PDMS composites depend on the initial applied pre-strain as in CNT/PDMS composites. Photomechanical stress change 2.4–3.6 times is greater for GNP/PDMS composites, compared to any other tested form of nanocarbon. These stress changes reported are usable and recoverable work achieved by the actuators. Energy conversion factor (ηM) of 7–9 MPa W^{-1} for optical-to-mechanical is obtained during testing. Until now, this is a largest energy conversion factor of an extraordinary photomechanical effect exhibited by any material so far.

Graphene in Piezoelectric Materials

Piezoelectric material has the property of converting mechanical movement into electrical movement and vice versa. Traditionally, piezoelectricity is considered to be an intrinsic property of a particular material phase. Piezoelectric materials have been frequently used in a wide variety of applications from pressure sensors (Pereira and Castro Neto, 2009), to acoustic transducers

(Guinea et al., 2010), to high voltage generators (Bunch et al., 2007) for dynamical control of material deformation by application of external electric force.

Luk'yanchuk et al. (2015) have found that piezoelectricity can be concocted into intrinsically non-piezoelectric materials such as graphene. This is a nanoscale new phenomenon and lacking a direct bulk analog (Luk'yanchuk et al., 2015). This new phenomenon has provided room for practical approach toward manipulation and dynamic control different concerns in nanodevices, such as optical, chemical, and electronic. Luk'yanchuk et al. (2015) have reported an extraordinary two dimensional piezoelectric effect, both on a strained and unstrained graphene junction. Interestingly, it has been formally attested that this 2D piezo effect is a direct consequence of the difference in the two work functions (of the two type of graphene) and hence to the charge transfer occurring from the biaxial strain when putting the two graphene types together (i.e., bend band structure). The effect has termed as the band-piezoelectric effect, which exhibits a massive magnitude due to the ultrathin structure of graphene (Luk'yanchuk et al., 2015).

Using this new type of piezo effect, a piezoelectric nanogenerator and a piezoresistive pressure sensor within a graphene nano-electro-mechanical-system (NEMS) platform have been demonstrated. In this novel device, the deformation caused by an AFM tip resulted in charge separation, due to the modified band structure of the bent membrane (Luk'yanchuk et al., 2015). Consequently, by using appropriate metal electrodes with suited work function, we can even collect the cumulated charge and produce an electrical voltage (Luk'yanchuk et al., 2015).

Graphene in Electrorheology Materials

Electrorheology (ER) material is a type of smart material where the rheological properties of the material can be reversibly transformed with the application of an external electric field (Zhang et al., 2010c, 2012; Yin et al., 2012; Zhang and Choi, 2014). An ER material is composed of polarizable particles suspended in an insulating medium and after applying an electric field the particles are polarized and form a column like structure. This modifies the rheological properties of the material, such as viscosity, shear stress, and dynamic modulus. Applications of such materials include damper systems, ER polishing, tactile displays, medical devices, and robotic actuators. Recently, graphene, r-GO, and GO, due to their unique properties, have gathered the interest of the scientific community as an additive for ER materials (Zhang and Choi, 2014). Zhang et al. (2010c) prepared colloidal r-GO using a modified Hummers method, which was used to prepare a nanocomposite material comprising of GO and polyaniline (PANI). The resulting material showed adjustable electrical conductivity, which has potential for use as an ER material. Additionally, Zhang et al. (2012) also prepared GO particles suspended in silicone oil, where the ER fluid is polarized and exhibits viscoelastic properties at various strains under an electric field.

Yin et al. (2012) prepared a novel ER suspension comprising of two dimensional PANI decorated GO sheets. With the application of an electric field, the suspension containing PANI and GO, shows higher ER effect when compared to pure PANI. The performance of the material is dependent on the thickness of the PANI coating applied to the GO sheets. Furthermore, Yin et al. (2013) prepared mesoporous silica-coated r-GO nanosheets as a dispersal for ER fluids. Silica coating limits the conductivity of graphene, which enables high polarization and ER response, with the application of an electric field particularly high-frequency AC electric fields. The silica shells also limit the restacking of graphene. Li et al. (2015) studied the ER effects of non-conducting GO and conducting r-GO comprise an insulating SiO_2 shell. The results show that the GO/SiO_2 has a higher ER response to DC and low-frequency AC electric fields, while the $r-GO/SiO_2$ has a higher ER response to high-frequency AC electric fields. The different behavior can be explained by the impact of the polarization rate with regards to the inter-particle interaction.

Multifunctional Graphene Nanocomposite Foams for Space Applications

Space is the new frontier and new materials, devices, and technologies for aerospace applications represent now an emerging sector with significant employment prospects and opportunities for profit. Materials and devices used in space applications (e.g., satellites) should combine functionality with low weight and reduced volume, to optimize cost effectiveness. The main cost of a satellite is its launch into orbit (estimated at about 7 M$), therefore reducing weight and volume can significantly decrease the overall cost. Weight savings can be achieved by replacing heavy copper wiring, which accounts for example up to 4000 lbs of weight on a Boeing 747 and about one-third of the weight of large satellites (Meador et al., 2010), with low density carbon-based wiring cables. In addition, Joule heating from metallic parts requires additional components for cooling (radiators), thus adding to the overall weight and cost. The combination of superior electronic and thermal properties of graphene materials could potentially revolutionize the design and fabrication of light weight electrical and electronic devices to be used in space applications. In this context, graphene is a promising candidate because its ballistic electron transport limits Joule heating and should allow reductions in weight, volume, and subsequently total cost.

Another potential space application of graphene is its use as nanofiller to fabricate light weight and robust nanocomposite, to be used, e.g., as thermal barrier. Recently, graphene nanoplatelets having FLG with 1–5 layers and typical diameters ranging from 1 to over 100 μm have been successfully grown by a team from Michigan State University[1]. These nanoplatelets show, overall, similar properties to a single graphene layer while being mechanically much robust. In addition, their cost production can be lowered to be competitive to other carbonaceous additives and fillers. These nanoplatelets could be used as a nanoadditive into various polymer foams, increasing thereby their thermal, mechanical, and electrical properties, while the foam still maintains its unique structure and low

---

[1] Space Technology missions directorate. The document is available at: http://www.nasa.gov/directorates/spacetech/strg/2012_nstrf_rollins.html

density. This combination of unique properties has direct potential application as flame resistant in space technology.

Conclusion and Outlooks

Research on graphene has considerably enriched our understandings and applications of two-dimensional atomic crystal. The combination of unprecedented physical and chemical properties, including its extremely high strength, thermal and electrical conductivity, and mechanical flexibility, have harried scientists to investigate in-depth its real potential in delivering improvements to different devices in many technological and scientific fields, ranging from materials science to physics, chemical engineering, and even biology. In this review, we have presented an overview of the main gas-phase synthesis of graphene and its applications as smart material systems. The possibility of successfully integrating graphene directly into device, could not only improve the electrical and/or mechanical properties but also enable the realization of a wide range of applications, such as actuation, thermoelectricity, shape memory, and self-healing.

However, this success is conditioned by the prior success to address the following issues: (i) developing a cost-effective growth process to synthesize functional graphene in a reasonable scale with acceptable degree of reproducibility, for the realization of practical applications; (ii) considering the unique electrical, mechanical, and optical properties of graphene, incorporation of these characteristics into the existing smart systems from an interdisciplinary point of view should be highly valued; (iii) finally, from practical view-point, its highly desirable to fabricate graphene multifunctional systems, that are responsive to multiple stimuli. Considering the current and ongoing achievements, it is believed that the smart applications of graphene systems with more functionality are expected to emerge as new complements – and/or even replacements – to existing conventional systems, especially in fields of electronics, energy, and space.

Acknowledgments

Authors thank the financial support of the Qatar Environment and Energy Research Institute and Qatar Foundation.

References

Ahir, S. V., and Terentjev, E. M. (2005). Photomechanical actuation in polymer-nanotube composites. *Nat. Mater.* 4, 491–495. doi:10.1038/nmat1391

Aïssa, B., Nechache, R., Haddad, E., Jamroz, W., Merle, P. G., and Rosei, F. (2012). Ruthenium Grubbs' catalyst nanostructures grown by UV-excimer-laser ablation for self-healing applications. *Appl. Surf. Sci.* 258, 9800–9804. doi:10.1510/icvts.2010.255588

Avouris, P. (2010). Graphene: electronic and photonic properties and devices. *Nano Lett.* 10, 4285–4294. doi:10.1021/nl102824h

Bae, S., Kim, H., Lee, Y., Xu, X., Park, J. S., Zheng, Y., et al. (2010). Roll-to-roll production of 30-inch graphene films for transparent electrodes. *Nat. Nanotechnol.* 5, 574–578. doi:10.1038/nnano.2010.132

Bai, J., Zhong, X., Jiang, S., Huang, Y., and Duan, X. (2010). Graphene nanomesh. *Nat. Nanotechnol.* 5, 190–194. doi:10.1038/nnano.2010.8

Balandin, A. A. (2011). Thermal properties of graphene and nanostructured carbon materials. *Nat. Mater.* 10, 569–581. doi:10.1038/nmat3064

Balooch, M. T., Tench, R. J., Siekhaus, W. J., Allen, M. J., Connor, A. L., and Olander, D. R. (1990). Deposition of SiC films by pulsed excimer laser ablation. *Appl. Phys. Lett.* 57, 1540–1542. doi:10.1063/1.103346

Basu, S., and Bhattacharyya, P. (2012). Recent developments on graphene and graphene oxide based solid state gas sensors. *Sens. Actuators B Chem.* 173, 1–21. doi:10.1016/j.snb.2012.07.092

Baughman, R. H., Cui, C., Zakhidov, A. A., Iqbal, Z., Barisci, J. N., Spinks, G. M., et al. (1999). Carbon nanotube actuators. *Science* 284, 1340. doi:10.1126/science.284.5418.1340

Bekaroglu, E., Topsakal, M., Cahangirov, S., and Ciraci, S. (2010). First-principles study of defects and adatoms in silicon carbide honeycomb structures. *Phys. Rev. B* 81, 075433. doi:10.1103/PhysRevB.81.075433

Bhaviripudi, S., Jia, X., Dresselhaus, M. S., and Kong, J. (2010). Role of kinetic factors in chemical vapor deposition synthesis of uniform large area graphene using copper catalyst. *Nano Lett.* 10, 4128–4133. doi:10.1021/nl102355e

Biunno, N., Narayan, J., Hofmeister, S. K., Srivatsa, A. R., and Singh, R. K. (1989). Low-temperature processing of titanium nitride films by laser physical vapor deposition. *Appl. Phys. Lett.* 54, 1519. doi:10.1063/1.101338

Bo, Z., Yang, Y., Chen, J., Yu, K., Yan, J., and Cen, K. (2013). Plasma-enhanced chemical vapor deposition synthesis of vertically oriented graphene nanosheets. *Nanoscale* 5, 5180–5204. doi:10.1039/c3nr33449j

Bonaccorso, F., Sun, Z., Hasan, T., and Ferrari, A. C. (2010). Graphene photonics and optoelectronics. *Nat. Photon* 4, 611–622. doi:10.1038/nphoton.2010.186

Boukhvalov, D. W., Katsnelson, M. I., and Lichtenstein, A. I. (2008). Hydrogen on graphene: electronic structure, total energy, structural distortions and magnetism from first-principles calculations. *Phys. Rev. B* 77, 035427. doi:10.1103/PhysRevB.77.035427

Bunch, J. S., van der Zande, A. M., Verbridge, S. S., Frank, I. W., Tanenbaum, D. M., Parpia, J. M., et al. (2007). Electromechanical resonators from graphene sheets. *Science* 315, 490–493. doi:10.1126/science.1136836

Bunch, J. S., Verbridge, S. S., Alden, J. S., van der Zande, A. M., Parpia, J. M., Craighead, H. G., et al. (2008). Impermeable atomic membranes from graphene sheets. *Nano Lett.* 8, 2458–2462. doi:10.1021/nl801457b

Casiraghi, C., Hartschuh, A., Lidorikis, E., Qian, H., Harutyunyan, H., Gokus, T., et al. (2007). Rayleigh imaging of graphene and graphene layers. *Nano Lett.* 7, 2711–2717. doi:10.1021/nl071168m

Casolo, S., Martinazzo, R., and Tantardini, G. F. (2011). Band engineering in graphene with superlattices of substitutional defects. *J. Phys. Chem. C* 115, 3250–3256. doi:10.1021/jp109741s

Chae, S. J., Güneş, F., Kim, K. K., Kim, E. S., Han, G. H., Kim, S. M., et al. (2009). Synthesis of large-area graphene layers on *poly*-nickel substrate by chemical vapor deposition: wrinkle formation. *Adv. Mater. Weinheim* 21, 2328–2333. doi:10.1002/adma.200803016

Chang, P.H. and Nikolic, B.K. (2007). Giant thermoelectric effect in graphene. *Appl. Phys. Lett.* 91, 203116. doi:10.1021/nl500755m

Chang, P. H., and Nikolic, B. K. (2012). Edge currents and nanopore arrays in zigzag and chiral graphene nanoribbons as a route toward high-ZT thermoelectrics. *Phys. Rev. B* 86, 041406. doi:10.1103/PhysRevB.86.041406

Chapin, D. M., Fuller, C. S., and Pearson, G. L. (1954). A new silicon p-n junction photocell for converting solar radiation into electrical power. *J. Appl. Phys.* 25, 676–677. doi:10.1063/1.1721711

Chen, Z., Berciaud, S., Nuckolls, C., Heinz, T. F., and Brus, L. E. (2010a). Energy transfer from individual semiconductor nanocrystals to graphene. *ACS Nano* 4, 2964–2968. doi:10.1021/nn1005107

Chen, Y., Jayasekera, T., Calzolari, A., Kim, K. W., and Nardelli, M. B. (2010b). Thermoelectric properties of graphene nanoribbons, junctions and superlattices. *J. Phys. Condens. Matter.* 22, 372202. doi:10.1088/0953-8984/22/37/372202

Choe, M., Lee, B. H., Jo, G., Park, J., Park, W., Lee, S., et al. (2010). Efficient bulk-heterojunction photovoltaic cells with transparent multi-layer graphene electrodes. *Org. Electron.* 11, 1864–1869. doi:10.1016/j.orgel.2010.08.018

Choi, W., Lahiria, I., and Seelaboyinaa, R. (2010). Synthesis of graphene and its applications: a review. *Crit. Rev. Solid State Mater. Sci.* 35, 52–71. doi:10.1080/10408430903505036

Chrisey, D. B., and Hubler, G. K. (1994). *Pulsed Laser Deposition of Thin Films*. New York, NY: John Wiley.

Curl, R. F., and Smalley, R. E. (1991). *Scientific American*, Vol. 33. 32.

Dato, A., Radmilovic, V., Lee, Z., Phillips, J., and Frenklach, M. (2008). Substrate-free gas-phase synthesis of graphene sheets. Nano Lett. 8, 2012–2016. doi:10.1021/nl8011566

Damjanovic, D. and Newnham, R. E. (1992). Electrostrictive and piezoelectric materials for actuator applications. J. Int. Mat. Sys. Struct. 3, 190–208. doi:10.1177/1045389X9200300201

Dean, C. R., Young, A. F., Meric, I., Lee, C., Wang, L., Sorgenfrei, S., et al. (2010). Boron nitride substrates for high-quality graphene electronics. Nat. Nanotechnol. 5, 722–726. doi:10.1038/nnano.2010.172

Dijkkamp, D., Venkatesan, T., Wu, X. D., Shaheen, S. A., Jisrawi, N., Min-Lee, Y. H., et al. (1987). Preparation of Y-Ba-Cu oxide superconductor thin films using pulsed laser evaporation from high Tc bulk materials. Appl. Phys. Lett. 51, 619. doi:10.1063/1.98366

Dong, J., Ding, J., Weng, J., and Dai, L. (2013). Graphene enhances the shape memory of poly (acrylamide-co-acrylic acid) grafted on graphene. Macromol. Rapid Commun. 34, 659–664. doi:10.1002/marc.201200814

Dreyer, D. R., Park, S., Bielawski, C. W., and Ruoff, R. S. (2010). The chemistry of graphene oxide. Chem. Soc. Rev. 39, 228–240. doi:10.1039/b917103g

Eason, R. (2006). Pulsed Laser Deposition of Thin Films: Applications-Led Growth of Functional materials. Hoboken, NJ: John Wiley & Sons, Inc.

Elias, D. C., Nair, R. R., Mohiuddin, T. M., Morozov, S. V., Blake, P., Halsall, M. P., et al. (2009). Control of graphene's properties by reversible hydrogenation: evidence for graphane. Science 323, 610–613. doi:10.1126/science.1167130

Eroms, J., and Weiss, D. (2009). Weak localization and transport gap in graphene antidot lattices. New J. Phys. 11, 095021. doi:10.1088/1367-2630/11/9/095021

Fogarassy, E., Fuchs, C., Slaoui, A., and Stoquert, J. P. (1990). SiO2 thin-film deposition by excimer laser ablation from SiO target in oxygen atmosphere. Appl. Phys. Lett. 57, 664–666. doi:10.1063/1.104253

Foster, C. M., Voss, K. F., Hagler, T. W., Mihailovi, D., Heeger, A. J., Eddy, M. M., et al. (1990). Infrared reflection of epitaxial Tl2Ba2CaCu2O8 thin films in the normal and superconducting states. Solid State Commun. 76, 651. doi:10.1016/0038-1098(90)90108-N

Geim, A. K. (2009). Graphene: status and prospects. Science 324, 1530–1534. doi:10.1126/science.1158877

Geim, A. K., and Novoselov, K. S. (2007a). The rise of graphene. Nat. Mater. 6, 183–191. doi:10.1038/nmat1849

Geim, A. K., and Novoselov, K. S. (2007b). The rise of graphene. Nat. Mater. 6, 183–191. doi:10.1038/nmat1849

Geurtsen, A. J. M., Kools, J.C.S., Wit, L., and Lodde, J.C. (1996). Pulsed laser deposition of permanent magnetic Nd2Fe14B thin films. Appl. Surf. Sci. 9698, 887–890. doi:10.1016/0169-4332(95)00541-2

Ghosh, S., Bao, W., Nika, D. L., Subrina, S., Pokatilov, E. P., Lau, C. N., et al. (2010). Dimensional crossover of thermal transport in few-layer graphene. Nat. Mater. 9, 555–558. doi:10.1038/nmat2753

Giannazzo, F., Raineri, V., and Rimini, E. (2011). "Transport properties of graphene with nanoscale lateral resolution," in Scanning Probe Microscopy in Nanoscience and Nanotechnology 2, ed. Bhushan B. (Berlin: Springer), 247–285.

Gomez De Arco, L. (2010). Continuous, highly flexible, and transparent graphene films by chemical vapor deposition for organic photovoltaics. ACS Nano 4, 2865–2873. doi:10.1021/nn901587x

Gómez-Navarro, C., Burghard, M., and Kern, K. (2008). Elastic properties of chemically derived single graphene sheets. Nano Lett. 8, 2045–2049. doi:10.1021/nl801384y

Green, A. A., and Hersam, M. C. (2009). Solution phase production of graphene with controlled thickness via density differentiation. Nano Lett. 9, 4031–4036. doi:10.1021/nl902200b

Guermoune, A., Chari, T., Popescu, F., Sabri, S. S., Guillemette, J., Skulason, H. S., et al. (2011). Chemical vapor deposition synthesis of graphene on copper with methanol, ethanol, and propanol precursors. Carbon 49, 4204–4210. doi:10.1016/j.carbon.2011.05.054

Guinea, F., Katsnelson, M. I., and Geim, A. K. (2010). Energy gaps and a zero-field quantum Hall effect in graphene by strain engineering. Nat. Phys. 6, 30–33. doi:10.1038/nphys1420

Gunst, T., Markussen, T., Jauho, A.-P., and Brandbyge, M. (2011). Thermoelectric properties of finite graphene antidot lattices. Phys. Rev. B 84, 155449. doi:10.1103/PhysRevB.84.155449

Guo, C. X., Guai, G. H., and Ming, C. (2011). Graphene based materials: enhancing solar energy harvesting. Adv. Energy Mater. 1, 448–452. doi:10.1002/aenm.201100119

Han, S., and Chun, B. C. (2014). Preparation of polyurethane nanocomposites via covalent incorporation of functionalized graphene and its shape

memory effect. Compos. Part A. Appl. Sci. Manuf. 58, 65–72. doi:10.1016/j.compositesa.2013.11.016

Hansen, S. G., and Robitaille, T. E. (1988). Formation of polymer films by pulsed laser evaporation. Appl. Phys. Lett. 52, 81–83. doi:10.1063/1.99332

Haskins, J., Kınacı, A., Sevik, C., Sevinçli, H., Cuniberti, G., and Çağın, T. (2009). A theoretical study on thermoelectric properties of graphene nanoribbons. Appl. Phys. Lett. 94, 263107. doi:10.1021/nn200114p

Hassan, R. (ed.) (2012). "Graphene nanoelectronics," in Metrology, Synthesis Properties and Applications (Berlin: Springer-Verlag), 598.

He, Z., Zhong, C., Huang, X., Wong, W.-Y., Wu, H., Chen, L., et al., (2011). Simultaneous enhancement of open-circuit voltage, short-circuit current density, and fill factor in polymer solar cells. Adv. Mater. Weinheim 23, 4636–4643. doi:10.1002/adma.201103006

Hernandez, Y., Nicolosi, V., Lotya, M., Blighe, F. M., Sun, Z., De, S., et al. (2008). High-yield production of graphene by liquid-phase exfoliation of graphite. Nat. Nanotechnol. 3, 563–568. doi:10.1038/nnano.2008.215

Hone, J., Ellwood, I., Muno, M., Mizel, A., Cohen, M. L., Zettl, A., et al., (1998). Thermoelectric power of single-walled carbon nanotubes. Phys. Rev. Lett. 80, 1042–1045. doi:10.1103/PhysRevLett.80.1042

Houssa, M., Pourtois, G., Afanas'ev, V. V., and Stesmans, A. (2010). Electronic properties of two-dimensional hexagonal germanium. Appl. Phys. Lett. 96, 082111. doi:10.1063/1.3332588

Huang, L., Yi, N., Wu, Y., Zhang, Y., Zhang, Q., Huang, Y., et al. (2013). Multichannel and repeatable self-healing of mechanical enhanced grapheme-thermoplastic polyurethane composites. Adv. Mater. Weinheim 25, 2224–2228. doi:10.1002/adma.201204768

Huang, W., Wang, J.-S., and Liang, G. (2011). Theoretical study on thermoelectric properties of kinked graphene nanoribbons. Phys. Rev. B 84, 045410. doi:10.1103/PhysRevB.84.045410

Inoue, K., Yanagisawa, R., Koike, E., Nishikawa, M., and Takano, H. (2010). Combustion synthesis of carbon nanotubes and related nano-structures. Prog. Energy Combust. Sci. 36, 696–727. doi:10.1016/j.freeradbiomed.2010.01.013

Ji, Y., Lee, S., Cho, B., Song, S., and Lee, T. (2011). Flexible organic memory devices with multilayer graphene electrodes. ACS Nano 5, 5995–6000. doi:10.1021/nn201770s

Jo, G., Na, S.-I., Oh, S.-H., Lee, S., Kim, T.-S., Wang, G., et al. (2012). The application of graphene as electrodes in electrical and optical devices. Nanotechnology 23, 112001. doi:10.1088/0957-4484/23/11/112001

Jo, G., et al. (2010). Tuning of a graphene-electrode work function to enhance the efficiency of organic bulk heterojunction photovoltaic cells with an inverted structure. Appl. Phys. Lett. 97, 213301. doi:10.1063/1.3514551

Kammler, H. K., Mäer, L. and Pratsinis, S. E. (2001). Flame synthesis of nanoparticles. Chem. Eng. Technol. 24, 583–596. doi:10.1002/1521-4125(200106)24:6<583::AID-CEAT583>3.0.CO;2-H

Karamitaheri, H., Pourfath, M., Faez, R., and Kosina, H. (2011). Geometrical effects on the thermoelectric properties of ballistic graphene antidot lattices. J. Appl. Phys. 110, 054506. doi:10.1063/1.3629990

Kellie, B. M., Silleck, A. C., Bellman, K., Snodgrass, R. and Prakash, S. (2013). Deposition of few-layered graphene in a microcombustor on copper and nickel substrates. RSC Adv. 3, 7100–7105. doi:10.1039/c3ra40632f

Kerbs, H. U., and Bremert, O. (1993). Pulsed-laser deposition of thin metallic alloys. Appl. Phys. Lett. 62, 2341. doi:10.1021/am3022976

Kidoh, H., Ogawa, T., Morimoto, A. and Shimizu, T. (1991). Ferroelectric properties of lead-zirconate-titanate films prepared by laser ablation. Appl. Phys. Lett. 58, 2910. doi:10.1063/1.104719

Kim, J., et al. (2011). Low-temperature synthesis of large-area graphene-based transparent conductive films using surface wave plasma chemical vapor deposition. Appl. Phys. Lett. 98, 091502. doi:10.1063/1.3561747

Kim, J.T., Kim, B.K., Kim, E.Y., Park, H.C., and Jeong, H.M. (2014). Synthesis and shape memory performance of polyurethane/graphene nanocomposites. React. Funct. Polym. 74, 16–21. doi:10.1016/j.reactfunctpolym.2013.10.004

Kim, K. S., Zhao, Y., Jang, H., Lee, S. Y., Kim, J. M., Kim, K. S., et al. (2009). Large-scale pattern growth of graphene films for stretchable transparent electrodes. Nature 457, 706–710. doi:10.1038/nature07719

Kim, M., Safron, N. S., Han, E., Arnold, M. S., and Gopalan, P. (2010). Fabrication and characterization of large-area, semiconducting nanoperforated graphene materials. Nano Lett. 10, 1125–1131. doi:10.1021/nl9032318

Koh, A. T. T., Foong, Y. M. and Chua, D. H. C. (2010). Cooling rate and energy dependence of pulsed laser fabricated graphene on nickel at reduced temperature. *Appl. Phys. Lett.* 97, 114102. doi:10.1063/1.3489993

Krebs, H. U., Weisheit, M., Faupel, J., Süske, E., Scharf, T., Fuhse, C., et al. (2003). Pulsed laser deposition (PLD), a versatile thin film technique. *Adv. Solid State Phys.* 505–518. doi:10.1007/978-3-540-44838-9_36

Krivchenko, V.A., Dvorkin, V.V., Dzbanovsky, N.N., Timofeyev, M.A., Stepanov, A.S., Rakhimov, A.T., et al., (2012). Evolution of carbon film structure during its catalyst-free growth in the plasma of direct current glow discharge. *Carbon N. Y.* 50, 1477–1487. doi:10.1016/j.carbon.2011.11.018

Kuribayashi, K., et al. (1989). Millimeter-sized joint actuator using a shape memory alloy. *Sens. Actuators* 20, 57–64. doi:10.1016/0250-6874(89)87102-1

Layek, R. K., and Nandi, A. K. (2013). A review on synthesis and properties of polymer functionalized graphene. *Polymer* 54, 5087–5103. doi:10.1016/j.polymer.2013.06.027

Lee, C., Wei, X., Kysar, J. W., and Hone, J. (2008). Measurement of the elastic properties and intrinsic strength of monolayer graphene. *Science* 321, 385–388. doi:10.1126/science.1157996

Lee, S.K. and Kim, B.K. (2014). Synthesis and properties of shape memory graphene oxide/polyurethane chemical hybrids. *Polym. Int.* 63, 1197–1202. doi:10.1002/pi.4617

Lee, W. H., Park, J., Sim, S. H., Jo, S. B., Kim, K. S., Hong, B. H., et al (2011). Transparent flexible organic transistors based on monolayer graphene electrodes on plastic. *Adv. Mater. Weinheim* 23, 1752–1756. doi:10.1002/adma.201004099

Li, J. F., et al. (2010d). High-performance nanostructured thermoelectric materials. *NPG Asia Mater.* 2, 152–158. doi:10.1038/asiamat.2010.138

Li, L., et al. (2015). Graphene oxide vs. reduced graphene oxide as core substrate for core/shell-structured dielectric nanoplates with different electro-responsive characteristics. *J. Mater. Chem. C* 3, 5098–5108. doi:10.1039/C5TC00474H

Li, S. S., Tu, K. H., Lin, C. C., Chen, C. W., and Chhowalla, M. (2010b). Solution-processable graphene oxide as an efficient hole transport layer in polymer solar cells. *ACS Nano* 4, 3169–3174. doi:10.1021/nn100551j

Li, X., Zhu, H., Wang, K., Cao, A., Wei, J., Li, C., et al. (2009). Large-area synthesis of high-quality and uniform graphene films on copper foils. *Science* 324, 1312–1314. doi:10.1126/science.1171245

Li, X., et al. (2010c). Graphene-on-silicon Schottky junction solar cells. *Adv. Mater. Weinheim* 22, 2743–2748. doi:10.1002/adma.200904383

Li, X., Zhu, H., Wang, K., Cao, A., Wei, J., Li, C., et al (2008). Chemically derived, ultrasmooth graphene nanoribbon semiconductors. *Science* 319, 1229–1232. doi:10.1126/science.1150878

Li, X., Zhu, H., Wang, K., Cao, A., Wei, J., Li, C., et al (2010a). Graphene-on-silicon Schottky junction solar cells. *Adv. Mater.* 22, 2743. doi:10.1002/adma.200904383

Li, J.-F., Liu, W.-S., Zhao, L.-D., and Zhou, M., (2011a). Ethanol flame synthesis of highly transparent carbon thin films. *Carbon N. Y.* 49, 237–241. doi:10.1016/j.carbon.2010.09.009

Li, Z., Zhu, H., Xie, D., Wang, K., Cao, A., Wei, J., et al. (2011b). Flame synthesis of few-layered graphene/graphite films. *Chem. Commun.* 47, 3520–3522. doi:10.1039/c0cc05139j

Liu, F., Ming, P., and Li, J. (2007). Ab initio calculation of ideal strength and phonon instability of graphene under tension. *Phys. Rev. B* 76, 064120. doi:10.1103/PhysRevB.76.064120

Loh, K. P., Bao, Q., Anga, P. K., and Yang, J. (2010). The chemistry of graphene. *J. Mater. Chem.* 20, 2277–2289. doi:10.1039/b920539j

Loomis, J., King, B., Burkhead, T., Xu, P., Bessler, N., Terentjev, E., et al. (2012). Graphene-nanoplatelet-based photomechanical actuators. *Nanotechnology* 23, 045501. doi:10.1088/0957-4484/23/4/045501

Losurdo, M., Giangregorio, M. M., Capezzuto, P., and Bruno, G. (2011). Graphene CVD growth on copper and nickel: role of hydrogen in kinetics and structure. *Phys. Chem. Chem. Phys.* 13, 20836–20843. doi:10.1039/c1cp22347j

Lotya, M., Hernandez, Y., King, P. J., Smith, R. J., Nicolosi, V., Karlsson, L. S., et al. (2009). Liquid phase production of graphene by exfoliation of graphite in surfactant/water solutions. *J. Am. Chem. Soc.* 131, 3611–3620. doi:10.1021/ja807449u

Luechinger, N. A., Athanassiou, E. K., and Stark, W. J. (2008). Graphene-stabilized copper nanoparticles as an air-stable substitute for silver and gold in low-cost ink-jet printable electronics. *Nanotechnology* 19, 445201. doi:10.1088/0957-4484/19/44/445201

Luk'yanchuk, I. A., Varlamov, A. A., and Kavokin, A. V. (2015). Observation of a giant two-dimensional band-piezoelectric effect on biaxial-strained graphene. *NPG Asia Mater* 7, e154. doi:10.1038/am.2014.124

Malesevic, A., Vitchev, R., Schouteden, K., Volodin, A., Zhang, L., Tendeloo, G. V., et al. (2008). Synthesis of few-layer graphene via microwave plasma-enhanced chemical vapour deposition. *Nanotechnology* 19, 305604. doi:10.1088/0957-4484/19/30/305604

Martin, J. A., Vazquez, L., Bernard, P., Comin, F. and Ferrer, S. (1990). Epitaxial growth of crystalline, diamond-like films on Si(100) by laser ablation of graphite. *Appl. Phys. Lett.* 57, 1742–1744. doi:10.1063/1.104053

Mattevi, C., Kima, H., and Chhowalla, M. (2011). A review of chemical vapour deposition of graphene on copper. *J. Mater. Chem.* 21, 3324–3334. doi:10.1039/C0JM02126A

Mayorov, A. S., Gorbachev, R. V., Morozov, S. V., Britnell, L., Jalil, R., Ponomarenko, L. A., et al. (2011). Micrometer-scale ballistic transport in encapsulated graphene at room temperature. *Nano Lett.* 11, 2396–2399. doi:10.1021/nl200758b

Mazzamuto, F., Hung Nguyen, V., Apertet, Y., Ca, C., Chassat, C., Saint-Martin, J., et al., (2011). Enhanced thermoelectric properties in graphene nanoribbons by resonant tunneling of electrons. *Phys. Rev. B* 83, 235426. doi:10.1103/PhysRevB.83.235426

Meador, M. A., et al. (2010). *DRAFT Nanotechnology Roadmap Technology Area 10. National Aeronautics and Space Administration (NASA)*. 5. Available at: http://www.nasa.gov/pdf/501325main_TA10-Nanotech-DRAFT-Nov2010-A.pdf

Memon, N. K., Tse, S. D., Al-Sharab, J. F., Yamaguchi, H., Goncalves, A.-M. B., Kear, B. H., et al., (2011). Flame synthesis of graphene films in open environments. *Carbon N. Y.* 49, 5064–5070. doi:10.1016/j.carbon.2011.07.024

Memon, N. K., Anjum, D. H., and Chung, S. H. (2013). Multiple-diffusion flame synthesis of pure anatase and carbon-coated titanium dioxide nanoparticles. *Combust. Flame* 160, 1848–1856. doi:10.1016/j.combustflame.2013.03.022

Morozov, S. V., Novoselov, K. S., Katsnelson, M. I., Schedin, F., Elias, D. C., Jaszczak, J. A., et al. (2008). Giant intrinsic carrier mobilities in graphene and its bilayer. *Phys. Rev. Lett.* 100, 016602. doi:10.1103/PhysRevLett.100.016602

Moser, J., Barreiro, A., and Bachtold, A. (2007). Current-induced cleaning of graphene. *Appl. Phys. Lett.* 91, 163513. doi:10.1063/1.2789673

Muge, A., and Chabal, Y. J. (2011). Nature of graphene edges: a review. *Jpn. J. Appl. Phys.* 50, 070101. doi:10.1021/ar500306w

Nair, R. R., Blake, P., Grigorenko, A. N., Novoselov, K. S., Booth, T. J., Stauber, T., et al. (2008). Fine structure constant defines visual transparency of graphene. *Science* 320, 1308. doi:10.1126/science.1156965

Nair, R. R., Ren, W., Jalil, R., Riaz, I., Kravets, V. G., Britnell, L., et al. (2010). Fluorographene: a two-dimensional counterpart of teflon. *Small* 6, 2877–2884. doi:10.1002/smll.201001555

Nakamura, K., Fukazawa, K., Yamada, K., and Saito, S. (1989). Hysteresis-free piezoelectric actuators using linbo3 plates with a ferroelectric inversion layer. *Ferroelectrics* 93, 211–216. doi:10.1080/00150198908017348

Nechache, R., Harnagea, C., Pignolet, A., Normandin, F., Veres, T., Carignan, L.-P. et al. (2006). Growth, structure, and properties of epitaxial thin films of first-principles predicted multiferroic Bi2FeCrO6. *Appl. Phys. Lett.* 89, 102902–102902–3. doi:10.1063/1.2346258

Neto, A. H. C., Guinea, F., Peres, N. M. R., Novoselov, K. S., and Geim, A. K. (2009). The electronic properties of graphene. *Rev. Mod. Phys.* 81, 109–162. doi:10.1103/RevModPhys.81.109

Ni, Z., Wang, Y., Yu, T., and Shen, Z. (2008). Raman spectroscopy and imaging of graphene. *Nano Res.* 1, 273–291. doi:10.1007/s12274-008-8036-1

Novoselov, K. S., Geim, A. K., Morozov, S. V., Jiang, D., Katsnelson, M. I., Grigorieva, I. V., et al. (2005). Two-dimensional gas of massless dirac fermions in graphene. *Nature* 438, 197–200. doi:10.1038/nature04233

Novoselov, K. S., Geim, A. K., Morozov, S. V., Jiang, D., Zhang, Y., Dubonos, S. V., et al. (2004). Electric field effect in atomically thin carbon films. *Science* 306, 666–669. doi:10.1126/science.1102896

Ohta, H., Kim, S., Mune, Y., Mizoguchi, T., Nomura, K., Ohta, S., et al. (2007). Giant thermoelectric Seebeck coefficient of a two-dimensional electron gas in SrTiO3. *Nat. Mater.* 6, 129–134. doi:10.1038/nmat1821

Ossler, F., Wagner, J., Canton, S., and Wallenberg, R. (2010). Sheet-like carbon particles with graphene structures obtained from a Bunsen flame. *Carbon N. Y.* 48, 4203–4206. doi:10.1016/j.carbon.2010.07.013

Ouyang, F., Peng, S., Liu, Z., and Liu, Z. (2011). Bandgap opening in graphene antidot lattices: the missing half. *ACS Nano* 5, 4023–4030. doi:10.1021/nn200580w

Ouyang, T., and Hu, M. (2010). Enhanced thermoelectric figure of merit in edge-disordered zigzag graphene nanoribbons. *Phys. Rev. B* 81, 113401. doi:10.1088/0957-4484/25/24/245401

Park, J. and Kim, B.K. (2014). Infrared light actuated shape memory effects in crystalline polyurethane/graphene chemical hybrids. *Smart Mater. Struct.* 23, 025038. doi:10.1088/0964-1726/23/2/025038

Park, J. B., Xiong, W., Gao, Y., Qian, M., Xie, Z. Q., Mitchell, M., et al., (2011). Fast growth of graphene patterns by laser direct writing. *Appl. Phys. Lett.* 98, 123109. doi:10.1063/1.3569720

Pedersen, T. G., Flindt, C., Pedersen, J., Mortensen, N. A., Jauho, A. P., and Pedersen, K. (2008). Graphene antidot lattices: designed defects and spin qubits. *Phys. Rev. Lett.* 100, 136804. doi:10.1103/PhysRevLett.100.136804

Pereira, V. M., and Castro Neto, A. H. (2009). Strain engineering of graphene's electronic structure. *Phys. Rev. Lett.* 103, 046801. doi:10.1103/PhysRevLett.103.046801

Petersen, R., Pedersen, T. G., and Jauho, A. P. (2011). Clar sextet analysis of triangular, rectangular, and honeycomb graphene antidot lattices. *ACS Nano* 5, 523–529. doi:10.1021/nn102442h

Ponnamma, D., Guo, Q., Krupa, I., Al-Maadeed, M. A., K, T. V., Thomas, S., et al. (2010). Recent advances in graphene based polymer composites. *Prog. Polym. Sci.* 35, 1350–1375. doi:10.1039/c4cp04418e

Potts, J. R., Dreyer, D. R., Bielawski, C. W., and Ruoff, R. S. (2011). Graphene-based polymer nanocomposites. *Polymer* 52, 5–25. doi:10.1016/j.polymer.2010.11.042

Rana, S., Cho, J. W. and Tan, L. P., (2013). Graphene-crosslinked polyurethane block copolymer nanocomposites with enhanced mechanical, electrical, and shape memory properties. *RSC Adv.* 3, 13796–13803. doi:10.1039/c3ra40711j

Reina, A., Jia, X., Ho, J., Nezich, D., Son, H., Bulovic, V., et al. (2008). Large area, few-layer graphene films on arbitrary substrates by chemical vapor deposition. *Nano Lett.* 9, 30–35. doi:10.1021/nl801827v

Reina, A., Jia, X., Ho, J., Nezich, D., Son, H., Bulovic, V., et al. (2009). Large area, few-layer graphene films on arbitrary substrates by chemical vapor deposition. *Nano Lett.* 9, 30–35. doi:10.1021/nl801827v

Ren, Z. F., Huang, Z. P., Xu, J. W., Wang, J. H., Bush, P., Siegal, M. P., et al. (1998). Synthesis of large arrays of well-aligned carbon nanotubes on glass. *Science* 282, 1105–1107. doi:10.1126/science.282.5391.1105

Saravanakumar, B., Mohan, R., Kim, S.-J. (2013). Facile synthesis of graphene/ZnO nanocomposites by low temperature hydrothermal method. *Mater. Res. Bull.* 48, 878–883. doi:10.1016/j.materresbull.2012.11.048

Sarma, S. D., Adam, S., Hwang, E. H., and Rossi, E. (2011). Electronic transport in two-dimensional graphene. *Rev. Mod. Phys.* 83, 407–470. doi:10.1103/RevModPhys.83.407

Schwierz, F. (2010). Graphene transistors. *Nat. Nanotechnol.* 5, 487–496. doi:10.1038/nnano.2010.89

Sevinçli, H., Sevik, C., Cain, T., and Cuniberti, G. (2009). Disorder enhances thermoelectric figure of merit in armchair graphane nanoribbons. *Appl. Phys. Lett.* 95, 192114. doi:10.1038/srep01228

Sharma, R., Alam, F., Sharma, A. K., Dutta, V. and Dhawan, S. K. (2014). ZnO anchored graphene hydrophobic nanocomposite-based bulk heterojunction solar cells showing enhanced short-circuit current. *J. Mater. Chem. C* 2, 8142–8151. doi:10.1039/C4TC01056F

Shen, T., Wu, Y. Q., Capano, M. A., Rokhinson, L. P., Engel, L. W. and Ye, P. D. (2008). Magneto-conductance oscillations in graphene antidot arrays. *Appl. Phys. Lett.* 93, 122102. doi:10.1063/1.2988725

Shinde, S. R., Ogale, S. B., Greene, R. L., Venkatesan, T., Canfield, P. C., Bud'ko, S. L., et al. (2001). Superconducting MgB2 thin films by pulsed laser deposition. *Appl. Phys. Lett.* 79, 227–229. doi:10.1063/1.1385186

Shim, J.-P., Choe, M., Jeon, S.-R., Seo, D., Lee, T. and Lee, D.-S., (2011). InGaN-based p-i-n solar cells with graphene electrodes. *Appl. Phys. Express* 4, 052302. doi:10.1143/APEX.4.052302

Singh, V., Joung, D., Zhai, L., Das, S., Khondaker, S. I., and Seal, S. (2011). Graphene based materials: past, present and future. *Prog. Mater. Sci.* 56, 1178–1271. doi:10.1016/j.pmatsci.2011.03.003

Smela, E. and Gadegaard, N. (1999). Surprising volume change in PPy(DBS): an atomic force microscopy study. *Adv. Mater.* 11, 953. doi:10.1002/(SICI)1521-4095(199908)11:11<953::AID-ADMA953>3.3.CO;2-8

Smith, H.M. and Turner, A.F. (1965). Vacuum deposited thin films using a ruby laser. *Appl. Opt.* 4, 147–148. doi:10.1364/AO.4.000147

Stankovich, S., Dikin, D. A., Dommett, G. H., Kohlhaas, K. M., Zimney, E. J., Stach, E. A., et al. (2006). Graphene-based composite materials. *Nature* 442, 282–286. doi:10.1038/nature04969

Suemitsu, M. and Fukidome, H. (2010). Epitaxial graphene on silicon substrates. *J. Phys. D Appl. Phys.* 43, 374012. doi:10.1088/0022-3727/43/37/374012

Suk, J. W., Kitt, A., Magnuson, C. W., Hao, Y., Ahmed, S., An, J., et al. (2011). Transfer of CVD-grown monolayer graphene onto arbitrary substrates. *ACS Nano* 5, 6916–6924. doi:10.1021/nn201207c

Sullivan, T. D., and Powers, J. M. (1977). Flexural disc piezoelectric polymer hydrophones. *J. Acoust. Soc. Am.* 60, S47–S47.

Sun, Z., Hasan, T., Torrisi, F., Popa, D., Privitera, G., Wang, F. et al., (2009). Graphene mode-locked ultrafast laser. *ACS Nano* 4, 803. doi:10.1021/nn901703e

Sun, Z., Yan, Z., Yao, J., Beitler, E., Zhu, Y., and Tour, J. M. (2010). Growth of graphene from solid carbon sources. *Nature* 468, 549–552. doi:10.1038/nature09579

Tang, S., and Cao, Z. (2010). Structural and electronic properties of the fully hydrogenated boron nitride sheets and nanoribbons: insight from first-principles calculations. *Chem. Phys. Lett.* 488, 67–72. doi:10.1016/j.cplett.2010.01.073

Thakur, S. and Karak, N. (2013). Bio-based tough hyperbranched polyurethane-graphene oxide nanocomposites as advanced shape memory materials. *RSC Adv.* 3, 9476–9482. doi:10.1039/c3ra40801a

Tung, V. C., Chen, L. M., Allen, M. J., Wassei, J. K., Nelson, K., Kaner, R. B., et al. (2009). Low-temperature solution processing of graphene– carbon nanotube hybrid materials for high-performance transparent conductors. *Nano Lett.* 9, 1949–1955. doi:10.1021/nl9001525

Tyler, W.W. and Wilson, A.C. (1953). Thermal conductivity, electrical resistivity, and thermoelectric power of graphite. *Phys. Rev.* 89, 870–875. doi:10.1103/PhysRev.89.870

Voon, L. C., Sandberg, E., Aga, R. S., and Farajian, A. A. (2010). Hydrogen compounds of group-IV nanosheets. *Appl. Phys. Lett.* 97, 163114. doi:10.1063/1.3495786

Wager, J.F. (2003). Transparent electronics. *Science* 300, 1245–1246. doi:10.1126/science.1085276

Wan, X., Long, G., Huang, L., and Chen, Y. (2011). Graphene – a promising material for organic photovoltaic cells. *Adv. Mater. Weinheim* 23, 5342–5358. doi:10.1002/adma.201102735

Wang, C., Liu, N., Allen, R., Tok, J. B., Wu, Y., Zhang, F., et al. (2013). A rapid and efficient self-healing thermo-reversible elastomer crosslinked with graphene oxide. *Adv. Mater. Weinheim* 25, 5785–5790. doi:10.1002/adma.201302962

Wang, J., Zhu, M., Outlaw, R. A., Zhao, X., Manos, D. M., and Holloway, B. C. (2004). Synthesis of carbon nanosheets by inductively coupled radio-frequency plasma enhanced chemical vapor deposition. *Carbon N. Y.* 42, 2867–2872. doi:10.1016/j.carbon.2004.06.035

Wang, X., Zhi, L., and Müllen, K. (2008). Transparent, conductive graphene electrodes for dye-sensitized solar cells. *Nano Lett.* 8, 323–327. doi:10.1021/nl072838r

Wang, Y., Zheng, Y., Xu, X., Dubuisson, E., Bao, Q., Lu, J., et al. (2011). Electrochemical delamination of CVD-grown graphene film: toward the recyclable use of copper catalyst. *ACS Nano* 5, 9927–9933. doi:10.1021/nn203700w

Wei, D. and Xu, X. (2012). Laser direct growth of graphene on silicon substrate. *Appl. Phys. Lett.* 100, 023110. doi:10.1063/1.3675636

Wei, D., Mitchell, J. I., Tansarawiput, C., Nam, W., Qi, M., Ye, P. D., et al. (2013). Laser direct synthesis of graphene on quartz. *Carbon N. Y.* 53, 374–379. doi:10.1016/j.carbon.2012.11.026

Wei, P., Bao, W., Pu, Y., Lau, C. N., and Shi, J. (2009). Anomalous thermoelectric transport of dirac particles in graphene. *Phys. Rev. Lett.* 102, 166808. doi:10.1103/PhysRevLett.102.166808

Yan, Y., Liang, Q.-F., Zhao, H., and Wu, C.-Q. (2012). Thermoelectric properties of hexagonal graphene quantum dots. *Phys. Lett. A* 376, 1154–1158. doi:10.1016/j.physleta.2012.02.013

Yannopoulos, S. N., Siokou, A., Nasikas, N. K., Dracopoulos, V., Ravani, F. and Papatheodorou, G. N. (2012). CO2-Laser-induced growth of epitaxial graphene on 6H-SiC(0001). *Adv. Funct. Mater.* 22, 113–120. doi:10.1002/adfm.201101413

Yin, J., Wang, X., Chang, R. and Zhao, X. (2012). Polyaniline decorated graphene sheet suspension with enhanced electrorheology. *Soft Matter* 8, 294–297. doi:10.1039/C1SM06728A

Yin, J., Chang, R., Kai, Y. and Zhao, X. (2013). Highly stable and AC electric field-activated electrorheological fluid based on mesoporous silica-coated graphene nanosheets. *Soft Matter* 9, 3910–3914. doi:10.1039/c3sm27835b

Yin, Z., Wu, S., Zhou, X., Huang, X., Zhang, Q., Boey, F., et al. (2010). Electrochemical deposition of ZnO nanorods on transparent reduced graphene oxide electrodes for hybrid solar cells. *Small* 6, 307–312. doi:10.1002/smll.200901968

Yokomizo, Y., Krishnamurthy, S., and Kamat, P. V. (2013). Photoinduced electron charge and discharge of graphene–ZnO nanoparticle assembly. *Catal. Today* 199, 36–41. doi:10.1016/j.cattod.2012.04.045

Yu, H., Wang, T., Wen, B., Lu, M., Xu, Z., Zhu, C., et al., (2012). Graphene/polyaniline nanorod arrays: synthesis and excellent electromagnetic absorption properties. *J. Mater. Chem.* 22, 21679–21685. doi:10.1039/c2jm34273a

Yu, Q., Lian, J., Siriponglert, S., Li, H., Chen, Y. P. and Pei, S.-S., (2008). Graphene segregated on Ni surfaces and transferred to insulators. *Appl. Phys. Lett.* 93, 113103. doi:10.1063/1.2982585

Yu, W. J., Lee, S. Y., Chae, S. H., Perello, D., Han, G. H., Yun, M., et al. (2011). Small hysteresis nanocarbon-based integrated circuits on flexible and transparent plastic substrate. *Nano Lett.* 11, 1344–1350. doi:10.1021/nl104488z

Yuan, G.D., Zhang, W.J., Yang, Y., Tang, Y.B., Li, Y.Q., Wang, J.X., et al., (2009). Graphene sheets via microwave chemical vapor deposition. *Chem. Phys. Lett.* 467, 361–364. doi:10.1016/j.cplett.2008.11.059

Zhang, H. and Feng, P. X. (2014). Environmentally responsive graphene systems. *Small* 10, 2151–2164. doi:10.1002/smll.201303080

Zhang, W. L., and Choi, H. J. (2014). Graphene oxide based smart fluids. *Soft Matter* 10, 6601–6608. doi:10.1039/c4sm01151a

Zhang, W. L., Liu, Y. D., Choi, H. J., and Kim, S. G. (2012). Electrorheology of graphene oxide. *ACS Appl. Mater. Interfaces* 4, 2267–2272. doi:10.1021/am300267f

Zhang, Y., Tang, T. T., Girit, C., Hao, Z., Martin, M. C., Zettl, A., et al. (2009). Direct observation of a widely tunable bandgap in bilayer graphene. *Nature* 459, 820–823. doi:10.1038/nature08105

Zhang, Y., Tiwarya, P., Scott Parenta, J., and Kontopoulou, M. (2013). Crystallization and foaming of coagent-modified polypropylene: nucleation effects of cross-linked nanoparticles. *Polymer* 54, 4814–4819. doi:10.1016/j.polymer.2013.07.020

Zhang, Y., Wu, S., Wen, Y.-H., and Zhu, Z. (2010a). Surface-passivation-induced metallic and magnetic properties of ZnO graphitic sheet. *Phys. Lett.* 96, 223113. doi:10.1063/1.3442507

Zhang, H. and Feng, P. X. (2010b). Fabrication and characterization of few-layer graphene. *Carbon N. Y.* 48, 359–364. doi:10.1016/j.carbon.2009.09.037

Zhang, W. L., Park, B. J., and Choi, H. J. (2010c). Colloidal graphene oxide/polyaniline nanocomposite and its electrorheology. *Chem. Commun.* 46, 5596–5598. doi:10.1039/c0cc00557f

Zhu, L., Ma, R., Sheng, L., Liu, M., and Sheng, D. N. (2009). Thermopower and Nernst effect in graphene in a magnetic field. *Phys. Rev. B* 80, 081413. doi:10.1103/PhysRevB.80.081413

Zhu, M., Wang, J., Holloway, B. C., Outlaw, R. A., Zhao, X., Hou, K., et al. (2007). A mechanism for carbon nanosheet formation. *Carbon N. Y.* 45, 2229–2234. doi:10.1016/j.carbon.2007.06.017

Zuev, Y. M., Chang, W., and Kim, P. (2009). Thermoelectric and magnetothermoelectric transport measurements of graphene. *Phys. Rev. Lett.* 102, 096807. doi:10.1103/PhysRevLett.102.096807

Conflict of Interest Statement: The authors declare that the research was conducted in the absence of any commercial or financial relationships that could be construed as a potential conflict of interest.

11

Polymer SU-8-based microprobes for neural recording and drug delivery

Ane Altuna[1]*, Javier Berganzo[2] and Luis J. Fernández[3,4]

[1] microLIQUID, Mondragon, Spain, [2] Microsystems Department, IK4-IKERLAN, Mondragon, Spain, [3] Centro de Investigación Biomédica en Red, Biomateriales y Nanomedicina (CIBER-BBN), Zaragoza, Spain, [4] Group of Structural Mechanics and Materials Modelling (GEMM), Aragón Institute of Engineering Research (I3A), Universidad de Zaragoza, Zaragoza, Spain

This manuscript makes a reflection about SU-8-based microprobes for neural activity recording and drug delivery. By taking advantage of improvements in microfabrication technologies and using polymer SU-8 as the only structural material, we developed several microprobe prototypes aimed to (a) minimize injury in neural tissue, (b) obtain high-quality electrical signals, and (c) deliver drugs at a micrometer precision scale. Dedicated packaging tools have been developed in parallel to fulfill requirements concerning electric and fluidic connections, size and handling. After these advances have been experimentally proven in brain using *in vivo* preparation, the technological concepts developed during consecutive prototypes are discussed in-depth now.

Keywords: SU-8 probes, drug delivery, neural activity recording, probe packaging, microfabrication techniques

Edited by:
Daniel J. Solis,
BioNano Genomics, USA

Reviewed by:
Jose L. Garcia-Cordero,
Center for Research and
Advanced Studies of the National
Polytechnic Institute, Mexico
Jungkyu Jay Kim,
Texas Tech University, USA

*Correspondence:
Ane Altuna,
Goiru 9, 20500, Arrasate, Gipuzkoa,
Spain
anealtuna@microliquid.com

Introduction

Microprobe development for neuronal applications has been refined to reach minimal invasiveness, high reproducibility, and monolithic integration with microelectronics. Reducing the invasive nature of neural probes is a major step while aiming to interfere with neural activity patterns at the cellular level. To this purpose, advances in microfabrication technology have been exploited to integrate high-density sensing sites into a single miniaturized probe for simultaneous discrimination of large neuronal ensembles. Hence, sophisticated fabrication methods have been developed in the microelectronics industry for over half a century. As a result, high reliability and high-volume production levels have been achieved. Furthermore, improvements in communication between microtechnologists and the neuroscientific community were updated in relation with the growth of connection sites, reduction of external signal influence, and autonomy of the device.

In addition to improvements of electronics for neural applications, microfluidic integration gained interest over the last years. In fact, precise and spatiotemporal controlled delivery by microprobes is considered to be one of the most promising directions for delivering pharmacological compounds directly into specific brain regions. From its beginning, silicon technology pioneered the integration of fluidic microchannels into a probe (Chen et al., 1997). Anisotropic silicon etching, boron etch stop, and thermal oxidation/low-pressure chemical vapor deposition sealing were key steps in putting together microelectrodes and fluidic channels for a first time. Changes in the response properties of isolated neurons during drug delivery confirmed the validity of this approach. Since then, efforts have been devised toward precisely control and monitor fluid movements (Lin and Pisano, 1999; McAllister et al., 2000; Rathnasingham et al., 2004; Neeves et al., 2006; Papageorgiou et al., 2006; Foley et al., 2009; Rohatgi et al., 2009; Pongrácz et al., 2013; Lee et al., 2015). Recently, multi-shank fluidic silicon probes have been developed for neuroscientific applications (Seidl et al., 2010; Frey et al., 2011; John et al., 2011;

Spieth et al., 2011). Fluidic channels have been successfully integrated into the silicon 3D configuration via isotropic etching and parylene resealing process.

In parallel, polymer-based microfabrication technology has been developed to deal with invasiveness and brittleness associated with silicon (Subbaroyan et al., 2005; Cheung and Renaud, 2006; Mercanzini et al., 2007, 2008). Parylene technology has already been reported as an option for fluidic applications (Takeuchi et al., 2005). A sacrificial photoresist was defined at specific thickness between two parylene layers and later dissolved to create the cavity. As an alternative to this method, thermal bonding was used (Ziegler et al., 2006). Its main advantages were simplicity, fast production, and the possibility to increase adhesion between the structural layers. Microprobes based on polyimide technology were also developed (Metz et al., 2004). Sputtering and dry etching technique were used to integrate electrodes on the polymer, and lamination technique was implemented to create the channel. It was concluded that this technology offers versatibility for the design of electrodes and channels in terms of size, number, and position. Also, benzocyclobutene (BCB) has been combined with silicon technology to include both, electrodes and fluidics (Lee et al., 2004).

Our group has already exploited polymer SU-8 microfabrication technology in order to integrate multiple electrodes and fluidic channels into a single probe. Over the last years, it has been demonstrated that SU-8 polymer provides optimal properties for neural applications (Lu et al., 2006; Cho et al., 2008; Fernández et al., 2009; Altuna et al., 2010, 2012, 2013). Its high-aspect ratio capability has enabled good dimensional control, and therefore, probes with vertical sidewalls provided a clean insertion into the soft tissue. In addition, high uniformity and adhesion properties ensured the positioning of the metallic tracks on top of the polymer. Recently, planar electrodes were integrated at probe surface and signal recording improved significantly (Altuna et al., 2012). Bonding at low temperature and pressure was first suggested as an optimal technique to create embedded microchannels (Blanco et al., 2004; Agirregabiria et al., 2005; Arroyo et al., 2007; Fernández et al., 2009). Now, even lower temperature combined with higher pressure has been used to avoid internal forces and to guarantee probe planarity. In all cases, the mechanical suitability of SU-8-based probes were experimentally tested for neural application (Lu et al., 2006; Cho et al., 2008; Fernández et al., 2009; Altuna et al., 2010, 2012, 2013). In addition, new achievements regarding probe packaging are addressed in this document. Packaging is known to be a major cost factor in the field of medical microdevices. Therefore, simple, robust, and re-usable housing becomes highly desirable, especially when life-limiting components are used. When electric and fluidic functions are integrated into a single microdevice, all connections have to fit into tight space restriction. Accordingly, our microfluidic SU-8 probes have been displayed every time by a dedicated packaging based on high-resolution 3D printing techniques (Fernández et al., 2009; Altuna et al., 2010, 2012, 2013).

Viability Study of SU-8 Microprobes for Neural Applications

Our first SU-8 prototype included novel design and material concepts to allow a flexible microprobe fabrication for a wide range of neural applications (**Figure 1A**). Polymer SU-8 was chosen as an exclusive structural material aimed to integrate miniaturized electrodes and fluidic microchannels. Insertion area was chosen to be considerably larger than typical silicon-based probes during the first stage of development (400 µm × 220 µm) to ensure mechanical stiffness of the SU-8 shank. Six electrodes with a size of 50 µm × 50 µm were placed close to the tip to facilitate recording of neuronal signals close to the delivery radius. Photolithography technique was used to define the pattern of the probe, and metallic layers were coated on top of a thick undeveloped SU-8 layer for the first time by a controlled sputtering process. In addition, SU-8 made microchannels were processed on top by thermocompression based on a recipe developed by Ikerlan/IK4 (Blanco et al., 2004; Agirregabiria et al., 2005; Arroyo et al., 2007). Preliminary electrical characterization verified the sensing capability of the probe and biological tests confirmed clean delivery into the brain (Fernández et al., 2009).

Microprobe Development: Minimal Tissue Damage and Improved Electrical Recording

Once functional viability of the first prototype was verified, the dimensions of the insertion area were reduced in an order of magnitude to limit injury of the neural tissue. Probe lengths and designs were adapted for *in vitro* experiments and a tetrode configuration was implemented at the tip in order to record activity from small neuronal groups and to discriminate independent neurons (**Figure 1B**). The electrode diameter was set at 20 µm, quite comparable to neuron size. In relation with the fabrication procedure, baking conditions and photolithography exposure dose were optimized to gain resolution of such reduced patterns. This design enabled the recording of extracellular action potentials with peak-to-peak amplitudes comparable to conventional tetrode (Altuna et al., 2010). Then, in an effort to improve the recording capability of the tetrode, a revolutionary fabrication sequence was developed (Altuna et al., 2012). The typical gap between the sensing site and tissue was fully eliminated and planar electrodes were located at the probe surface level (**Figure 1C**). In addition, the thickness of the nearest SU-8 layer was reduced to minimize injury at depth brain *in vivo* recordings. This advances positively impacted on recording capability, as tested experimentally in the rat hippocampus *in vivo*. Action potentials emerging from individual neurons with high peak-to-peak amplitudes were measured using SU-8 probes.

At this stage, a novel fabrication procedure was developed to better integrate planar electrodes and well-defined microchannels in reduced probe sizes. Once the technological scope was demonstrated (**Figure 1D**), we moved forward in order to increase specificity in the application field. The position of the outlets was

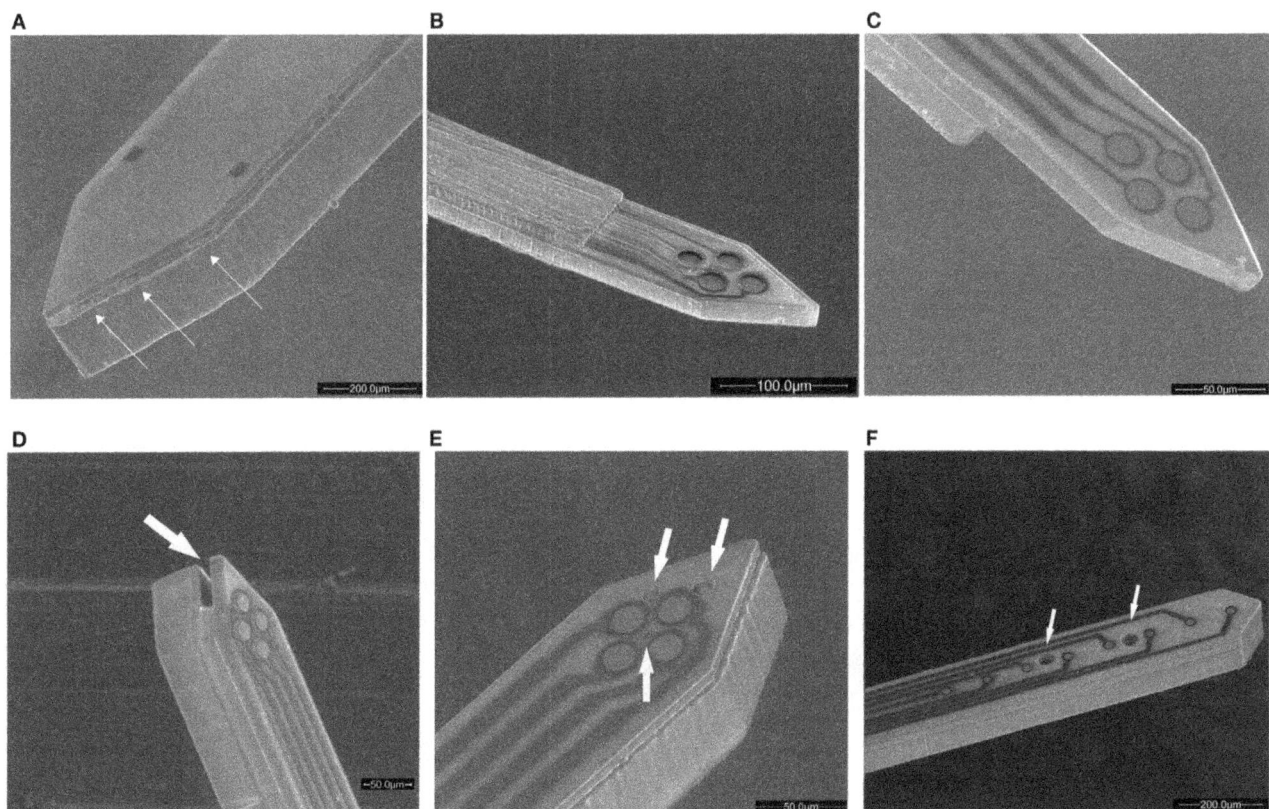

FIGURE 1 | Evolution of the SU-8 microprobe shown by scanning electron microscopy (SEM) figures: (A) initial electrode configuration and a fluidic channel with multiple lateral outlets, **(B)** first tetrode version for *in vitro* applications, **(C)** the tetrode at the probe surface level, **(D)** fluidic implementation to the tetrode version, **(E)** the tetrode with three fluidic outlets in the same face of the probe, **(F)** eight electrodes in a row and two fluidic outlets in the same face of the probe. The outlets are indicated with arrows.

changed from the lateral side of the probe to the top side in order to place the outlets in the same level as the electrodes (**Figure 1E**). Additionally, a linear electrode configuration was developed to facilitate recording from several brain layers simultaneously. In relation to fluidics, two channels with internal dimensions of $40\,\mu m \times 20\,\mu m$ were added. Both channels had independent inlet and outlet ports in order to control drug delivery individually (**Figure 1F**).

Development of Packaging Tools

In parallel to the progress of probe design and microfabrication, packaging was adjusted to each probe prototype. The first packaging was aimed to integrate multiple sensing sites and a single fluidic connection (Fernández et al., 2009). A proper fixation of the probe into the capsule and an improved sealing ensured pressure application up to 4 bar (**Figure 2A**). Next probe design required exclusively electric connection and to facilitate handling in typical experimental setups available at neuroscience research laboratories, therefore, a printed circuit board (PCB) with 16 electric contacts was developed (Altuna et al., 2010). A conductive adhesive between the microelectrode and electric pad

of the PCB enabled a robust connection for electric data transfer (**Figure 2B**). The third design added complexity due to its eight electric and fluidic ports in such a reduced device (**Figure 2C**). With this prototype we recorded for the first time neural ensemble activity in response to local drug delivery at a microscale resolution (Altuna et al., 2013).

Specific details of early SU-8 microprobe prototypes (**Figures 1A–D**) and their corresponding packaging (**Figures 2A–C**) have been already published. Since the last probe and packaging prototype provided us reliable experimental results, both probe and packaging designs are now under a commercialization program by microLIQUID (www.microliquid.com).

Discussion and Future

This manuscript summarizes the evolution of our SU-8 microprobe's design and packaging since its beginning up to the present. The first probe prototype was the master key, since it was demonstrated the viability of polymer SU-8 as structural material for neural applications at fine scale. In each consecutive prototype, electrode design was improved for better signal-to-noise ratio and the fluidic configuration evolved to multiple independent

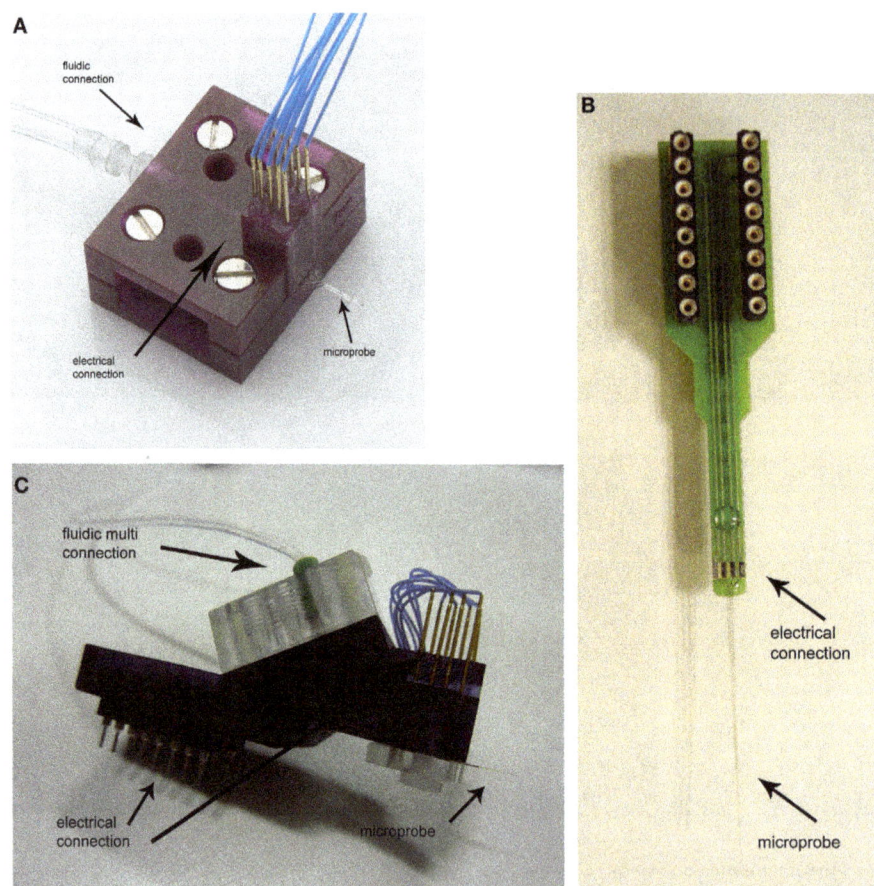

FIGURE 2 | SU-8 probe-packaging designs: (A) first packaging design with eight electric contacts and a fluidic connection, (B) PCB with tetrode connections, and (C) novel packaging with multiple electric and fluidic connections.

microchannels facing the same side of the electrode. Accordingly, our three packaging designs not only conceived to ensure reliable connections but also easy manipulation. Just as technological aspects of the probe and packaging were improved over time, the complexity of the experiments also increased until simultaneous functionality for depth electrical recording and drug delivery in the brain at a scale of few hundreds of micrometers was demonstrated. Future steps will be focused on chronic neural applications aimed to gain better spatiotemporal control of drug delivery and recording during long-term experiments. In this direction, specific packaging tools will be manufactured with the aim to get a compact probe-packaging ensemble.

Acknowledgments

The authors would like to thank the Biomedical Application Group at CNM, Barcelona and the work team of Liset Menendez de la Prida at Instituto Cajal, Madrid.

References

Agirregabiria, M., Blanco, F. J., Berganzo, J., Arroyo, M. T., Fullaondo, A., Mayora, K., et al. (2005). Fabrication of SU-8 multilayer microstructures based on successive CMOS compatible adhesive bonding and releasing steps. *Lab. Chip* 5, 545–552. doi:10.1039/b500519a

Altuna, A., Bellistri, E., Cid, E., Aivar, P., Gal, B., Berganzo, J., et al. (2013). SU-8 based microprobes for simultaneous neural depth recording and drug delivery in the brain. *Lab. Chip* 13, 1422–1430. doi:10.1039/c3lc41364k

Altuna, A., Gabriel, G., De La Prida, L. M., Tijero, M., Guimerá, A., Berganzo, J., et al. (2010). SU-8-based microneedles for in vitro neural applications. *J. Micromech. Microeng.* 20. doi:10.1088/0960-1317/20/6/064014

Altuna, A., Menendez de la Prida, L., Bellistri, E., Gabriel, G., Guimerá, A., Berganzo, J., et al. (2012). SU-8 based microprobes with integrated planar electrodes for enhanced neural depth recording. *Biosens. Bioelectron.* 37, 1–5. doi:10.1016/j.bios.2012.03.039

Arroyo, M. T., Fernández, L. J., Agirregabiria, M., Ibáez, N., Aurrekoetxea, J., and Blanco, F. J. (2007). Novel all-polymer microfluidic devices monolithically integrated within metallic electrodes for SDS-CGE of proteins. *J. Micromech. Microeng.* 17, 1289–1298. doi:10.1088/0960-1317/17/7/011

Blanco, F. J., Agirregabiria, M., Garcia, J., Berganzo, J., Tijero, M., Arroyo, M. T., et al. (2004). Novel three-dimensional embedded SU-8 microchannels fabricated using a low temperature full wafer adhesive bonding. *J. Micromech. Microeng.* 14, 1047–1056. doi:10.1088/0960-1317/14/7/027

Chen, J., Wise, K. D., Hetke, J. F., and Bledsoe, S. C. Jr. (1997). A multichannel neural probe for selective chemical delivery at the cellular level. *IEEE Trans. Biomed. Eng.* 44, 760–769. doi:10.1109/10.605435

Cheung, K. C., and Renaud, P. (2006). BioMEMS for medicine: on-chip cell characterization and implantable microelectrodes. *Solid State Electron.* 50, 551–557. doi:10.1016/j.sse.2006.03.023

Cho, S. H., Lu, H. M., Cauller, L., Romero-Ortega, M. I., Lee, J. B., and Hughes, G. A. (2008). Biocompatible SU-8-based microprobes for recording neural spike signals from regenerated peripheral nerve fibers. *IEEE Sens. J.* 8, 1830–1836. doi:10.1109/JSEN.2008.2006261

Fernández, L. J., Altuna, A., Tijero, M., Gabriel, G., Villa, R., Rodríguez, M. J., et al. (2009). Study of functional viability of SU-8-based microneedles for neural applications. *J. Micromech. Microeng.* 19. doi:10.1088/0960-1317/19/2/025007

Foley, C. P., Nishimura, N., Neeves, K. B., Schaffer, C. B., and Olbricht, W. L. (2009). Flexible microfluidic devices supported by biodegradable insertion scaffolds for convection-enhanced neural drug delivery. *Biomed. Microdevices* 11, 915–924. doi:10.1007/s10544-009-9308-6

Frey, O., Holtzman, T., McNamara, R. M., Theobald, D. E. H., Van Der Wal, P. D., De Rooij, N. F., et al. (2011). Simultaneous neurochemical stimulation and recording using an assembly of biosensor silicon microprobes and SU-8 microinjectors. *Sens. Actuators B Chem.* 154, 96–105. doi:10.1016/j.snb.2010.01.034

John, J., Li, Y., Zhang, J., Loeb, J. A., and Xu, Y. (2011). Microfabrication of 3D neural probes with combined electrical and chemical interfaces. *J. Micromech. Microeng.* 21. doi:10.1088/0960-1317/21/10/105011

Lee, H. J., Son, Y., Kim, J., Lee, C. J., Yoon, E. S., and Cho, I. J. (2015). A multichannel neural probe with embedded microfluidic channels for simultaneous in vivo neural recording and drug delivery. *Lab. Chip* 15, 1590–1597. doi:10.1039/c4lc01321b

Lee, K., He, J., Clement, R., Massia, S., and Kim, B. (2004). Biocompatible benzocyclobutene (BCB)-based neural implants with micro-fluidic channel. *Biosens. Bioelectron.* 20, 404–407. doi:10.1016/j.bios.2004.02.005

Lin, L., and Pisano, A. P. (1999). Silicon-processed microneedles. *J. Microelectromech. Syst.* 8, 78–84. doi:10.1109/84.749406

Lu, H., Cho, S.-H., Lee, J.-B., Romero-Ortega, M., Cauller, L., and Hughes, G. (2006). "SU8-based micro neural probe for enhanced chronic in-vivo recording of spike signals from regenerated axons," in *Proceedings of IEEE Sensors, Art. No. 4178557*, 66–69.

McAllister, D. V., Allen, M. G., and Prausnitz, M. R. (2000). Microfabricated microneedles for gene and drug delivery. *Annu Rev Biomed Eng* 2, 289–313.

Mercanzini, A., Cheung, K., Buhl, D. L., Boers, M., Maillard, A., Colin, P., et al. (2008). Demonstration of cortical recording using novel flexible polymer neural probes. *Sens. Actuators A Phys.* 143, 90–96. doi:10.1016/j.sna.2007.07.027

Metz, S., Bertsch, A., Bertrand, D., and Renaud, P. (2004). Flexible polyimide probes with microelectrodes and embedded microfluidic channels for simultaneous drug delivery and multi-channel monitoring of bioelectric activity. *Biosens. Bioelectron.* 19, 1309–1318. doi:10.1016/j.bios.2003.11.021

Neeves, K. B., Lo, C. T., Foley, C. P., Saltzman, W. M., and Olbricht, W. L. (2006). Fabrication and characterization of microfluidic probes for convection enhanced drug delivery. *J. Control. Release* 111, 252–262. doi:10.1016/j.jconrel.2005.11.018

Papageorgiou, D. P., Shore, S. E., Bledsoe, S. C. Jr., and Wise, K. D. (2006). A shuttered neural probe with on-chip flowmeters for chronic in vivo drug delivery. *J. Microelectromech. Syst.* 15, 1025–1033. doi:10.1109/JMEMS.2005.863733

Pongrácz, A., Fekete, Z., Márton, G., Bérces, Z., Ulbert, I., and Fürjes, P. (2013). Deep-brain silicon multielectrodes for simultaneous in vivo neural recording and drug delivery. *Sens. Actuators B Chem.* 189, 97–105. doi:10.1039/c4lc01321b

Rathnasingham, R., Kipke, D. R., Bledsoe, S. C. Jr., and McLaren, J. D. (2004). Characterization of implantable microfabricated fluid delivery devices. *IEEE Trans. Biomed. Eng.* 51, 138–145. doi:10.1109/TBME.2003.820311

Rohatgi, P., Langhals, N. B., Kipke, D. R., and Patil, P. G. (2009). In vivo performance of a microelectrode neural probe with integrated drug delivery laboratory investigation. *Neurosurg. Focus* 27. doi:10.3171/2009.4.FOCUS0983

Seidl, K., Spieth, S., Herwik, S., Steigert, J., Zengerle, R., Paul, O., et al. (2010). In-plane silicon probes for simultaneous neural recording and drug delivery. *J. Micromech. Microeng.* 20. doi:10.1088/0960-1317/20/10/105006

Spieth, S., Brett, O., Seidl, K., Aarts, A. A. A., Erismis, M. A., Herwik, S., et al. (2011). A floating 3D silicon microprobe array for neural drug delivery compatible with electrical recording. *J. Micromech. Microeng.* 21. doi:10.1088/0960-1317/21/12/125001

Subbaroyan, J., Martin, D. C., and Kipke, D. R. (2005). A finite-element model of the mechanical effects of implantable microelectrodes in the cerebral cortex. *J. Neural Eng.* 2, 103–113. doi:10.1088/1741-2560/2/4/006

Takeuchi, S., Ziegler, D., Yoshida, Y., Mabuchi, K., and Suzuki, T. (2005). Parylene flexible neural probes integrated with microfluidic channels. *Lab. Chip* 5, 519–523. doi:10.1039/b417497f

Ziegler, D., Suzuki, T., and Takeuchi, S. (2006). Fabrication of flexible neural probes with built-in microfluidic channels by thermal bonding of Parylene. *J. Microelectromech. Syst.* 15, 1477–1482. doi:10.1109/JMEMS.2006.879681

Conflict of Interest Statement: The authors declare that the research was conducted in the absence of any commercial or financial relationships that could be construed as a potential conflict of interest.

12

Development of New Smart Materials and Spinning Systems Inspired by Natural Silks and Their Applications

Jie Cheng[1] and Sang-Hoon Lee[1,2,3]*

[1] Department of Biomedical Engineering, College of Health Science, Korea University, Seoul, South Korea, [2] Department of Bio-Convergence Engineering, College of Health Science, Korea University, Seoul, South Korea, [3] KU-KIST Graduate School of Converging Science and Technology, Korea University, Seoul, South Korea

Silks produced by spiders and silkworms are charming natural biological materials with highly optimized hierarchical structures and outstanding physicomechanical properties. The superior performance of silks relies on the integration of a unique protein sequence, a distinctive spinning process, and complex hierarchical structures. Silks have been prepared to form a variety of morphologies and are widely used in diverse applications, for example, in the textile industry, as drug delivery vehicles, and as tissue engineering scaffolds. This review presents an overview of the organization of natural silks, in which chemical and physical functions are optimized, as well as a range of new materials inspired by the desire to mimic natural silk structure and synthesis.

Keywords: silk, protein primary sequence, spinning, hierarchical structure, self-assembly, biomimetic, silk-inspired functional materials

Edited by:
Yongmei Zheng,
Beihang University, China

Reviewed by:
Zhongbing Huang,
Sichuan University, China
Liping Wen,
Chinese Academy of Sciences, China

*Correspondence:
Sang-Hoon Lee
dbiomed@korea.ac.kr

INTRODUCTION

Over the last 5000 years, silk has been widely studied and used in the textile industry; however, many lessons may still be learned from this outstanding natural biological material. A remarkable feature of natural silk is its outstanding toughness, which is superior to that of any of other artificial fiber (Liu et al., 2005; Keten et al., 2010; Omenetto and Kaplan, 2010; Porter et al., 2013; Tokareva et al., 2014). Most researchers believe that the toughness of silk originates from the combined action of the alternating hydrophobic and hydrophilic pattern in the protein primary sequence and the unique spinning process used to prepare the silk fibers. Spiders and silkworms generate a variety of silk fibers. Among the known silk variants, the silk of *Bombyx mori* (*B. mori*) and the orb web spider dragline silks of *Nephila clavipes* and *Araneus diadematus* are the most widely used one. These particular silks have inspired researchers to develop new smart materials, which is the main topic of this review. Although *B. mori* silks are not as tough as spider dragline silks, the high yield of these silks through domesticated production has facilitated the use of these silks in a variety of fields, from luxury fabrics to biomedicine, since their development in China thousands of years ago. Spider dragline silk is thought to be the toughest biological material yet identified; however, limitations on methods for mass production of spider silk have contributed to a gap between applications of silkworm silk and spider silk. Researchers recently developed a series of recombinant DNA technologies that enable spider silk production of the large scale (Tokareva et al., 2013). Recombinant technologies permit genetic modifications that can be imparted on the functionalities of spider silks in addition to their extraordinary mechanical properties (Schacht and Scheibel, 2014). Such advances in biotechnology have led to a flourishing of spider silk applications that have narrowed the gap with silkworm silk applications (Schacht and Scheibel, 2014).

The hierarchical structures of natural silks are crucial to the silks' strength. The design of materials with hierarchical structures similar to that of silk may contribute to the development of novel high-performance functional materials. Silk proteins are hierarchical amphiphilic block proteins composed of alternating hydrophobic and hydrophilic blocks. Hierarchical amphiphilic patterns drive silk proteins to self-assemble into micelles and play a critical role in silk protein storage inside the glands (Jin and Kaplan, 2003; Askarieh et al., 2010; Hagn et al., 2010; He et al., 2012; Schwarze et al., 2013). Therefore, the synthesis of silk proteins similar to the natural proteins is a key issue in material science. Another critical aspect of silk engineering is the spinning system used to prepare the fibers. Such systems are typically designed to mimic natural spinning mechanisms. The tapered spinning duct of silkworm or spider provides an elongational field during spinning process, which orients the silk protein micelles along the direction of the duct via a shear force, leading to the formation of β-sheet crystals. The fibers are then drawn away from the spinning ducts either by the legs of the spider or by the "figure eight" movement of the silkworm head to further orient the silk proteins and form β-sheets. The as-spun silk fibers are composed a hierarchical system of β-sheet crystal domains and α-helix/turn non-crystalline domains, which convey to the silk fibers non-linear material properties and outstanding mechanical properties. Although several reviews of synthetic silk protein morphologies have been published (Altman et al., 2003; Vepari and Kaplan, 2007; Kim et al., 2010, 2014; Rockwood et al., 2011; Tao et al., 2012; Kundu et al., 2014), new materials inspired by silk protein and various spinning methods mimicking spider and silkworm have not been reviewed, despite their importance and extensive applications. In addition, how silk fibers assemble into 3D hierarchical macroscale structures (cocoons and webs) to meet various needs by silkworms and spiders are not summarized until now.

This review presents an overview of the beautiful organizational structures of natural silks from the nanoscale (protein sequence), microscale (spinning) to macroscale (cocoons and webs) aspect and introduces new materials and processes inspired by nature. As shown in **Figure 1**, the natural silk is produced from the synthesis of silk protein, spinning by shear force and formation of web and cocoon, and understanding of sequence–structure–property interplay is critical in developing smart materials inspired by the spider and silkworm. Here, we described a variety of new materials and applications inspired by silk's hierarchical structures and silk spinning process.

NEW SYNTHETIC NANOMATERIALS INSPIRED BY THE SILK PRIMARY SEQUENCE AND DIRECTED SELF-ASSEMBLY

Nanomaterials are potentially useful in a variety of applications (Zhang, 2003). As scientists have dreamed of designing nanoscale

material macrostructures through bottom-up methods, they have studied mimics of the self-assembly processes used by natural systems, such as silkworms or spiders. The self-assembly of silk inside the gland and spinning duct is a marvelous process. Silk proteins are produced inside glands and stored in relatively high concentrations of up to ~50% w/w without aggregation. Self-assembly among silk proteins into highly ordered micelle cluster structures protects the silk proteins from aggregation. The silk proteins are amphiphilic and consist of alternating hydrophobic and hydrophilic domains with highly conserved C- and N-terminal domains. The hydrophobic interactions between hydrophobic domains constitute the main force driving self-assembly (Jin and Kaplan, 2003). The small hydrophilic domains may remain hydrated inside the micelles, and the terminal domains define the micelle edges, stabilizing the solubilized proteins (Hagn et al., 2010). The hydrophobic domains are dominated by a hydrophobic sequence (GA)n or An, which enables protein chain self-assembly to form β-sheet-rich crystal structure and gives silk fibers their outstanding mechanical properties. Stimuli-responsive C- and N-terminal domains play a role in fiber formation under shear force. The primary silk protein sequence has been introduced into artificial polypeptides/peptides and polymer–peptide hybrids in an effort to direct self-assembly and prepare supramolecular hierarchical architectures (**Table 1**). This section describes the design of diverse supramolecules based on silk proteins.

Polypeptide/Peptide Design Inspired by Silk Proteins
Pure Silk-Like Polypeptides with a Lower Molecular Weight

Recombinant forms of the major native ampullate dragline spidroin 1-like proteins, with a molecular weight range of 250–320 kDa, have been successfully overexpressed in the host *Escherichia coli* BL21 (DE3), and strong fibers have been fabricated using these synthesized proteins (Xia et al., 2010). However, the synthesis of silk-like proteins comparable in size to the native silk protein still poses significant challenges to many material research groups. The most famous spider silk protein, the dragline silk protein of *A. diadematus* fibroin and *N. clavipes* fibroin, has received its fame through intensive investigations and recombinant genetic engineering redesign. Spider dragline silk proteins are great models and guides for the design of novel polypeptide. The spider dragline silk of *N. clavipes* contains at least two proteins: the major ampullate dragline spidroin 1 (MaSp1) and the major ampullate dragline spidroin 2 (MaSp2). The spider dragline silk of *A. diadematus* also contains two major proteins: *A. diadematus* fibroins 3 and 4 (ADF-3 and ADF-4). These dragline spidroins are composed of two alternating peptide motifs: a crystalline [An or (GA)n] motif and a less crystalline (GGX or GPGXX) motif (Scheibel, 2004). The An and (GA)n motifs form β-sheet crystal structures to provide a high tensile strength and stiffness in the dragline silk. On the other hand, the GPGGX motifs form type II β-turn structures, and the GGX motif forms a 3_1-helix, both of which convey extensibility to the dragline fiber (Scheibel, 2004; Tokareva et al., 2013). With this understanding of the interplay between the sequence and the secondary structure

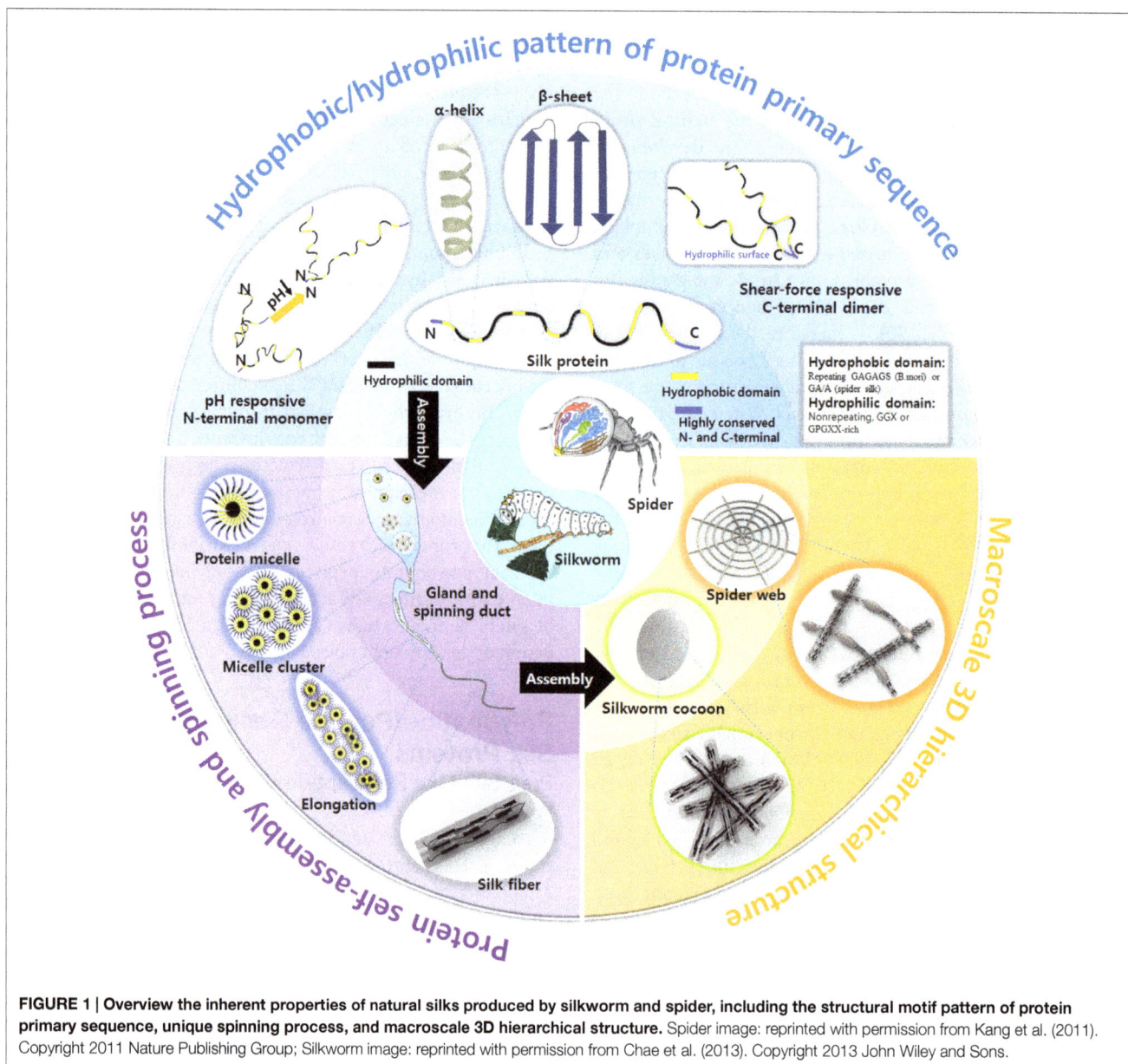

FIGURE 1 | Overview the inherent properties of natural silks produced by silkworm and spider, including the structural motif pattern of protein primary sequence, unique spinning process, and macroscale 3D hierarchical structure. Spider image: reprinted with permission from Kang et al. (2011). Copyright 2011 Nature Publishing Group; Silkworm image: reprinted with permission from Chae et al. (2013). Copyright 2013 John Wiley and Sons.

properties, scientists have designed a series of spider silk-like polypeptides and peptides. The repeating motifs of ADF-3 and ADF-4 have been multimerized through recombinant DNA technology to obtain an ADF-like polypeptide (Huemmerich et al., 2004; Rammensee et al., 2006, 2008). The ADF-4-like polypeptide (called C16) with a molecular mass of 48 kDa self-assembles into nanofibers in the presence of methanol. These nanofibers then further self-assemble to form a hydrogel (Rammensee et al., 2006). The same artificial polypeptide may be used to fabricate both microspheres and nanofibers by inducing self-assembly in the presence of a kosmotropic agent, such as potassium phosphate (Slotta et al., 2007). C16 also can self-assemble to form a thin film at an oil–water interface and encapsulate active agents inside a water phase (Hermanson et al., 2007). C16 adsorbs onto oil–water interface, and a film or microcapsule is formed rapidly due to a secondary structure transition undertaken by C16 at the interface (**Figure 2A**). These encapsulates can be released rapidly by digesting the C16 film with proteinase K (Hermanson et al., 2007). The ADF-3-like polypeptide and ADF-3's conserved non-repeating region at the C-terminus has been recombinantly produced in an effort to probe the natural spider spinning process (Rammensee et al., 2008; Hagn et al., 2010).

Silk Peptide Fusion with Other Functional Peptide Sequences

Spider silk proteins have been fused with a variety of functional peptides, including the integrin-binding motif RGD (Leal-Egana and Scheibel, 2012; Leal-Egaña et al., 2012; Wohlrab et al., 2012), the cell-binding peptide IKVAV (Graf et al., 1987; Grant et al., 1989; Widhe et al., 2013), antimicrobial peptides

TABLE 1 | Summary of the self-assembled nano/microstructures based on silk peptide sequence and their potential applications.

	Sequence[a–c]	Structures	Media	Potential applications	Reference
Peptide/ polypeptide	(GSSAAAAAAAASGPGGYGPENQGPSGPG GYGPGGP)$_{16}$	Nanofiber/hydrogel	Methanol/water	Tissue engineering	Rammensee et al. (2006)
	(GSSAAAAAAAASGPGGYGPENQGPSGPG GYGPGGP)$_{16}$	Sphere	Potassium phosphate	Drug delivery	Slotta et al. (2007)
	(GSSAAAAAAAASGPGGYGPENQGPSG PGGYGPGGP)$_{16}$	Capsule	Water/oil interface	Encapsulation	Hermanson et al. (2007)
	(SGRGGLGGQGAGAAAAAGGAGQG GYGGLGSQGT)$_6$-30K-ppTG1	Nanoparticle	Water	Gene delivery	Numata and Kaplan (2010)
	(SGRGGLGGQGAGAAAAAGGAGQGGYG GLGSQGT)$_6$-30K-CGKRK or -(CGKRK)$_2$	Nanoparticle	Water	Gene delivery	Numata et al. (2011)
	[(GVGVP)$_4$(GYGVP)(GVGVP)$_3$(GAGAGS)]$_{14}$	Nanoparticle	Water	Controlled release	Xia et al. (2011) and Xia et al. (2014)
	[(GVGVP)$_4$(GYGVP)(GVGVP)$_3$(GAGAGS)$_2$]$_{12}$	Nanoparticle/ nanofiber	Water	Controlled release	Xia et al. (2011) and Xia et al. (2014)
	[(GVGVP)$_4$(GYGVP)(GVGVP)$_3$(GAGAGS)$_4$]$_9$	Nanoparticle/ nanofiber	Water	Controlled release	Xia et al. (2011) and Xia et al. (2014)
	H$_{10}$-[(GVGVP)$_4$(GGGVP)(GVGVP)$_3$(GAGAGS)$_4$]$_{n = 3, 8, 13, 18}$	Micelle	Water	Biosensor/drug delivery	Lin et al. (2014)
	(GAGSGAGAGS)[(GVGVP)$_4$(GKGVP) (GVGVP)$_3$(GAGAGS)$_4$]$_{12}$(GVGVP)$_4$(GKGVP) (GVGVP)$_3$(GAGAGS)$_2$GAGA	Injectable hydrogel	Water/in vivo	Gene delivery	Megeed et al. (2002) and Megeed et al. (2004)
	C$_2$-(GAGAGAGX)$_{n = 8, 16, 24, 48}$-C$_2$, X = H, E, K; C = collagen triplet G**XY**	Micelle/nanofiber/ nanotape/hydrogel	Water	Target delivery/tissue engineering	Martens et al. (2009), Beun et al. (2012), Beun et al. (2014), and Golinska et al. (2014)
	A$_n$K, n = 3, 4, 6, 8, 9, 10	Nanotube/nematic lamellar/nanorod/ nanofibril	Water	Nanofabrication/ protein misfolding research/antibacterial peptide	von Maltzahn et al. (2003), Bucak et al. (2009), Qiu et al. (2009), Wang et al. (2009), Castelletto et al. (2010), Chen et al. (2010), Cenker et al. (2011), and Cenker et al. (2014)
	A$_6$D	Nanotube/nanofibril	Water	Protein misfolding research/membrane proteins stabilization	Vauthey et al. (2002) and Zhao et al. (2006)
	A$_6$H	Nanotape/ nanosheet	Acidic/neutral ZnCl$_2$ aqueous solutions	Ions chelation	Castelletto et al. (2014)
	A$_{12}$R$_2$	Twisted fibrils	Water	Nanobiotechnology	Hamley et al. (2013)
Peptide– polymer hybrid	PEG-[(AG)$_3$EG]$_n$-PEG, N = 10 or 20	Fibril/gel	Formic acid/ methanol	Nanofabrication	Smeenk et al. (2005) and Smeenk et al. (2006)
	(PI$_n$-A$_8$-spacer-A$_8$)m, n = 31 or 72	Micelle	Chloroform/2-chloroethanol	Nanofabrication	Zhou et al. (2006)
	A$_x$-(O-CH$_2$-CH$_2$)-A$_x$-PEG, X = 4 or 6	Nanoparticle	Trifluoroethanol	Nanofabrication	Rathore and Sogah (2001a,b)
	(α)tetrathiophene-(GA)$_3$/(GA)$_3$-(α)tetrathiophene-(GA)$_3$	Bilayer/fibril	Dichloromethane/ graphite surface	Nanoelectronics	Gus'kova et al. (2008)
	Poly(isobutylene)-A$_n$/A$_n$-Poly(isobutylene)-A$_n$, n = 0–5	Tape/nanofibril/ interpenetrating supramolecular network	Tetrachlorethane/ chloroform	Elastomer	Croisier et al. (2014)
	HSNGLPLGGGSEEEAAAVVV(K)-CO(CH$_2$)$_{10}$CH$_3$	Nanofiber/hydrogel		Drug delivery/tissue engineering	Shah et al. (2010)

[a]Red-bolds denote silk protein sequence.
[b]Single amino acid abbreviations used: A, Ala; C, Cys; D, Asp; E, Glu; G, Gly; H, His; K, Lys; L, Leu; M, Met; N, Asn; P, Pro; Q, Gln; R, Arg; S, Ser; and V, Val.
[c]Non-functional amino acid sequence domains at N- and C-terminus are omitted for simplicity.
G**XY**: X and Y positions can be any residue but are frequently occupied by Pro and Hyp.

(Gomes et al., 2011), silver ion-binding peptides (Currie et al., 2011), and silica-binding peptides (Wong Po Foo et al., 2006). These hybrids may be applied as a film and coating material to improve the surface properties of biomaterials. Silk protein-based biomaterials prepared via top-down methods have been reviewed extensively elsewhere. Here, we discuss supramolecular assemblies formed by silk-functional peptide fusions. Numata and Kaplan (2010) reported the use of a silk–polylysine–ppTG1

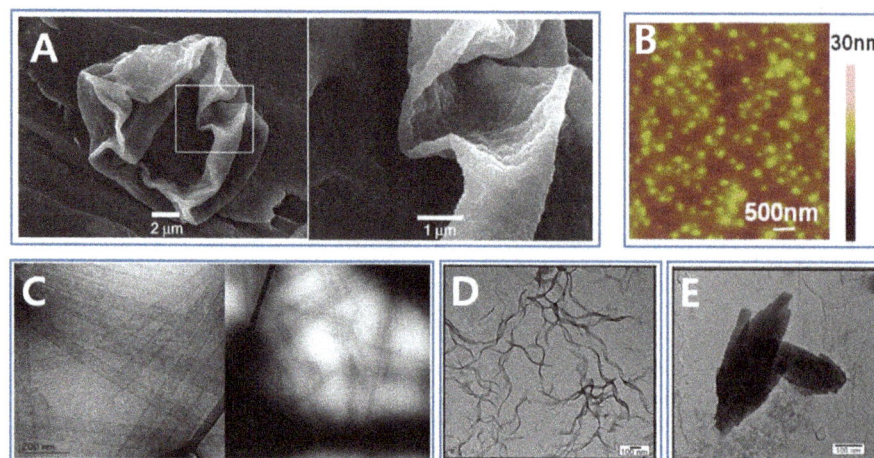

FIGURE 2 | Nano/microstructures formed through molecular self-assembly of silk-based peptide and peptide/polymer hybrid. (A) Microcapsule [Reprinted with permission from Hermanson et al. (2007). Copyright 2007 John Wiley and Sons]. **(B)** Nanoparticle [Reprinted with permission from Numata and Kaplan (2010). Copyright 2010 American Chemical Society]. **(C)** Nanotube [Reprinted with permission from Cenker et al. (2011). Copyright 2011 Royal Society of Chemistry]. **(D)** Nanotape [Reprinted with permission from Castelletto et al. (2014). Copyright 2014 American Chemical Society]. **(E)** Nanosheet [Reprinted with permission from Castelletto et al. (2014). Copyright 2014 American Chemical Society].

fusion polypeptide as gene carrier. The consensus repeat sequence (SGRGGLGGQGAGAAAAAGGAGQGGYGGLGSQGT) of the native MaSp1 was chosen to prepare the fusion polypeptide (silk6mer-30lysine-ppTG1), with 6 contiguous copies of MaSp1 repeating domains, 30 lysines, and a ppTG1 peptide domain fused at its C-terminus. The silk–polylysine–ppTG1 protein trapped pDNA to form micelle-like nanoparticles with an average hydrodynamic diameter of 99 nm (**Figure 2B**), and this structure displayed a high transfection efficiency (Numata and Kaplan, 2010). The profile of the released genes could be further regulated by selectively controlling the β-sheet content of the silk domains in the gene carrier nanoparticle. The same group also fused this basic silk-30 lysine polypeptide to tumor-homing peptides to form gene delivery nanoparticles 150–250 nm in size. This gene delivery system showed significant target specificity to tumor cells (Numata et al., 2011). Silk crystalline protein domains have been genetically fused with other fibrous protein domains, including the elastin repeating hydrophobic domain (GVGVP) and the collagen repeating triplet domain. Silk–elastin polypeptides (SELPs) consist of silkworm silk protein domains (GAGAGS) and environmentally sensitive elastin domains (GXGVP, X could be any amino acid other than proline). SELPs undergo a two-step self-assembly process in aqueous solutions (Xia et al., 2011). The first assembly step is driven by the silk domains. Because the elastin domains are more hydrophilic below their transition temperature than the silk domains, the silk domains drive SELPs to self-assemble into micelles through hydrophobic associations among the silk domains in water. The second self-assembly step is driven by the elastin domains at higher temperatures (Cheng et al., 2013, 2014a). In the second step, the SELP micelles associate to form higher-order nanostructures, including hydrogels and nanofibers (Xia et al., 2011). These SELPs may be used to encapsulate the hydrophobic antitumor drug doxorubicin into the core of the SELP nanoparticles to promote significant uptake by the cells (Xia

et al., 2014). SELPs are good candidates for use in the design of thermo-responsive photonic/electronic systems (Lin et al., 2014). Lin et al. (2014) designed a series of SELPs with a polyhistidine tag at the N-terminus, and these SELPs were shown to bind to the surfaces of nickel chelate nitrilotriacetate-functionalized gold nanoparticles through the nickel affinity of the polyhistidine tag. SELPs drive the gold nanoparticles to aggregate through hydrophobic interactions among the elastin domains at high temperatures, leading to a red shift in the Au nanoparticle surface plasmon resonance. The SELP-Au nanoparticle system may be used as a self-adaptive plasmonic nanodevice for biosensing. Megeed et al. (2002, 2004) designed SELPs-based injectable hydrogels for gene delivery applications. An SELP-47K was designed based on four silk domains and eight elastin domains. A large number of lysine amino acids were coded into elastin domains, and negatively charged DNA bound to the positively charged lysine groups to form a DNA/SELP complex. The SELP/DNA complex formed a hydrogel at body temperature and could be used for injectable gene delivery. Cohen Stuart's group designed a series of stimuli-responsive silk collagen-like polypeptides $\left(C_2S_n^XC_2\right.$ composed of a middle silkworm silk-like peptide domain (S) flanked by two collagen-like peptide domains (C) (Martens et al., 2009; Beun et al., 2012, 2014; Golinska et al., 2014). The middle silk domains consisted of the sequence GAGAGAGX, and pH-responsive amino residues were introduced at position X, resulting in the pH-triggered self-assembly of $C_2S_n^XC_2$ into various nanomaterials (micelles, nanofibers, nanotapes, and hydrogels). The $C_2S_n^XC_2$ sequence was shown to mimic the properties of natural collagen, lending its potential utility in tissue engineering applications.

Small Surfactant-Like Peptide Based on the Silk Protein Domain

The crystalline An domains of the native silk proteins drive self-assembly via hydrophobic interactions and stabilize the assembly via inter- or intramolecular hydrogen bonding. This sequence has been attached to a head group bearing one or more charged amino acids [e.g., histidine (H), arginine (R), lysine (K), or aspartic acid (D)] to prepare surfactant-like small peptides (SLSPs). SLSPs have properties very similar to those observed in biological surfactant molecules and show particular promise in bionanotechnology and biomedicine applications. A variety of silk domain-based SLSPs have been reported, including A_6H, A_6D, AnK, and $A_{12}R_2$, and these SLSPs undergo self-assembly to form nanostructures in an aqueous environment.

The A_nK family is the most extensively studied SLSP, where n may be 3, 4, 6, 8, 9, or 10. The self-assembly of A_nK is concentration dependent. A_6K was found to self-assemble above a critical concentration in aqueous solutions and form hollow nanotubes with a radius of 26 nm (**Figure 2C**), followed by a transition to a nematic lamellar phase with increasing peptide concentration (Bucak et al., 2009; Castelletto et al., 2010; Cenker et al., 2011). A_4K could not self-assemble to form nanostructures. On the other hand, A_8K and $A_{10}K$ self-assembled to form thin rod-like aggregates, possibly due to the increased hydrophobicity as more alanine amino residues were added (Cenker et al., 2014). The A_nK peptides designed by Zhang et al. (von Maltzahn et al., 2003) and later by Zhao et al. (Qiu et al., 2009; Wang et al., 2009) were capped either at both ends or at the N-terminus. These peptides formed nanofibrils other than nanotubes, because the caps influenced the hydrophobicity of the peptides. The A_6D, A_6H, and $A_{12}R_2$ constructs are self-assembled to form various nanostructures, including nanotubes, nanovesicles, nanotapes (**Figure 2D**), nanosheets (**Figure 2E**), or twisted fibrils (Vauthey et al., 2002; Zhao et al., 2006; Hamley et al., 2013; Castelletto et al., 2014).

Polymer–Peptide Hybrids Inspired by Silk Proteins

The incorporation of silk's β-sheet-forming peptide domain (GA)n or (A)n to form a synthetic polymer offers an effective approach to design supramolecular materials via self-assembly. Smeenk et al. (2005, 2006) reported the development of a triblock polymer consisting of a central silk β-sheet domain [(AG)₃EG] n with one PEG block at both its N- and C-termini. The central silk β-sheet domain was overexpressed in *E. coli* and purified via nickel–nitrilotriacetate (Ni–NTA) affinity chromatography. One cysteine amino residue at each peptide terminus was used to selectively alkylate the maleimide-functionalized PEG. The PEG blocks prevented the central silk β-sheet domains from forming random macroscopic aggregates. Instead, the central β-sheet domains stacked together in an orderly fashion to form well-defined fibrils (Smeenk et al., 2005).

Zhou et al. (2006) reported a series of silk-like multiblock peptide–polymer hybrids containing A_5 blocks and polyisoprene oligomers (PI). The PI oligomer improved the solubility of the hybrid polymer, just as glycine-rich domains do in natural silk proteins. The A_5 blocks directed the hybrid polymer to self-assemble into micelles through the formation of an antiparallel β-sheet core in the presence of chloroform or 2-chloroethanol. A similar design concept was applied by Rathore et al., who replaced amorphous domains of the silk proteins with PEG. The remaining β-sheet-forming domain of the silk proteins were then included

in the synthesized PEG–peptide copolymer, which self-assembled to form nanostructures (Rathore and Sogah, 2001a,b).

The GAGAG pentapeptide was conjugated to tetrathiophene to form diblock or triblock peptide/polymer hybrids that self-assembled into supramolecular structures at the graphite surface and in organic solvents due to β-sheet formation among the peptide blocks (Gus'kova et al., 2008). The thiophene blocks further stabilized the supramolecule through the formation of π–π stacks, which conveyed conductivity to the supramolecular structures (Gus'kova et al., 2008).

Croisier et al. (2014) synthesized a series of peptide/ poly(isobutylene) hybrid elastomers. In these hybrids, silk peptides were conjugated to one terminus or both termini of an poly(isobutylene) unit to form a diblock hybrid (monofunctional hybrid M) or a triblock hybrid (difunctional hybrid D). The silk peptide domains directed M and D to self-assemble into various nanostructures. The length of the silk peptide domain determined the morphology of the nanostructures formed through self-assembly. Both the M and D units comprising short peptide domains formed small aggregates. As the length of the peptide domain increased, single β-sheet tapes and stacked nanofibrils formed. Mixing M and D together gave rise to inherently reinforced thermoplastic elastomers and interpenetrating supramolecular networks (Croisier et al., 2014).

Self-assembly in the various silk-inspired peptide–polymer hybrids mentioned above required an organic solvent environment, and hydrogen bonds formed among the silk peptides to strengthen and direct the self-assembly process. Stupp's group described the use of a series of well-known diblock polymers, the peptide amphiphiles (PA), consisting of a peptide segment and a more hydrophobic alkyl tail that self-assemble to form supramolecular nanofibers in an aqueous solution (Hartgerink et al., 2001, 2002; Claussen et al., 2003; Behanna et al., 2005; Shah et al., 2010). The silk GA domains were incorporated into the PAs to provide stability to the nanofibers by forming β-sheets in the fibers (Shah et al., 2010). The PAs provide a powerful tool for drug delivery and tissue engineering.

NOVEL SPINNING PROCESSES AND MICROPLATFORMS INSPIRED BY NATURE

The outstanding mechanical properties of silk fibers originate from both the silk protein's unique primary sequence and the distinctive spinning process used during fiber fabrication. The success of spider silk spinning lies in the various silk proteins produced in its multiple glands, the extensional field formed inside its tapered duct, and the post-spin drawing action of the spider's legs. This section discusses a variety of spinning methods using microfluidic platforms that provide excellent quality fibers through the control over multiphase flows in small amounts of liquid.

Spinning and the Mimicking of Multiple Glands

Individual spiders can spin a variety of silks. For example, the female orb-weaving spiders spin up to seven different types of silks to meet different needs using seven different glands. The multigland design of the orb-weaving spider was implemented by our group to continuously spin fibers with spatiotemporal variations in chemical composition and morphology using polydimethylsiloxane (PDMS) microfluidic chips (**Figure 3A**) (Kang et al., 2011). The microfluidic chip consisted of several individually controllable inlets that mimicked the multiple glands of the spider. Switching among these inlets was precisely controlled using a computer-controlled pneumatic valve that enabled the preparation of fibers with spatially organized morphologies and compositions. Multiple compositions and topographies, including gas microbubbles, nanoporous spindle knots, and a variety of cells, could be coded into a single fiber (Kang et al., 2011). Similar design concepts were implemented by Qin et al. to prepare hybrid microfibers with bamboo-like or pearl-like morphologies by combining the droplet microfluidic technique with microfluidic wet spinning (Yu et al., 2014). Hydrophobic droplets containing polymer microspheres or multicellular spheroids were encapsulated in a continuous alginate hydrophilic phase using "T" junction channels. The Gu's group replaced the PDMS microfluidic chip and mimicked spider's multiple glands using a capillary coaxial microfluidic device (Cheng et al., 2014b). Multiple tapered barrel capillaries were coaxially inserted into a single collection capillary to fabricate two- to six-compartment calcium alginate microfibers. Capillaries have been inserted into a single collection capillary in a hierarchical manner to prepare multilayers microfibers. Different cells could then be encapsulated into different alginate compartments or layers.

Although silkworms, such as *B. mori*, have only two identical glands, three partitions are present in each gland. Silk fibroins are secreted in the posterior gland and transported to the middle gland, where sericins are secreted. In the middle gland, the silk fibroins phase separate to form microscale micelle globules, which improve silk fibroin alignment under physical shear stress (Jin and Kaplan, 2003). A PDMS platform was used to simulate the silkworm spinning process, and a series of micro/nanometer-scale alginate fibers with highly aligned alginate polymers were fabricated by inducing alginate phase separation inside the PDMS channel (Chae et al., 2013). Isopropyl alcohol (IPA) in a sheath flow diffused into the alginate core flow, leading to alginate phase separation to form small alginate particles. The shear forces in the microfluidic channels aligned these alginate particles to form highly crystallized alginate fibers (Chae et al., 2013).

Spinning via a Tapered Duct-Based Elongational Mechanism

The outstanding mechanical properties of spider silk are closely related to the spider's tapered spinning duct, which applies an elongational field that orients the short spidroin filaments along the direction of the elongational field (Vollrath and Knight, 2001). Capillaries with tapered orifices provide a simple approach to create elongational flows. This approach has been used to align small nanofibrils and form long microfibers. We reported the first example of a microfluidic spinning process using a microfluidic chip that integrated a tapered glass capillary (Jeong et al., 2004). Diverse hydrogel-based solids and hollow fibers were fabricated

using variations on these chip designs (Shin et al., 2007; Hwang et al., 2008, 2009; Lee et al., 2009, 2010). Stupp's group used a tapered pipet to prepare noodle-like strings in which peptide amphiphile nanofibers aligned and were bundled along the longitudinal direction of the string (Zhang et al., 2010). The strings could be used to encapsulate cells and direct the orientations of the cells in a 3D environment. Human mesenchymal stem cells (hMSCs) were found to remain viable during the string preparation process. Both the cell bodies and filopodia were aligned along the peptide amphiphile nanofibers due to contact guidance by the oriented matrix (Zhang et al., 2010). Takeuchi's group designed a series of coaxial microfluidic glass capillaries with tapered orifices to control the assembly of diverse small structures under flow (Kiriya et al., 2012a,b; Hirayama et al., 2013; Onoe et al., 2013). In the elongational flow, small nanofibers formed the core flow and were oriented along the flow stream as the sheath flow velocity exceeded the core flow (Kiriya et al., 2012b). The elongational flow overcame the hydrophobic/hydrophilic interactions and oriented the lipid-type monomers to form meter-long fibers. The fibers were then used as templates to synthesize poly(aniline) conductive nanofibers with a high conductivity and potential utility as sensor materials (Kiriya et al., 2012a). Fast sheath flows and slower core flows can create an elongational field and such fields have been used to induce alignment of carbon nanotubes (CNTs) (Davis et al., 2009). Single-wall CNTs in pure chlorosulfonic acid at a concentration of 8.5 vol% were spun into a faster-flowing 96% sulfuric acid sheath flow to form CNT fibers uniformly composed of oriented CNT fibrils. These fibers showed great mechanical properties and a high conductivity (Davis et al., 2009).

Our group has fabricated PDMS-only cylindrical coaxial flow channels with a tapered orifice to mimic the spider's tapered spinning duct (**Figure 3B**) (Kang et al., 2010, 2011). The PDMS-based mold was prepared using soft lithography techniques. A thin PDMS membrane was subsequently bonded to the PDMS base mold, and a negative pressure was applied to form a concave hemicylindrical channel. The hemicylindrical channel was replicated with SU-8, and a convex hemicylindrical SU-8 master was obtained. The convex hemicylindrical SU-8 master was used to prepare a concave hemicylindrical PDMS channel. The PDMS chip used to prepare the fibers was constructed by aligning and bonding two concave hemicylindrical PDMS channels using oxygen plasma treatment (Kang et al., 2010, 2011). This PDMS chip was used to continuously prepare monodisperse microfibers for cell encapsulation and tissue engineering (Kang et al., 2012, 2014; Jun et al., 2013, 2014a,b).

Post-Spin Drawing Process

During the natural silk spinning process, the fibers are drawn out of the spinning duct either using the legs of the spider or through the "figure eight" movements of the silkworm head (Omenetto and Kaplan, 2010). This drawing process orients the silk proteins along the longitudinal axis of the fibers and improves the fibers' mechanical properties. The post-spin drawing process used by silkworms and spiders has been replicated in conventional processes to improve the mechanical properties of the artificial

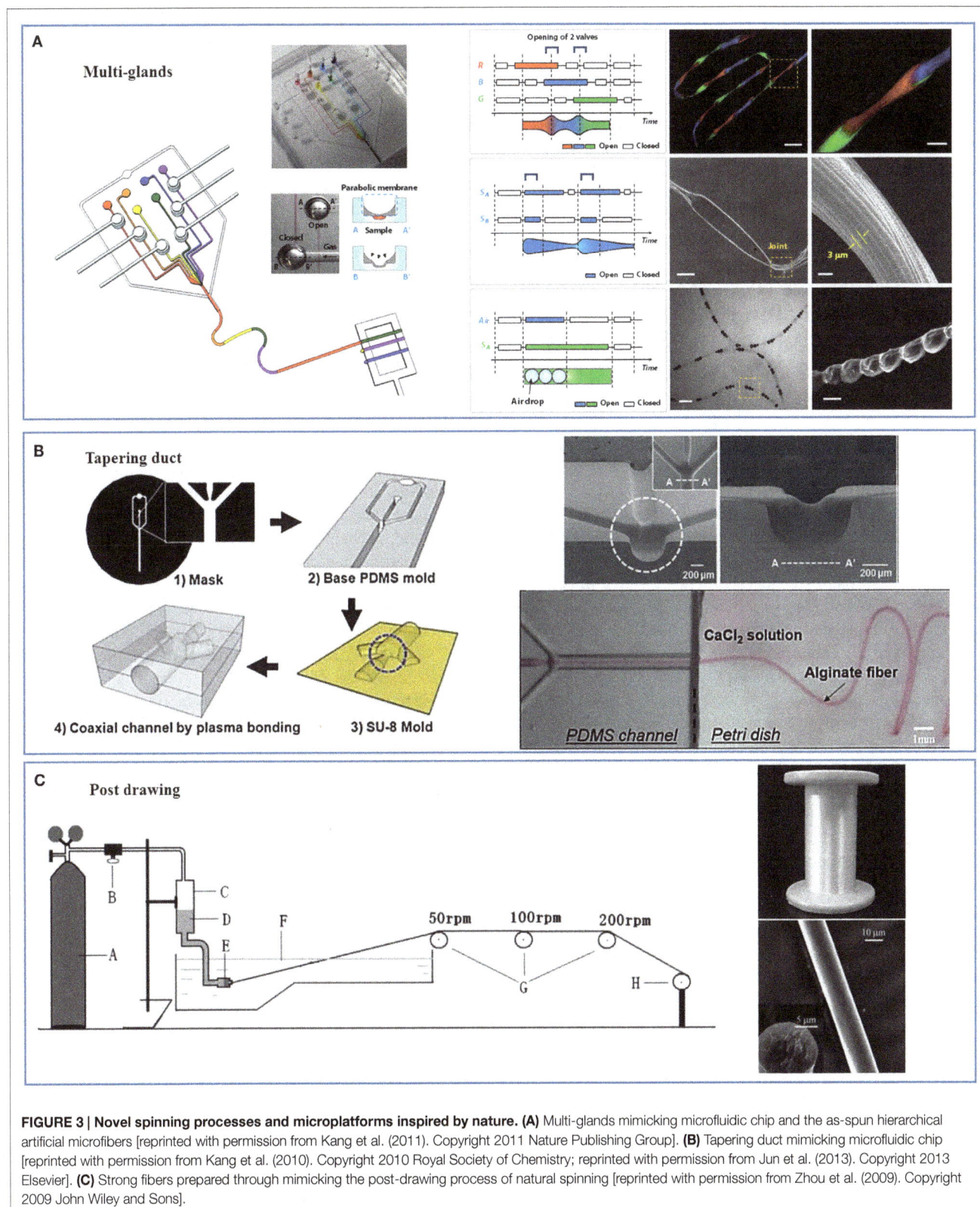

FIGURE 3 | Novel spinning processes and microplatforms inspired by nature. (A) Multi-glands mimicking microfluidic chip and the as-spun hierarchical artificial microfibers [reprinted with permission from Kang et al. (2011). Copyright 2011 Nature Publishing Group]. **(B)** Tapering duct mimicking microfluidic chip [reprinted with permission from Kang et al. (2010). Copyright 2010 Royal Society of Chemistry; reprinted with permission from Jun et al. (2013). Copyright 2013 Elsevier]. **(C)** Strong fibers prepared through mimicking the post-drawing process of natural spinning [reprinted with permission from Zhou et al. (2009). Copyright 2009 John Wiley and Sons].

fibers. Chen et al. prepared strong artificial silk fibroin fibers by continuously drawing the fibers through a series of plastic rollers with different rotating speeds, the post-spun drawn fibers showed mechanical properties comparable to those of the natural silkworm silk fibers (**Figure 3C**) (Zhou et al., 2009; Yan et al., 2010). The same spinning process was applied by Chen's group to a spinning dope composed of a silk fibroin–CNT composite to prepare strong artificial fibers with breaking energies comparable to that of the spider dragline silk fibers (Fang et al., 2015). Lewis's group used post-spin drawing to prepare strong artificial spider silk fibers (An et al., 2011). Post-spin drawing significantly increased the fiber tensile strength due to the improved protein molecule rearrangement and β-sheet formation in the polyalanine region (An et al., 2011). We recently successfully spun microfibers using silk fibroin and alginate, and the mechanical properties of the fibers were significantly enhanced by the post-spin drawing process (unpublished data).

DEVELOPMENT OF NEW SMART MATERIALS AND APPLICATIONS INSPIRED BY COMPLEX HIERARCHICAL ARCHITECTURES CONSTRUCTED BY SILKWORM AND SPIDER

The advanced functions of most natural materials rely on the complex hierarchical structures of the materials (Sanchez et al., 2005). Recent studies have revealed that the strength of silk fibers also relies on its hierarchical structure (Cranford et al., 2012; Koski et al., 2013; Qin and Buehler, 2013; Silva and Rech, 2013; Qin et al., 2015). The structure of a spider's web is hierarchical in two aspects: the protein secondary structure inside a single silk fiber is hierarchical and the geometric structure of the whole web is hierarchical. Together, these structures contribute to the outstanding properties of natural spider webs (Cranford et al., 2012; Qin et al., 2015). The hierarchical protein secondary structure (semi-amorphous and β-sheet nanocrystals) in a single spider silk fiber leads to non-linear stress–strain behaviors of the silk fiber, resulting in small web deformations under the application of an external force to a single fiber (Cranford et al., 2012). Orb web spiders produce up to seven different types of silk fibers using their different glands to weave a hierarchical web net, and any insects trapped in this thin fiber net have no chance of escape. The framework of the web is constructed using dragline silk produced by the major ampullate gland (MA gland). The capture spiral is jointly constructed by minor ampullate spidroin silk fibers produced in the minor ampullate gland (MI gland) and by flagelliform silk fibers produced in the flagelliform gland. The auxiliary spiral is first constructed using ampullate spidroin silk fibers to stabilize the emerging web structure and to serve as a template in support of the construction of the capture structure using the flagelliform silk fibers. The whole web structure is fixed by a gluey substance called attachment cement, which is produced in the piriform gland (Heim et al., 2010).

Spiders spin a variety of silk fibers to meet their different needs, such as prey capture, prey swathing, or egg protection. Although the silkworm *B. mori* produces only one type of silk fiber to meet its exclusive need for cocoon weaving as a defense mechanism, the single silkworm silk fibers and the cocoon are hierarchical structural materials. Silkworm silk fibers consist of a fibroin core and a sericin shell, and the cocoon wall is a multilayer hierarchical architecture randomly weaved using the silk fibers.

Cocoon Silk's Core/Shell Composite Structure Has Inspired the Development of Strong Fibers

Cocoon silk fibers consist of a fibroin core and a sericin coating layer. The sericin layer allows for the efficient transfer of an external loading force to the fibroin core and improves the mechanical properties of the silk fiber. Inspired by this core/shell design, Jiang's group enforced lotus fibers by applying a poly(vinyl alcohol) (PVA) coating layer (**Figure 4A**) (Wu et al., 2014). In addition to improving the mechanical properties, the PVA layers acted as a glue to hold the individual fibers together as a fiber bundle. The mechanical properties of the fibers were further improved by introducing cross-links through heat treatment and glutaraldehyde.

Spider Silk-Inspired CNT-Based Fiber Design

The outstanding mechanical properties of spider silk rely on the plentiful intra- and intermolecular hydrogen bonds among the structural proteins. These features suggested a simple approach to strengthening artificial fibers. CNTs are extremely stiff and strong nanomaterials and have been widely used in nanocomposites to improve the mechanical properties of the materials. Scientists and engineers believe that the best approach to fully utilizing the outstanding mechanical properties of CNTs is to spin them into continuous fibers on the macroscale (Ericson et al., 2004; Zhang et al., 2005; Peng et al., 2008). The continuous spinning of CNTs has remained challenging due to the atomically smooth surfaces and cylindrical profile of the CNTs, which provide a low shear load transfer and facilitate sliding among aligned CNTs (Kis et al., 2004; Peng et al., 2008). Inspired by the silk fibers, hydrogen bonding functional groups were introduced onto the pristine CNT surfaces to improve the shear load transfer and prepare continuous CNT fibers. In addition to small functional groups, such as OH, NH_n may be introduced onto the CNT surfaces to form bridging hydrogen bonds among CNTs. Ultra-short polymers may also be covalently conjugated to the CNT surfaces to create a microstructural environment similar to that present in natural spider silk. Naraghi et al. (2013) functionalized CNT bundles with poly(methyl methacrylate)-like oligomers using chemical vapor deposition methods (**Figure 4B**). The van der Waals interactions increased the shear strength of the bundle junctions by an order of magnitude compared to the shear strength of the pristine CNT bundles.

Spider Silk Adhesion Glue

The viscid fibers of web-weaving spiders are coated with glycoprotein-based glue droplets that facilitate prey capture (**Figure 4C**) (Opell and Hendricks, 2009, 2010; Sahni et al., 2010; Stellwagen et al., 2014). The properties of the glue droplets depend on the rate at which they are extended. The glue droplets act as a viscous material under rapid extension, for example, when snared by a fast-flying insect, or as a viscoelastic material under slow extension, for example, during the attempts by trapped prey to escape (Sahni et al., 2010). The adhesion of the spider's glue droplet is

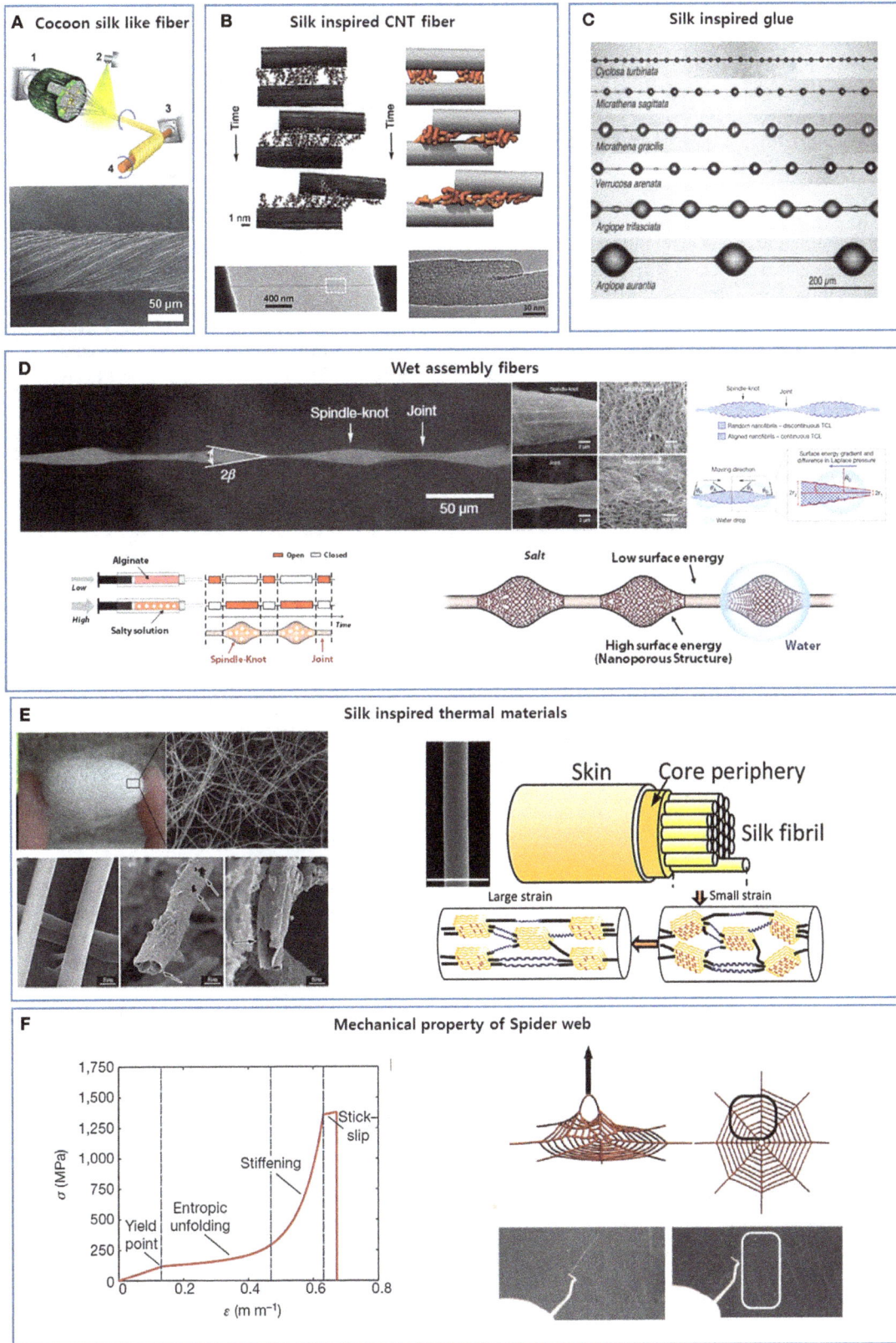

A Cocoon silk like fiber

B Silk inspired CNT fiber

C Silk inspired glue

D Wet assembly fibers

E Silk inspired thermal materials

F Mechanical property of Spider web

FIGURE 4 | Development of new smart materials by mimicking silk's complex hierarchical structure.

(Continued)

humidity responsive and increases with humidity. These properties are exactly opposite the properties displayed by synthetic adhesive systems. Although glue droplets appear to consist of a glycoprotein core and a viscous aqueous salt coating (Vollrath et al., 1990; Opell and Hendricks, 2009, 2010; Sahni et al., 2010; Stellwagen et al., 2014), an in-depth examination of the relationship between the chemical composition and the physical properties has not yet been conducted. Synthetic adhesive systems that display adhesion properties similar to those of spider silk have not yet been developed.

Wet Assembly Fibers

Aside from their outstanding mechanical properties, the water-collecting properties of spider silk can provide critical insights into the design of advanced materials (Wen et al., 2015). The water-collecting properties of silk result from the hierarchical spindle knot-joint structure formed during wetting in humid air (the upper panel in **Figure 4D**) (Zheng et al., 2010). Dry spider silk consists of puffs separated by joints. As the dry spider fibers encounter humid air, tiny water drops condense on the semitransparent puffs, and the puffs transform into periodic spindle knots as water condensation continues (Zheng et al., 2010). The spindle knots and joints are composed of random and aligned nanofibrils, respectively, which results in a surface energy gradient and a difference in the Laplace pressure between the spindle knots and the joints (Zheng et al., 2010). As a result, water continuously condenses and collects around the spindle knots.

Mechanistic insights gleaned from the water collection properties of spider silk have informed the development of spider silk-like artificial fibers based on a range of materials. Water collection properties similar to those of natural spider silk have been achieved. Zheng et al. prepared an artificial spider silk-like fiber using a simple dip-coating method. Uniform nylon fibers were immersed into a poly(methyl methacrylate)/*N,N*-dimethylformamide ethanol (PMMA/DMF-EtOH) solution and then horizontally drawn out quickly (Zheng et al., 2010). After drying, the polymers formed periodic spindle knots on the surfaces of the nylon fibers (Zheng et al., 2010). The artificial polymer spindle knots were composed of random pores and artificial joints prepared from the stretched pores. The differences between the surface morphologies of the artificial spindle knots and joints gave rise to a surface energy gradient and difference in the Laplace pressure, yielding directional water collection around the spindle knots (Zheng et al., 2010).

Coaxial electrospinning methods offer a good approach to preparing spider silk-like fibers with periodic spindle knot/joint patterns (Tian et al., 2011; Dong et al., 2012). A viscous polystyrene (PS) solution and a less viscous PMMA or poly(ethylene glycol) (PEG) solution were used as the inner and outer solutions, respectively, in a coaxial system. Under a high-voltage electric field, the inner PS solution spun to form a thin thread, whereas the outer PMMA or PEG solutions formed a thin film on the surface of the PS thread. The PMMA or PEG films broke to form periodic droplets and solidified into spindle knots on the surface of the PS thread (Tian et al., 2011; Dong et al., 2012). These electrospun fibers demonstrated capacities for directional water collection that resembled the corresponding capacity of natural spider silk fibers.

Our group used a multigland spider-mimicking microfluidic chip to prepare artificial spider silk-like fibers containing a spindle knot/joint pattern that directionally transported water toward the spindle knots (Kang et al., 2011). This microfluidic chip continuously created microfibers with strictly coded morphological and chemical features using a digital programmable flow controller. An alginate solution containing salt was injected into one of sample channels at a high flow rate to construct porous spindle knots (the lower left image in **Figure 4D**). An alginate solution without salt was injected into another sample channel at a low flow rate to construct joints (the lower left image in **Figure 4D**). As the salt leached from one flow to the other through diffusion, nanoporous spindle knots were generated. These spindle knots provided a higher surface energy than the joints. The tapered morphologies of the spindle knots generated a difference in the Laplace pressures of the spindle knots and joints. As expected, water droplets were directionally collected around the spindle knots under the combined action of the surface energy and the Laplace pressure differential (the lower right image in **Figure 4D**).

The relationship between directional water transport and differences in the surface energies and Laplace pressures was used to prepare artificial fibers with the capacity to transport water droplets in the reverse direction, that is, from the spindle knots toward the joints. The relative wettability of the spindle knots and joints was tuned by changing the spindle knots' chemical composition or the roughness (Bai et al., 2010). Interestingly, artificial fibers were prepared to enable *in situ* switching of the directional water droplet movement by introducing a stimulus-responsive material onto the spindle knots (Feng et al., 2013; Hou et al., 2013). Thermo-responsive copolymer poly(methyl methacrylate)-b-poly(*N*-isopropylacrylamide) (PMMA-b-PNIPAAm) or photosensitive azobenzene polymers have been introduced into spindle knots. PMMA-b-PNIPAAm-modified spindle knots displayed a wettability that could be tuned by adjusting the environmental temperature below or above the lower critical solution temperature (LCST) of PNIPAAm. This stimulus switched the direction of the water droplets' movement (Hou et al., 2013). Photosensitive azobenzene polymer-modified spindle knots permitted control over the wettability and, therefore, the droplet movement direction via UV–vis irradiation, which transformed the molecular configuration of the azobenzene polymer from a less hydrophilic to a more hydrophilic state (Feng et al., 2013).

Silk-Inspired Thermal Materials

Silkworms weave cocoons to protect the pupae from extremely cold weather. The cocoon wall is a multilayer hierarchical fiber-based structure that traps air within the cocoon to provide thermal insulation. The still air is retained within the cocoon wall, even in the presence of wind, by depositing a calcium oxalate crystal layer on the outer surface of the cocoon wall. This layer reduces wind penetration (Zhang et al., 2013; Tao et al., 2015). With this understanding in mind, Wang et al. (2013) fabricated thermally insulating hollow biomorphic fibers using silkworm silk as the templates (the left image in **Figure 4E**). The silk fibers were immersed in an $AlCl_3$ solution and dried at 70°C over 24 h. An alumina layer formed on the surfaces of the silk fibers during an oxidation reaction. The silk fibers themselves were removed by heating, yielding free-standing alumina fibers with a hollow core (Wang et al., 2013). The fibers showed better thermo-insulating properties than traditional alumina fibers due to their hollow structures and infrared absorbing properties (Wang et al., 2013).

In general, bulk polymers have a very low thermal conductivity owing to strong photon scattering within the material (Chae and Kumar, 2008). Wang et al. recently found that dragline silk is a thermally conductive material with a thermal conductivity of 340 W^{-1} K^{-1} in its relaxed state (0% strain). The thermal conductivity may be further improved to 416 W^{-1} K^{-1} simply by stretching the fiber by 19.7% (Huang et al., 2012). The inherently high thermal conductivity of dragline silk is largely attributed to its crystal/non-crystal hierarchical internal structure. The nanofibrils are intensely packed in the dragline silk fibers with few defects, leading to a high thermal conductivity. The silk's thermal conductivity enhances after stretching due to an increase in the alignment of both the β-sheet crystal structure and the helical non-crystal structure (the right image in **Figure 4E**), thereby further enhancing photon conduction (Huang et al., 2012).

Prior to Wang's work, researchers recognized the effects of polymer chain orientation on the thermal conductivity of the polymer fibers (Shen et al., 2010). Chen et al. fabricated high-quality ultra-drawn polyethylene nanofibers with a thermal conductivity as high as ~10^4 W m^{-1} K^{-1}, larger than the thermal conductivities of half of the pure metals, including platinum, iron, and nickel (Shen et al., 2010). The high thermal conductivity was attributed to the restructuring of the polyethylene chains to form large crystals along the drawing direction after stretching, which decreased the defect density in the fibers and made the fibers ideal single crystalline fibers (Shen et al., 2010).

Spider Web Network-Inspired Materials

Buehler's group examined the mechanism by which the non-linear mechanical characteristics of single silk fibers contribute to the integrity and performance of a spider web through web deformation experiments and atomistic simulations (**Figure 4F**)

(Cranford et al., 2012). The non-linear mechanical response to strain derives from the inherent semi-amorphous/β-sheet nanocrystal hierarchical molecular structure of the silk fibers. Under relatively low stress, semi-amorphous domains permit entropic unfolding, which softens the stress–strain curves. Once unfolding is complete, stress is transferred to the rigid β-sheet crystal domains, leading to a sharp increase in the stress. These unique mechanical properties of single silk fibers localize load-induced deformations by sacrificing the loaded fibers while maintaining the integrity of the web. This model offers a good explanation of why most natural webs are not intact in structure but remain functional for a spider's use (Cranford et al., 2012). The sacrifice of individual silk fibers to avoid damaging the whole web system has provided crucial inspiration for the design of new advanced 3D structure.

Recently, the same group prepared PDMS-based artificial spider webs using 3D printing to study how the geometrical arrangement of the silk fibers in a web influenced the web's mechanical performance (Qin et al., 2015). They found that the ratio of the radial silk diameter to the spiral silk diameter played a key role in tuning the mechanical response of the whole web to load. A homogeneous distribution of fibers with a radial/spiral ratio of ~1 tended to localize loading, whereas a heterogeneous distribution with much thicker radial silk fibers tended to distribute loading. This finding agrees well with the designs implemented naturally by spiders. Small webs built by garden spiders consist of radial and spiral silk fibers that are similar in size, which provides the maximum web strength against point loading by small prey. By contrast, giant webs built by spiders in rainforests are used to catch much larger prey; therefore, the spiders weave their webs with much thicker radial silk fibers to maximize the web strength against distributed loading (Qin et al., 2015).

Non-Web Spider Silk Materials

Besides aforementioned web silk, spiders also produce other types of silk, including pyriform silk, aciniform silk, and tubuliform/cylindriform silk for different uses, which also provide a lot of inspiration to fabricate new materials. Spiders produce attachment disks spun from pyriform silk to anchor their webs to diverse substrates. Cobweb spiders produce two types of attachment disks with distinct architectures using the same pyriform silk: "staple-pin" and "dendritic" disks (Sahni et al., 2012). "Stable-pin" disk firmly attaches the dragline silk to the substrate and "dendritic" disk weakly attaches the gumfoot silk to the substrate, the reason behind the difference in adhesion strengths of the two disks is the peeling angles. Small peeling angle of "stable-in" disk is the secret of high adhesion strength; on the other hand, dendritic disk is peeled at much higher angles, resulting in its much lower adhesion strength. With this understating, synthetic adhesive attachment disks inspired by spider have been prepared. Jain et al. (2014) fabricated artificial attachment disk with a similar architecture to spider's "stable-pin" disks using polyurethane electrospun fibers. The adhesion energy and peeling forces can be controlled by adjusting the spacing of fibers and the surface energy of the substrate (Jain et al., 2014).

Spiders fabricate egg case using tubuliform/cylindriform silks to protect the egg from predator, because of their high toughness. At least three different proteins have been found in tubuliform silk: tubuliform spidroin 1 (TuSp1), egg case protein 1 and 2 (ECP-1 and ECP-2). Tutuliform silks is the second most studied spider fiber following the dragline silk and the primary sequence and structure of tobuliform silk proteins have been clearly demonstrated (Lin et al., 2009, 2013; Gnesa et al., 2012). Lin et al. (2013) produced TuSp1 consisting of both repetitive and conserved terminal domains using recombinant DNA technology, and the artificial fibers spun from the recombinant TuSp1 showed outstanding mechanical properties.

Spiders produce aciniform silk to wrap and immobilize prey because aciniform silks possess the highest toughness out of all the spider silks including the famous dragline silk. Acriniform silk is also used to form the soft inner layer of egg case. Tremblay et al. (2015) found that the high toughness of aciniform silk is related to the primary sequence of spidroin and the distinctive mixture of α-helical, β-sheet, and non-canonical secondary structures. We expect that well understanding of the hierarchical architecture about these silks will inspire to the designs and development of advanced materials.

Artificial Silk Fiber Prepared by Biomimetic Spinning

Artificial spinning platforms have been designed to mimic the geometric construction of silkworm/spider spinning ducts, and artificial silk fibers have been prepared. Rammensee et al. (2008) designed a biomimetic microfluidic device in which ions and pH gradients were established along the elongational flow. Silk formation of a MaSp2 analog has been achieved using this device under similar condition to natural spider spinning duct. Kinahan et al. (2011) designed a microfluidic chip with a cross channel to prepare strong artificial silk fibroin fiber. PEO solution was chosen as the outer stream to apply shear stress on the silk fibroin core stream, and methanol bath was used to collect, dehydrate, and crystallize silk fiber. Because silk fibroin chains were aligned under the shear stress, the mechanical strength of the fibers could

be relatively high, and this strength could be further improved by post-drawing (Kinahan et al., 2011). Luo et al. (2014) designed a biomimetic microfluidic chip with the shear and elongation conditions similar to the spinning duct of spider and silkworm. Regenerated silk fibroin solutions with a very high concentration (50 wt%) was prepared and injected into the microchip as a spinning dope. Fibers were extruded out of the outlet and reeled on a rotating roller in air. The fibers were post-drawn in ethanol solution to further improve the mechanical property. The artificial fiber was tougher than the degummed natural silkworm silk (Luo et al., 2014).

CONCLUSION AND OUTLOOK

This review has outlined the sequence–structure–property interplay observed in natural silks and overviewed how smart materials are fabricated by mimicking natural silks. Significant progress in understanding the silk proteins' self-assembly, natural spinning process, and structure–property interplay has been achieved over the past decade, and many hidden secrets lie behind silk production and structure have been unveiled. This progress enabled the development of new materials inspired by natural silk. We expect that a much deeper understanding of the outstanding properties of silks will inspire researchers to design novel functional materials and to apply in diverse fields.

AUTHOR CONTRIBUTIONS

Prof. S-HL is the corresponding author and contributed to writing. Dr. JC contributed to data collecting and writing.

FUNDING

This work was supported by National Research Foundation of Korea (NRF-2015R1A2A1A09004998), Republic of Korea.

REFERENCES

Altman, G. H., Diaz, F., Jakuba, C., Calabro, T., Horan, R. L., Chen, J., et al. (2003). Silk-based biomaterials. *Biomaterials* 24, 401–416. doi:10.1016/S0142-9612(02)00353-8

An, B., Hinman, M. B., Holland, G. P., Yarger, J. L., and Lewis, R. V. (2011). Inducing β-sheets formation in synthetic spider silk fibers by aqueous post-spin stretching. *Biomacromolecules* 12, 2375–2381. doi:10.1021/bm200463e

Askarieh, G., Hedhammar, M., Nordling, K., Saenz, A., Casals, C., Rising, A., et al. (2010). Self-assembly of spider silk proteins is controlled by a pH-sensitive relay. *Nature* 465, 236–238. doi:10.1038/nature08962

Bai, H., Tian, X., Zheng, Y., Ju, J., Zhao, Y., and Jiang, L. (2010). Direction controlled driving of tiny water drops on bioinspired artificial spider silks. *Adv. Mater.* 22, 5521–5525. doi:10.1002/adma.201003169

Behanna, H. A., Donners, J. J. J. M., Gordon, A. C., and Stupp, S. I. (2005). Coassembly of amphiphiles with opposite peptide polarities into nanofibers. *J. Am. Chem. Soc.* 127, 1193–1200. doi:10.1021/ja044863u

Beun, L. H., Beaudoux, X. J., Kleijn, J. M., De Wolf, F. A., and Cohen Stuart, M. A. (2012). Self-assembly of silk-collagen-like triblock copolymers resembles a supramolecular living polymerization. *ACS Nano* 6, 133–140. doi:10.1021/nn203092u

Beun, L. H., Storm, I. M., Werten, M. W. T., De Wolf, F. A., Cohen Stuart, M. A., and De Vries, R. (2014). From micelles to fibers: balancing self-assembling

and random coiling domains in ph-responsive silk-collagen-like protein-based polymers. *Biomacromolecules* 15, 3349–3357. doi:10.1021/bm500826y

Bucak, S., Cenker, C., Nasir, I., Olsson, U., and Zackrisson, M. (2009). Peptide nanotube nematic phase. *Langmuir* 25, 4262–4265. doi:10.1021/la804175h

Castelletto, V., Hamley, I. W., Segarra-Maset, M. D., Berdugo Gumbau, C., Miravet, J. F., Escuder, B., et al. (2014). Tuning chelation by the surfactant-like peptide A(6)H using predetermined pH values. *Biomacromolecules* 15, 591–598. doi:10.1021/bm401640j

Castelletto, V., Nutt, D. R., Hamley, I. W., Bucak, S., Cenker, C., and Olsson, U. (2010). Structure of single-wall peptide nanotubes: in situ flow aligning X-ray diffraction. *Chem. Commun.* 46, 6270–6272. doi:10.1039/C0CC00212G

Cenker, C. C., Bucak, S., and Olsson, U. (2011). Nanotubes and bilayers in a model peptide system. *Soft Matter* 7, 4868–4875. doi:10.1039/C0SM01186J

Cenker, C. C., Bucak, S., and Olsson, U. (2014). Aqueous self-assembly within the homologous peptide series AnK. *Langmuir* 30, 10072–10079. doi:10.1021/la5016324

Chae, H. G., and Kumar, S. (2008). Making strong fibers. *Science* 319, 908–909. doi:10.1126/science.1153911

Chae, S.-K., Kang, E., Khademhosseini, A., and Lee, S.-H. (2013). Micro/nanometer-scale fiber with highly ordered structures by mimicking the spinning process of silkworm. *Adv. Mater.* 25, 3071–3078. doi:10.1002/adma.201300837

Chen, C., Pan, F., Zhang, S., Hu, J., Cao, M., Wang, J., et al. (2010). Antibacterial activities of short designer peptides: a link between propensity for nanostructuring and capacity for membrane destabilization. *Biomacromolecules* 11, 402–411. doi:10.1021/bm901130u

Cheng, J., Park, M., and Hyun, J. (2014a). Simple conditions for large morphological variations in thermoresponsive biopolymeric microstructures. *Chem. Commun.* 50, 2954–2957. doi:10.1039/C3CC49044K

Cheng, Y., Zheng, F., Lu, J., Shang, L., Xie, Z., Zhao, Y., et al. (2014b). Bioinspired multicompartmental microfibers from microfluidics. *Adv. Mater.* 26, 5184–5190. doi:10.1002/adma.201400798

Cheng, J., Park, M., Lim, D. W., and Hyun, J. (2013). Polypeptide microgel capsules as drug carriers. *Macromol. Res.* 21, 1163–1166. doi:10.1007/s13233-013-1167-6

Claussen, R. C., Rabatic, B. M., and Stupp, S. I. (2003). Aqueous self-assembly of unsymmetric peptide bolaamphiphiles into nanofibers with hydrophilic cores and surfaces. *J. Am. Chem. Soc.* 125, 12680–12681. doi:10.1021/ja035882r

Cranford, S. W., Tarakanova, A., Pugno, N. M., and Buehler, M. J. (2012). Nonlinear material behaviour of spider silk yields robust webs. *Nature* 482, 72–76. doi:10.1038/nature10739

Croisier, E., Liang, S., Schweizer, T., Balog, S., Mionic, M., Snellings, R., et al. (2014). A toolbox of oligopeptide-modified polymers for tailored elastomers. *Nat. Commun.* 5, 4728. doi:10.1038/ncomms5728

Currie, H. A., Deschaume, O., Naik, R. R., Perry, C. C., and Kaplan, D. L. (2011). Genetically engineered chimeric silk-silver binding proteins. *Adv. Funct. Mater.* 21, 2889–2895. doi:10.1002/adfm.201100249

Davis, V. A., Parra-Vasquez, A. N. G., Green, M. J., Rai, P. K., Behabtu, N., Prieto, V., et al. (2009). True solutions of single-walled carbon nanotubes for assembly into macroscopic materials. *Nat. Nanotechnol.* 4, 830–834. doi:10.1038/nnano.2009.302

Dong, H., Wang, N., Wang, L., Bai, H., Wu, J., Zheng, Y., et al. (2012). Bioinspired electrospun knotted microfibers for fog harvesting. *Chemphyschem* 13, 1153–1156. doi:10.1002/cphc.201100957

Ericson, L. M., Fan, H., Peng, H., Davis, V. A., Zhou, W., Sulpizio, J., et al. (2004). Macroscopic, neat, single-walled carbon nanotube fibers. *Science* 305, 1447–1450. doi:10.1126/science.1101398

Fang, G., Zheng, Z., Yao, J., Chen, M., Tang, Y., Zhong, J., et al. (2015). Tough protein-carbon nanotube hybrid fibers comparable to natural spider silks. *J. Mater. Chem. B* 3, 3940–3947. doi:10.1039/C5TB00448A

Feng, S., Hou, Y., Xue, Y., Gao, L., Jiang, L., and Zheng, Y. (2013). Photo-controlled water gathering on bio-inspired fibers. *Soft Matter* 9, 9294–9297. doi:10.1039/C3SM51517F

Gnesa, E., Hsia, Y., Yarger, J. L., Weber, W., Lin-Cereghino, J., Lin-Cereghino, G., et al. (2012). Conserved C-terminal domain of spider tubuliform spidroin 1 contributes to extensibility in synthetic fibers. *Biomacromolecules* 13, 304–312. doi:10.1021/bm201262n

Golinska, M. D., Wlodarczyk-Biegun, M. K., Werten, M. W. T., Cohen Stuart, M. A., De Wolf, F. A., and De Vries, R. (2014). Dilute self-healing hydrogels of silk-collagen-like block copolypeptides at neutral pH. *Biomacromolecules* 15, 699–706. doi:10.1021/bm401682n

Gomes, S. C., Leonor, I. B., Mano, J. F., Reis, R. L., and Kaplan, D. L. (2011). Antimicrobial functionalized genetically engineered spider silk. *Biomaterials* 32, 4255–4266. doi:10.1016/j.biomaterials.2011.02.040

Graf, J., Iwamoto, Y., Sasaki, M., Martin, G. R., Kleinman, H. K., Robey, F. A., et al. (1987). Identification of an amino acid sequence in laminin mediating cell attachment, chemotaxis, and receptor binding. *Cell* 48, 989–996. doi:10.1016/0092-8674(87)90707-0

Grant, D. S., Tashiro, K.-I., Segui-Real, B., Yamada, Y., Martin, G. R., and Kleinman, H. K. (1989). Two different laminin domains mediate the differentiation of human endothelial cells into capillary-like structures in vitro. *Cell* 58, 933–943. doi:10.1016/0092-8674(89)90945-8

Gus'kova, O. A., Khalatur, P. G., Bäuerle, P., and Khokhlov, A. R. (2008). Silk-inspired 'molecular chimeras': atomistic simulation of nanoarchitectures based on thiophene-peptide copolymers. *Chem. Phys. Lett.* 461, 64–70. doi:10.1016/j.cplett.2008.06.058

Hagn, F., Eisoldt, L., Hardy, J. G., Vendrely, C., Coles, M., Scheibel, T., et al. (2010). A conserved spider silk domain acts as a molecular switch that controls fibre assembly. *Nature* 465, 239–242. doi:10.1038/nature08936

Hamley, I. W., Dehsorkhi, A., Castelletto, V., Seitsonen, J., Ruokolainen, J., and Iatrou, H. (2013). Self-assembly of a model amphiphilic oligopeptide incorporating an arginine headgroup. *Soft Matter* 9, 4794–4801. doi:10.1039/c3sm50303h

Hartgerink, J. D., Beniash, E., and Stupp, S. I. (2001). Self-assembly and mineralization of peptide-amphiphile nanofibers. *Science* 294, 1684–1688. doi:10.1126/science.1063187

Hartgerink, J. D., Beniash, E., and Stupp, S. I. (2002). Peptide-amphiphile nanofibers: a versatile scaffold for the preparation of self-assembling materials. *Proc. Natl. Acad. Sci. U.S.A.* 99, 5133–5138. doi:10.1073/pnas.072699999

He, Y.-X., Zhang, N.-N., Li, W.-F., Jia, N., Chen, B.-Y., Zhou, K., et al. (2012). N-terminal domain of *Bombyx mori* fibroin mediates the assembly of silk in response to pH decrease. *J. Mol. Biol.* 418, 197–207. doi:10.1016/j.jmb.2012.02.040

Heim, M., Romer, L., and Scheibel, T. (2010). Hierarchical structures made of proteins. The complex architecture of spider webs and their constituent silk proteins. *Chem. Soc. Rev.* 39, 156–164. doi:10.1039/B813273A

Hermanson, K. D., Huemmerich, D., Scheibel, T., and Bausch, A. R. (2007). Engineered microcapsules fabricated from reconstituted spider silk. *Adv. Mater.* 19, 1810–1815. doi:10.1002/adma.200602709

Hirayama, K., Okitsu, T., Teramae, H., Kiriya, D., Onoe, H., and Takeuchi, S. (2013). Cellular building unit integrated with microstrand-shaped bacterial cellulose. *Biomaterials* 34, 2421–2427. doi:10.1016/j.biomaterials.2012.12.013

Hou, Y., Gao, L., Feng, S., Chen, Y., Xue, Y., Jiang, L., et al. (2013). Temperature-triggered directional motion of tiny water droplets on bioinspired fibers in humidity. *Chem. Commun.* 49, 5253–5255. doi:10.1039/C3CC41060A

Huang, X., Liu, G., and Wang, X. (2012). New secrets of spider silk: exceptionally high thermal conductivity and its abnormal change under stretching. *Adv. Mater.* 24, 1482–1486. doi:10.1002/adma.201104668

Huemmerich, D., Helsen, C. W., Quedzuweit, S., Oschmann, J., Rudolph, R., and Scheibel, T. (2004). Primary structure elements of spider dragline silks and their contribution to protein solubility†. *Biochemistry* 43, 13604–13612. doi:10.1021/bi048983q

Hwang, C. M., Khademhosseini, A., Park, Y., Sun, K., and Lee, S.-H. (2008). Microfluidic chip-based fabrication of PLGA microfiber scaffolds for tissue engineering. *Langmuir* 24, 6845–6851. doi:10.1021/la800253b

Hwang, C. M., Park, Y., Park, J. Y., Lee, K., Sun, K., Khademhosseini, A., et al. (2009). Controlled cellular orientation on PLGA microfibers with defined diameters. *Biomed. Microdevices* 11, 739–746. doi:10.1007/s10544-009-9287-7

Jain, D., Sahni, V., and Dhinojwala, A. (2014). Synthetic adhesive attachment discs inspired by spider's pyriform silk architecture. *J. Polym. Sci. B Polym. Phys.* 52, 553–560. doi:10.1002/polb.23453

Jeong, W., Kim, J., Kim, S., Lee, S., Mensing, G., and Beebe, D. J. (2004). Hydrodynamic microfabrication via "on the fly" photopolymerization of microscale fibers and tubes. *Lab. Chip* 4, 576–580. doi:10.1039/B411249K

Jin, H.-J., and Kaplan, D. L. (2003). Mechanism of silk processing in insects and spiders. *Nature* 424, 1057–1061. doi:10.1038/nature01809

Jun, Y., Kang, A. R., Lee, J. S., Park, S.-J., Lee, D. Y., Moon, S.-H., et al. (2014a). Microchip-based engineering of super-pancreatic islets supported by adipose-derived stem cells. *Biomaterials* 35, 4815–4826. doi:10.1016/j.biomaterials.2014.02.045

Jun, Y., Kang, E., Chae, S., and Lee, S.-H. (2014b). Microfluidic spinning of micro- and nano-scale fibers for tissue engineering. *Lab. Chip* 14, 2145–2160. doi:10.1039/C3LC51414E

Jun, Y., Kim, M. J., Hwang, Y. H., Jeon, E. A., Kang, A. R., Lee, S.-H., et al. (2013). Microfluidics-generated pancreatic islet microfibers for enhanced immunoprotection. *Biomaterials* 34, 8122–8130. doi:10.1016/j.biomaterials.2013.07.079

Kang, A., Park, J., Ju, J., Jeong, G. S., and Lee, S.-H. (2014). Cell encapsulation via microtechnologies. *Biomaterials* 35, 2651–2663. doi:10.1016/j.biomaterials.2013.12.073

Kang, E., Choi, Y. Y., Chae, S.-K., Moon, J.-H., Chang, J.-Y., and Lee, S.-H. (2012). Microfluidic spinning of flat alginate fibers with grooves for cell-aligning scaffolds. *Adv. Mater.* 24, 4271–4277. doi:10.1002/adma.201201232

Kang, E., Jeong, G. S., Choi, Y. Y., Lee, K. H., Khademhosseini, A., and Lee, S.-H. (2011). Digitally tunable physicochemical coding of material composition and topography in continuous microfibres. *Nat. Mater.* 10, 877–883. doi:10.1038/nmat3108

Kang, E., Shin, S.-J., Lee, K. H., and Lee, S.-H. (2010). Novel PDMS cylindrical channels that generate coaxial flow, and application to fabrication of microfibers and particles. *Lab. Chip* 10, 1856–1861. doi:10.1039/C002695F

Keten, S., Xu, Z., Ihle, B., and Buehler, M. J. (2010). Nanoconfinement controls stiffness, strength and mechanical toughness of [beta]-sheet crystals in silk. *Nat. Mater.* 9, 359–367. doi:10.1038/nmat2704

Kim, D.-H., Viventi, J., Amsden, J. J., Xiao, J., Vigeland, L., Kim, Y.-S., et al. (2010). Dissolvable films of silk fibroin for ultrathin conformal bio-integrated electronics. *Nat. Mater.* 9, 511–517. doi:10.1038/nmat2745

Kim, S., Marelli, B., Brenckle, M. A., Mitropoulos, A. N., Gil, E.-S., Tsioris, K., et al. (2014). All-water-based electron-beam lithography using silk as a resist. *Nat. Nanotechnol.* 9, 306–310. doi:10.1038/nnano.2014.47

Kinahan, M. E., Filippidi, E., Köster, S., Hu, X., Evans, H. M., Pfohl, T., et al. (2011). Tunable silk: using microfluidics to fabricate silk fibers with controllable properties. *Biomacromolecules* 12, 1504–1511. doi:10.1021/bm1014624

Kiriya, D., Ikeda, M., Onoe, H., Takinoue, M., Komatsu, H., Shimoyama, Y., et al. (2012a). Meter-long and robust supramolecular strands encapsulated in hydrogel jackets. *Angew. Chem. Int. Ed.* 51, 1553–1557. doi:10.1002/anie.201104043

Kiriya, D., Kawano, R., Onoe, H., and Takeuchi, S. (2012b). Microfluidic control of the internal morphology in nanofiber-based macroscopic cables. *Angew. Chem. Int. Ed.* 51, 7942–7947. doi:10.1002/anie.201202078

Kis, A., Csanyi, G., Salvetat, J. P., Lee, T.-N., Couteau, E., Kulik, A. J., et al. (2004). Reinforcement of single-walled carbon nanotube bundles by intertube bridging. *Nat. Mater.* 3, 153–157. doi:10.1038/nmat1076

Koski, K. J., Akhenblit, P., Mckiernan, K., and Yarger, J. L. (2013). Non-invasive determination of the complete elastic moduli of spider silks. *Nat. Mater.* 12, 262–267. doi:10.1038/nmat3549

Kundu, B., Kurland, N. E., Bano, S., Patra, C., Engel, F. B., Yadavalli, V. K., et al. (2014). Silk proteins for biomedical applications: bioengineering perspectives. *Prog. Polym. Sci.* 39, 251–267. doi:10.1016/j.progpolymsci.2013.09.002

Leal-Egaña, A., Lang, G., Mauerer, C., Wickinghoff, J., Weber, M., Geimer, S., et al. (2012). Interactions of fibroblasts with different morphologies made of an engineered spider silk protein. *Adv. Eng. Mater.* 14, B67–B75. doi:10.1002/adem.201180072

Leal-Egana, A., and Scheibel, T. (2012). Interactions of cells with silk surfaces. *J. Mater. Chem.* 22, 14330–14336. doi:10.1039/C2JM31174G

Lee, K. H., Shin, S. J., Kim, C.-B., Kim, J. K., Cho, Y. W., Chung, B. G., et al. (2010). Microfluidic synthesis of pure chitosan microfibers for bio-artificial liver chip. *Lab. Chip* 10, 1328–1334. doi:10.1039/B924987G

Lee, K. H., Shin, S. J., Park, Y., and Lee, S.-H. (2009). Synthesis of cell-laden alginate hollow fibers using microfluidic chips and microvascularized tissue-engineering applications. *Small* 5, 1264–1268. doi:10.1002/smll.200801667

Lin, Y., Xia, X., Wang, M., Wang, Q., An, B., Tao, H., et al. (2014). Genetically programmable thermoresponsive plasmonic gold/silk-elastin protein core/shell nanoparticles. *Langmuir* 30, 4406–4414. doi:10.1021/la403559t

Lin, Z., Deng, Q., Liu, X.-Y., and Yang, D. (2013). Engineered large spider eggcase silk protein for strong artificial fibers. *Adv. Mater.* 25, 1216–1220. doi:10.1002/adma.201204357

Lin, Z., Huang, W., Zhang, J., Fan, J.-S., and Yang, D. (2009). Solution structure of eggcase silk protein and its implications for silk fiber formation. *Proc. Natl. Acad. Sci. U.S.A.* 106, 8906–8911. doi:10.1073/pnas.0813255106

Liu, Y., Shao, Z., and Vollrath, F. (2005). Relationships between supercontraction and mechanical properties of spider silk. *Nat. Mater.* 4, 901–905. doi:10.1038/nmat1534

Luo, J., Zhang, L., Peng, Q., Sun, M., Zhang, Y., Shao, H., et al. (2014). Tough silk fibers prepared in air using a biomimetic microfluidic chip. *Int. J. Biol. Macromol.* 66, 319–324. doi:10.1016/j.ijbiomac.2014.02.049

Martens, A. A., Portale, G., Werten, M. W. T., De Vries, R. J., Eggink, G., Cohen Stuart, M. A., et al. (2009). Triblock protein copolymers forming supramolecular nanotapes and pH-responsive gels. *Macromolecules* 42, 1002–1009. doi:10.1021/ma801955q

Megeed, Z., Cappello, J., and Ghandehari, H. (2002). Controlled release of plasmid DNA from a genetically engineered silk-elastinlike hydrogel. *Pharm. Res.* 19, 954–959. doi:10.1023/a:1016406120288

Megeed, Z., Haider, M., Li, D. Q., O'Malley, B. W., Cappello, J., and Ghandehari, H. (2004). In vitro and in vivo evaluation of recombinant silk-elastinlike hydrogels for cancer gene therapy. *J. Control. Release* 94, 433–445. doi:10.1016/j.jconrel.2003.10.027

Naraghi, M., Bratzel, G. H., Filleter, T., An, Z., Wei, X., Nguyen, S. T., et al. (2013). Atomistic investigation of load transfer between DWNT bundles "crosslinked" by PMMA oligomers. *Adv. Funct. Mater.* 23, 1883–1892. doi:10.1002/adfm.201201358

Numata, K., and Kaplan, D. L. (2010). Silk-based gene carriers with cell membrane destabilizing peptides. *Biomacromolecules* 11, 3189–3195. doi:10.1021/bm101055m

Numata, K., Reagan, M. R., Goldstein, R. H., Rosenblatt, M., and Kaplan, D. L. (2011). Spider silk-based gene carriers for tumor cell-specific delivery. *Bioconjug. Chem.* 22, 1605–1610. doi:10.1021/bc200170u

Omenetto, F. G., and Kaplan, D. L. (2010). New opportunities for an ancient material. *Science* 329, 528–531. doi:10.1126/science.1188936

Onoe, H., Okitsu, T., Itou, A., Kato-Negishi, M., Gojo, R., Kiriya, D., et al. (2013). Metre-long cell-laden microfibres exhibit tissue morphologies and functions. *Nat. Mater.* 12, 584–590. doi:10.1038/nmat3606

Opell, B. D., and Hendricks, M. L. (2009). The adhesive delivery system of viscous capture threads spun by orb-weaving spiders. *J. Exp. Biol.* 212, 3026–3034. doi:10.1242/jeb.030064

Opell, B. D., and Hendricks, M. L. (2010). The role of granules within viscous capture threads of orb-weaving spiders. *J. Exp. Biol.* 213, 339–346. doi:10.1242/jeb.036947

Peng, H., Jain, M., Li, Q., Peterson, D. E., Zhu, Y., and Jia, Q. (2008). Vertically aligned pearl-like carbon nanotube arrays for fiber spinning. *J. Am. Chem. Soc.* 130, 1130–1131. doi:10.1021/ja077767c

Porter, D., Guan, J., and Vollrath, F. (2013). Spider silk: super material or thin fibre? *Adv. Mater.* 25, 1275–1279. doi:10.1002/adma.201204158

Qin, Z., and Buehler, M. J. (2013). Spider silk: webs measure up. *Nat. Mater.* 12, 185–187. doi:10.1038/nmat3578

Qin, Z., Compton, B. G., Lewis, J. A., and Buehler, M. J. (2015). Structural optimization of 3D-printed synthetic spider webs for high strength. *Nat. Commun.* 6, 7038. doi:10.1038/ncomms8038

Qiu, F., Chen, Y., and Zhao, X. (2009). Comparative studies on the self-assembling behaviors of cationic and catanionic surfactant-like peptides. *J. Colloid Interface Sci.* 336, 477–484. doi:10.1016/j.jcis.2009.04.014

Rammensee, S., Huemmerich, D., Hermanson, K. D., Scheibel, T., and Bausch, A. R. (2006). Rheological characterization of hydrogels formed by recombinantly produced spider silk. *Appl. Phys. A* 82, 261–264. doi:10.1007/s00339-005-3431-x

Rammensee, S., Slotta, U., Scheibel, T., and Bausch, A. R. (2008). Assembly mechanism of recombinant spider silk proteins. *Proc. Natl. Acad. Sci. U.S.A.* 105, 6590–6595. doi:10.1073/pnas.0709246105

Rathore, O., and Sogah, D. Y. (2001a). Nanostructure formation through β-sheet self-assembly in silk-based materials. *Macromolecules* 34, 1477–1486. doi:10.1021/ma001553x

Rathore, O., and Sogah, D. Y. (2001b). Self-assembly of β-sheets into nanostructures by poly(alanine) segments incorporated in multiblock copolymers inspired by spider silk. *J. Am. Chem. Soc.* 123, 5231–5239. doi:10.1021/ja004030d

Rockwood, D. N., Preda, R. C., Yucel, T., Wang, X., Lovett, M. L., and Kaplan, D. L. (2011). Materials fabrication from *Bombyx mori* silk fibroin. *Nat. Protoc.* 6, 1612–1631. doi:10.1038/nprot.2011.379

Sahni, V., Blackledge, T. A., and Dhinojwala, A. (2010). Viscoelastic solids explain spider web stickiness. *Nat. Commun.* 1, 19. doi:10.1038/ncomms1019

Sahni, V., Harris, J., Blackledge, T. A., and Dhinojwala, A. (2012). Cobweb-weaving spiders produce different attachment discs for locomotion and prey capture. *Nat. Commun.* 3, 1106. doi:10.1038/ncomms2099

Sanchez, C., Arribart, H., and Giraud Guille, M. M. (2005). Biomimetism and bioinspiration as tools for the design of innovative materials and systems. *Nat. Mater.* 4, 277–288. doi:10.1038/nmat1339

Schacht, K., and Scheibel, T. (2014). Processing of recombinant spider silk proteins into tailor-made materials for biomaterials applications. *Curr. Opin. Biotechnol.* 29, 62–69. doi:10.1016/j.copbio.2014.02.015

Scheibel, T. (2004). Spider silks: recombinant synthesis, assembly, spinning, and engineering of synthetic proteins. *Microb. Cell Fact.* 3, 14. doi:10.1186/1475-2859-3-14

Schwarze, S., Zwettler, F. U., Johnson, C. M., and Neuweiler, H. (2013). The N-terminal domains of spider silk proteins assemble ultrafast and protected from charge screening. *Nat. Commun.* 4, 2815. doi:10.1038/ncomms3815

Shah, R. N., Shah, N. A., Lim, M. M. D. R., Hsieh, C., Nuber, G., and Stupp, S. I. (2010). Supramolecular design of self-assembling nanofibers for cartilage regeneration. *Proc. Natl. Acad. Sci. U.S.A.* 107, 3293–3298. doi:10.1073/pnas.0906501107

Shen, S., Henry, A., Tong, J., Zheng, R., and Chen, G. (2010). Polyethylene nanofibres with very high thermal conductivities. *Nat. Nanotechnol.* 5, 251–255. doi:10.1038/nnano.2010.27

Shin, S.-J., Park, J.-Y., Lee, J.-Y., Park, H., Park, Y.-D., Lee, K.-B., et al. (2007). "On the fly" continuous generation of alginate fibers using a microfluidic device. *Langmuir* 23, 9104–9108. doi:10.1021/la700818q

Silva, L. P., and Rech, E. L. (2013). Unravelling the biodiversity of nanoscale signatures of spider silk fibres. *Nat. Commun.* 4, 3014. doi:10.1038/ncomms4014

Slotta, U., Hess, S., Spiess, K., Stromer, T., Serpell, L., and Scheibel, T. (2007). Spider silk and amyloid fibrils: a structural comparison. *Macromol. Biosci.* 7, 183–188. doi:10.1002/mabi.200600201

Smeenk, J. M., Otten, M. B. J., Thies, J., Tirrell, D. A., Stunnenberg, H. G., and Van Hest, J. C. M. (2005). Controlled assembly of macromolecular beta-sheet fibrils. *Angew. Chem. Int. Ed.* 44, 1968–1971. doi:10.1002/anie.200462415

Smeenk, J. M., Schön, P., Otten, M. B. J., Speller, S., Stunnenberg, H. G., and Van Hest, J. C. M. (2006). Fibril formation by triblock copolymers of silklike β-sheet polypeptides and poly(ethylene glycol). *Macromolecules* 39, 2989–2997. doi:10.1021/ma0521654

Stellwagen, S. D., Opell, B. D., and Short, K. G. (2014). Temperature mediates the effect of humidity on the viscoelasticity of glycoprotein glue within the droplets of an orb-weaving spider's prey capture threads. *J. Exp. Biol.* 217, 1563–1569. doi:10.1242/jeb.097816

Tao, H., Kaplan, D. L., and Omenetto, F. G. (2012). Silk materials – a road to sustainable high technology. *Adv. Mater.* 24, 2824–2837. doi:10.1002/adma.201104477

Tao, P., Shang, W., Song, C., Shen, Q., Zhang, F., Luo, Z., et al. (2015). Bioinspired engineering of thermal materials. *Adv. Mater.* 27, 428–463. doi:10.1002/adma.201401449

Tian, X., Bai, H., Zheng, Y., and Jiang, L. (2011). Bio-inspired heterostructured bead-on-string fibers that respond to environmental wetting. *Adv. Funct. Mater.* 21, 1398–1402. doi:10.1002/adfm.201002061

Tokareva, O., Jacobsen, M., Buehler, M., Wong, J., and Kaplan, D. L. (2014). Structure-function-property-design interplay in biopolymers: spider silk. *Acta Biomater.* 10, 1612–1626. doi:10.1016/j.actbio.2013.08.020

Tokareva, O., Michalczechen-Lacerda, V. A., Rech, E. L., and Kaplan, D. L. (2013). Recombinant DNA production of spider silk proteins. *Microb. Biotechnol.* 6, 651–663. doi:10.1111/1751-7915.12081

Tremblay, M.-L., Xu, L., Lefèvre, T., Sarker, M., Orrell, K. E., Leclerc, J., et al. (2015). Spider wrapping silk fibre architecture arising from its modular soluble protein precursor. *Sci. Rep.* 5, 11502. doi:10.1038/srep11502

Vauthey, S., Santoso, S., Gong, H. Y., Watson, N., and Zhang, S. G. (2002). Molecular self-assembly of surfactant-like peptides to form nanotubes and nanovesicles. *Proc. Natl. Acad. Sci. U.S.A.* 99, 5355–5360. doi:10.1073/pnas.072089599

Vepari, C., and Kaplan, D. L. (2007). Silk as a biomaterial. *Prog. Polym. Sci.* 32, 991–1007. doi:10.1016/j.progpolymsci.2007.05.013

Vollrath, F., Fairbrother, W. J., Williams, R. J. P., Tillinghast, E. K., Bernstein, D. T., Gallagher, K. S., et al. (1990). Compounds in the droplets of the orb spider's viscid spiral. *Nature* 345, 526–528. doi:10.1038/345526a0

Vollrath, F., and Knight, D. P. (2001). Liquid crystalline spinning of spider silk. *Nature* 410, 541–548. doi:10.1038/35069000

von Maltzahn, G., Vauthey, S., Santoso, S., and Zhang, S. U. (2003). Positively charged surfactant-like peptides self-assemble into nanostructures. *Langmuir* 19, 4332–4337. doi:10.1021/la026526+

Wang, J., Han, S., Meng, G., Xu, H., Xia, D., Zhao, X., et al. (2009). Dynamic self-assembly of surfactant-like peptides A(6)K and A(9)K. *Soft Matter* 5, 3870–3878. doi:10.1039/b901653h

Wang, T., Kong, S., Jia, Y., Chang, L., Wong, C., and Xiong, D. (2013). Synthesis and thermal conductivities of the biomorphic Al2O3 fibers derived from silk template. *Int. J. Appl. Ceram. Tech.* 10, 285–292. doi:10.1111/j.1744-7402.2012.02841.x

Wen, L., Tian, Y., and Jiang, L. (2015). Bioinspired super-wettability from fundamental research to practical applications. *Angew. Chem. Int. Ed.* 54, 3387–3399. doi:10.1002/anie.201409911

Widhe, M., Johansson, U., Hillerdahl, C.-O., and Hedhammar, M. (2013). Recombinant spider silk with cell binding motifs for specific adherence of cells. *Biomaterials* 34, 8223–8234. doi:10.1016/j.biomaterials.2013.07.058

Wohlrab, S., Müller, S., Schmidt, A., Neubauer, S., Kessler, H., Leal-Egaña, A., et al. (2012). Cell adhesion and proliferation on RGD-modified recombinant spider silk proteins. *Biomaterials* 33, 6650–6659. doi:10.1016/j.biomaterials.2012.05.069

Wong Po Foo, C., Patwardhan, S. V., Belton, D. J., Kitchel, B., Anastasiades, D., Huang, J., et al. (2006). Novel nanocomposites from spider silk-silica fusion (chimeric) proteins. *Proc. Natl. Acad. Sci. U.S.A.* 103, 9428–9433. doi:10.1073/pnas.0601096103

Wu, M., Shuai, H., Cheng, Q., and Jiang, L. (2014). Bioinspired green composite lotus fibers. *Angew. Chem. Int. Ed.* 53, 3358–3361. doi:10.1002/anie.201310656

Xia, X.-X., Qian, Z.-G., Ki, C. S., Park, Y. H., Kaplan, D. L., and Lee, S. Y. (2010). Native-sized recombinant spider silk protein produced in metabolically engineered *Escherichia coli* results in a strong fiber. *Proc. Natl. Acad. Sci. U.S.A.* 107, 14059–14063. doi:10.1073/pnas.1003366107

Xia, X.-X., Wang, M., Lin, Y., Xu, Q., and Kaplan, D. L. (2014). Hydrophobic drug-triggered self-assembly of nanoparticles from silk-elastin-like protein polymers for drug delivery. *Biomacromolecules* 15, 908–914. doi:10.1021/bm4017594

Xia, X.-X., Xu, Q., Hu, X., Qin, G., and Kaplan, D. L. (2011). Tunable self-assembly of genetically engineered silk-elastin-like protein polymers. *Biomacromolecules* 12, 3844–3850. doi:10.1021/bm201165h

Yan, J., Zhou, G., Knight, D. P., Shao, Z., and Chen, X. (2010). Wet-spinning of regenerated silk fiber from aqueous silk fibroin solution: discussion of spinning parameters. *Biomacromolecules* 11, 1–5. doi:10.1021/bm900840h

Yu, Y., Wen, H., Ma, J., Lykkemark, S., Xu, H., and Qin, J. (2014). Flexible fabrication of biomimetic bamboo-like hybrid microfibers. *Adv. Mater.* 26, 2494–2499. doi:10.1002/adma.201304974

Zhang, J., Rajkhowa, R., Li, J. L., Liu, X. Y., and Wang, X. G. (2013). Silkworm cocoon as natural material and structure for thermal insulation. *Mater. Des.* 49, 842–849. doi:10.1016/j.matdes.2013.02.006

Zhang, M., Fang, S., Zakhidov, A. A., Lee, S. B., Aliev, A. E., Williams, C. D., et al. (2005). Strong, transparent, multifunctional, carbon nanotube sheets. *Science* 309, 1215–1219. doi:10.1126/science.1115311

Zhang, S. (2003). Fabrication of novel biomaterials through molecular self-assembly. *Nat. Biotechnol.* 21, 1171–1178. doi:10.1038/nbt874

Zhang, S., Greenfield, M. A., Mata, A., Palmer, L. C., Bitton, R., Mantei, J. R., et al. (2010). A self-assembly pathway to aligned monodomain gels. *Nat. Mater.* 9, 594–601. doi:10.1038/nmat2778

Zhao, X., Nagai, Y., Reeves, P. J., Kiley, P., Khorana, H. G., and Zhang, S. (2006). Designer short peptide surfactants stabilize G protein-coupled receptor bovine rhodopsin. *Proc. Natl. Acad. Sci. U.S.A.* 103, 17707–17712. doi:10.1073/pnas.0607167103

Zheng, Y., Bai, H., Huang, Z., Tian, X., Nie, F.-Q., Zhao, Y., et al. (2010). Directional water collection on wetted spider silk. *Nature* 463, 640–643. doi:10.1038/nature08729

Zhou, C., Leng, B., Yao, J., Qian, J., Chen, X., Zhou, P., et al. (2006). Synthesis and characterization of multiblock copolymers based on spider dragline silk proteins. *Biomacromolecules* 7, 2415–2419. doi:10.1021/bm060199t

Zhou, G., Shao, Z., Knight, D. P., Yan, J., and Chen, X. (2009). Silk fibers extruded artificially from aqueous solutions of regenerated *Bombyx mori* silk fibroin are tougher than their natural counterparts. *Adv. Mater.* 21, 366–370. doi:10.1002/adma.200800582

Conflict of Interest Statement: The authors declare that the research was conducted in the absence of any commercial or financial relationships that could be construed as a potential conflict of interest.

13

Dodecylamine-Loaded Halloysite Nanocontainers for Active Anticorrosion Coatings

Jesus Marino Falcón[1], Tiago Sawczen[2] and Idalina Vieira Aoki[1]*

[1] *Polytechnic School, University of São Paulo, São Paulo, Brazil,* [2] *Nanocorr – Aditivos Inteligentes e Soluções Contra Corrosão Ltd., Paraná, Brazil*

Currently, the most promising approach in the corrosion protection of smart coatings is the use of nanoreservoirs loaded with corrosion inhibitors. Nanocontainers are filled with anti-corrosive agents and are embedded into a primer coating. Future prospective containers are halloysite nanotubes (HNTs) due to their low price, availability, durability, with high mechanical strength and biocompatibility. The aim of this work is to study the use of HNTs as nanocontainers for encapsulated dodecylamine for active corrosion protection of carbon steel. Halloysite clay was characterized by X-ray diffraction and TGA – thermogravimetric analysis techniques. HNTs were loaded with dodecylamine and were embedded into an alkyd primer with a weight ratio of 10 wt.%. The anti-corrosive performance of the alkyd primer doped with 10 wt.% of entrapped-dodecylamine halloysite was tested on coated carbon steel by direct exposure of the coated samples with a provoked defect into 0.01 mol/L NaCl corrosive media using electrochemical impedance spectroscopy (EIS) and scanning vibrating electrode technique (SVET). EIS and SVET measurements showed the self-healing properties of the doped alkyd coating. Coated samples were also evaluated in a salt spray chamber and the self-healing effect was unequivocally noticed.

Keywords: nanocontainers, self-healing, halloysite, corrosion inhibitor, dodecylamine, smart coatings

Edited by:
Wolfram Fürbeth,
DECHEMA-Forschungsinstitut,
Germany

Reviewed by:
Ole Øystein Knudsen,
SINTEF and Norwegian University of Science and Technology, Norway
Flavio Deflorian,
University of Trento, Italy

*Correspondence:
Idalina Vieira Aoki
idavaoki@usp.br

INTRODUCTION

Corrosion of metals is one of the serious technological problems, resulting in huge economic losses, especially in the aerospace, automotive, and petroleum industries. For this reason, a variety of methods such as cathodic protection, metallic coatings, and polymeric coating systems were developed to overcome it. However, if this barrier is partially disrupted, the coating itself cannot stop the corrosion process. One of the possible solutions to achieve active corrosion protection is to introduce an ecofriendly corrosion inhibitor directly into the coating, providing release of the inhibitor, and termination of the corrosion propagation at already damaged corrosion defects (Shchukin et al., 2008). Expected undesirable reactions between inhibitor molecule and constituents of the coating polymeric matrix are the main reasons for avoiding the direct introduction of inhibitors into coatings and trying to encapsulate them (Zheludkevich et al., 2007).

Accordingly, nanocontainers are loaded with corrosion inhibitors and dispersed in organic coatings applied as primers. These nanomaterials have the ability to release encapsulated inhibitors in a controlled manner, which can be tuned to coincide with an increase of the aggressiveness in the surrounding environment or corrosion initiation on the metallic substrates (Hu et al., 2005; Zheludkevich et al., 2005, 2007; Suryanarayana et al., 2008; Wu et al., 2008; Tedim et al., 2010; Evaggelos et al., 2011; Nesterova et al., 2011). When the local environment undergoes changes or if the corrosion process is started on the metal surface, the nanocontainers release encapsulated active material (inhibitor) directly into the damaged area, thus preventing undesirable leakage of the inhibitor and reduction of the barrier properties of the whole applied film (Zheludkevich et al., 2007; Andreeva et al., 2008, 2010; Shchukin et al., 2008; Borisova et al., 2011). Recently, a review on smart coatings stresses the use of smart nanocontainers for the enhancement of active anticorrosion performance in protective coatings (Montemor, 2014). Different types of inorganic corrosion inhibitors include chromates, phosphates, molybdates, and nitrites. One of the disadvantages of inorganic inhibitors is that some of them are toxic. For example, chromates are proven to cause several diseases, including cancer and are forbidden in Europe since 2007, for common usage. Therefore, introduction of ecofriendly corrosion inhibitors for protective coatings is a very intriguing challenge (Zheludkevich et al., 2007). For instance, the large-scale implementation of the nanocontainer-based coatings is limited by the high nanocontainers price. This calls for the finding of low-cost nanocontainers, which can be employed successfully in self-healing or smart coatings. One of the prospective future nanocontainers that can be industrially mined is halloysite clay nanotubes (Yuan et al., 2015). Often, long and thin tubular systems are highly desirable, since for the same amount of cargo (compared to spherical capsules), they exhibit superior aero and hydrodynamic properties and thus better processability (Suh et al., 2011). Thus, the use of halloysite nanotubes (HNTs) has become probably one of the best proposals due to their low cost and ability to encapsulate a range of active agents within their structure followed by their retention and triggered release (Shi et al., 2011; Kamble et al., 2012; Rawtani and Agrawal, 2012; Yuan et al., 2012; Joshi et al., 2013; Lvov et al., 2014). Halloysite occurs mainly in two different six polymorphs, with a interlayer where water molecules are entrapped; the hydrated form (basal distance around 10 Å) with the minimal formula of $Al_2Si_2O_5 (OH)_4 \cdot 2H_2O$, and the dehydrated form (basal distance around 7 Å) with the minimal formula of $Al_2Si_2O_5(OH)_4$ which is identical to that of kaolinite. The hydrated form converts irreversibly into the dehydrated form when dried at temperatures below 100°C (Nicolini et al., 2009). Various inorganic and organic species can be used in the intercalation of halloysite into its interlayer spaces, such as formamide, dimethylsulfoxide, urea, potassium acetate, aniline, and hydrazine. When intercalation takes place with larger molecules, the basal distance between interlayers is higher than 10 Å (Frost et al., 2010; Wilson, 2013).

Different types of inhibitors can be loaded within halloysite lumen and used for the doping of coatings. Fix et al. (2009) investigated the corrosion protection efficiency of aluminum AA2024 alloy covered with a sol–gel matrix, which was doped with inhibitor-filled HNTs. The corrosion protection efficiency was monitored via scanning vibrating electrode technique (SVET). A self-healing effect of the sol–gel doped with inhibitor-loaded HNTs was demonstrated. Abdullayev et al. (2009) evaluated the

anti-corrosive performance of the sol–gel coating and paint loaded with 2–5 wt.% of halloysite-entrapped benzotriazole applied on copper and 2024-allumium alloy by direct exposure to corrosive media. To characterize the halloysite structure scanning electron microscope and transmission electron microscope were used. The corrosion inhibition efficiency of halloysite nanocontainers for aluminum and copper samples was monitored by SVET. A good corrosion protection was observed for halloysite-loaded samples in comparison to samples coated with only sol–gel. Shchukin et al. (2008) studied the anticorrosion properties of the sol–gel films doped with HNTs loaded with corrosion inhibitor (2-mercaptobenzothiazole) and outer surfaces layer-by-layer covered with polyelectrolyte multilayers. An increase in the anticorrosion properties were reached for the samples coated with sol–gel doped with HNTs. The use of dodecylamine encapsulated in halloysite was proposed to provide corrosion protection to coatings (Joshi, 2014). Kinetics of releasing of corrosion inhibitors encapsulated within halloysite lumen was also studied. Lvov et al. (2014) studied kinetics of releasing of drugs and proteins from HNTs in water. The results showed that halloysite inner lumen can store and release molecules in a controllable manner resulting useful for applications in drug delivery, antimicrobial materials, and self-healing polymeric composites. Yuan et al. (2012) studied the loading and the release of an anionic dye compound from functionalized HNTs. The results showed that functionalization of HNTs makes pH an external trigger for controlling the loading and the subsequent release of the encapsulated species. The main objective of this article is to encapsulate dodecylamine as corrosion inhibitor into halloysite nanoclay and dope an alkyd primer with them and to evaluate the corrosion inhibitor release kinetics and the self-healing effect.

MATERIALS AND METHODS

Materials
Plates of AISI 1020 carbon steel were used in this study, which were previously treated with CSi emery papers from 120 to 600 grit, sequentially, and then rinsed with distilled water, alcohol, and acetone. Samples were cut in different dimensions depending on the specific test. The halloysite used was purchased from Sigma-Aldrich, and it is mainly formed by SiO_2 and Al_2O_3, with 30–70 nm diameter and 4 μm long with surface area of 64 m²/g what is very typical for tubular halloysite (Wilson, 2013). Pure dodecylamine was purchased from Sigma-Aldrich and used as corrosion inhibitor. A commercial alkyd paint modified with phenolic resin with 65% solids was used.

Methods
Characterization of Halloysite Nanotubes
TG was performed using a thermogravimetric (TGA)/DSC1 from Mettler Toledo, in the temperature range from 30 to 800°C, at the heating rate of 10 K/min. The samples were placed on a platinum plate in nitrogen flux of 100 mL/min.

X-ray diffraction (XRD) patterns of the powdered samples were performed using a Panalytical X'Pert diffractometer using Cu Kα radiation, tube power 40 kV, current of 40 mA at room temperature. The angle measurements were performed in the range $2\theta = 4°$ to $70°$ with a step of $0.02°$.

Halloysite Treatment, Loading, and Addition into a Coating

Before loading the inhibitor into HNTs, they were exposed to a treatment with 2 mol/L sulfuric acid during 6 and 12 h in order to enhance the tube-loading efficiency (two to three times) (Abdullayev et al., 2009), increasing the inner space available for encapsulation of dodecylamine inhibitor. The embedment of dodecylamine inside the inner gallery of the HNTs was performed according to the adapted procedure described by Price et al. (2001). In the first step, 6 mL of dodecylamine ethanolic solution with a concentration of 10 mg/mL was prepared. In the second step, 50 mg of HNTs was added to the dodecylamine solution. The vial containing the mixture was transferred into a vacuum jar and then evacuated, which deaerates the halloysite lumen. Slight fizzing of the suspension indicates the air is being removed from the halloysite inner part. After the fizzing was stopped, the vial was sealed for 30 min to reach equilibrium in relation to dodecylamine distribution between the inner volume and the surrounding solution. Furthermore, due to a rapid evaporation of the solvent, the inhibitor concentration increases that improves the loading efficiency. During the complete loading procedure, the dispersion was stirred. The vacuum treatment was followed by water washing and centrifugation. The aqueous upper phase was removed and the precipitate was dried at 60°C in an oven overnight. The halloysite particles were dispersed in an alkyd paint diluent (50 wt.%) before addition into the paint.

Coated samples were obtained applying the paint with the help of a brush in two layers of approximately 200 μm total dry film thickness. A commercial alkyd paint modified with phenolic resin with 65% solids containing dispersed HNTs was prepared to coat carbon steel panels (dimensions of 100 mm × 150 mm × 2 mm) as a primer (first layer of about 94 μm dry thickness) with 10 wt.% of dodecylamine-loaded HNTs and a second layer of about 99 μm dry film thickness without HNTs was also applied. Dry coating thickness measurements were made using a Fisher Model DualScope® MP40.

Electrochemical Measurements

Electrochemical impedance spectroscopy (EIS) measurements were employed to evaluate two different types of systems: at first, for determining indirectly the release of corrosion inhibitor from halloysite lumen by monitoring of the corrosion behavior of carbon steel samples in 0.01 mol/L NaCl solution at different pHs (2, 6.2, and 9) and containing 1 wt.% of nanocontainers loaded with and without dodecylamine as corrosion inhibitor and, at second, for evaluating the corrosion protection performance of coated samples with alkyd paint doped with HNTs loaded with dodecylamine corrosion inhibitor in 0.01 mol/L NaCl solution. In order to accelerate the corrosion process of the coated samples, a small defect (about 130 μm diameter) was made on the coating sample by an indenter just before starting the experiment. Kinetic curves of dodecylamine inhibitor release from HNTs were obtained plotting the values of the ratio (|Z| with HNTs loaded

with inhibitor/|Z| with HNTs without inhibitor) vs. immersion time for each condition of pH. In order to calculate the impedance modulus values (|Z| with HNTs loaded with inhibitor and |Z| with HNTs without inhibitor) for different immersion times was necessary to find the values of log|Z| for each condition from Bode plots (log (|Z|) vs. log f) and at a fixed frequency $f = 31.5$ mHZ. This value of 31.5 mHz was chosen due to the fact that at low frequencies, it is possible to detect the charge transfer phenomena in the metal/solution interface.

Electrochemical impedance spectroscopy measurements were performed on duplicate samples at open circuit potential for different immersion times using a Gamry Reference 600 potentiostat/galvanostat/frequency analyzer and controlled by Gamry Framework software. A frequency range from 50 kHz to 5 mHz with a sinusoidal potential amplitude perturbation of 10 mV rms and 10 measurements for each frequency decade was adopted.

Scanning vibrating electrode technique measurements were performed on duplicate samples by using the Applicable Electronics equipment controlled by ASET (Sciencewares) software. Samples were prepared for SVET measurements by cutting into 1 cm × 1 cm area plates. The probe was located 100 μm above the surface and vibrated in the perpendicular direction to the surface (Z) with 20 μm amplitude. The frequency of vibration of the probe was 164 Hz. The scanned area was around 2 mm × 5 mm. In order to accelerate the corrosion process and evaluate the corrosion resistance of these samples, small scratches on the coated sample using a sharp tool were made of approximately 3–4 mm long and about 200 μm wide. Periodical measurements were taken during the course of immersion of the coated samples in a 0.01 mol/L NaCl solution used as electrolyte.

Accelerated Corrosion Test in Salt Spray Chamber

Salt spray experiments were performed on triplicate samples by using the Bass USC-ISSO-(PLUS) Model equipment following the prescriptions of ASTM B 117-11 standard. Scribes with 9 cm long were made on the coated samples using a sharp tool.

RESULTS

Thermogravimetric Measurements

Thermogravimetric curves for HNTs without inhibitor and with inhibitor are shown in **Figures 1A,B**, respectively. The red lines indicate the inflection points of the TG curve for each sample. **Figure 1A** presents the TGA results for sulfuric acid treated halloysite for 6 h without loading with dodecylamine. The thermogram shows the mass loss % for increasing temperatures starting from 30 to 800°C. As indicated in the curve, there are four different slopes or four inflection points (in the middle of red lines) where each one can be related to a different thermal-activated process occurrence. It starts losing adsorbed water (slope 1) from 30 to 100–150°C (Frost et al., 2010; Liu et al., 2013; Carrillo et al., 2014) summing 2.3% of mass loss in the dehydration process and from 150 to 380°C (slope 2) occurs the loss of structural water or so-called bound water (Frost et al., 2010; Liu et al., 2013; Carrillo et al., 2014) contributing to 1.4% of mass loss. For temperatures higher than 400–450°C (slopes 3 and 4) another

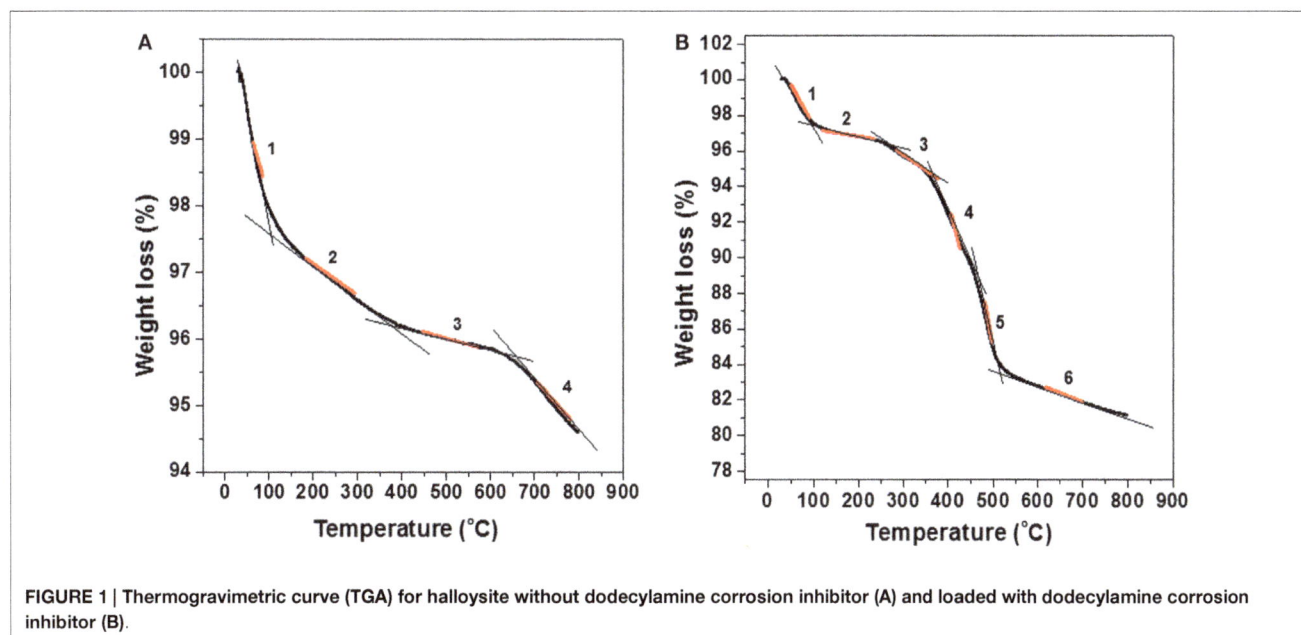

FIGURE 1 | Thermogravimetric curve (TGA) for halloysite without dodecylamine corrosion inhibitor (A) and loaded with dodecylamine corrosion inhibitor (B).

process is occurring and sums up to only 0.5% of mass loss. In the literature, transformations at temperatures beyond 450–600°C are assigned to dehydroxylation process during degradation of clays like halloysite or kaolinite (Yuan et al., 2008, 2013; Frost et al., 2010; Bodeepong et al., 2011; Carrillo et al., 2014; Luo et al., 2014). So, slopes 3 and 4 can be attributed to the dehydroxylation of halloysite.

At **Figure 1B**, one can find TGA results for halloysite loaded with dodecylamine inhibitor. In the thermogravimetric curve, it is possible to identify six slopes (inflection points) that can be related to different mass loss processes. It is important to stress the need for comparing both thermograms for loaded (**Figure 1B**) and for non-loaded (**Figure 1A**) halloysite. Since the clay is the same, the additional slopes or inflection points that appear in **Figure 1B** can be attributed to the release or degradation of the loaded organic compound, dodecylamine. From 30 to 120°C and from 150 to 250°C, the mass losses can be assigned to the loss of adsorbed water (2.5%) (slope 1) and loss of structural water (1.9%) (slope 2), respectively. Slope 3 in the range of 250–370°C could be related to the release of adsorbed dodecylamine molecules from the external walls of halloysite, which sums a mass loss of 0.8%. Slope 4 from 370 to 470°C range could be related to the release of dodecylamine molecules from the inner part of the lumen of halloysite and their thermal degradation corresponding to a significant mass loss of 6.0%. In the literature (Frost et al., 2010; Chen and Fu, 2012a,b; Liu et al., 2013), many articles report the initial release/degradation of organic compounds entrapped in inorganic structures at temperatures higher than 250°C. The very steep slope 5 starts about 470°C and goes to 530°C corresponding to a 4.8% of mass loss and can be a sum of two concomitant processes: thermal degradation of the remaining dodecylamine in the inner halloysite structure and the beginning of dihydroxylation of the halloysite clay.

TABLE 1 | Percentage of weight loss for different stages of thermal behavior of halloysite not loaded with dodecylamine inhibitor.

Slope	Range of temperature (°C)	% Weight loss
1	30–120	2.3
2	120–370	1.4
3	370–650	0.5
4	650–800	1.2
	Total mass loss (wt.%)	5.4

The final slope 6, from 550 to 800°C is related to the dehydroxylation of AlOH groups in the clay structure. It is reasonable to infer about concomitant process after comparison of **Figures 1A,B** in the range 470 to 800°C. Slope 3 in **Figure 1A** is less abrupt than the corresponding slope 5 in **Figure 1B**, denoting the presence of another process besides the beginning of dehydroxylation process in slope 5 of **Figure 1B**. If the sum of mass losses assigned to slopes 3, 4, and 5 is calculated in **Figure 1B**, a good estimate of the total amount dodecylamine loaded in halloysite is obtained. In this case, the sum of 1.9, 6.0, and 4.8% reaches 12.7%. If 0.5% is discounted from the total (mass loss related to slope 3 in **Figure 1A**), 12.2% is obtained. This value makes sense, because in literature, a range of 5–10% of cargo is always reported for different nanocontainers (Lamaka et al., 2008; Abdullayev et al., 2009; Fix et al., 2009; Skorb et al., 2009). These findings indicate that TGA technique is a good tool to estimate the inhibitor load in such inorganic nanocontainers.

Tables 1 and **2** show the different stages of mass loss of halloysite without and with encapsulated dodecylamine corrosion inhibitor, respectively. It is possible to deduce that dodecylamine load was in the range of a maximum of 13.4 wt.% (obtained from the difference of total mass loss in the presence and in the absence of dodecylamine from **Tables 1** and **2**) and a minimum of 12.2 wt.%, as discussed above. These estimates are reasonable and supported by other published papers (Lamaka et al., 2008; Abdullayev et al., 2009; Fix et al., 2009; Skorb et al., 2009).

Characterization of Halloysite Nanotubes by X-Ray Diffraction

Figure 2 shows the diffractograms for the samples of halloysite without and with treatment for 6 h in 2 mol/L sulfuric acid. The X-ray diffraction patterns in **Figure 2** show that the samples are all dominated by halloysite [$Al_2Si_2O_5(OH)_4$] and also contain other minerals as impurities. Beyond halloysite, minerals as alunite (aluminum sulfate), quartz, kaolinite, and gibbsite [$CaSO_4.2H_2O$] were also found. As shown in **Figure 2**, natural halloysite shows a sharp peak at 2θ of 20.1°, which is the characteristic (1 1 0) peak of tubular halloysite (Bodeepong et al., 2011; Abdullayev et al., 2012; Zhang et al., 2012; Carrillo et al., 2014). No intercalation of sulfuric acid into interlayer space is observed, as the (0 0 1) reflection does not shift to lower angles. This indicates that etching of alumina takes place from halloysite lumen and proceeds

toward the outermost layer (Abdullayev et al., 2012). For the halloysite treated for 6 h with 2 mol/L sulfuric acid is possible to observe an increase in the intensity of the peak corresponding to quartz (0 1 1) due to decrease of peak of halloysite (0 0 1) as consequence of the process of dealumination suffered during the treatment with sulfuric acid. Furthermore, when halloysite is treated in sulfuric acid for 6 h is possible to see that the intensity of peak (1 1 3) corresponding to alunite [$KAl_3(SO_4)_2(OH)_6$] increases due to a greater presence of aluminum sulfate caused by the attack provoked by sulfuric acid. In the pattern of halloysite after treatment does not appear a peak around 6°–8° that usually appears when there is adsorbed specie in the interlayer of the halloysite structure (Wilson, 2013). This could indicate that during treatment some organic species present was destroyed by treatment with sulfuric acid. Unfortunately, it was not possible to get XRD pattern of the dodecylamine-loaded halloysite that could prove the appearing of a basal interlayer peak at lower angles related to a higher basal distance (with intercalation compound) instead of that related to a peak at 10° related to a basal distance of 10 A (without intercalation compound) (Frost et al., 2010; Wilson, 2013).

Kinetics of Releasing of Dodecylamine Inhibitor from Halloysite Lumen for Different Values of pH

Figures 3–5 show the impedance diagrams for carbon steel in 0.1 mol/L NaCl solution containing 1 wt.% of halloysite nanocontainers loaded with dodecylamine inhibitor after different immersion times and pH values 2.0, 6.2, and 9.0.

TABLE 2 | Percentage of weight loss for different stages of thermal behavior of halloysite loaded with dodecylamine inhibitor.

Slope	Range of temperature (°C)	% Weight loss
1	30–120	2.5
2	120–270	0.8
3	270–370	1.9
4	370–470	6.0
5	470–530	4.8
6	510–800	2.8
	Total mass loss (wt.%)	18.8

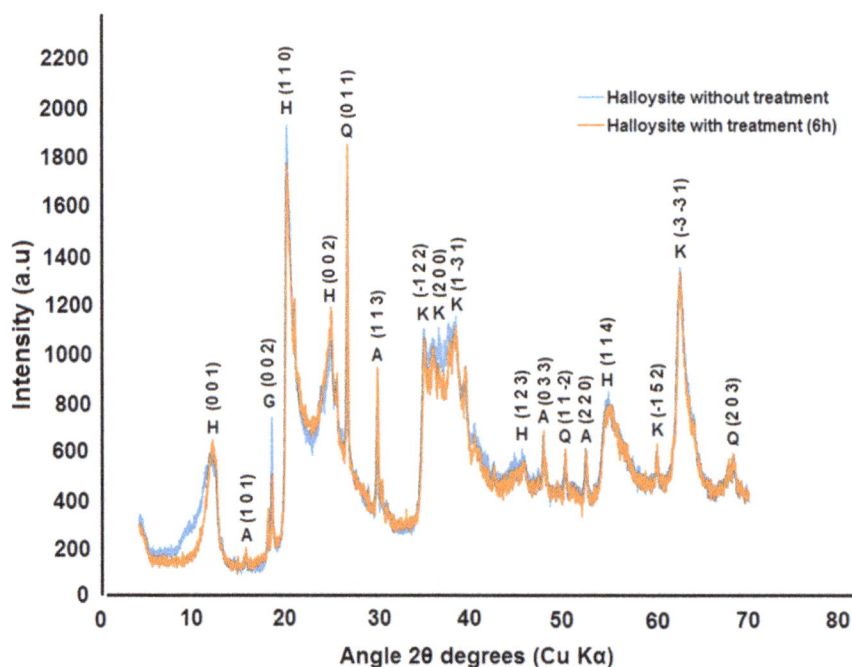

FIGURE 2 | X-ray diffraction diffractograms of powdered samples for the halloysite without treatment and with treatment in 2 mol/L sulfuric acid during 6 h. H, halloysite; A, alunite; Q, quartz; K, kaolinite; G, gibbsite.

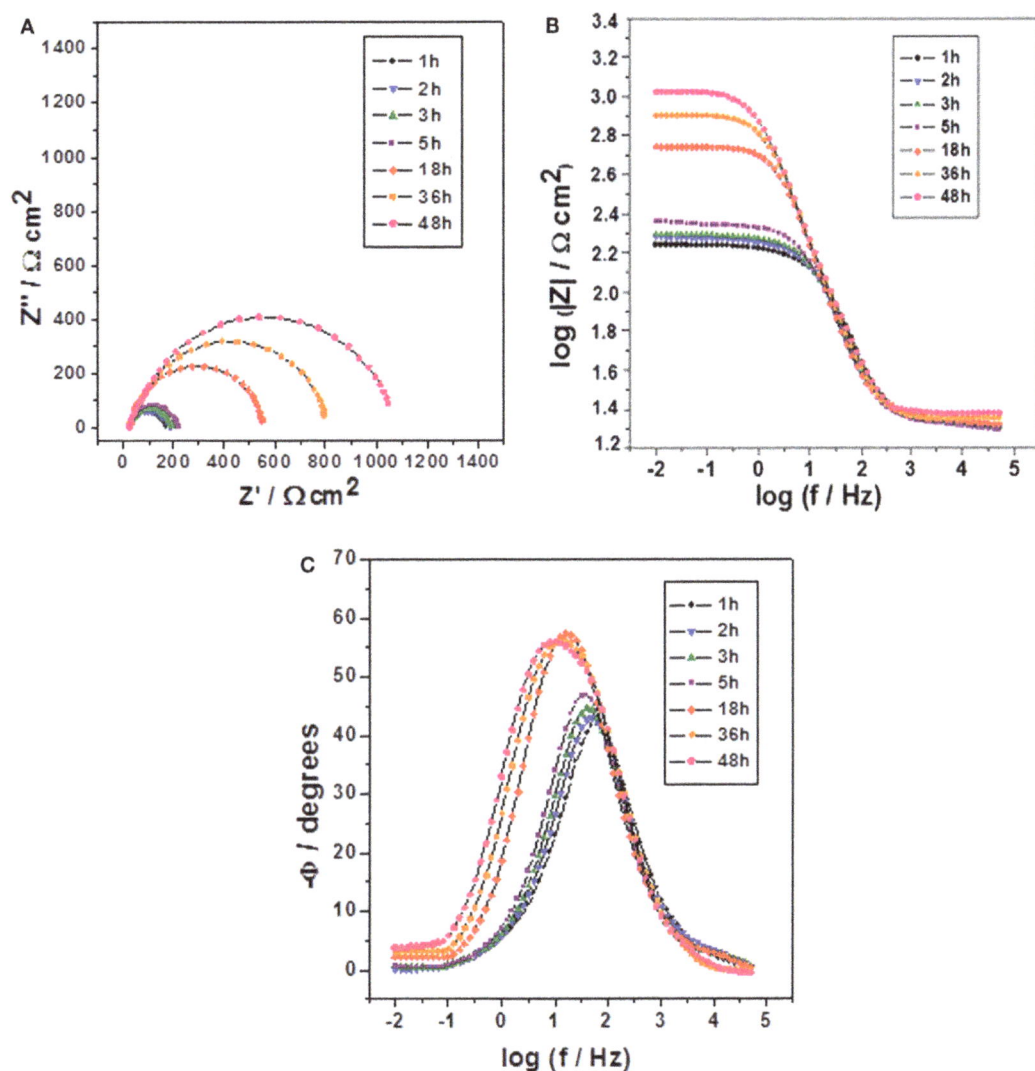

FIGURE 3 | Impedance diagrams: Nyquist (A) and Bode (B,C) plots for carbon steel after different immersion times in 0.1 mol/L NaCl at pH 2 and containing 1 wt.% of halloysite nanotubes with encapsulated dodecylamine.

Figure 3 shows the impedance diagrams for carbon steel for various immersion times in 0.1 mol/L NaCl solution at pH 2.0 and containing 1 wt.% of HNTs loaded with dodecylamine inhibitor, where it can be observed that for initial immersion times (1, 2, 3, and 5 h), there was a small increase in the capacitive arcs diameter shown in the Nyquist diagrams whose values oscillate around 200 Ω cm^2, but for long contact period (18, 36, and 48 h), this diameter increases significantly up to a value of 1100 Ω cm^2, which can be confirmed in the Bode diagram (θ vs. log f) where there is a displacement of higher phase angles from higher frequencies to lower frequencies indicating a more protective character of the dodecylamine to the metal for long immersion times. These results confirm that the release of the inhibitor is accelerated at low pH after long immersion times (Zheludkevich et al., 2007; Andreeva et al., 2008) what guarantees the release of corrosion inhibitor in defective areas.

Figure 4 shows the impedance diagrams for carbon steel for different immersion times in 0.1 mol/L NaCl solution at pH 6.2 and containing 1 wt.% of HNTs loaded with dodecylamine inhibitor. In this case, is possible to observe that for short immersion times (1, 2, 3, and 5 h) the capacitive arc diameter was around 1700 Ω cm^2. For longer periods of immersion (18, 36, and 48 h), a slight increase in the capacitive loop diameter till 2000 Ω cm^2 can be observed. From these results, it can be conclude that for this condition of neutral pH, the inhibitor release shows a slower kinetics for longer immersion time what means that an amount of inhibitor load will last for longer time in the case of a local mechanical defect occurs, what is good scenery.

Figure 5 depicts the results of impedance for carbon steel for various immersion times in 0.1 mol/L NaCl solution at pH 9.0 and containing 1 wt.% of HNTs loaded with dodecylamine inhibitor. Similarly, for this pH, it is also possible to observe a

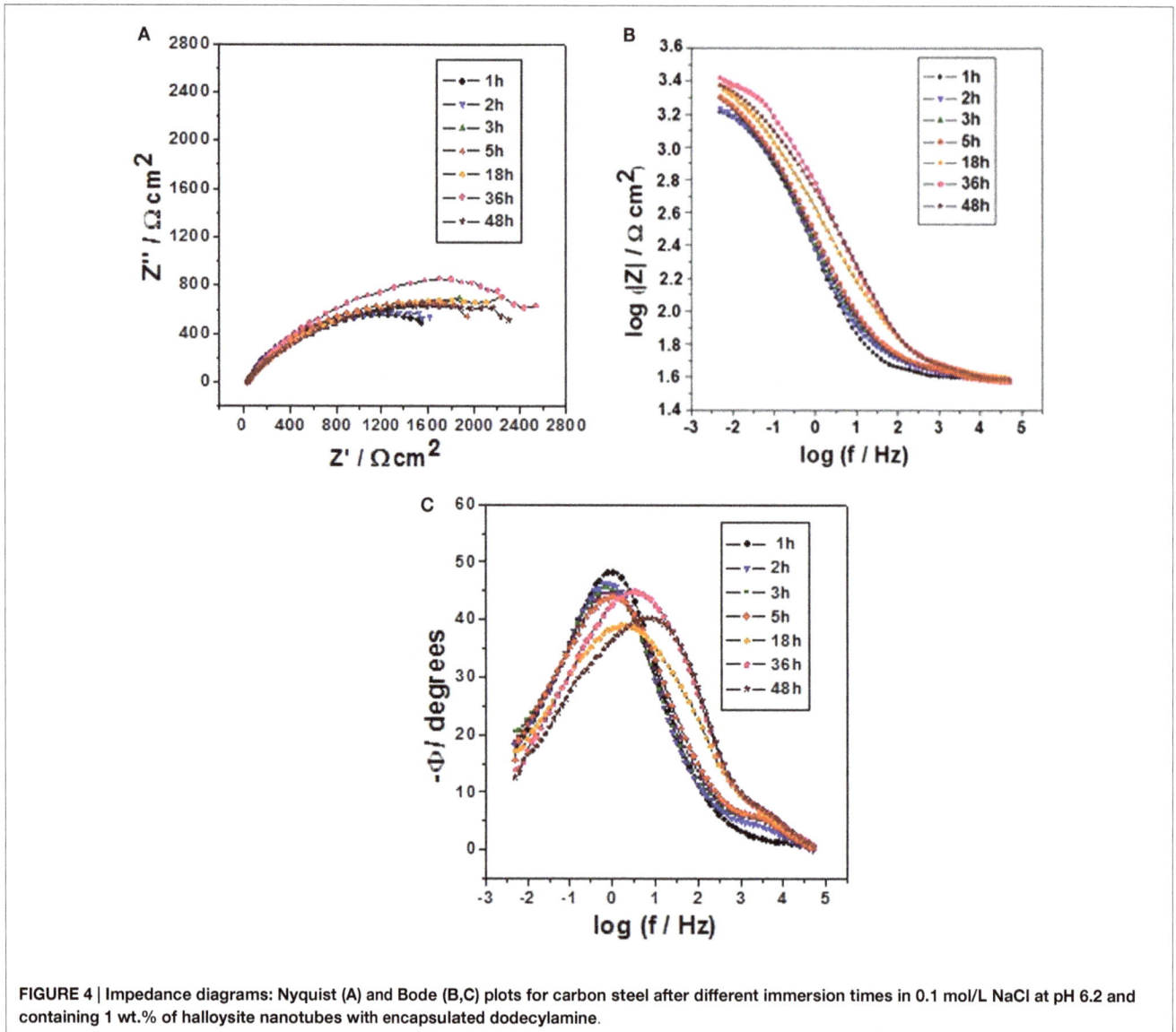

FIGURE 4 | Impedance diagrams: Nyquist (A) and Bode (B,C) plots for carbon steel after different immersion times in 0.1 mol/L NaCl at pH 6.2 and containing 1 wt.% of halloysite nanotubes with encapsulated dodecylamine.

slight increase in the capacitive arc diameter for long immersion times (18, 36, and 48 h) in comparison to the values obtained for short immersion times (1, 2, 3, and 5 h) due to a slower kinetics of release of dodecylamine from halloysite lumen what means that in this conditions the load of corrosion inhibitor will also last for longer time, shown also a good perspective, since, in a coating defect, higher pH values have been detected after long exposure time. So, if a local low pH is developed, the inhibitor will be released more slowly for short immersion times and faster for long immersion times and, if the pH is locally high, the inhibitors still will be released, but in a slower kinetics, for longer immersion times, lasting for longer periods.

Figure 6 shows the Nyquist diagrams for carbon steel after different immersion times in 0.1 mol/L NaCl at different pH values and containing 1 wt.% of HNTs without encapsulated dodecylamine. In these curves, it is possible to observe that for pH 2 condition the values of capacitive arc diameters are not constant

and increase slightly until reaching a value of 475 Ω cm². On the other hand, for the conditions of pH 6.2 and pH 9, the values of capacitive arc diameters do not present a significant increase. The values of $|Z|$ at 31.5 mHz for each pH and different immersion time will be the reference point for the determination of the curves of corrosion inhibitor release kinetics.

According to results obtained in **Figures 3–6** kinetic curves of dodecylamine inhibitor release from halloysite lumen were plotted for different immersion times in 0.1 mol/L NaCl solution at different pH values following the procedure explained in the Section "Materials and Methods" and are presented in **Figure 7**. It can be seen that for pH 6.2 and 9.0, there is a steep increase in the ratio between the impedance modulus for the condition with and without HNTs for shorter immersion times in comparison with values obtained for pH 2 and for longer immersion times a less pronounced release is noticed providing global logarithmic kinetics curves. For pH 2, the release kinetics is linear and after

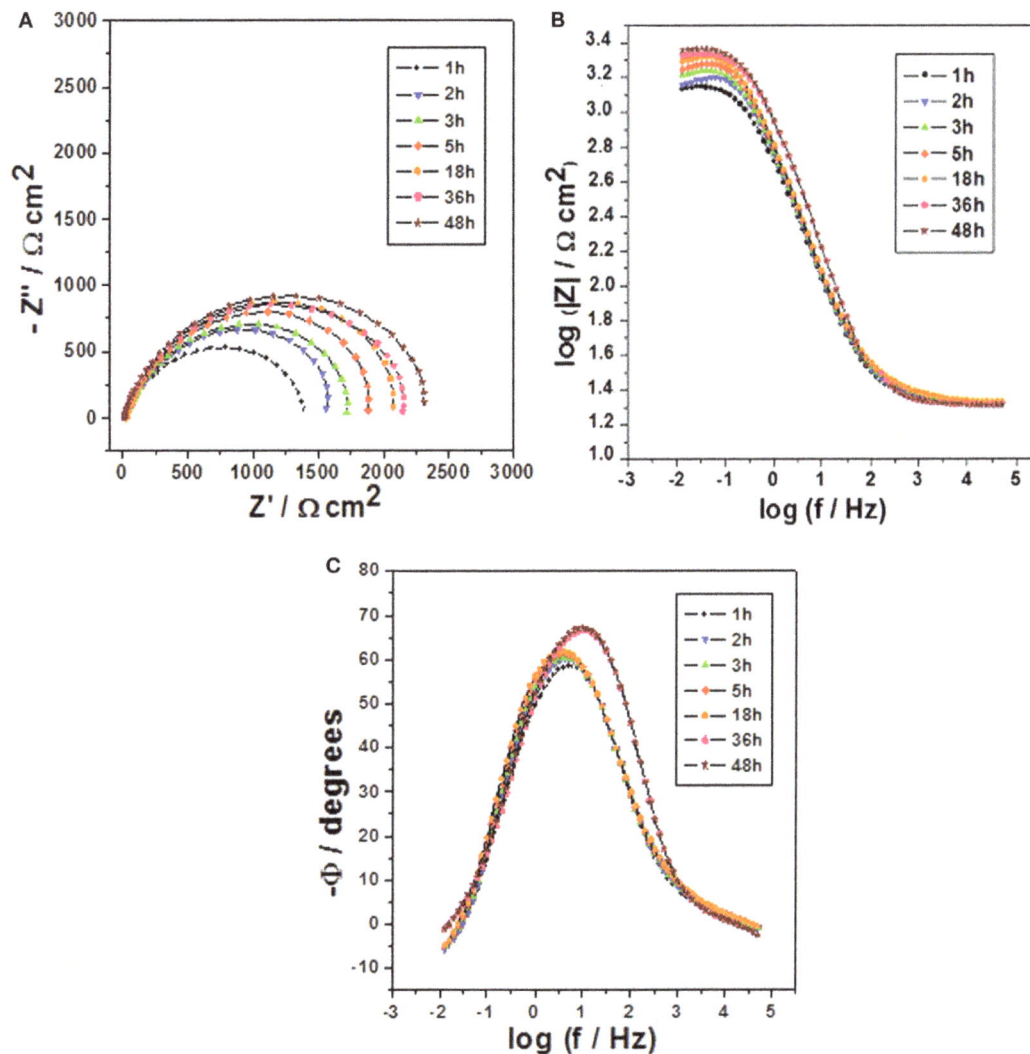

FIGURE 5 | Impedance diagrams: Nyquist (A) and Bode (B,C) plots for carbon steel after different immersion times in 0.1 mol/L NaCl at pH 9 and containing 1 wt.% of halloysite nanotubes with encapsulated dodecylamine.

48 h (high immersion times) and it is possible to observe that the release kinetics of pH 2 is very close to the release kinetics at pH 6.2 and 9.

Self-Healing Effect for Carbon Steel Samples Coated with Alkyd Primers Doped with Halloysite Nanotubes Loaded with Dodecylamine Inhibitor by EIS and SVET

EIS Measurements

The Bode diagrams (log $|Z|$ vs. log f) corresponding to the coated sample with two layers of alkyd paint containing 0 and 10 wt.% of HNTs in the primer are presented in **Figures 8A,B**, respectively. In both figures, it is possible to observe a capacitive response for the condition without defect, which extends from

the high frequencies to the medium frequencies. HNTs cause a small decrease in impedance modulus at low frequencies. This behavior indicates that the addition of the HNTs did not affect markedly the coating barrier properties due to their good dispersion in the paint and small size, which prevented the agglomeration and allowed a more uniform distribution on the coating layers.

In the case of defective coatings, it can be observed that after 4 h of immersion (**Figure 8A**), the addition of 10 wt.% of HNTs improved protection properties against corrosion due to the release of the inhibitor from the halloysite lumen into the damaged area. These results are in agreement with those already reported in the literature (Shchukin and Möhwald, 2007; Lvov et al., 2008; Shchukin et al., 2008; Abdullayev et al., 2013; Snihirova et al., 2013). After 8 h of immersion (**Figure 8B**), the coating self-healing has occurred and achieved the initial coating protective properties due to the increasing of impedance

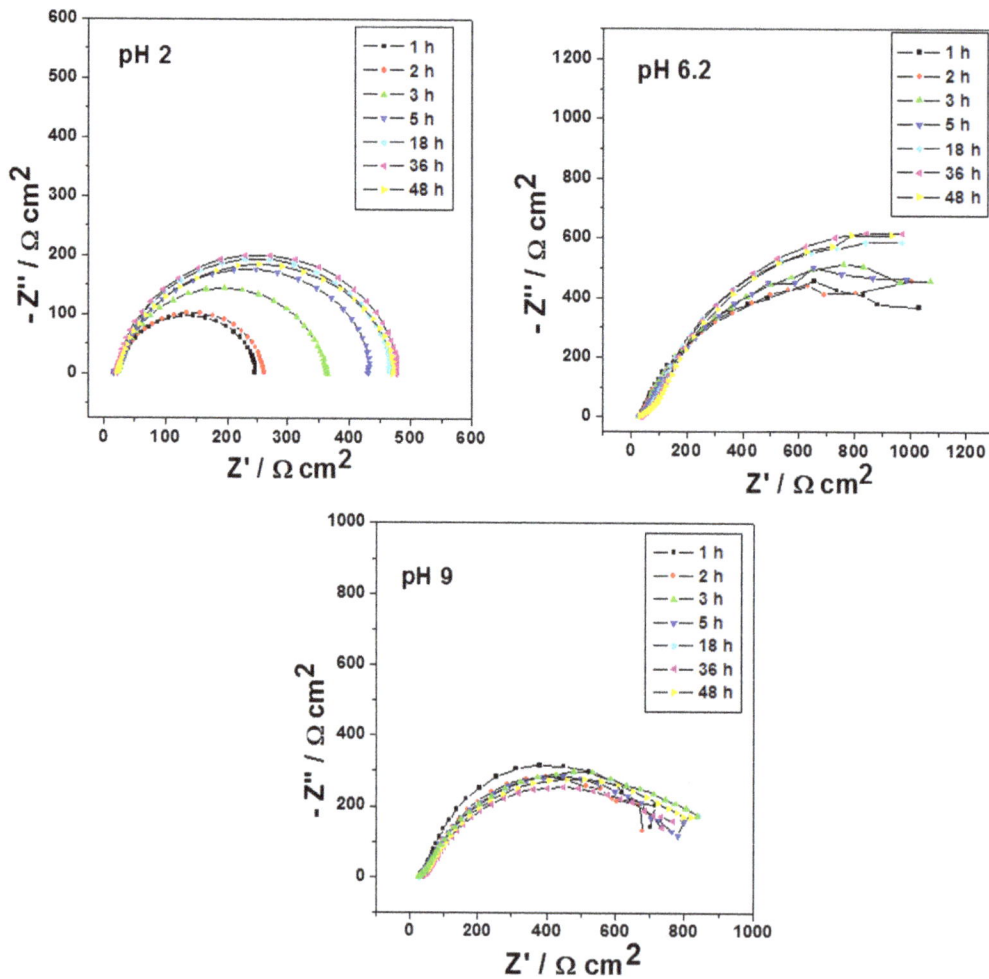

FIGURE 6 | Impedance diagrams: Nyquist plots for carbon steel after different immersion times in 0.1 mol/L NaCl at pH 2, 6.2, and 9 and containing 1 wt.% of halloysite nanotubes without encapsulated dodecylamine.

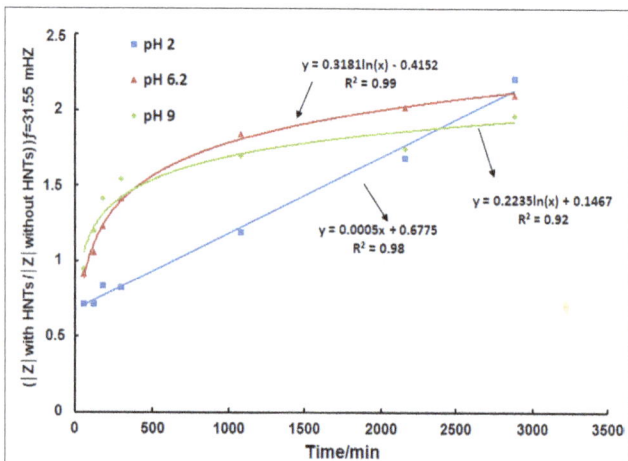

FIGURE 7 | Curves of the kinetics of the release of encapsulated dodecylamine inhibitor from halloysite for different immersion times in 0.1 mol/L NaCl solution at different pH values.

modulus values at low frequencies caused by the action of the dodecylamine inhibitor in the defect site, which provides a long and effective protection against carbon steel corrosion.

Scanning Vibrating Electrode Technique Measurements

Scanning vibrating electrode technique measurements for carbon steel coated with two layers of alkyd paint, with the *primer without HNTs*, for different immersion times of 1, 3, 4, and 12 h in 0.01 mol/L NaCl are shown in **Figure 9**. The results showed the presence of a high anodic ionic current density (383 μA/cm^2) after 1 h of immersion, which decreases over time until reaching a value of ionic current density of 325 μA/cm^2 after 12 h of immersion. This is due to the initial concentration of corrosion sites in a small region but over time, this region migrates along scratch causing the decrease of anodic ionic current density (Fix et al., 2009), and the precipitation of corrosion products provokes the slowdown of corrosion process as usual.

Figure 10 shows the SVET measurements and the resulting ionic current densities map obtained for carbon steel panels coated with two layers of alkyd paint and the *primer containing 10 wt.% of HNTs*, for different immersion times of 1, 3, 4, and 12 h in 0.01 mol/L NaCl. For these conditions, the results of SVET in **Figure 10** show that initially the ionic current density values start to increase until a value of 552 μA/cm^2 (after 3 h of immersion). After this time, the activity around the defect decrease sharply until reaching a value of anodic current density of only 72 μA/cm^2 (4 h of immersion) showing that the corrosion process was

stopped due to the action of dodecylamine inhibitor that was released from the HNTs on the provoked defect area (Lamaka et al., 2008; Lvov et al., 2008; Abdullayev et al., 2009, 2013; Fix et al., 2009; Snihirova et al., 2013). After 4 h of immersion, there was a slight decrease of the values of anodic current density until 68 μA/cm^2 (12 h of immersion) and then a slight increase for 24 h of immersion (data not shown), which indicate some depletion in the amount of inhibitor released from the nanocontainers.

As the release of the corrosion inhibitor is significant at all the pHs as shown above, the corroding site locally triggers the action

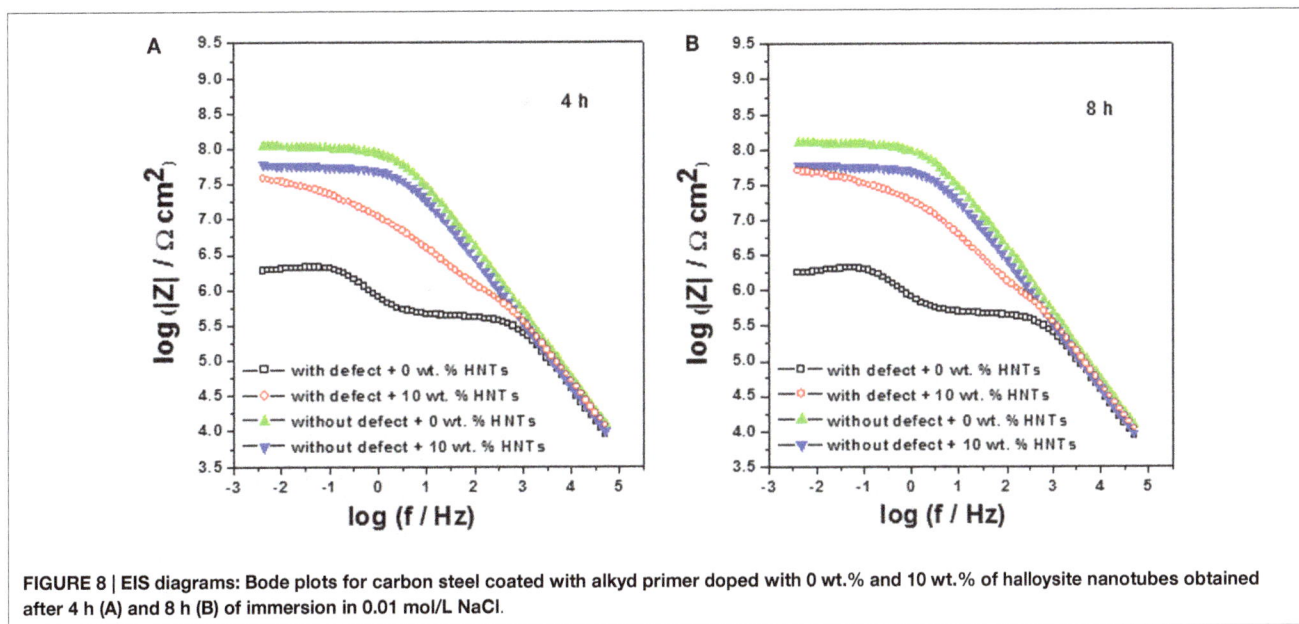

FIGURE 8 | EIS diagrams: Bode plots for carbon steel coated with alkyd primer doped with 0 wt.% and 10 wt.% of halloysite nanotubes obtained after 4 h (A) and 8 h (B) of immersion in 0.01 mol/L NaCl.

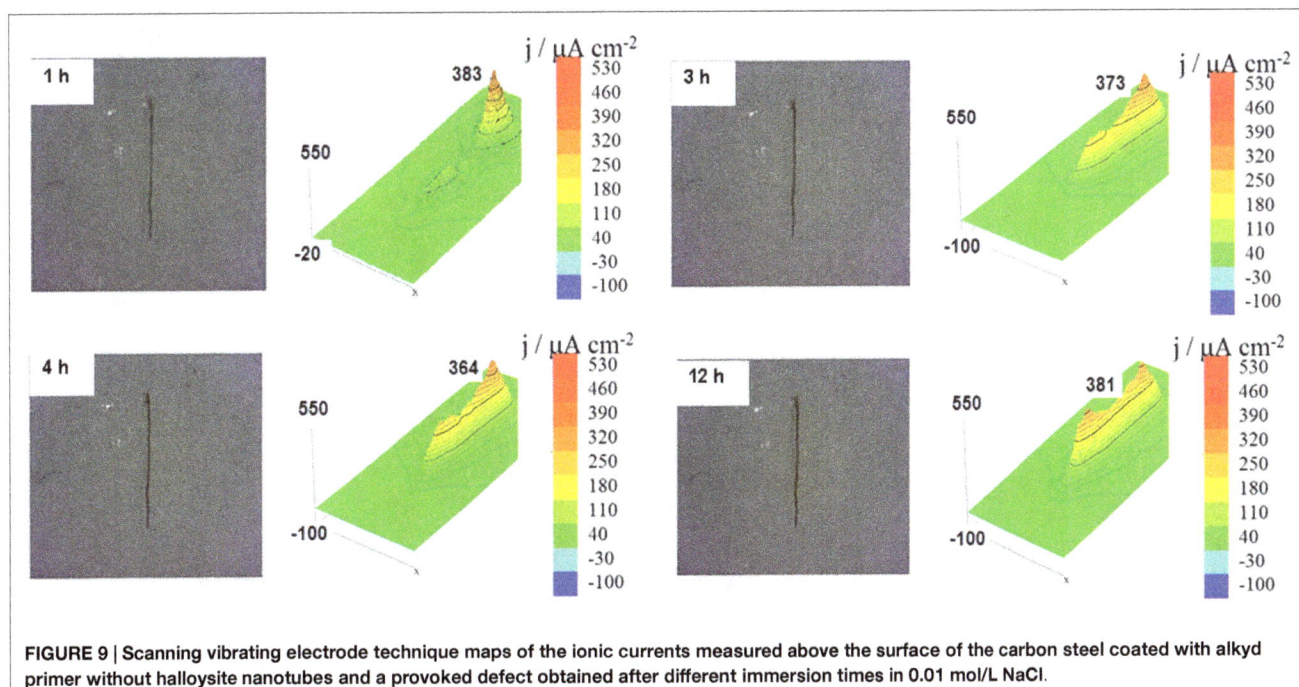

FIGURE 9 | Scanning vibrating electrode technique maps of the ionic currents measured above the surface of the carbon steel coated with alkyd primer without halloysite nanotubes and a provoked defect obtained after different immersion times in 0.01 mol/L NaCl.

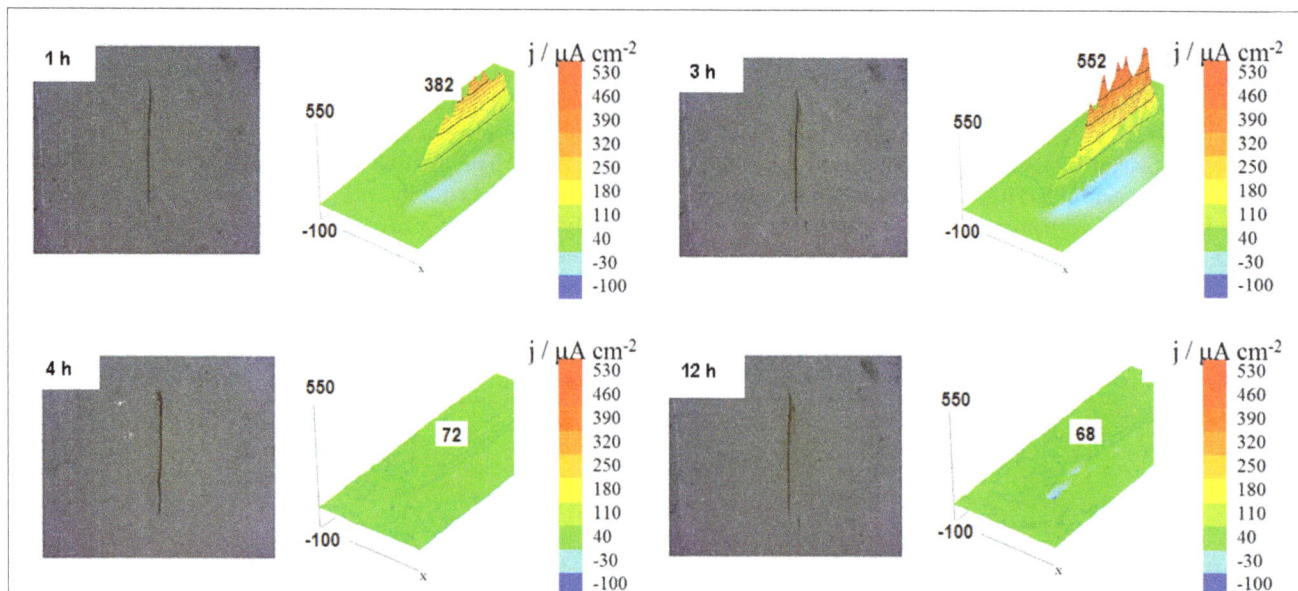

FIGURE 10 | Scanning vibrating electrode technique maps of the ionic currents measured above the surface of the carbon steel coated with alkyd primer doped with 10 wt.% of halloysite nanotubes loaded with dodecylamine inhibitor obtained after different immersion times in 0.01 mol/L NaCl.

of entrapped inhibitor, which blocks the initial corrosion process (Lvov et al., 2008; Fix et al., 2009) and the release continues as pH is getting higher as evidenced from kinetics curves of inhibitor release in **Figure 7**.

Salt Spray Tests

Figure 11 presents the results obtained for coated samples in the salt spray tests for exposure times of 0 and 720 h, where it is possible to see that after 720 h a severe corrosion process is presented especially for the coated samples without halloysite. On the other hand, for the coated samples with 10 wt.% of dodecylamine-loaded halloysite (on the primer layer), there was less corrosion products and no blistering around scribe compared to coated samples without nanocontainers, indicating a better protection against ingress of aggressive species at the scribe and preventing the spread of corrosion under the paint.

CONCLUSION

The XRD studies showed that the halloysite sample contain other types of minerals as alunite, quartz, kaolinite, and gibbsite, which may have affected the performance of the halloysite.

Thermogravimetric curves indicated more than one simple step for the release and/or degradation of dodecylamine from halloysite structure. It was possible to estimate the amount of inhibitor loaded in the halloysite nanoclay.

Electrochemical impedance spectroscopy and SVET results for the carbon steel coated with a primer layer doped with 10 wt.% of HNTs loaded with dodecylamine inhibitor showed self-healing ability by release of the encapsulated inhibitor triggered by pH changes around provoked defect inhibiting the kinetics of corrosion process.

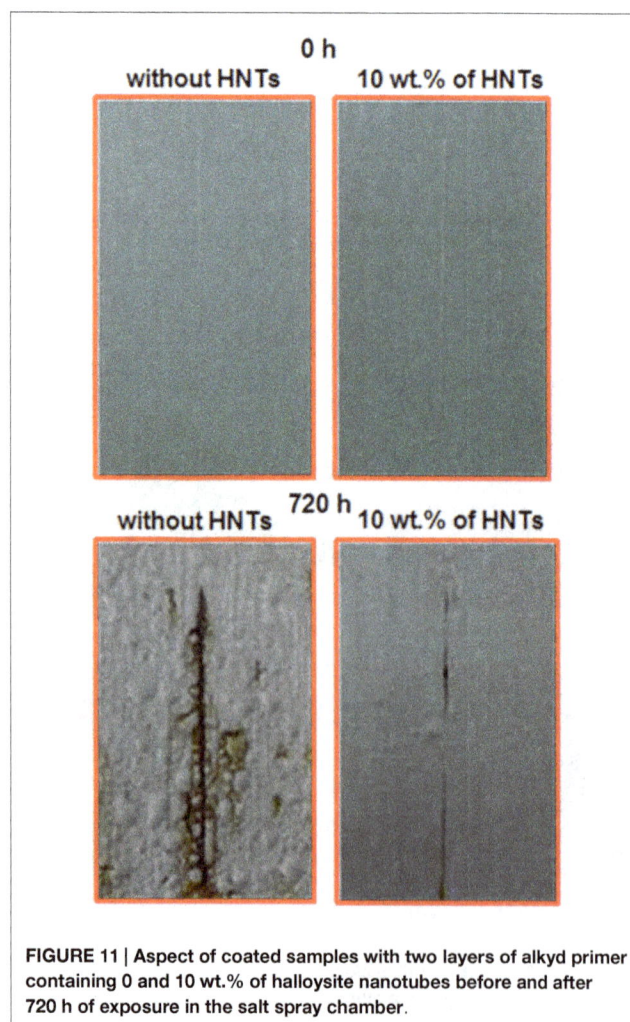

FIGURE 11 | Aspect of coated samples with two layers of alkyd primer containing 0 and 10 wt.% of halloysite nanotubes before and after 720 h of exposure in the salt spray chamber.

The salt spray tests showed that the addition of dodecylamine-loaded HNTs on the primer of coated samples provided self-healing properties, with the release of inhibitor from the lumen of halloysite inhibiting the attack provoked by the corrosion process.

These results unquestionably prove the self-healing capacity of the alkyd primer doped with halloysite loaded with dodecylamine and prescribes the possibility of enhancing the protective properties of such coating system that is well known as not the best for high performance coatings.

FUNDING

Authors would like to express sincere gratitude to CNPq for the financial support for developing this research (Process no. 141051/2010-08).

REFERENCES

Abdullayev, E., Abbasov, V., Tursunbayeva, A., Portnov, V., Ibrahimov, H., Mukhtarova, G., et al. (2013). Self-healing coatings based on halloysite clay polymer composites for protection of copper alloys. *ACS Appl. Mater. Interfaces* 5, 4464–4471. doi:10.1021/am400936m

Abdullayev, E., Joshi, A., Wei, W., Zhao, Y., and Lvov, Y. (2012). Enlargement of halloysite clay nanotube lumen by selective etching of aluminium oxide. *ACS Nano* 6, 7216–7226. doi:10.1021/nn302328x

Abdullayev, E., Price, R., Shchukin, D. G., and Lvov, Y. M. (2009). Halloysite tubes as nanocontainers for anticorrosion coating with benzotriazole. *ACS Appl. Mater. Interfaces.* 1, 1437–1443. doi:10.1021/am9002028

Andreeva, D. V., Fix, D., Möhwald, H., Hack, T., and Shchukin, D. (2008). Self-healing anticorrosion coatings based on pH-sensitive polyelectrolyte/inhibitor sandwichlike nanostructures. *Adv. Mater* 20, 2789–2794. doi:10.1002/adma.200800705

Andreeva, D. V., Skorb, E. V., and Shchukin, D. (2010). Layer-by-layer polyelectrolyte/inhibitor nanostructures for metal corrosion protection. *ACS Appl. Mater. Interfaces* 2, 1954–1962. doi:10.1021/am1002712

Bodeepong, S., Bhongsuwan, D., Pungrassami, T., and Bhongsuwan, T. (2011). Characterization of halloysite from Thung Yai Distric, Nackon Si Thammarat Province, in Shouthern Thailand. *Songklanakarin J. Sci. Technol.* 33, 599–607. doi:10.1080/09243046.2014.915116

Borisova, D., Mohwald, H., and Shchukin, D. (2011). Mesoporous silica nanoparticles for active corrosion protection. *J. Mater. Chem.* 5, 1939–1946. doi:10.1021/nn102871v

Carrillo, A. M., Urruchurto, C. M., Carriazo, J. G., Moreno, S., and Molina, R. A. (2014). Structural and textural characterization of a Colombian halloysite. *Rev. Mex. Chem. Eng* 13, 563–571. doi:10.1016/j.clay.2009.11.029

Chen, T., and Fu, J. (2012a). An intelligent anticorrosion coating based on pH-responsive supramolecular nanocontainers. *J. Nanotechnol.* 23, 1–12. doi:10.1088/0957-4484/23/50/505705

Chen, T., and Fu, J. (2012b). pH-responsive nanovalves base on hollow mesoporous silica spheres for controlled release of corrosion inhibitor. *J. Nanotechnol.* 23, 1–8. doi:10.1088/0957-4484/23/23/235605

Evaggelos, M., Ioannis, K., George, P., and George, K. (2011). Release studies of corrosion inhibitors from cerium titanium oxide nanocontainers. *J. Nanopart. Res.* 13, 541–554. doi:10.1007/s11051-010-0044-x

Fix, D., Andreeva, D. V., Shchukin, D., and Mohwald, H. (2009). Application of inhibitor-loaded halloysite nanotubes in active-corrosive coatings. *Adv. Funct. Mater.* 19, 1720–1727. doi:10.1002/adfm.200800946

Frost, R. L., Yang, J., Hongfei, C., Liu, Q., and Zhang, J. (2010). Thermal analysis and infrared emission spectroscopic study of halloysite potassium acetate intercalation compound. *Thermochim. Acta.* 511, 124–128. doi:10.1007/s10973-014-3692-8

Hu, Y., Chen, Y., Zhang, L., Jiang, X., and Yang, C. (2005). Synthesis and stimuli-responsive properties of chitosan/poly(acrylic acid) hollow nanospheres. *Polymer* 46, 12703–12710. doi:10.1016/j.polymer.2005.10.110

Joshi, A. (2014). *Applications of Halloysite Nanocontainers for Functional Protective Coating.* Ph.D. thesis, Louisiana Tech University, Ruston, LA.

Joshi, A., Abdullayev, E., Vasiliev, A., Volkova, O., and Lvov, Y. (2013). Interfacial modification of clay nanotubes for the sustained release of corrosion inhibitors. *Langmuir* 29, 7439–7448. doi:10.1021/la3044973

Kamble, R., Ghag, M., Gaikawad, S., and Panda, B. K. (2012). Halloysite nanotubes and applications: a review. *J. Adv. Sci. Res.* 3, 25–29.

Lamaka, S. V., Shchukin, D., Andreeva, D. V., Zheludkevich, M. L., Möhwald, H., and Ferreira, M. G. S. (2008). Sol-gel/polyelectrolyte active corrosion protection system. *Adv. Funct. Mater.* 18, 3137–3147. doi:10.1002/adfm.200800630

Liu, M., Wu, C., Jiao, Y., Xiong, S., and Zhou, C. (2013). Chitosan-halloysite nanotubes nanocomposite scaffolds for tissue engineering. *J. Mater. Chem. B.* 1, 2078–2089. doi:10.1039/c3tb20084a

Luo, Z., Wang, A., Wang, C., Qin, W., Zhao, N., Song, H., et al. (2014). Liquid crystalline phase behavior and fiber spinning of cellulose/ionic liquid/halloysite nanotubes dispersions. *J. Mater. Chem. A* 2, 7327–7336. doi:10.1039/c4ta00225c

Lvov, Y. M., Aerov, A., and Fakhrullin, R. (2014). Clay nanotube encapsulation for functional biocomposites. *Adv. Colloid Interface Sci.* 207, 189–198. doi:10.1016/j.cis.2013.10.006

Lvov, Y. M., Shchukin, D., Mohwald, H., and Price, R. R. (2008). Halloysite clay nanotubes for controlled release of protective agents. *ACS Nano* 2, 814–820. doi:10.1021/nn800259q

Montemor, M. F. (2014). Functional and smart coatings for corrosion protection: a review of recent advances. *Surf. Coat. Tech.* 258, 17–37. doi:10.1016/j.surfcoat.2014.06.031

Nesterova, T., Dam-Johansen, K., and Kill, S. (2011). Syntheses of durable microcapsules for self-healing coatings: a comparison of selected methods. *Prog. Org. Coat* 70, 325–342. doi:10.1016/j.porgcoat.2010.09.032

Nicolini, K. P., Fukamachi, C. R. B., Wypych, F., and Mangrich, A. S. (2009). Dehydrated halloysite intercalated mechanochemically with urea: thermal behavior and structural aspects. *J. Colloid Interface Sci.* 338, 474–479. doi:10.1016/j.jcis.2009.06.058

Price, R. R., Gaber, B. P., and Lvov, Y. M. (2001). In-vitro release characteristics of tetracycline HCl, khellin and nicotinamide adenine dineculeotide from halloysite; a cylindrical mineral. *J. Microencapsul.* 18, 713–722. doi:10.1080/02652040010019532

Rawtani, D., and Agrawal, Y. K. (2012). Multifarious applications of halloysite nanotubes: a review. *Rev. Adv. Mater. Sci.* 30, 282–295.

Shchukin, D., Lamaka, S. V., Yasakau, K. A., Zheludkevich, M. L., Ferreira, M. G. S., and Möhwald, H. (2008). Active anticorrosion coatings with halloysite nanocontainers. *J. Phys. Chem. C* 112, 958–964. doi:10.1021/jp076188r

Shchukin, D., and Möhwald, H. (2007). Surface-engineered nanocontainers for entrapment of corrosion inhibitors. *Adv. Funct. Mater* 17, 1451–1458. doi:10.1002/adfm.200601226

Shi, Y. F., Tian, Z., Zhang, Y., Shen, H. B., and Jia, N. Q. (2011). Functionalized halloysite nanotube-based carrier for intracellular delivery of antisense oligonucleotides. *Nanoscale Res. Lett.* 6, 1–7. doi:10.1186/1556-276X-6-608

Skorb, E. V., Fix, D., Andreeva, D. V., Möhwald, H., and Shchukin, D. (2009). Surface-modified mesoporous SiO$_2$ containers for corrosion protection. *Adv. Funct. Mater* 19, 2373–2379. doi:10.1002/adfm.200801804

Snihirova, D., Liphardt, L., Grundmeier, G., and Montemor, F. (2013). Electrochemical study of the corrosion inhibition ability of "smart" coatings applied on AA2024. *J. Solid State Electrochem.* 17, 2183–2192. doi:10.1007/s10008-013-2078-3

Suh, Y. J., Kil, D. S., Chung, K. S., Abdullayev, E., Lvov, Y. M., and Mongayt, D. (2011). Natural nanocontainer for the controlled delivery of glycerol as a moisturizing agent. *J. Nanosci. Nanotechnol.* 11, 661–665. doi:10.1166/jnn.2011.3194

Suryanarayana, C., Rao, K. Z., and Kumar, D. (2008). Preparation and characterization of microcapsules containing linseed oil and its use in self-healing coatings. *Prog. Org. Coat.* 63, 72–78. doi:10.1007/s11434-010-4133-0

Tedim, J., Poznyak, S. K., Kuznetsova, A., Raps, D., Hack, T., Zheludkevich, M. L., et al. (2010). Enhancement of active corrosion protection via combination of inhibitor-loaded nanocontainers. *ACS Appl. Mater. Interfaces* 2, 1528–1535. doi:10.1021/am100174t

Wilson, M. J. (2013). *Rock-Forming Minerals: Sheet Silicates – Clay Minerals*, 2nd Edn. London: Geological Society of London.

Wu, D. Y., Meure, S., and Solomom, D. (2008). Self-healing polymeric materials: a review of recent developments. *Prog. Polym. Sci.* 33, 479–522. doi:10.1016/j.progpolymsci.2008.02.001

Yuan, P., Southon, P. D., Liu, Z., Green, M. E. R., Hook, J. M., Antill, S. J., et al. (2008). Functionalization of halloysite clay nanotubes by grafting with γ-aminopropyltriethoxysilane. *J. Phys. Chem. C* 112, 15742–15751. doi:10.1021/jp805657t

Yuan, P., Southon, P. D., Liu, Z., and Kepert, C. J. (2012). Organosilane functionalization of halloysite nanotubes for enhanced loading and controlled release. *Nanotechnology.* 23, 1–5. doi:10.1088/0957-4484/23/37/375705

Yuan, P., Tan, D., and Annabi-Bergaya, F. (2015). Properties and applications of halloysite nanotubes: recent research advances and future prospects. *Appl. Clay Sci.* 11, 75–93. doi:10.1016/j.clay.2015.05.001

Yuan, P., Tan, D., Annabi-Bergaya, F., Yan, W., Liu, D., and Liu, Z. (2013). From platy kaolinite to aluminosilicate nanoroll via one-step delamination of kaolinite: effect of the temperature of intercalation. *Appl. Clay Sci.* 8, 68–76. doi:10.1016/j.clay.2015.05.001

Zhang, A. B., Pan, L., Zhang, H. L., Liu, S. T., Ye, Y., Xia, M. S., et al. (2012). Effects of acid treatment on the physic-chemical and pore characteristics of halloysite. *Colloids Surf. A.* 396, 182–188. doi:10.1016/j.colsurfa.2011.12.067

Zheludkevich, M. L., Serra, R., Montemor, M. F., and Ferreira, M. G. S. (2005). Oxide nanoparticle reservoirs for storage and prolonged release of corrosion inhibitors. *Electrochem. Commun.* 7, 836–840. doi:10.1016/j.elecom.2005.04.039

Zheludkevich, M. L., Shchukin, D., Yasakau, K. A., Möhwald, H., and Ferreira, M. G. S. (2007). Anticorrosion coatings with self-healing effect based on nanocontainers impregnated with corrosion inhibitor. *Chem. Mater.* 19, 402–411. doi:10.1021/cm062066k

Conflict of Interest Statement: The authors declare that the research was conducted in the absence of any commercial or financial relationships that could be construed as a potential conflict of interest.

14

Sporopollenin, the least known yet toughest natural biopolymer

Grahame Mackenzie[1,2]*, Andrew N. Boa[1], Alberto Diego-Taboada[1,2], Stephen L. Atkin[3] and Thozhukat Sathyapalan[4]

[1] *Department of Chemistry, University of Hull, Hull, UK,* [2] *Sporomex Ltd., Driffield, UK,* [3] *Weill Cornell Medical College Qatar, Doha, Qatar,* [4] *Hull York Medical School, University of Hull, Hull, UK*

Sporopollenin is highly cross-linked polymer composed of carbon, hydrogen, and oxygen that is extraordinarily stable and has been found chemically intact in sedimentary rocks some 500 million years old. It makes up the outer shell (exine) of plant spores and pollen and when extracted it is in the form of an empty exine or microcapsule. The exines resemble the spores and pollen from which they are extracted, in size and morphology. Also, from any one plant such characteristics are incredible uniform. The exines can be used as microcapsules or simply as micron-sized particles due to the variety of functional groups on their surfaces. The loading of a material into the chamber of the exine microcapsule is via multi-directional nano-diameter sized channels. The exines can be filled with a variety of polar and non-polar materials. Enzymes can be encapsulated within the shells and still remain active. *In vivo* studies in humans have shown that an encapsulated active substance can have a substantially increased bioavailability than if it is taken alone. The sporopollenin exine surface possesses phenolic, alkane, alkene, ketone, lactone, and carboxylic acid groups. Therefore, it can be derivatized in a number of ways, which has given rise to applications in areas, such as solid supported for peptide synthesis, catalysis, and ion-exchange chromatography. Also, the presence of the phenolic groups on sporopollenin endows it with antioxidant activity.

Keywords: sporopollenin, pollen, spores, exines, microcapsules

Edited by:
José Alejandro Heredia-Guerrero,
Istituto Italiano di Tecnologia, Italy

Reviewed by:
Frederik Roman Wurm,
Max Planck Institute for Polymer
Research, Germany
Wesley Toby Fraser,
Oxford Brookes University, UK

*Correspondence:
Grahame Mackenzie
g.mackenzie@hull.ac.uk

The Oxford English Dictionary defines a biopolymer as "any of various polymers occurring in living organisms." They are often categorized simply as carbohydrates, proteins, nucleic acids, polyiso-prenoid lipids, and lignin. Sporopollenin seems to fall off the list or appears much lower in profile, despite its extraordinary properties that can exceed those of more widely known biopolymers. Sporopollenin makes up the fabric of the outer shell (exine) of plant spores and pollens and is arguably the toughest plant substance known. It has been referred to as the "diamond of the plant world" (Kesseler, 2004), as "probably one of the most extraordinary resistant materials known in the organic world" (Feagri and Iverson, 1964) and, in the plural sense, "the most resistant organic materials of direct biological origin found in nature and in geological samples" (Brooks and Shaw, 1978). In support of these statements, intact exines have been found in some of the most ancient sedimentary rocks found on the Earth's surface (Wellman et al., 2003). In 1814, John (1814) was first to comment on the inertness of tulip pollen wall material, which he called "pollenin." Such inertness was later reported in 1829 by Braconnot (1829) in the pollen wall of bullrush (*Scirpoides holoschoenus*), which he named "sporonin." Almost a century later Zetzsche et al. made a systematic study to characterize the material from the asexual spores of *Lycopodium clavatum* L. (club moss) (Zetzsche and Kälin, 1931; Zetzsche and Vicari, 1931). He combined the previous names to form the word "sporopollenin," as the generic term for the resistant exine material forming both spore and pollen grain walls, since they appeared to be of the same or have a very similar chemical character. More simply, sporopollenin has been defined as "the resistant non-soluble material left after acetolysis" (Heslop-Harrison and Shaw, 1971; Shaw, 1971) (using a hot mixture acetic anhydride and concentrated sulfuric acid) (Erdtman, 1960) since it readily digests most other biopolymers. Perhaps because of this resilience, the structure and biosynthesis of sporopollenin has yet to be fully elucidated although significant progress has been made over the years (Brooks and Shaw, 1978; Hemsley et al., 1992; Shaw and Apperley, 1996; Ahlers et al., 2000; Fraser et al., 2012; Watson et al., 2012). There is evidence (Hemsley et al., 1998; Gabarayeva and Hemsley, 2006) for the involvement of self-assembly in the formation of exines and more recently evidence for genetic and molecular regulation in sporopollenin biosynthesis (Basketter et al., 1997; Ariizumi and Toriyama, 2011). Also, there is strong support for sporopollenin being built up via catalytic enzyme reactions present in the tapetum. In spite of the widely divergent exine morphology and patterning across plant taxa, the developmental processes of sporopollenin formation appear, surprisingly, not to vary significantly, which suggests the preservation of a common molecular mechanism (Fraser et al., 2012). The quest to unravel the structure of sporopollenin has attracted several workers and a variety of approaches over the years (Shaw, 1971; Schulze Osthoff and Wiermann, 1987; Guilford et al., 1988; Wehling et al., 1989; Shaw and Apperley, 1996). Cumulative evidence is in support of sporopollenin being a highly cross-linked polymer composed of carbon, hydrogen, and oxygen. The building blocks appear to be varied and complex, comprising straight- and branched-aliphatic chains, some of which are saturated, unsaturated, and polyhydroxylated (Ahlers et al., 2000). Other important building blocks involve oxygenated aromatic rings and phenylpropanoid moieties. Cross-linking of the blocks has been shown to involve ether cross-linking (Domínguez et al., 1998; Ahlers et al., 2000) but carbon–carbon cross-linking is likely. Exines can be extracted from spores or pollen by a variety of methods, all of which rely upon and demonstrate the extreme stability of sporopollenin to harsh chemical treatment and a wide range of digestive enzymes. The early extraction work was performed to provide a source material for structural characterization. Chemical methods have included successive treatments with hot acetone, potassium hydroxide, and phosphoric acid (Zetzsche and Huggler, 1928; Zetzsche and Kälin, 1931; Zetzsche and Vicari, 1931; Zetzsche et al., 1937) although sulfuric acid was suggested (Shaw and Apperley, 1996) to be a better alternative to phosphoric acid. Acetolysis, introduced by Erdtman (1960) is often used as a cleaning method for palynological sample preparation and has been used frequently since it is rapid and efficient. Interestingly, when used under relatively mild conditions acetolysis has relatively little impact on sporopollenin chemistry (Jardin et al., 2015). Domínguez et al. (1998) developed a method involving anhydrous hydrofluoric acid in pyridine in one-step and claimed that it isolated unaltered sporopollenin. An alternative one-step method used aqueous 4-methylmorpholine-N-oxide and sucrose in alkaline conditions (Loewus et al., 1985; Espelie et al., 1989; Couderchet et al., 1996). Most of the extraction methods produce sporopollenin that is nitrogen free by combustion elemental analysis, which is

indicative of sporopollenin being protein free. However, exines treated with sequential hot alkali and acid treatments were also shown to be protein free by mass spectrometry (MALDI-TOF-MS and ESI-QqToF-MS) and electrophoresis (SDS-PAGE), which are techniques in keeping with the Advisory Committee on Novel Foods and Processes (ACNFP) guidelines for the detection of trace protein in foods. It was proposed by Wiermann and Gubatz (1992) that all the harsh chemical methods denature sporopollenin in spore or pollen wall. For this reason, they embarked upon a detailed investigation using an extensive sequence of digestive enzymes to remove all biopolymers and leave only the natural sporopollenin intact (Schulze Osthoff and Wiermann, 1987). Clearly, it is important that sporopollenin be biologically degraded in nature "otherwise we would quite simply be submerged in these materials," as Faegri (1971) stated. Various bacteria are known to degrade sporopollenin under certain conditions (e.g., pH or aerobic milieu) (Faegri, 1971; Stanley and Linskens, 1974; Edlund et al., 2004). There is surprisingly little in the literature pertaining to characterized enzymes that are capable of degrading sporopollenin. Following pollination, the intine secretes an enzyme cocktail (acid phosphatase, ribonuclease, esterase, and amylase) to break up the exine, but the precise characterization of such enzymes not been made (Edlund et al., 2004). Similarly, an uncharacterized pollen esterase has been found in *Hordeum vulgare* L. (barley) pollen that hydrolyzes sporopollenin, in pore formation (Ahokas, 1976).

Exines are unique as microcapsules due to their resilience, consistency in size and characteristic morphology, the latter two features being peculiar to the plant taxa. The exines (**Figure 1**, left) are penetrated by a myriad of nano-diameter sized channels that provide access to the internal chamber (**Figure 1**, centre) for hydration and dehydration processes (Rowley et al., 2003). The biological role of the exines is to protect and transport the delicate genetic and nutrient payloads as part of the plant replication process. Therefore, the exines have evolved to be light shielding (Rozema et al., 2001; Atkin et al., 2011), antioxidant, elastic, and resilient, both chemically and physically. Such a wide range of properties endowed upon sporopollenin exines has attracted interest toward a wide range of applications, particularly where the raw pollen or spores can be obtained in bulk commercial amounts. Examples include club moss (*L. clavatum*) (**Figure 1**), sunflower (*Helianthus annuus*), pine (*Pinus*), and green algae (*Chlorella vulgaris*). The foremost of these is the club moss, which is perhaps the most robust and plentiful since it is used in herbal remedies (Orhan et al., 2007; Mandal et al., 2010) and pyrotechnics (Ellern, 1968). However, the use of pollen from cereal crops, such as rye, maize, and wheat, holds great potential due to the large amounts of such crops being grown worldwide. Exines have served in two roles: (i) as simple "microparticles" that are resilient, uniform, and have a multifunctional and highly decorated surface and (ii) resilient microcapsules with an accessible chamber that can be filled with a variety of polar and non-polar actives with relatively favorable weight-for-weight loadings (**Figure 1**, right). Initially, applications used only the former capability.

Applications as microparticles: the first reported application used exines extracted from *L. clavatum* using sequential treatment with hot alkali and acid (Mackenzie and Shaw, 1980). The aromatic moieties of the exines were chloromethylated and used as a

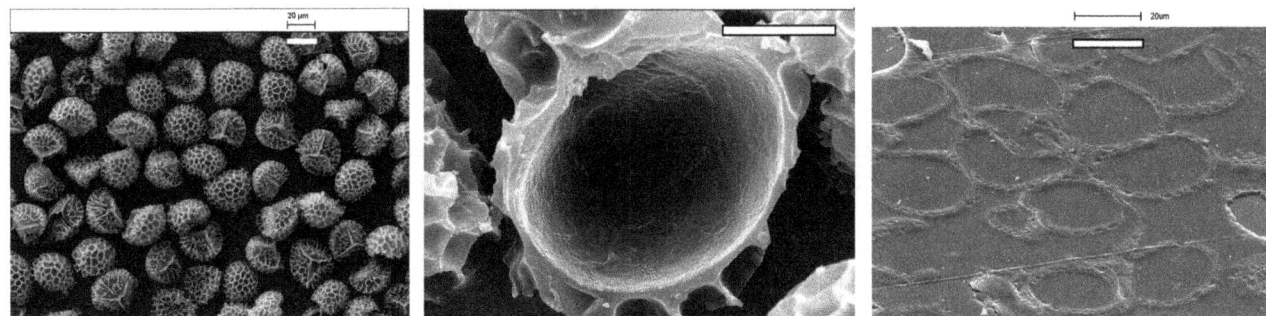

FIGURE 1 | Scanning electron microscopy (SEM) micrographs of *Lycopodium clavatum* L exines [left (bar = 20 μm)] exterior view of a cluster of exines [center (bar = 10 μm)], microtome section of a single exine showing the empty internal cavity, and [left (bar = 20 μm)] a microtome section of exines embedded in acrylic resin with internal cavities filled with the resin.

Merrifield-type resin to synthesize a tripeptide. Conveniently, the loading of the chloromethyl linker was similar to that found on commercial synthetic resins. The chemical resilience and consistency in particle size of exines, coupled with available aromatic and carboxylic acid functions, were put to use as an immobile phase in chromatography and ion-exchange processes (Shaw et al., 1988). The former could be sulfonated with chlorosulfonic acid to produce a strongly acid material and the carboxylic acid could be converted into basic amino functionalities, either by using different types of diamines or reductive amination using ammonia (Barrier et al., 2010). Various chemically modified and unmodified exines from *L. clavatum* have been used for metal remediation (Archibald et al., 2014). Modifications have also been reported to demonstrate the use of sporopollenin exines as a support for use in heterogeneous catalysis. Examples include sporopollenin-supported Schiff bases (Sahin et al., 2012) and palladium catalysts for use in Heck coupling reactions (Keles, 2013). More recently, the exines have been demonstrated to act as an efficient support for enzyme-catalyzed reactions (de Souza et al., 2015). The exine surface possesses ionizable groups (phenols and carboxylic acids) that become increasingly negatively charged with an increase in pH (Binks et al., 2011). This has facilitated their being able to adsorb at both air–water and oil–water interfaces, forming stabilized liquid marbles and emulsions, respectively. The surface conjugated phenol functionalities have been shown by cyclic voltammetry to cause either a two-electron two-proton, or a two-electron one-proton process, which suggests that a rapid electron transfer takes place over the over the surface. This is in accordance with the observed antioxidant behavior of sporopollenin for protecting encapsulated omega-3 oils. Furthermore, the exines have been shown to adsorb ultra-violet and visible light (Atkin et al., 2011; Lomax and Fraser, 2015), which might be expected in terms of their biological role of protecting the genetic material within the pollen or spore.

Applications as microcapsules: the capability of exines from *L. clavatum* to act as microcapsules (**Figure 1**) was demonstrated (Barrier et al., 2011) for a range of polar and non-polar materials, over a wide range of molecular weights, from water to proteins up to 2000 kDa (Atwe et al., 2014). Notably, neither the sporopollenin nor the encapsulation procedure was significantly deleterious to the enzymes' activity, thus offering the possibility of the exines to act as micro-reaction vessels or transporters of enzymes. This concept was also shown to be effective for the encapsulation of a lipase and subsequent use as catalyst in a variety of reactions (Tutar et al., 2009; Yilmaz, 2012). *L. clavatum* exines were also shown to work as micro-reactors for inorganic reactions. Magnetite nanoparticles and low-soluble salts, such as calcium phosphates, were formed within the exines' chambers. The relatively large available space in the *L. clavatum* exine's chamber, coupled with the exine's amphiphilic surface properties have shown application by being able to sequester oil very efficiently from oil-in-water emulsions. Not surprisingly, the efficiency could be controlled by modifying the polarity of the exines by converting the exine's surface hydroxyl groups (alcohols, phenols, and carboxylic acids), into polar salts (Na^+ and K^+) or non-polar derivatives (ethers, esters, and acetates). The ester forms were shown to sequester oils in near quantitative fashion. The elasticity of the exines permitted encapsulated oils to be released from the microcapsules in a stepwise manner simply by repeated rubbing. It is of note that the exines can be obtained with the cellulose intine (under) layer remaining intact, by hydrolysis of the spores or pollen with strong base, such as aqueous potassium hydroxide (Diego-Taboada et al., 2012). These double layered microcapsules (intine plus exine) can also be alkylated and acylated to sequester oil efficiently from emulsions. Normally, encapsulation into the exine's available chamber is *via* the nano-diameter channels that penetrate them. However, it would appear that compression of the dry exines under some 5–10 tonnes/cm^2 can open the trilete scar feature of the exines, allowing entry of living yeast cells into the exines' chambers (Barrier et al., 2011; Hamad et al., 2011). After this process, the cells still remain viable within the chamber and the exines undamaged, which further illustrate the remarkable elasticity and physical robustness of the exines. The potential for sporopollenin exines to be used in food and pharmaceutical applications has been demonstrated by their having taste-masking properties to the tongue when encapsulated with, such as fish oil and ibuprofen (Diego-Taboada et al., 2013). Also, the exines were shown to encapsulate a commercial gadolinium(III) MRI contrast agent and allow slow release over 8 h in blood plasma (Lorch et al., 2009). *In vitro* studies have shown exines to have the potential to act as an effective drug delivery vector in which release can be triggered by pH (Beckett et al., 2009; Diego-Taboada et al., 2013).

However, more importantly, the potential for exines to be used to deliver specialized foods and drugs has been demonstrated by the significantly enhanced bioavailability of the omega oil eicosapentaenoic acid (EPA) in a double cross-over study in human volunteers. It was shown that the taking the EPA in the encapsulated form provided a *circa* 10-fold increase in bioavailability over taking the EPA alone. The mechanism involved may be associated with bioadhesion of the exine capsules in the gastrointestinal tract. A later *in vivo* study in mouse showed that the sporopollenin exines filled with ovalbumin as a model antigen could be used for oral immunization and which demonstrates the potential of the exines to be used for oral vaccination (Atwe et al., 2014). The authors offered evidence by confocal microscopy that the exines translocate across the mouse intestinal epithelium, which, the authors proposed, might be involved

in the stimulation of the immune system. This work further supports the potential of such microcapsules to have universality toward the delivery of polar and non-polar payloads. The unique three-dimensional shapes and surface topographies of pollens have attracted interest to synthesis inorganic mimics for use in drug delivery and adhesive microparticles with tuneable short- and long-range attractive forces (Goodwin et al., 2013).

For many years, sporopollenin has been studied as a curiosity due its robust nature and role in nature; however, it is only since 1980 that there has been a steady rise of interest in the potential uses of sporopollenin, particularly from *L. clavatum*. The authors believe that there will be more applications and uses in the years to come of this unique renewable polymer that is plentiful in nature from a wide variety of plant sources.

REFERENCES

Ahlers, F., Bubert, H., Steuernagel, S., and Wiermann, R. (2000). The nature of oxygen in sporopollenin from the pollen of typha angustifolia L. *Z. Naturforsch. C* 55, 129–136. doi:10.1515/znc-2000-3-401

Ahokas, H. (1976). Evidence of a pollen esterase capable of hydrolysing sporopollenin. *Experientia* 32, 175–177. doi:10.1007/BF01937750

Archibald, S. J., Atkin, S. L., Bras, W., Diego-Taboada, A., Mackenzie, G., Mosselmans, J. F. W., et al. (2014). How does iron interact with sporopollenin exine capsules? An X-ray absorption study including microfocus XANES and XRF imaging. *J. Mater. Chem. B* 2, 945–959. doi:10.1039/C3TB21523G

Ariizumi, T., and Toriyama, K. (2011). Genetic regulation of sporopollenin synthesis and pollen exine development. *Annu. Rev. Plant Biol.* 62, 437–460. doi:10.1146/annurev-arplant-042809-112312

Atkin, S. L., Barrier, S., Cui, Z., Fletcher, P. D. I., Mackenzie, G., Panel, V., et al. (2011). UV and visible light screening by individual sporopollenin exines derived from *Lycopodium clavatum* (club moss) and *Ambrosia trifida* (giant ragweed). *J. Photochem. Photobiol. B* 102, 209–217. doi:10.1016/j.jphotobiol.2010.12.005

Atwe, S. U., Ma, Y., and Gill, H. S. (2014). Pollen grains for oral vaccination. *J. Control. Release* 194, 45–52. doi:10.1016/j.jconrel.2014.08.010

Barrier, S., Diego-Taboada, A., Thomasson, M. J., Madden, L., Pointon, J. C., Wadhawan, J. D., et al. (2011). Viability of plant spore exine capsules for microencapsulation. *J. Mater. Chem.* 21, 975–981. doi:10.1039/C0JM02246B

Barrier, S., Loebbert, A., Boasman, A. J., Boa, A. N., Lorch, M., Atkin, S. L., et al. (2010). Access to a primary aminosporopollenin solid support from plant spores. *Green Chem.* 12, 234–240. doi:10.1039/B913215E

Basketter, D. A., Chamberlain, M., Griffiths, H. A., Rowson, M., Whittle, E., and York, M. (1997). The classification of skin irritants by human patch test. *Food Chem. Toxicol.* 35, 845–852. doi:10.1016/S0278-6915(97)00053-7

Beckett, S. T., Atkin, S. L., and Mackenzie, G. (2009). *Dosage Form.* US 07608270.

Binks, B. P., Boa, A. N., Kibble, M. A., Mackenzie, G., and Rocher, A. (2011). Sporopollenin capsules at fluid interfaces: particle-stabilised emulsions and liquid marbles. *Soft Matter* 7, 4017–4024. doi:10.1039/c0sm01516d

Braconnot, H. (1829). Recherches chimiques sur le pollen de *Typha latifolia* L., famille des typhacées. *Ann. Chim. Phys.* 2, 91–105.

Brooks, J., and Shaw, G. (1978). Sporopollenin a review of its chemistry paleochemistry and geochemistry. *Grana* 17, 91–98. doi:10.1080/00173137809428858

Couderchet, M., Schmalfuß, J., and Böger, P. (1996). Incorporation of oleic acid into sporopollenin and its inhibition by the chloroacetamide herbicide metazachlor. *Pestic. Biochem. Physiol.* 55, 189–199. doi:10.1006/pest.1996.0048

de Souza, S. P., Bassut, J., Marquez, H. V., Junior, I. I., Miranda, L. S. M., Huang, Y., et al. (2015). Sporopollenin as an efficient green support for covalent immobilization of a lipase. *Catal. Sci. Technol.* 5, 3288–3295. doi:10.1039/C4CY01682C

Diego-Taboada, A., Cousson, P., Raynaud, E., Huang, Y., Lorch, M., Binks, B. P., et al. (2012). Sequestration of edible oil from emulsions using new single and double layered microcapsules from plant spores. *J. Mater. Chem.* 22, 9767–9773. doi:10.1039/c2jm00103a

Diego-Taboada, A., Maillet, L., Banoub, J. H., Lorch, M., Rigby, A. S., Boa, A. N., et al. (2013). Protein free microcapsules obtained from plant spores as a model for drug delivery: ibuprofen encapsulation, release and taste masking. *J. Mater. Chem. B* 1, 707–713. doi:10.1039/C2TB00228K

Domínguez, E., Mercado, J. A., Quesada, M. A., and Heredia, A. (1998). Isolation of intact pollen exine using anhydrous hydrogen fluoride. *Grana* 37, 93–96. doi:10.1080/00173139809362649

Edlund, A. F., Swanson, R., and Preuss, D. (2004). Pollen and stigma structure and function: the role of diversity in pollination. *The Plant Cell* 16, S84–S97. doi:10.1105/tpc.015800

Ellern, H. (1968). *Military and Civilian Pyrotechnics*. New York, NY: Chemical Publishing Company Inc.

Erdtman, G. (1960). The acetolysis method. A revised description. *Svensk Botanisk Tidskrift* 54, 561–564.

Espelie, K. E., Loewus, F. A., Pugmire, R. J., Woolfenden, W. R., Baldi, B. G., and Given, P. H. (1989). Structural-analysis of lilium-longiflorum sporopollenin by C-13 NMR-spectroscopy. *Phytochemistry* 28, 751–753. doi:10.1016/0031-9422(89)80108-6

Faegri, K. (1971). "The preservation of sporopollenin membranes under natural conditions," in *Sporopollenin*, eds J. Brooks, P. R. Grant, M. Muir, P. Van Gijzel, and G. Shaw (London: Academic Press), 256–272.

Feagri, I., and Iverson, J. (1964). *Textbook of Pollen Analysis*. London: Blackwell.

Fraser, W. T., Scott, A. C., Forbes, A. E. S., Glasspool, I. J., Plotnick, R. E., Kenig, F., et al. (2012). Evolutionary stasis of sporopollenin biochemistry revealed by unaltered Pennsylvanian spores. *New Phytol.* 196, 397–401. doi:10.1111/j.1469-8137.2012.04301.x

Gabarayeva, N., and Hemsley, A. R. (2006). Merging concepts: the role of self-assembly in the development of pollen wall structure. *Rev. Palaeobot. Palynol.* 138, 121–139. doi:10.1016/j.revpalbo.2005.12.001

Goodwin, W. B., Gomez, I. J., Fang, Y., Meredith, J. C., and Sandhage, K. H. (2013). Conversion of pollen particles into three-dimensional ceramic replicas tailored for multimodal adhesion. *Chem. Mater.* 25, 4529–4536. doi:10.1021/cm402226w

Guilford, W. J., Schneider, D. M., Labovitz, J., and Opella, S. J. (1988). High-resolution solid-state C-13 NMR-spectroscopy of sporopollenins from different plant taxa. *Plant Physiol.* 86, 134–136. doi:10.1104/pp.86.1.134

Hamad, S. A., Dyab, A. F. K., Stoyanov, S. D., and Paunov, V. N. (2011). Encapsulation of living cells into sporopollenin microcapsules. *J. Mater. Chem.* 21, 18018–18023. doi:10.1039/c1jm13719k

Hemsley, A. R., Chaloner, W. G., Scott, A. C., and Groombridge, C. J. (1992). C-13 Solid-state nuclear-magnetic-resonance of sporopollenins from modern and fossil plants. *Ann. Bot.* 69, 545–549.

Hemsley, A. R., Vincent, B., Collinson, M., and Griffiths, P. C. (1998). Simulated self-assembly of spore exines. *Ann. Bot.* 82, 105–109. doi:10.1006/anbo.1998.0653

Heslop-Harrison, J., and Shaw, G. (1971). "Sporopollenin in the biological context," in *Sporopollenin*, eds J. Brooks, P. R. Grant, M. Muir, and P. Van Gijzel (London: Academic Press), 1–30.

Jardin, P. E., Fraser, W. T., Lomax, B. H., and Gosling, W. D. (2015). The impact of oxidation on spore and pollen chemistry. *J. Micropalaeontolog* 34, 139–149. doi:10.1144/jmpaleo2014-022

John, J. F. (1814). Über den befruchtungsstaub, nebst einer analyse des tulpenpollens. *J. Chem. Phys.* 12, 244–252.

Keles, M. (2013). Preparation of heterogeneous palladium catalysts supported on sporopollenin for heck coupling reactions. *Synthesis and Reactivity in Inorganic Metal-Organic and Nano-Metal Chemistry* 43, 575–579. doi:10.1080/15533174. 2012.749895

Kesseler, I. H. M. (2004). *Pollen, The Hidden Sexuality of Flowers*. London: Papadakis Publisher.

Loewus, F. A., Baldi, B. G., Franceschi, V. R., Meinert, L. D., and McCollum, J. J. (1985). Pollen sporoplasts – dissolution of pollen walls. *Plant Physiol.* 78, 652–654. doi:10.1104/pp.78.3.652

Lomax, B. H., and Fraser, W. T. (2015). Palaeoproxies: botanical monitors and recorders of atmospheric change. *Palaeontology* 58, 759–768. doi:10.1111/pala. 12180

Lorch, M., Thomasson, M. J., Diego-Taboada, A., Barrier, S., Atkin, S. L., Mackenzie, G., et al. (2009). MRI contrast agent delivery using spore capsules: controlled release in blood plasma. *Chem. Commun.* 42, 6442–6444. doi:10.1039/b909551a

Mackenzie, G., and Shaw, G. (1980). Sporopollenin. A novel, naturally occurring support for solid phase peptide synthesis. *Int. J. Pept. Protein Res.* 15, 298–300. doi:10.1111/j.1399-3011.1980.tb02580.x

Mandal, S. K., Biswas, R., Bhattacharyya, S. S., Paul, S., Dutta, S., Pathak, S., et al. (2010). Lycopodine from *Lycopodium clavatum* extract inhibits proliferation of HeLa cells through induction of apoptosis via caspase-3 activation. *Eur. J. Pharmacol.* 626, 115–122. doi:10.1016/j.ejphar.2009.09.033

Orhan, I., Kupeli, E., Sener, B., and Yesilada, E. (2007). Appraisal of anti-inflammatory potential of the clubmoss, *Lycopodium clavatum* L. *J. Ethnopharmacol.* 109, 146–150. doi:10.1016/j.jep.2006.07.018

Rowley, J. R., Skvarla, J. J., and El-Ghazaly, G. (2003). Transfer of material through the microspore exine – from the loculus into the cytoplasm. *Can. J. Bot.* 81, 1070–1082. doi:10.1139/b03-095

Rozema, J., Broekman, R. A., Blokker, P., Meijkamp, B. B., de Bakker, N., van de Staaij, J., et al. (2001). UV-B absorbance and UV-B absorbing compounds (para-coumaric acid) in pollen and sporopollenin: the perspective to track historic UV-B levels. *J. Photochem. Photobiol. B* 62, 108–117. doi:10.1016/S1011-1344(01)00155-5

Sahin, M., Gubbuk, I. H., and Kocak, N. (2012). Synthesis and characterization of sporopollenin-supported schiff bases and ruthenium(III) sorption studies. *J. Inorg. Organomet. Polym. Mater.* 22, 1279–1286. doi:10.1007/s10904-012-9739-z

Schulze Osthoff, K., and Wiermann, R. (1987). Phenols as integrated compounds of sporopollenin from *Pinus* pollen. *J. Plant Physiol.* 131, 5–15. doi:10.1016/S0176-1617(87)80262-6

Shaw, G. (1971). "The chemistry of sporopollenin," in *Sporopollenin*, eds J. Brooks, P. R. Grant, M. Muir, P. V. Gijzel, and G. Shaw (London: Academic Press), 305–334.

Shaw, G., and Apperley, D. C. (1996). C-13-NMR spectra of *Lycopodium clavatum* sporopollenin and oxidatively polymerised beta-carotene. *Grana* 35, 125–127. doi:10.1080/00173139609429483

Shaw, G., Sykes, M., Humble, R. W., Mackenzie, G., Marsden, D., and Pehlivan, E. (1988). The use of modified sporopollenin from *Lycopodium clavatum* as a novel ion- or ligand-exchange medium. *React. Polym. Ion Exchangers Sorbents* 9, 211–217. doi:10.1016/0167-6989(88)90034-7

Stanley, R. G., and Linskens, H. F. (eds) (1974). *Pollen: Biology, Biochemistry, Management*. Berlin: Springer-Verlag, 307.

Tutar, H., Yilmaz, E., Pehlivan, E., and Yilmaz, M. (2009). Immobilization of *Candida rugosa* lipase on sporopollenin from *Lycopodium clavatum*. *Int. J. Biol. Macromol.* 45, 315–320. doi:10.1016/j.ijbiomac.2009.06.014

Watson, J. S., Fraser, W. T., and Sephton, M. A. (2012). Formation of a polyalkyl macromolecule from the hydrolysable component within sporopollenin during heating/pyrolysis experiments with *Lycopodium* spores. *J. Anal. Appl. Pyrolysis* 95, 138–144. doi:10.1016/j.jaap.2012.01.019

Wehling, K., Niester, C., Boon, J. J., Willemse, M. T. M., and Wiermann, R. (1989). *p*-Coumaric acid – a monomer in the sporopollenin skeleton. *Planta* 179, 376–380. doi:10.1007/BF00391083

Wellman, C. H., Osterloff, P. L., and Mohiuddin, U. (2003). Fragments of the earliest land plants. *Nature* 425, 282–285. doi:10.1038/nature01884

Wiermann, R., and Gubatz, S. (1992). Pollen wall and sporopollenin. *Int. Rev. Cytol.* 140, 35–72. doi:10.1016/S0074-7696(08)61093-1

Yilmaz, E. (2012). Enantioselective enzymatic hydrolysis of racemic drugs by encapsulation in sol-gel magnetic sporopollenin. *Bioprocess Biosyst. Eng.* 35, 493–502. doi:10.1007/s00449-011-0622-z

Zetzsche, F., and Huggler, K. (1928). Untersuchungen über die membran des sporen und pollen I. 1. *Lycopodium clavatum* L. *Justus Liebigs. Ann. Chem.* 461, 89–108. doi:10.1002/jlac.19284610105

Zetzsche, F., and Kälin, O. (1931). Untersuchungen über die membran der sporen und pollen v. 4. Zur autoxydation der sporopollenine. *Helv. Chim. Acta* 14, 517–519. doi:10.1002/hlca.19310140151

Zetzsche, F., Kalt, P., Lietchi, J., and Ziegler, E. (1937). Zur konstitution des lycopodium-sporonins, des Tasmanins und des lange-sporonins. *J. Prakt. Chem.* 148, 267–286. doi:10.1002/prac.19371480903

Zetzsche, F., and Vicari, H. (1931). Untersuchungen über die membran der sporen und pollen. III. 2. *Picea orientalis, Pinus sylvestris* L., *Corylus avellana* L. *Helv. Chim. Acta* 14, 62–67. doi:10.1002/hlca.19310140105

Conflict of Interest Statement: Some of the work cited was sponsored by the company Sporomex Ltd. Grahame Mackenzie is a Director of Sporomex Ltd., and Dr. Alberto Diego Taboada has received sponsorship from the company.

15

Fatigue Performance of Laser Additive Manufactured Ti–6Al–4V in Very High Cycle Fatigue Regime up to 10^9 Cycles

Eric Wycisk[1]*, Shafaqat Siddique[2], Dirk Herzog[1], Frank Walther[2] and Claus Emmelmann[1]

[1] *Institute of Laser and System Technologies (iLAS), Hamburg University of Technology, Hamburg, Germany,* [2] *Department of Materials Test Engineering (WPT), TU Dortmund University, Dortmund, Germany*

Edited by:
Han-Yong Jeon,
Inha University, South Korea

Reviewed by:
Alberto D'Amore,
Second University of Naples, Italy
Sriramya Duddukuri Nair,
The University of Texas at Austin,
USA

*Correspondence:
Eric Wycisk
eric.wycisk@tuhh.de

Additive manufacturing technologies are in the process of establishing themselves as an alternative production technology to conventional manufacturing, such as casting or milling. Especially laser additive manufacturing (LAM) enables the production of metallic parts with mechanical properties comparable to conventionally manufactured components. Due to the high geometrical freedom in LAM, the technology enables the production of ultra-light weight designs, and therefore gains increasing importance in aircraft and space industry. The high quality standards of these industries demand predictability of material properties for static and dynamic load cases. However, fatigue properties especially in the very high cycle fatigue (VHCF) regime until 10^9 cycles have not been sufficiently determined yet. Therefore, this paper presents an analysis of fatigue properties of laser additive manufactured Ti–6Al–4V under cyclic tension–tension until 10^7 cycles and tension–compression load until 10^9 cycles. For the analysis of laser additive manufactured titanium alloy Ti–6Al–4V, Woehler fatigue tests under tension–tension and tension–compression were carried out in the high cycle and VHCF regime. Specimens in stress-relieved as well as hot-isostatic-pressed conditions were analyzed regarding crack initiation site, mean stress sensitivity, and overall fatigue performance. The determined fatigue properties show values in the range of conventionally manufactured Ti–6Al–4V with particularly good performance for hot-isostatic-pressed additive-manufactured material. For all conditions, the results show no conventional fatigue limit but a constant increase in fatigue life with decreasing loads. No effects of test frequency on life span could be determined. However, independently of testing principle, a shift of crack initiation from surface to internal initiation could be observed with increasing cycles to failure.

Keywords: laser additive manufacturing, very high cycle fatigue, Ti–6Al–4V, selective laser melting, 3D printing, titanium alloy

INTRODUCTION

Additive manufacturing technologies are no longer restricted to prototyping but establish themselves as manufacturing technologies for functional parts with properties comparable to conventionally manufactured components. Since the digital process chain and layer wise manufacturing principle of additive manufacturing technologies enable an economic production of small lot sizes and complex geometries additive manufacturing is commonly used in the production of synthetic components from polyamide as well as dental prostheses and implants from metal (Emmelmann et al., 2011c; Gebhardt, 2007; Wohlers, 2013). Especially technologies, such as laser additive manufacturing (LAM), are capable of producing metal components with properties necessary for serial production. In LAM, the energy of a laser beam is used to selectively melt metal powder of a single composition alloy layer by layer. The high energy of the laser beam enables a full melting of the powder particles and consequently creates solid material with a density larger than 99.5% (Meiners, 1999; Over, 2003). Additionally to economic production of small lot sizes, LAM enables the production of innovative lightweight structures with weight savings of up to 50% compared to conventional designs by utilizing new design approaches and the use of numerical optimization algorithms combined with the design flexibility of additive manufacturing (Emmelmann et al., 2011a,b). This potential recently accelerated the introduction of LAM as an alternative to conventional manufacturing technologies in the aircraft and space industry (Wohlers, 2013). Due to its high specific strength and corrosion resistance, the titanium alloy Ti–6Al–4V is of particular interest for lightweight designs in aerospace applications. The high demand on quality and structural integrity in this industry, however, requires, particularly with regard to dynamic load cases, a detailed investigation of the fatigue behavior of laser additive manufactured Ti–6Al–4V. Regardless of the need for reliable fatigue data, current state-of-the-art is predominantly marked by investigations on process stability and capability (Wirtz, 2005; Yasa, 2011), reduction of residual stresses during manufacturing (Munsch, 2013), and studies on static mechanical properties (Wirtz, 2005; Vlcek, 2006, 2007; Facchini et al., 2010; Kausch, 2013). Only recently investigations on fatigue behavior came into the focus of research (Vlcek, 2007; Brandl, 2010; Van Hooreweder et al., 2012; Wycisk et al., 2012a,b; Kausch, 2013; Leuders et al., 2013; Edwards and Ramulu, 2014). The existing studies, however, have been concentrating on fatigue properties until 10^7 cycles neglecting the investigation of the material behavior in the very high cycle fatigue (VHCF) regime until 10^9 cycles. To broaden the understanding of the fatigue behavior of laser additive manufactured Ti–6Al–4V beyond conventional investigations, this paper presents an analysis of fatigue properties under cyclic tension–compression load until 10^9 cycles. For the analysis of laser additive manufactured titanium alloy Ti–6Al–4V, Woehler fatigue tests under tension–tension and tension–compression were carried out in the high cycle fatigue (HCF) and VHCF regime. Specimens in stress-relieved as well as hot-isostatic-pressed conditions were analyzed regarding crack initiation site, mean stress sensitivity, and overall fatigue performance.

LASER ADDITIVE MANUFACTURED Ti–6Al–4V AND EXPERIMENTAL METHODS

Production of all specimens analyzed in this study was done on a commercially available manufacturing system M270 from EOS. Frequent machine maintenance and laser power measurements ensure process reproducibility across all manufacturing batches. The specimens were built under Argon atmosphere in an upright standing position (90°) with 200 W laser power, a layer thickness of 30 μm, and an energy density $E = 45.33$ J/mm³. Powder material used for the production of specimen was obtained from EOS with a grain size distribution of 20–63 μm. To relief process-induced internal stresses and improve fatigue properties, the specimens were either treated with a stress relief treatment at 650°C for 3 h in vacuum or with a hot-isostatic-pressing process (HIP) at 920°C and 1000 bar pressure for 2 h. All specimens were manufactured with a material allowance and finished with machining processes to final geometry after the heat treatment. Machining was performed with constant cutting parameters for all batches and completed by manual grinding and polishing. The manual finish was performed in three steps with 2 min each. For finishing, grinding paper size 2000 and 5000 and a diamond polishing paste with grain size of 3 ± 0.5 μm were used.

The fatigue behavior is characterized with constant amplitude tests under tension–tension ($R = 0.1$) and tension–compression ($R = -1$) load. The experimental planning and statistical analysis of the fatigue data are performed in accordance to ISO 12107:2012 (2012). Investigations at $R = 0.1$ were carried out on a servo-hydraulic test rig from MTS at 50 Hz. Tension–compression tests in the HCF regime were carried out on a resonance testing machine by Rumul at 59 Hz. VHCF tests were done on an ultrasonic fatigue testing system (USF-2000 by Shimadzu) at a frequency of $f = 20$ kHz. To eliminate effects from elevated temperatures due to the high test frequency at VHCF testing, the specimens were tested with pulse-pause mode and cooled with air during testing. The pulse-pause mode was set to 1:1 with the testing system set in resonance for 200 ms and stopped testing for 200 ms to let the specimen cool down.

RESULTS AND DISCUSSION

Microstructure and Mechanical Properties

Figure 1 displays the microstructure of the manufactured specimens after stress relieve treatment at 650°C in the building direction (z) and in plane with the manufacturing layers (x,y). The additive manufactured material shows columnar grains of prior β – phase in orientation of the building direction. With approximately 100 μm the grain thickness corresponds to the melting width in the manufacturing process. The microstructure of additive manufactured Ti–6Al–4V is predominantly controlled by the high cooling rate in the manufacturing process generating a very fine lamellar microstructure consisting completely of α – phase and α′ – martensite. The size of the α – lamellae was measured using a light microscope Olympus GX51 and the corresponding Olympus Stream software. Single α – lamellae

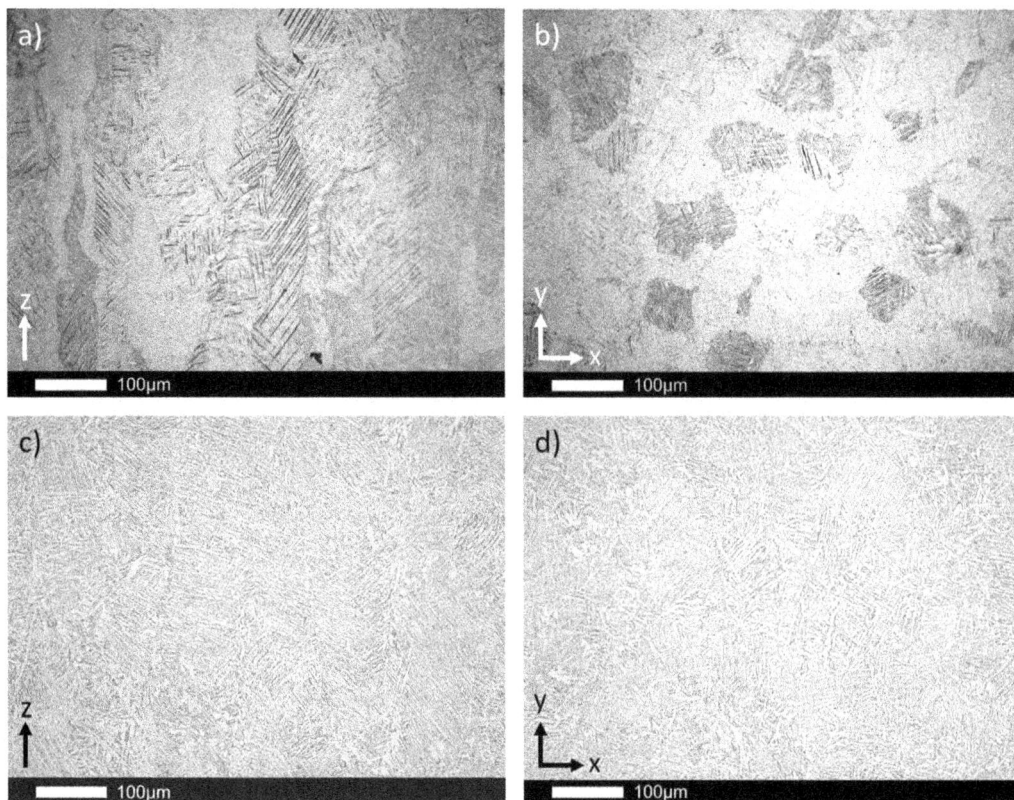

FIGURE 1 | Microstructure of laser additive manufactured Ti–6Al–4V in stress-relieved (A,B) and hot-isostatic-pressed (C,D) condition.

showing a thickness of <1 µm. Hot-isostatic-pressing results in a coarsening of the lamellar microstructure with an α – lamellae thickness of approximately 4 µm and the formation of small amounts of β – phase in between single α – lamellae (compare **Figure 1**). The defined structure of columnar grains is disappearing with the heat treatment at elevated temperatures. Also, prior observed gas porosity and process inherent defects from insufficient bonding are minimized by the heat treatment at high pressure. The observed microstructure corresponds to findings in other investigations of laser additive manufactured Ti–6Al–4V (Facchini et al., 2010; Leuders et al., 2013).

In agreement with the observed microstructure the stress-relieved material shows high tensile strength with reduced ductility. The post treatment at elevated temperatures of 920°C results in an increase of ductility with additional decrease in tensile strength. The values for tensile and ultimate strength as well as the elongation at break are listed in **Table 1**. The results show that residual porosity and process inherent defects cause an increase of variance for stress-relieved specimens. By contrast, tensile test of specimen after hot-isostatic-pressing show very little scatter.

High Cycle Fatigue up to 10⁷ Cycles

Preceding research investigated the effect of process inherent defects on the fatigue properties of laser additive manufactured Ti–6Al–4V under tension–tension loading. Results showed large scatter of fatigue life depending on defect kind, size, and location.

TABLE 1 | Static mechanical properties of laser additive manufactured Ti–Al6–4V.

Heat treatment	$R_{p0.2}$ (MPa)	R_m (MPa)	A (%)
650°C	1076 ± 14	1189 ± 16	13.6 ± 1.3
920°C @ 1000 bar	907 ± 4	1022 ± 5	17.7 ± 0.8

The investigations exhibited crack initiation at internal defects for all specimens. It was concluded that the Murakami criteria for surface interaction was applicable to describe the scatter in the fatigue data with failure agglomeration at 10^5 cycles for defects with increased stress concentration due to surface interaction and specimen failure around 3×10^6 cycles for defects without surface interaction (Wycisk et al., 2014).

Figure 2 shows the results from preceding tension–tension testing compared to the current fatigue investigation at $R = -1$. Identical to previous investigations crack initiation for tension–compression only starts at internal defects; however, these data show significantly less scatter and no visible agglomeration of fatigue life. Analyzing the crack initiating defects regarding increased stress concentration due to surface interaction the results indicate that for tension–compression the surface interaction of defects plays a minor role in fatigue life. The finding is illustrated in **Figure 3** showing two specimens with crack initiation at bonding defects under identical stress amplitude

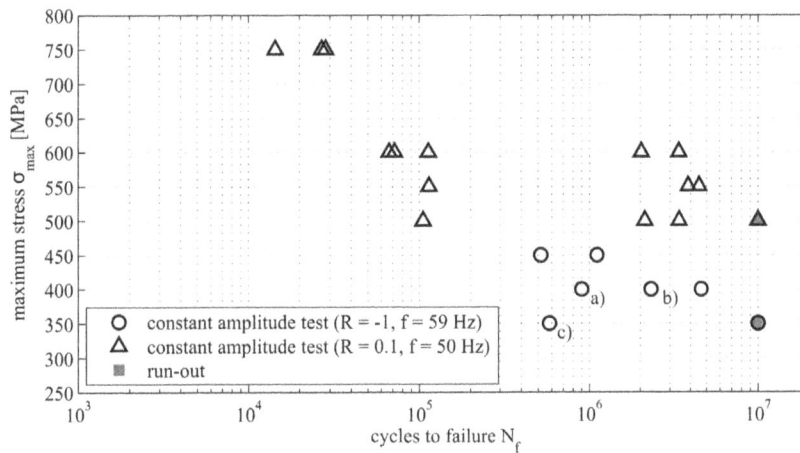

FIGURE 2 | High cycle fatigue properties of stress-relieved Ti–6Al–4V under tension–tension and tension–compression loading.

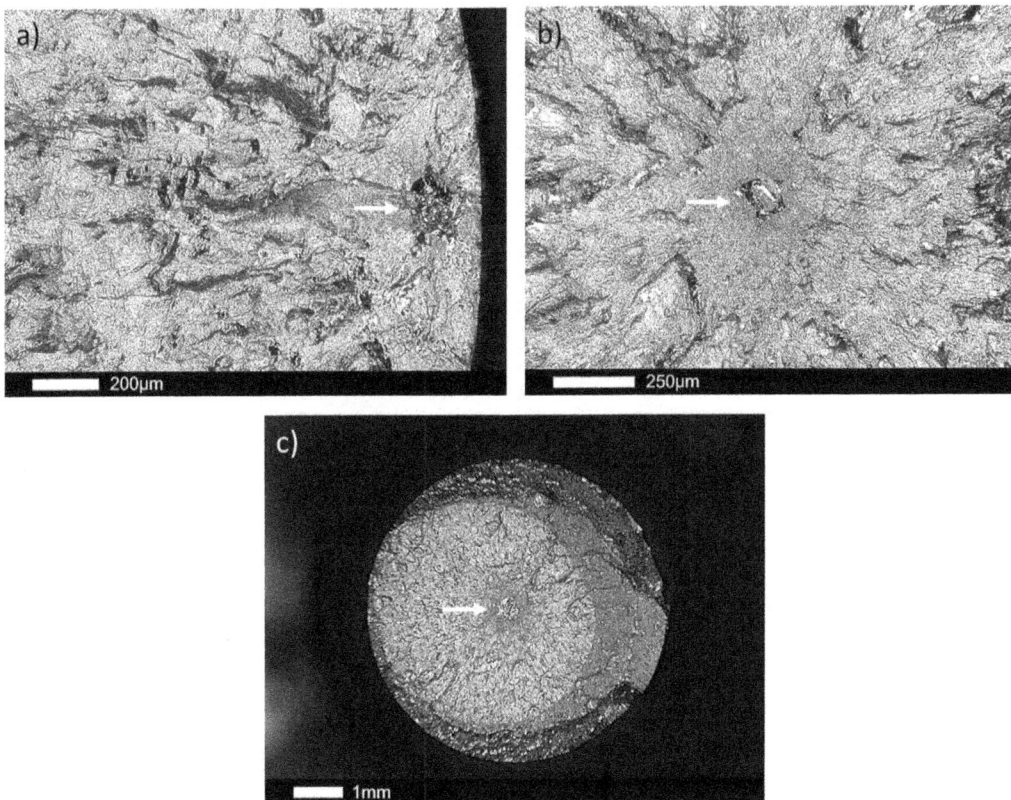

FIGURE 3 | Crack initiation at process inherent defects for stress-relieved specimen under tension–compression loading; defect with (A) and without (B,C) surface interaction.

of 400 MPa. Defect (a) fulfills the Murakami criteria for surface interaction and increased stress concentration causing failure at 9×10^5 cycles; whereas defect (b) without surface interaction causes failure at 2.3×10^6 cycles. A significant reduction of lifetime to failure can only be observed for one very large defect. Defect (c) with a diameter larger than 500 μm reduces the specimen cross

section significantly and leads to increased stress concentration and early failure

Using hot-isostatic-pressing defects, such as gas porosity and insufficient layer bonding, are cured in additive manufactured Ti–6Al–4V. In **Figure 4**, the Woehler curve for HIPed specimens under tension–tension loading ($R = 0.1$) is shown. The results

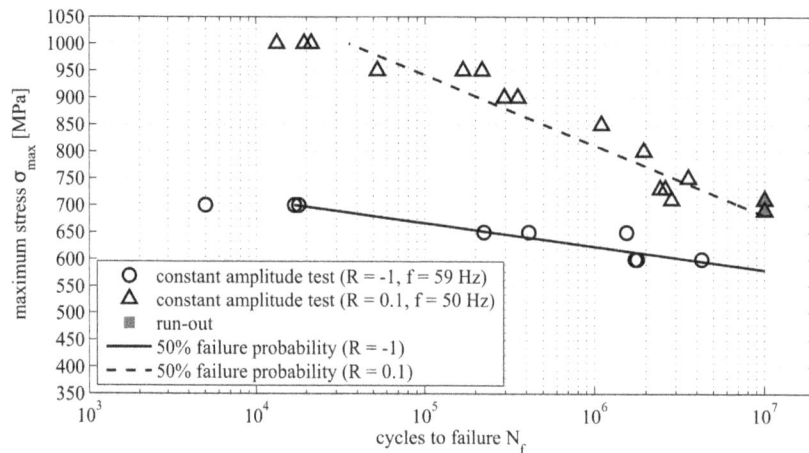

FIGURE 4 | High cycle fatigue properties of hot-isostatic-pressed Ti–6Al–4V under tension–tension and tension–compression loading.

FIGURE 5 | Crack initiation site for hot-isostatic-pressed specimen under tension–tension loading; surface crack initiation at $N_f = 2.9 \times 10^5$ (left) and internal crack initiation at $N_f = 2.4 \times 10^6$ (right).

indicate a strong reduction of scatter compared to additive manufactured material in stress-relieved conditions and an increased fatigue strength of $\sigma_{max} = 680 \pm 35$ MPa at 10^7 cycles. Electron and light microscopic analysis of the fracture surface show no sign of crack initiation at process inherent material defects. However, depending on cycles to failure different sites of crack initiation can be observed. For cycles to failure of $N_f < 10^6$ crack initiation is only observed at the surface of the specimen with one exception. With increasing fatigue life and $N_f > 10^6$ all cracks initiate from the inside of the specimen. **Figure 5** shows surface and internal crack initiation for specimen at $N_f = 2.9 \times 10^5$ and $N_f = 2.4 \times 10^6$ cycles.

Additionally, **Figure 4** shows the results of tension-compression investigations of HIPed specimen. The results show an equally good fatigue performance with a fatigue strength of approximately $\sigma_{max} = 575$ MPa at 10^7 cycles and little scatter. Contrary to the tests at $R = 0.1$ the specimen under tension-compression exclusively show surface crack initiation until 10^7 cycles.

Very High Cycle Fatigue up to 10^9 Cycles

The results of HCF and VCHF tests at $R = -1$ for stress-relieved specimen are presented in **Figure 6**. These data show no conventional fatigue limit for $N_f > 10^7$ but a steady decrease in fatigue strength with increasing lifetime to failure. The typical scatter due to crack initiation at process inherent material defects of stress-relieved additive manufactured Ti–6Al–4V can be observed. Crack initiation of all specimens is located at internal defects. Additionally, these data show no apparent influence of test frequency on fatigue life.

In comparison to stress-relieved specimen, **Figure 7** depicts the HCF and VHCF results of hot-isostatic-pressed specimens. Identical to the previous results, the investigation shows no fatigue limit but a steady decrease in strength and no influence of test frequency on fatigue strength. Similar to investigations at $R = 0.1$ no crack initiation at process inherent defects could be observed. The location of crack formation changes under tension–compression at 10^7 cycles from surface crack to internal crack initiation.

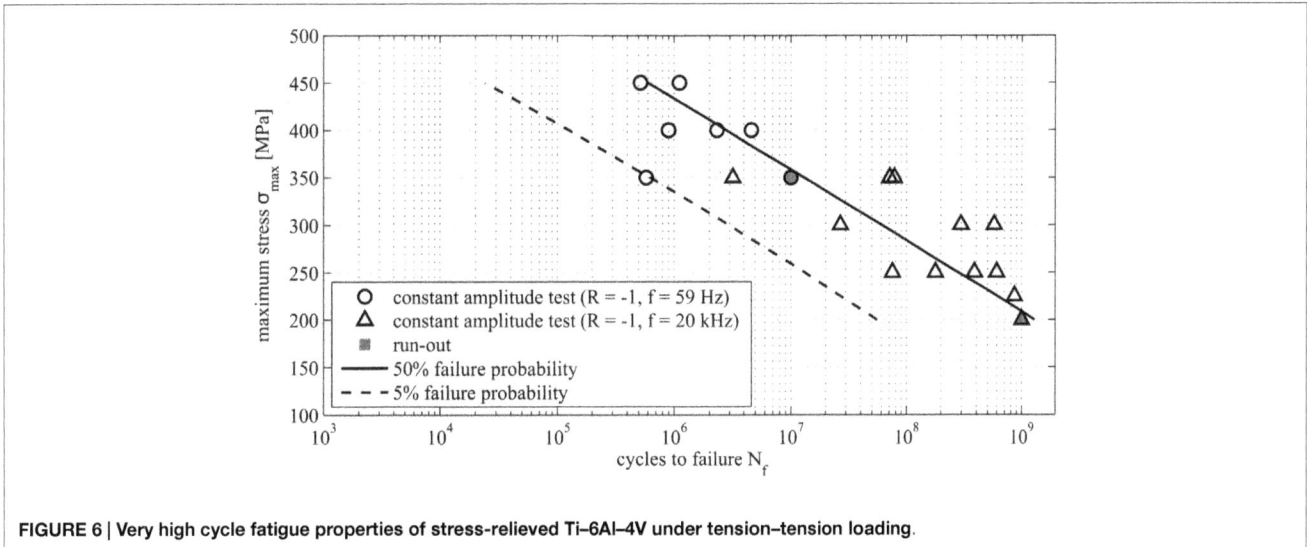

FIGURE 6 | Very high cycle fatigue properties of stress-relieved Ti–6Al–4V under tension–tension loading.

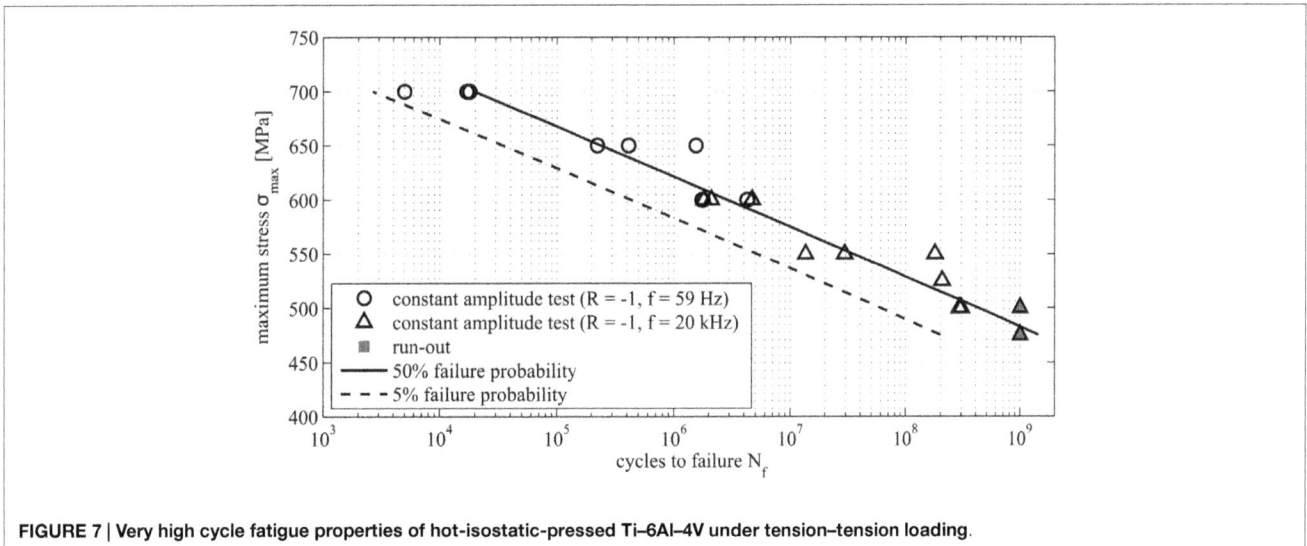

FIGURE 7 | Very high cycle fatigue properties of hot-isostatic-pressed Ti–6Al–4V under tension–tension loading.

Figure 8 shows surface crack formation for $N_f < 10^7$ and internal crack initiation at $N_f = 1.8 \times 10^8$ cycles. The hot-isostatic-pressed Ti-6Al-4V under tension-compression loading exhibits a very high fatigue strength of $\sigma_{max} = 483 \pm 16$ MPa at 10^9 cycles.

Mean Stress Sensitivity

In **Figure 9**, the experimentally determined fatigue strength at 10^7 cycles for tension–tension and tension–compression testing of stress-relieved and hot-isostatic-pressed Ti–6Al–4V is illustrated in a Smith diagram. Primarily, the diagram illustrates the large negative effect of process inherent defects on the fatigue performance. Under quasi-static load, the strength of stress-relieved material excels the HIPed material but stress concentration at process inherent material defects causes a strong decline in strength under cyclic loading. Regarding the mean stress, the results show only small apparent sensitivity

for both material conditions. In previous research, mean stress sensitivity was mainly observed in globular or bi-modal titanium alloys. Though the causing mechanisms are not fully understood mean stress sensitivity in globular and bi-modal microstructure is generally said to be driven by the crack initiation stage (Adachi et al., 1985; Lindemann and Wagner, 1997). In lamellar micro-structure of conventional Ti-6Al-4V, mean stress sensitivity has not been observed (Adachi, 1987; Notkina, 2005). In the case of the lamellar microstructure of laser additive manufactured Ti-6Al-4V, the slight mean stress sensitivity for stress-relieved material, however, might be explained by the higher influence of stress concentration at defects on the fatigue strength under tension–tension compared to tension–compression loading as observed in the HCF testing. Lacking the existence of defects, the mean stress sensitivity of hot-isostatic-pressed specimen could be explained due to earlier change of crack initiation site under

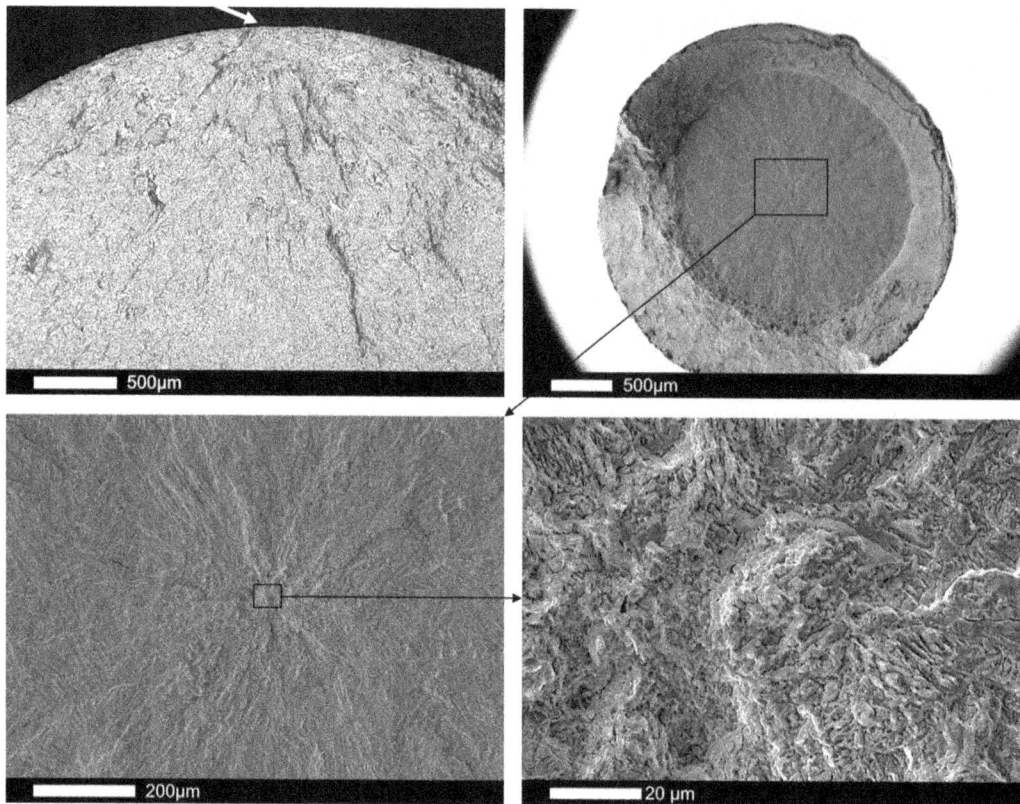

FIGURE 8 | Crack initiation site for hot-isostatic-pressed specimen under tension–compression loading; surface crack initiation at $N_f < 10^7$ (top left), internal crack initiation at $N_f = 1.8 \times 10^8$ (top right), and magnification of internal crack initiation (bottom left and right).

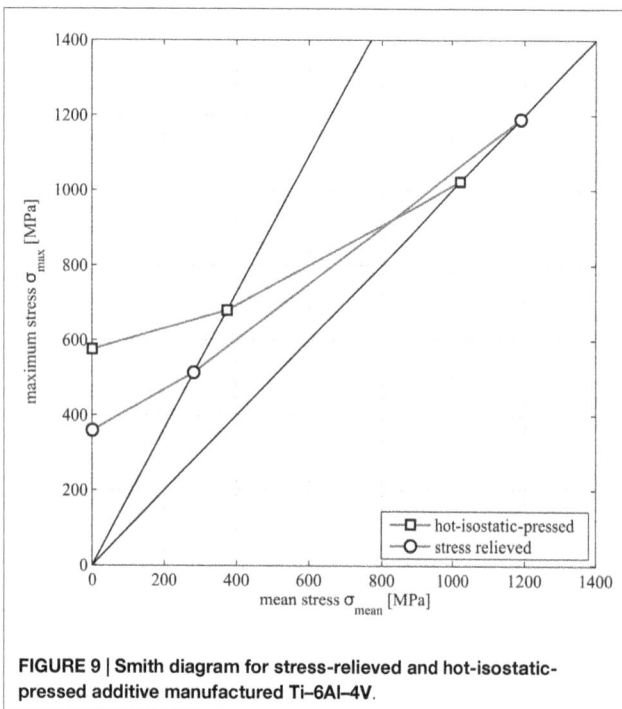

FIGURE 9 | Smith diagram for stress-relieved and hot-isostatic-pressed additive manufactured Ti–6Al–4V.

tension-tension loading. As the analysis of fracture surfaces of HIPed specimens show, the shift of crack initiation from surface to internal occurs at 10^6 cycles for $R = 0.1$ but not until 10^7 cycles for $R = -1$.

CONCLUSION

From the described results following conclusions can be drawn:

- Process inherent defects have a major influence on the fatigue performance of laser additive manufactured Ti–6Al–4V. Regardless of the applied load ratio, all specimens fail due to crack initiation at process inherent defects. Depending on size and location of the defect, the lifetime to failure can vary greatly.
- Under tension–compression crack initiation shows less influence on fatigue life compared to tension–tension loading. Therefore, increased stress concentration at defects due to surface interaction causes significantly less scatter in fatigue data at $R = -1$.
- Post treatment by hot-isostatic-pressing cures process inherent defects and improves the fatigue performance significantly. Crack initiation at defects can no longer be observed.

- Depending on the applied load ratio crack initiation for HIPed specimen shifts from surface to internal crack initiation at 10^6 cycles for $R = 0.1$ and at 10^7 cycles for $R = -1$.
- Testing until 10^9 cycles shows no conventional fatigue limit but decreasing fatigue strength with increasing fatigue life for stress-relieved and hot-isostatic-pressed material.
- An influence of test frequency on fatigue life cannot be observed.
- Laser additive manufactured Ti–6Al–4V shows a slight mean stress sensitivity in stress-relieved as well as HIPed condition. In stress-relieved condition, mean stress sensitivity can be explained by increased sensitivity for crack initiation due

to increased stress concentration at defects under tension–tension loading. Decrease in fatigue strength for $R = 0.1$ at 10^7 cycles for HIPed material is explained by earlier shift of crack initiation site compared to tension–compression loading.

ACKNOWLEDGMENTS

This publication was supported by the German Research Foundation (DFG) and the Hamburg University of Technology (TUHH) in the funding program "Open Access Publishing."

REFERENCES

Adachi, S. (1987). *Mean Stress Dependence of Fatigue Strength in Titanium Alloys.* Hamburg: TU Hamburg – Harburg.

Adachi, S., Wagner, L., and Lütjering, G. (1985). "Influence of mean stress on fatigue strength of Ti-6A1-4V," in *Proceedings of the 7th International Conference on the Strength of Metals and Alloys*, Montreal, 2117–2122.

Brandl, E. (2010). *Microstructural and Mechanical Properties of Additive Manufactured Titanium (Ti-6Al-4V) Using Wire.* Cottbus: TU Cottbus.

Edwards, P., and Ramulu, M. (2014). Fatigue performance evaluation of selective laser melted Ti-6Al-4V. *Mater Sci Eng* 598, 327–337. doi:10.1016/j.msea.2014.01.041

Emmelmann, C., Petersen, M., Kranz, J., and Wycisk, E. (2011a). "Bionic lightweight design by laser additive manufacturing (LAM) for aircraft industry," in *Proceedings of SPIE Eco-photonics 2011: Sustainable Design, Manufacturing, and Engineering Workforce Education for a Green Future*, Vol. 8065, Strasbourg.

Emmelmann, C., Sander, P., Kranz, J., and Wycisk, E. (2011b). Laser additive manufacturing and bionics: redefining lightweight design. *Phys. Procedia* 12, 364–368. doi:10.1016/j.phpro.2011.03.046

Emmelmann, C., Scheinemann, P., Munsch, M., and Seyda, V. (2011c). Laser additive manufacturing of modified implant surfaces with osseointegrative characteristics. *Phys. Procedia* 12, 375–384. doi:10.1016/j.phpro.2011.03.048

Facchini, L., Magalini, E., Robotti, P., Molinari, A., Höges, S., and Wissenbach, K. (2010). Ductility of a Ti-6Al-4V alloy produced by selective laser melting of prealloyed powders. *Rapid Prototyp. J.* 16, 450–459. doi:10.1108/13552541011083371

Gebhardt, A. (2007). *Generative Fertigungsverfahren.* München: Carl Hanser Verlag.

ISO 12107:2012. (2012). *Metallic Materials – Fatigue Testing – Statistical Planning and Analysis of Data.* Berlin: Beuth Verlag.

Kausch, M. (2013). *Entwicklung Hochbelasteter Leichtbaustrukturen Aus Lasergenerierten Metallischen Komponenten Mit Faserverbundverstärkung.* Auerbach: TU Chemnitz, Wissenschaftliche Scripten.

Leuders, S., Thöne, M., Riemer, A., Niendorf, T., Tröster, T., Richard, H. A., et al. (2013). On the mechanical behaviour of titanium alloy TiAl6V4 manufactured by selective laser melting: Fatigue resistance and crack growth performance. *Int. J. Fatigue* 48, 300–307. doi:10.1016/j.ijfatigue.2012.11.011

Lindemann, J., and Wagner, L. (1997). Mean stress sensitivity in fatigue of α, (α+β) and β titanium alloys. *Mater. Sci. Eng.* 234–326, 1118–1121. doi:10.1016/S0921-5093(97)00347-X

Meiners, W. (1999). *Direktes Selektives Laser Sintern Einkomponentiger Metallischer Werkstoffe.* Aachen: RWTH Aachen, Shaker Verlag.

Munsch, M. (2013). *Reduzierung von Eigenspannungen und Verzug in der Laseradditiven Fertigung.* Göttingen: TU Hamburg – Harburg, Cuvillier Verlag.

Notkina, E. (2005). *Einfluss der Mikrostruktur aus das Ermüdungsrissbildungs- und Ermüdungsrissausbreitungsverhalten der Legierung Ti-6Al-4V.* Hamburg: TU Hamburg – Harburg.

Over, C. (2003). *Generative Fertigung von Bauteilen aus Werkzeugstahl X38CrMoV5-1 und Titan TiAl6V4 mit "Selective Laser Melting".* Aachen: RWTH Aachen, Shaker Verlag.

Van Hooreweder, B., Moens, D., Boonen, R., Kruth, J.-P., and Sas, P. (2012). Analysis of fracture toughness and crack propagation of Ti6Al4V produced by selective laser melting. *Adv. Eng. Mater.* 14, 92–97. doi:10.1002/adem.201100233

Vlcek, J. (2006). *Property Investigation of Laser Cladded, Laser Melted and Electron Beam Melted Ti-6Al-4V. Tech. Rept. RTO-AVT-139.* Amsterdam: NATO-RTO Specialist Meeting on Netshape Manufacturing.

Vlcek, J. (2007). "Property investigation of Ti-6Al-4V produced by additive manufacturing," in *Proceedings of the 136th TMS Annual Meeting*, Orlando.

Wirtz, T. P. (2005). *Herstellung von Knochenimplantaten aus Titanwerkstoffen durch Laserformen.* Aachen: RWTH Aachen.

Wohlers, T. (2013). *Wohlers Report 2013.* Fort Collins, CO: Wohlers Associates Inc.

Wycisk, E., Kranz, J., and Emmelmann, C. (2012a). "Fatigue strength of light weight structures produced by laser additive manufacturing in TiAl6V4," in *Proceedings of 1st International Conference of the International Journal of Structural Integrity*, Porto.

Wycisk, E., Kranz, J., and Emmelmann, C. (2012b). "Influence of surface properties on fatigue strength of lightweight structure produced by laser additive manufacturing in TiAl6V4," in *Proceedings of Direct Digital Manufacturing Conference*, Berlin.

Wycisk, E., Solbach, A., Siddique, S., Herzog, D., Walther, F., and Emmelmann, C. (2014). Effects of defects in laser additive manufactured Ti-6Al-4V on fatigue properties. *Phys. Procedia* 56, 371–378. doi:10.1016/j.phpro.2014.08.120

Yasa, E. (2011). *Manufacturing by Combining Selective Laser Melting and Selective Laser Erosion/Laser Re-Melting.* Leuven: Katholieke Universiteit Leuven.

Conflict of Interest Statement: The authors declare that the research was conducted in the absence of any commercial or financial relationships that could be construed as a potential conflict of interest.

16

SU-8 based microdevices to study self-induced chemotaxis in 3D microenvironments

Jose Maria Ayuso [1,2,3], Rosa Monge [1,2,3], Guillermo A. Llamazares [1,2,3], Marco Moreno [1,2,3], Maria Agirregabiria [4], Javier Berganzo [5], Manuel Doblaré [1,2,3], Iñaki Ochoa [1,2,3†] and Luis J. Fernández [1,2,3†*]

[1] Group of Structural Mechanics and Materials Modelling (GEMM), Centro de Investigación Biomédica en Red en Bioingeniería, Biomateriales y Nanomedicina (CIBER-BBN), Zaragoza, Spain, [2] Aragón Institute of Engineering Research (I3A), University of Zaragoza, Zaragoza, Spain, [3] Aragon Institute of Biomedical Research, Instituto de Salud Carlos III, Zaragoza, Spain, [4] MEMS/MST Department, Ikerlan S. Coop., Mondragón, Spain, [5] Department of Design and Manufacturing Engineering, University of Zaragoza, Zaragoza, Spain

Edited by:
Daniel J. Solis,
BioNano Genomics, USA

Reviewed by:
Eduardo Fernandez,
University Miguel Hernandez, Spain
Verena Charwat,
University of Natural Resources and
Life Sciences, Austria

*Correspondence:
Luis J. Fernández,
Group of Structural Mechanics and
Materials Modelling (GEMM), I3A,
Zaragoza University, Marinano
Esquillor, Zaragoza 50018, Spain
luisf@unizar.es

[†] Iñaki Ochoa and Luis J. Fernández
have contributed equally to this work

Tissues are complex three-dimensional structures in which cell behavior is frequently guided by chemotactic signals. Although starvation and nutrient restriction induce many different chemotactic processes, the recreation of such conditions *in vitro* remains difficult when using standard cell culture equipment. Recently, microfluidic techniques have arisen as powerful tools to mimic such physiological conditions. In this context, microfluidic three-dimensional cell culture systems require precise control of cell/hydrogel location because samples need to be placed within a microchamber without obstruction of surrounding elements. In this article, SU-8 is studied as structural material for the fabrication of complex cell culture microdevices due to its good mechanical properties and sensor integration capacity. Moreover, SU-8 physical properties and their effect on a successful design for precise control of hydrogel location within microfluidic devices are studied. In particular, this manuscript presents a SU-8 based microdevice designed to create "self-induced" medium starvation, based on the combination of nutrient restriction and natural cell metabolism. Results show a natural migratory response toward nutrient source, showing how cells adapt to their own microenvironment modifications. The presented results demonstrate the SU-8 potential for microdevice fabrication applied to cell culture.

Keywords: SU-8 photoresist, microfluidic devices, capillarity, microfluidic cell culture, hydrogel confinement, self-induced chemotaxis, cell migration

Introduction

Living tissues are composed of embedded cells within a heterogeneous extracellular matrix, which plays a key role in three-dimensional cell migration and intercellular communication (Raines, 2000; Savino et al., 2004; Korpos et al., 2010; Franco and Muller, 2011). Several previous studies report that cell culture in three-dimensional systems produces multiple changes in cell behavior when compared to the use of classical two-dimensional systems (Baker and Chen, 2012; Huang et al., 2012; Breslin and O'Driscoll, 2013; Page et al., 2013). In migration, such behavioral differences are caused not only by the availability of a three-dimensional environment affecting cell migration but also by generation of chemotactic processes. For example, cancer cells have been reported to migrate and invade other tissues following chemotaxis induced by SDF-1, FGF, PDGF, or TGF (Roussos et al., 2011). Examples of chemotactic processes are angiogenesis (Guo et al., 2012), immune response (Nieto et al., 1997), and axon elongation during embryonic development (Bhattacharjee et al., 2010). Numerous reports have shown the influence of nutrient on cell migration (Chen et al., 2011). In this context, tumor cells often proliferate much faster than non-malignant cells, causing nutrient starvation in tumor microenvironments (Noman et al., 2011). These conditions trigger tumor cell migration and invasiveness, worsening patient prognosis (Nagelkerke et al., 2013). Therefore, there is increasing interest among the biological community in study cell response under these nutrient restricted conditions (Kim et al., 2013a; Zhang et al., 2013). Various strategies have been developed for the study of cell migration in response to chemotaxis. Perhaps one of the most well-known assays involves the use of a Boyden chamber, in which cells migrate through a polycarbonate membrane in response to chemoattractants (Falasca et al., 2011). However, as this system does not allow direct observation of cellular migration, alternative tools were developed such as the under-agarose gel (Mousseau et al., 2007), Zigmond chamber (Zigmond, 1977), Dunn chamber (Zicha et al., 1997), and Insall chamber assays (Muinonen-Martin et al., 2010). Although each of these assays has its own advantages, they generally struggle to maintain stable gradients for long periods of time (Keenan and Folch, 2008; Muinonen-Martin et al., 2010) and more important: the chemotaxis conditions are externally imposed. In all of these assays, study how cells response to their own nutrient consumption remains challenging. Recently, microfabrication and microfluidic technologies have arisen as interesting alternatives for creating high-performance cell culture systems (El-Ali et al., 2006; Kim et al., 2010; Lesher-Perez et al., 2013). In general, such devices possess a culture microchamber housing a hydrogel and lateral microchannels delimited by a series of pillars (**Figure 1**). These lateral microchannels are usually kept hydrogel-free to allow for medium perfusion (Huang et al., 2009; Farahat et al., 2012; Song et al., 2012). Depending on the experimental protocol, cells could either be embedded within the hydrogel or located in the lateral microchannels. Injecting different media on each lateral microchannel chemical gradients across the culture chamber can be imposed (Huang et al., 2009; Kothapalli et al., 2011; Farahat et al., 2012; Funamoto et al., 2012). The majority of recently

designed microfluidic devices are fabricated using polydimethylsiloxane (PDMS), which possesses excellent optical properties. However, using PDMS has some disadvantages, including poor mechanical properties (Jeon et al., 2011) and sensor integration is challenging. Here, we present a SU-8 based microfluidic device for three-dimensional cell cultures capable to establish nutrient restricted conditions. SU-8 material was selected due to the wide range of sensors and actuators already reported to be available to be monolithically integrated: valves (Ezkerra et al., 2007; Calvo et al., 2011), pumps (Ezkerra et al., 2011), flow sensors (Vilares et al., 2010), microneedles (Fernández et al., 2009), polymerase chain reaction (PCR) systems (Verdoy et al., 2012), cell sorting devices (González et al., 2012), electrophoresis (Castaño-Álvarez et al., 2009), and Mach–Zender interferometer based sensors (Duval et al., 2012) among others. Additionally, SU-8 have been reported previously for cell culture applications (Kotzar et al., 2002; Nemani et al., 2013; Torrejon et al., 2013; Rigat-Brugarolas et al., 2014), showing some advantages over PDMS for the fabrication of cell cultures systems, including broader chamber design possibilities (Ni et al., 2009), good mechanical properties (Lorenz et al., 1997), and gas impermeability (Gerhardt and Betsholtz, 2005). Huang et al. described extensively that when using PDMS microdevices, the hydrogel location can be achieved creating a specific geometry based on a series of pillars and taking advantage of PDMS surface hydrophobicity (Huang et al., 2009). They also demonstrated how when hydrophilic material are used, hydrogel cannot be confined using the same geometry. As SU-8 is hydrophilic (Jokinen et al., 2012), those described designs are not applicable. Although SU-8 can be chemically treated to render its surface hydrophobic, this requires aggressive and dangerous chemical agents such as strong acids and hydrogen peroxide (Schumacher et al., 2008). In this study, a new design capable of confining hydrogels within hydrophilic microdevices was successfully developed and used with SU-8-based microfluidic devices. The biocompatibility of SU-8-based microdevices was evaluated at different cell densities and collagen concentration using the pre-osteoblasts cell line, MC3T3. Finally, cells were long-term cultured within the microdevice under two different conditions: (1) unrestricted conditions, medium was refreshed once a day through both lateral microchannels; (2) Restricted conditions, one lateral microchannel was sealed while continuous medium perfusion was enabled on the other lateral microchannel. After 1 week, in culture cell distribution within the central microchamber was analyzed, showing an intense self-induced chemotactic response under restricted conditions due to cell metabolism. Contrary to previous cell culture studies under artificially generated gradients (Zicha et al., 1997; Zimmermann et al., 2007; Bhattacharjee et al., 2010; Kim et al., 2013b, 2010; Chen et al., 2011; Funamoto et al., 2012) the proposed device and protocol allows the study of cell response under self-induced nutrient gradient conditions. The results presented in this paper validate SU-8 based microfluidic devices as a novel and robust tool for future three-dimensional self-induced cell migration assays and allow studying cell behavior under nutrient starvation. In this article, this methodology is validated in the context of bone tissue repair, since blood vessels are absent in bone scaffolds and nutrient support is compromised. This is the reason why MC3T3 cells were selected. However,

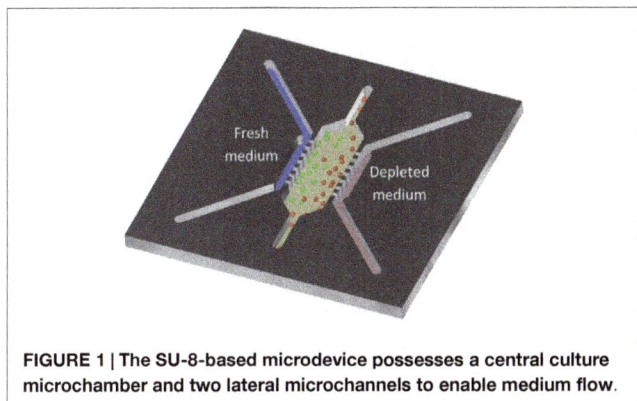

FIGURE 1 | The SU-8-based microdevice possesses a central culture microchamber and two lateral microchannels to enable medium flow.

the described methodology could be applied in many different biological scenarios.

Materials and Methods

Microfluidic Chip Fabrication

Microfluidic devices were fabricated using SU-8 photolithography combined with an SU-8 to SU-8 bonding process (Blanco et al., 2004). **Figure 2** summarizes the fabrication process, which yielded a stand-alone SU-8 microfluidic chip that was without any substrate material such as glass or silicon. The fabrication process was inspired by previously reported work describing the fabrication of SU-8 microneedles (Altuna et al., 2012, 2013). First, a polyimide film (Kapton) was temporarily bonded onto a Pyrex wafer (**Figure 2A**). Then, a 60 μm-thick layer of SU-8 was spun onto the Kapton film and soft baked at 65°C for 30 min. After the wafer was cooled down to room temperature, a spinning protocol for the deposition of a 20 μm-thick SU-8 layer was performed, followed by an additional soft bake. The total thickness of obtained SU-8 layers was 90 μm instead of the expected 80 μm (**Figure 2B**). This can be explained by the higher frictional forces present when layering on an SU-8 surface compared to Kapton. The wafer was then exposed to a 140 mJ/cm^2 dose of UV light filtered with a mask that defines the bottom layer of the device. The exposure was followed by a post-bake step which consisted of heating the wafer to 65°C for 15 min. Three new depositions of SU-8 layers were then performed (60-, 60-, and 20-μm thick), followed by corresponding soft bakes, producing a total height of 150 μm. The wafer was again exposed to UV light (140 mJ/cm^2) to pattern microchannels and microchambers, post baked (**Figure 2C**), immersed in SU-8 developer, and finally rinsed in DI H$_2$O (**Figure 2D**).

An additional Pyrex wafer was processed with temporary bonding to a Kapton film (**Figure 2E**). The previously described spinning procedures and soft bake processes were applied to this wafer to obtain a 90-μm-thick SU-8 layer. Inlets and outlets were defined by standard photolithography using a 140 mJ/cm^2 dose of UV light. Then, a 150 μm-thick layer of SU-8 was deposited and patterned with the same mask used to obtain microchannels and microchambers on the first wafer (**Figure 2H**). Both wafers were then aligned and bonded by heating to 90°C for 15 min under a pressure of 3 bar (**Figure 2I**). The weak adhesion between SU-8 and Kapton allowed the devices to be released manually (**Figure 2J**). A picture of a finished chip is illustrated in **Figure 2K**.

Packaging Tool Fabrication

In order to recreate restricted conditions, a regular flow of medium must be provided through the microdevice. A dedicated package was designed and fabricated to provide proper housing and hermetic connections to the microdevice inlets and outlets, allowing for automated cellular culture. The package body was made of polycarbonate to withstand cell culture humidity and temperature conditions. Flat bottom Upchurch fluidic connectors (Idex M-644-03x) were used to reduce dead volume and provide a smooth fluid tube-to-package transition. The package is divided into two parts: a microdevice holder with an optical observation window, and an upper manifold cover that holds inlet sealing o-rings (Barnwell BS1806-001) and tube connectors. When the manifold cover is screwed down over the chip holder, the microdevice is sealed and connected to the tubes. Both parts were fabricated with precision machining, taking special care to avoid roughness on critical surfaces such as sealing elements, fluidic connectors, and o-ring grooves.

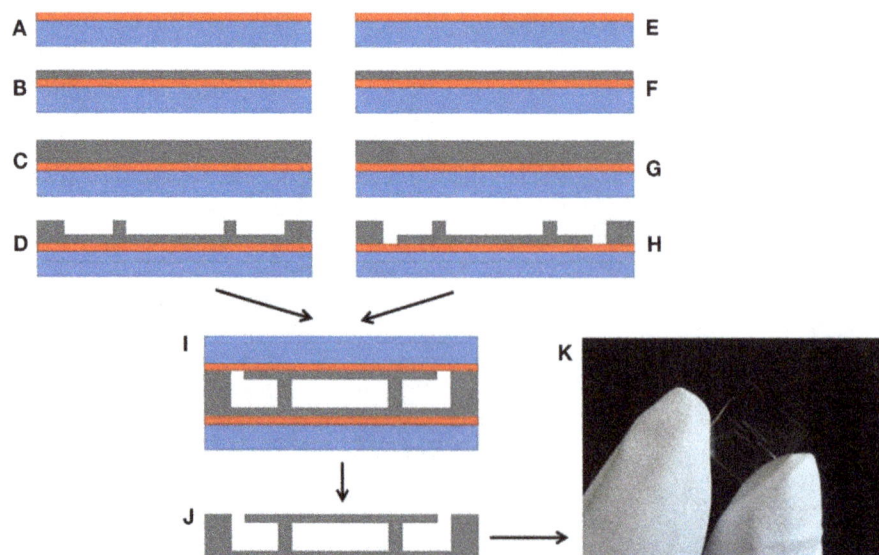

FIGURE 2 | Chip fabrication process: (A) Kapton film bonding to a Pyrex wafer, **(B)** processing of a 90 μm-thick SU-8 layer, **(C)** processing of chamber and microchannels layer, **(D)** wafer development, **(E)** Kapton film bonding to a Pyrex wafer, **(F)** processing of a 90 μm-thick SU-8 layer, **(G)** processing of chamber and microchannels layer, **(H)** wafer development, **(I)** SU-8 to SU-8 bonding, **(J)** SU-8 device release, and **(K)** finished SU-8 microdevice.

Cell Culture

MC3T3-E1 cells, mouse pre-osteoblasts (ATCC, CRL-2593, kindly donated by Doctor Izal from the University of Navarra, Spain), were routinely grown in -minimal essential media (MEM) (Lonza BE12-169) supplemented with 10% v/v fetal calf serum (Sigma F7524) and penicillin/streptomycin (DE 17-602E). For three-dimensional cultures, all needed reagents, microdevices included, were placed on ice. Cells were trypsinized and resuspended in a calculated volume of medium (-MEM supplemented with 10% fetal bovine serum) to reach the desired concentration of cells in the final hydrogel solution. About 100 µl of hydrogel mixture (for a final collagen concentration of 1.2 mg/ml) were prepared as follows: 17.25 µl of sterile and distilled water; 12 µl of DMEM 5X with L-glutamine (Sigma D5523); 0.75 µl of NaOH 1N (Sigma 221465); 30 µl of collagen 4 mg/ml (Corning, 356236); and finally 40 µl of cell suspension were added. If another collagen concentration was used, reagent amounts were modified accordingly. Using a chilled tip, solutions were injected into the device using a micropipette (detailed protocol in Section "Microdevice Design"). Then, droplets of 5–10 µl were placed on top of each inlet to prevent hydrogel leakage and evaporation. Afterwards, the microfluidic device was placed into an incubator (37°C and 5% CO_2) for 15 min to allow collagen polymerization. For use as a macroscopic control, 100 µl of solution was allowed to polymerize on Petri dishes.

Cell Viability

Stock solutions of 5 mg/ml fluorescein diacetate (FDA) (Sigma F7378) and 2 mg/ml propidium iodide (PI) (Sigma P4170) were prepared following supplier instructions. To test cell viability within microfluidic devices and in Petri dishes, stock solutions of FDA and PI were diluted to 5 and 4 µg/ml, respectively, in phosphate-buffered saline (PBS) (Lonza BE17-516F). Microdevices and Petri controls were washed once with PBS, and then filled with FDA/PI. Confocal images were immediately taken using a Nikon Eclipse Ti microscope. Images were collected at three different focal planes within each microdevice and Petri control. Experiments were repeated at least three times.

Long-Term Cultures Under Restricted Conditions

In order to study cell behavior during long-term experiments, cells were seeded into microfluidic devices at a concentration of 4 million cells/ml, using 1.2 mg/ml collagen hydrogel. Microdevices were incubated for 24 h in a CO_2 incubator, and then placed inside the packaging tool. To enable medium flow to the packaging tool and microdevice assembly, the device was connected to a pressurized medium reservoir. A pressure regulator (Camozzi MC104-R20) was used to apply 0.1 bar and establish medium flow. Using a flow sensor (Sensirion LG16-0430), medium flow was monitored during experiments. Connections were made with fluorinated ethylene propylene (FEP) tubes (IDEX 1650 and 1684). A GL 45 screw cap (Schott Duran 11 297 51) was placed onto the reservoir bottle along with corresponding connectors (Schott Duran 11 298 14, Schott Duran 11 298 15, Schott Duran 11 562 92) that ensure good coupling of tubing. The outlet flow from the packaging tool was directed into a 15 ml conical tube for collection of waste (**Figure 3A**). Flow data were recorded during

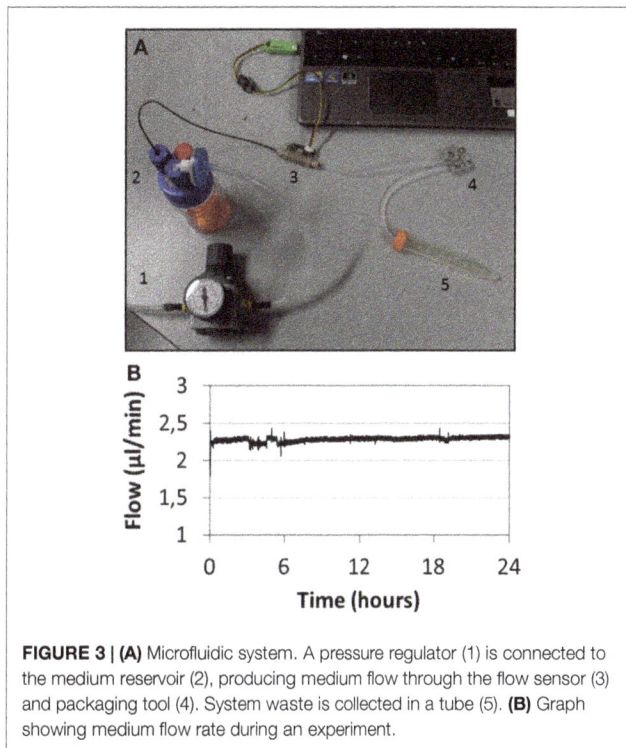

FIGURE 3 | (A) Microfluidic system. A pressure regulator (1) is connected to the medium reservoir (2), producing medium flow through the flow sensor (3) and packaging tool (4). System waste is collected in a tube (5). (B) Graph showing medium flow rate during an experiment.

experiments to verify flow stability (**Figure 3B**). Reservoir bottle culture medium was changed every 2 days to maintain stable concentration of media components. All the system was located within the incubator to maintain 37°C and 5% CO_2. Experiments were repeated three times.

Image Analysis

In order to analyze cell viability in different culture conditions, viable and dead cells were counted within the central microchamber using automated Fiji software (http://fiji.sc/Fiji). Due to the high cell densities reached in some microdevice regions during medium restriction experiments, cell counting algorithms were not reliable. In order to reduce error associated with experiments, we measured the area occupied by living or dead cells comparing live/dead occupied area distribution under restricted and unrestricted conditions. Data were analyzed using PSPP software, and statistical significance was set at $p < 0.05$. Results are presented as mean $\pm$ SE.

Results

Microdevice Design

In order to establish three-dimensional cell culture, microfluidic devices are usually designed with a central culture microchamber (including a dedicated inlet and outlet) and lateral microchannels at both sides that are delimited with a series of pillars (**Figure 1**). In this design, hydrogel, with or without cells embedded, should be confined within the culture chamber while lateral microchannels remain empty. To ensure unobstructed medium flow, it is important to avoid any hydrogel leakage from the microchamber into the lateral microchannels. During microdevice filling different interfaces appear between lateral pillars (50 µm spacing distance) and one main interface across the central microchamber (1000 μm width). The liquid behavior during microdevice filling has been deduced from geometry of these interfaces and material properties using the Young–Laplace equation (Huang et al., 2009):

$$\Delta P = \frac{}{R} \qquad (1)$$

where "ΔP" is the pressure difference between air and liquid, " " is the liquid surface tension, and "R" is the interface curvature radius. Within microfluidic devices, Eq. 1 can be rewritten as (Huang et al., 2009):

$$\Delta P = -2\left(\frac{\cos}{W} + \frac{\cos}{H}\right) \qquad (2)$$

Pressure difference at the interface is therefore expressed only as a function of the native contact angle between the liquid and solid ( ), the microchannel width (W), and microchannel height (H). Therefore, in hydrophilic materials ($< 90°$) such as SU-8 (range between $72°$ and $86°$) (Schumacher et al., 2008;

Jokinen et al., 2012), this pressure difference is negative and spontaneous microdevice filling started (**Figure 4A**). When liquid spontaneously reached the central microchamber two lateral interfaces and one main front interface appeared, reaching a new interface curvature radius of 0 (**Figure 4B**). Under these circumstances capillary forces disappeared and interfaces got pinned, stopping microdevice filling. In order to stop this spontaneous filling, square-shape pillars were critical and showed a robust interface halting. To resume advancing external pressure had to be exerted, increasing the curvature radius of all interfaces until at least one of them reached a curvature equal to the native angle (**Figure 4C**). Due to their different geometry, when an external pressure was applied this curvature increase was lower on lateral interfaces compared to the main front interface. Therefore, main front interface reached first a curvature equal to the contact angle, resuming advancing trough central microchamber but without liquid leakage on lateral microchannels (**Figure 4D**). When design with such geometry and square shaped pillars, hydrogel confinement was achieved just locating a hydrogel droplet on the central inlet (10 μl). Due to its inherent curvature, this droplet exerted the needed pressure to trigger fluid advancement through the

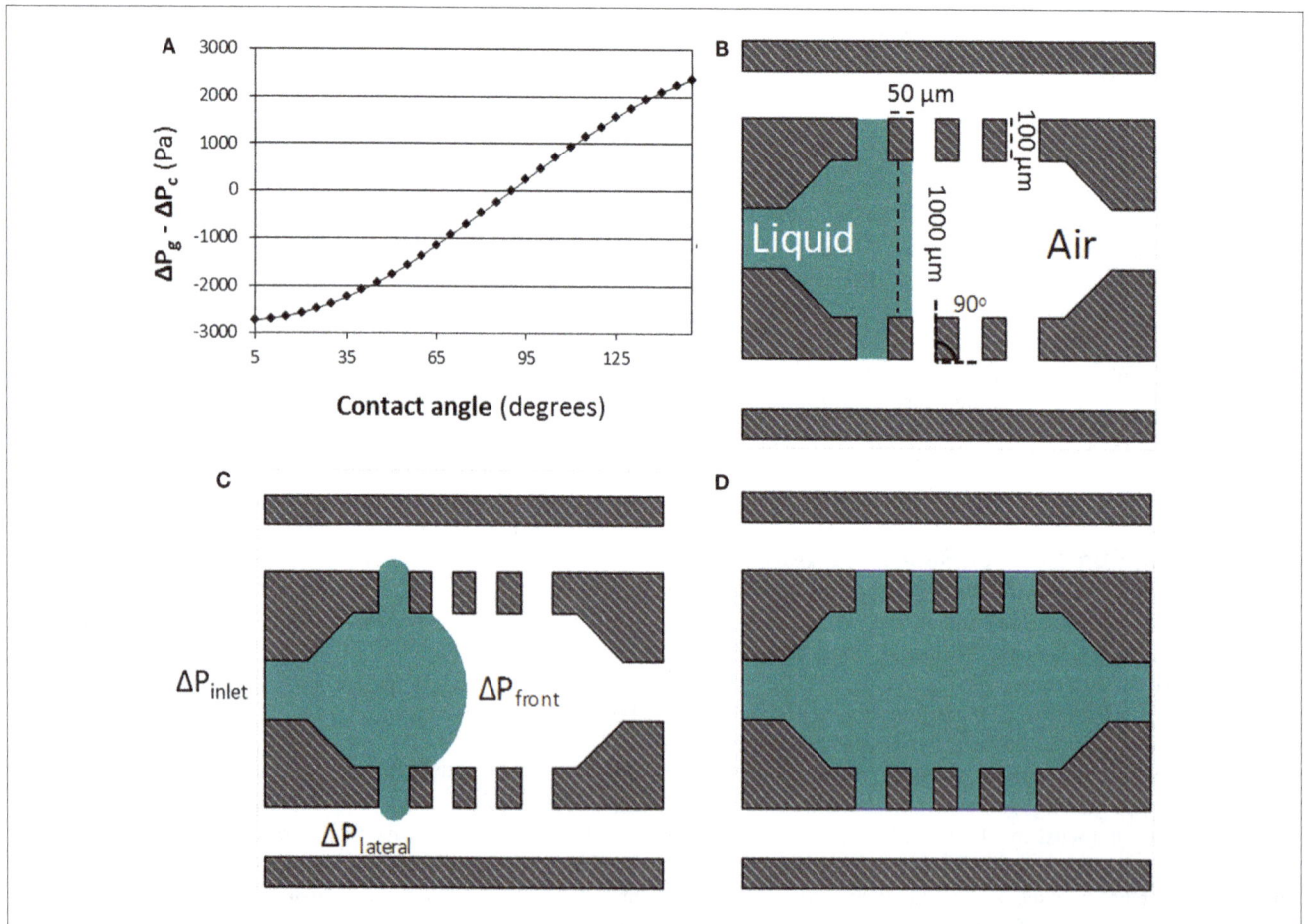

FIGURE 4 | (A) shows how the difference between $\Delta P_{lateral}$ and ΔP_{front} changes with the contact angle between the liquid and the material. In this model, water surface tension (0.072 N/m^2), gap distance (50 μm), microchamber width (1000 μm), and microdevice height (300 μm) were kept constant, while contact angle ranged between 5° and 150° in 5° steps. **(B–D)** shows the confinement process within hydrophilic SU-8-based microdevices. Capillary valves are created at the posts due to the 90° change in post wall direction, illustrated in **(B)**. Lateral capillary valves can support a higher pressure than the central one ($\Delta P_{lateral} > \Delta P_{front}$). This configuration allows hydrogel confinement within a hydrophilic microdevice, as shown in **(D)**.

central microchamber, but not on lateral channels. Blue-colored water and either gelatine (1 and 4 mg/ml) or collagen (1.2 and 2.4 mg/ml) hydrogel was used to fill the microdevice employing the above-described method, and resulted in no leakage into lateral microchannels (**Figure 5A**). Once hydrogel, with cells embedded, was polymerized, additional solutions could be introduced to the microdevice through one or both lateral microchannels without any mixing except diffusion (**Figure 5B**).

Influence of Culture Conditions on Cell Viability

Biocompatibility of microdevices was evaluated using different cellular and collagen densities. Cell densities of 1×10^6, 3×10^6, and 10×10^6 cells/ml in collagen hydrogels of 1.2 and 2.4 mg/ml were tested. Cell viability for each condition was evaluated within microfluidic devices as well as in macroscopic controls on cell culture dishes. After 24 h, cell viability was evaluated by FDA/PI labeling. All culture conditions yielded viability higher than 90% (**Figure 6**), and similar values were obtained for both microdevices and macroscopic controls. In these experiments, measurements were made after 24 h in culture to minimize possible nutrient starvation within the microdevice. Each condition was tested at least three times. Taken together with the aforementioned design considerations, these results support the use of SU-8 as suitable material for the development of three-dimensional cell culture microfluidic devices.

Self-Induced Chemotaxis Under Restricted Conditions

Cells are capable to adapt their behavior in response to their own microenvironment modifications. To investigate this notion, 1.2 mg/ml collagen hydrogel holding 4 million cells/ml was confined into the central microchamber. Microfluidic devices were left in static conditions on a Petri dish for 24 h to allow cell attachment and spreading within the hydrogel. One microdevice was then enclosed in a packaging tool, and all inlets were sealed with the exception of one pair to allow medium perfusion through one lateral microchannel (restricted conditions). After medium flow was enabled, the system was incubated for 7 days. The other microdevice was left on the Petri dish and media was refreshed once a day through both lateral microchannels (unrestricted conditions). Post-incubation images of cell distribution within the culture microchamber show cell distribution in the

different conditions (**Figure 7**). Cell distribution just after hydrogel confinement was apparently homogeneous across the entire microchamber (**Figure 7A**). After 1 week in culture under unrestricted conditions cell distribution seemed to remain unchanged, showing a homogeneous cell distribution (**Figure 7B**). However, under restricted conditions the result was completely different, showing an asymmetric cell distribution within the central microchamber (**Figure 7C**). Under restricted condition cells seemed to concentrate nearby to the perfused lateral microchannel. This asymmetric distribution under restricted conditions could be explained by an asymmetric proliferation or asymmetric mortality within the central microchamber or also by a migratory process. To address this question, the taken images in the different conditions were split vertically into two symmetric halves and area occupied by living and dead cells was analyzed (**Figure 8**). As expected, viable cells distribution after hydrogel injection (0 days) and after 7 days under unrestricted condition was nearly identical in both halves. On the other hand, after 7 days under restricted conditions, cell distribution dramatically changed. In this case, the area occupied by viable cells in the microchamber half closer to the nutrient source was significantly higher compared to the other half (p-value <0.05). Additionally, the total area occupied by live cells in all the microchambers remained almost constant between the different conditions (p-value >0.05). This suggests that there was no cell proliferation under restricted or unrestricted conditions. Furthermore, the area occupied by dead cells remained also constant in all conditions, showing medium restriction caused any cell mortality (p-value >0.05). Taking together, these results demonstrated that under restricted conditions cell metabolism generated a medium depletion, which led to a self-induced chemotactic process toward the nutrient source.

Discussion

When working with microdevices, material properties play a critical role during the critical step of liquid filling, as it has been described previously. In most of the microsystems reported in literature made of PDMS including pillars, hydrogel confinement is achieved by taking advantage of the substrate inherent hydrophobicity. Due to this hydrophobicity, interface pressure barriers are always positive, allowing the use external pressure with the pipette to push the hydrogel within the microdevice. As liquid injection

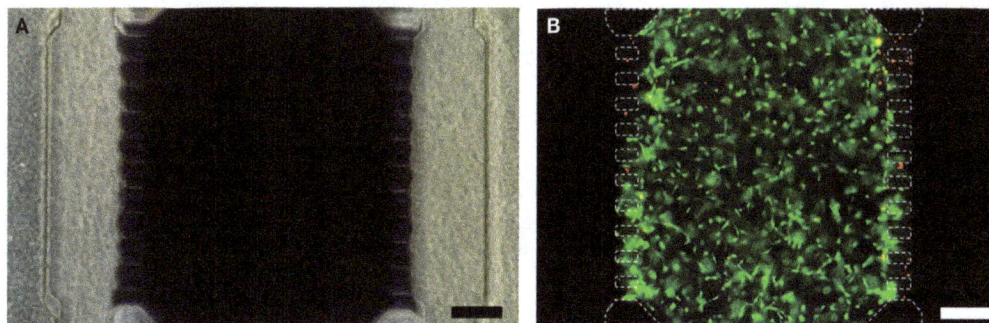

FIGURE 5 | (A) Blue-colored water is confined in the central microchamber with no leakage observed into lateral microchannels. **(B)** Fluorescence image of cells (4×10^6 cells/ml) embedded in a collagen hydrogel (1.2 mg/ml) confined for 24 h. Viable cells are shown in green while dead cells are in red. Scale bar is 200 μm.

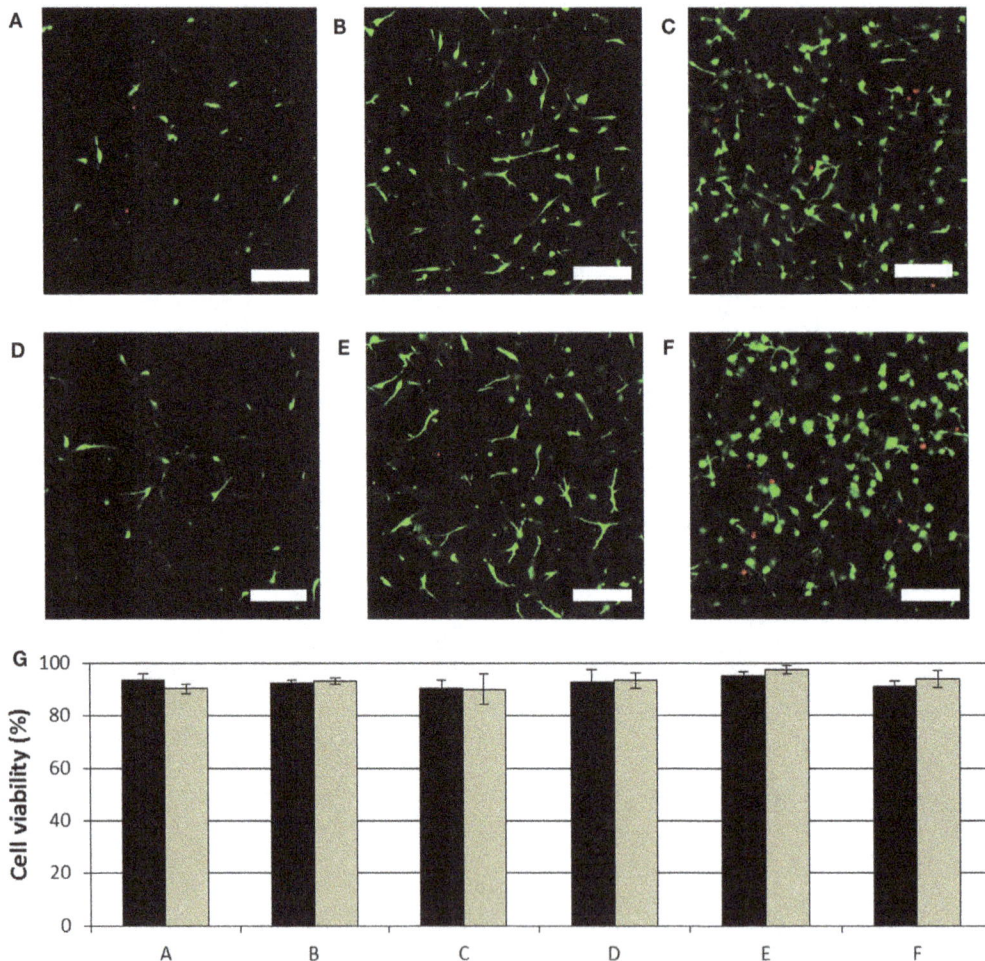

FIGURE 6 | Cell viability at 1×10^6, 3×10^6, and 10×10^6 cells/ml in 1.2 mg/ml collagen hydrogels is shown in (A–C), respectively. Cell viability at 1×10^6, 3×10^6, and 10×10^6 cells/ml in 2.4 mg/ml collagen hydrogels is shown in **(D–F)**, respectively. Scale bar is 100 μm. **(G)** Graph shows cell viability analysis under **(A–F)** culture conditions within the microdevice and on a Petri dish. No statistical difference (*p*-value >0.05) was found between microdevices and hydrogels on Petri dishes.

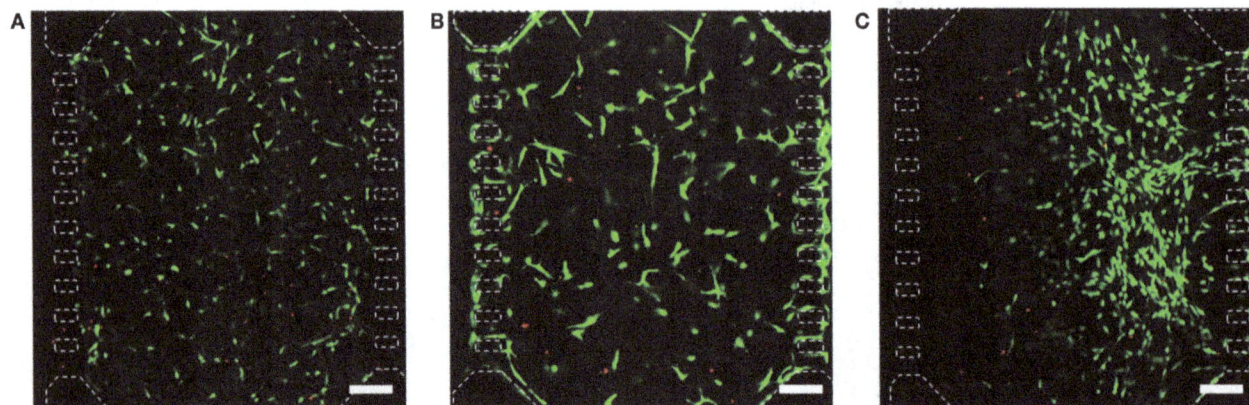

FIGURE 7 | Confocal images showing cell viability and distribution for 4×10^6 cells/ml in 1.2 mg/ml collagen hydrogel. **(A)** Cell distribution after hydrogel injection. **(B)** Cell distribution after 7 days in unrestricted conditions. **(C)** Cell distribution in restricted conditions, medium flow was enabled only through the right microchannel. Scale bar is 200 μm.

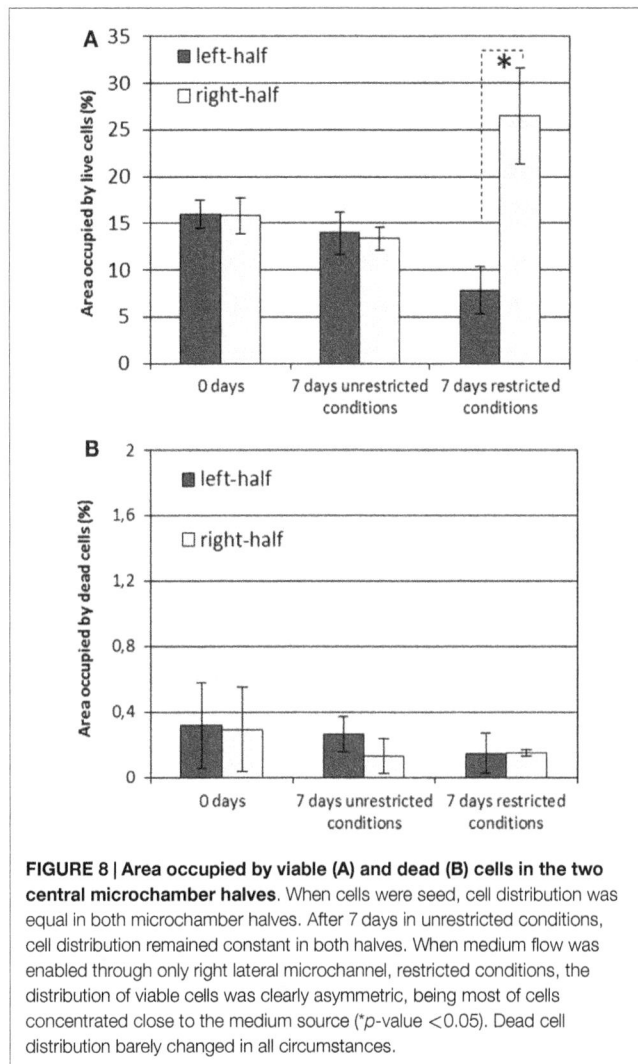

FIGURE 8 | Area occupied by viable (A) and dead (B) cells in the two central microchamber halves. When cells were seed, cell distribution was equal in both microchamber halves. After 7 days in unrestricted conditions, cell distribution remained constant in both halves. When medium flow was enabled through only right lateral microchannel, restricted conditions, the distribution of viable cells was clearly asymmetric, being most of cells concentrated close to the medium source (*p-value <0.05). Dead cell distribution barely changed in all circumstances.

radius becomes 0, and spontaneous filling is stopped. In order to design capillary valves in hydrophilic microdevices, the change in pillar wall direction ( ) needs to satisfy $\geq 90° -$ (for SU-8, $\geq 18°$). As increases, capillary valves can withstand higher pressures before failure (in our square shaped pillars $= 90°$). In principle, the main front interface created across the central microchamber could be halted at the first series of pillars, since they create also a capillary valve. Due to its larger width, however, lower external pressure is required to reach a curvature equal to the native angle (P_{front}) as compared to the lateral interface ($P_{lateral}$) (Safavieh and Juncker, 2013). As long as the pressure applied to the inlet (P_{inlet}) is kept between these two values (i.e., $P_{front} < P_{inlet} < P_{lateral}$), microchamber filling will occur without leakage into lateral microchannels. When a SU-8 based microdevice is designed with the described geometry, location of a droplet ($\approx 10\ \mu l$) exerted a pressure that spanned between these ranges, allowing proper hydrogel confinement.

The microdevice presented in this article has been then tested on self-induced chemotaxis experiments. In contrast with previous reported chemotaxis experiments, the nutrient chemotaxis is created by the activity of cells, a situation is to be considered closer to natural gradient environments. During the course of the experiment, metabolism of cells located close to the perfused channel created a nutrient starvation for cells located far away, which led to a strong migratory process toward the perfused lateral microchannel. Interestingly, in this long-term culture, experiments cells seemed to enclose around SU-8 pillars. Proliferation rate reduction was quite expected since many reports have demonstrated how cell proliferation is normally lower in 3D systems compared with 2D cultures (Smalley et al., 2006; Wells, 2008; Ali et al., 2014; Antoni et al., 2015). Furthermore, an inverse relationship between the collagen hydrogel stiffness and proliferation rate has been shown in bone cells (Jha et al., 2014). Previous results obtained in imposed chemotactic assays show that cell migration occurs after few minutes or hours (Zicha et al., 1997; Zimmermann et al., 2007; Bhattacharjee et al., 2010; Kim et al., 2013b, 2010; Chen et al., 2011; Funamoto et al., 2012). In contrast, in the presented paper, a more complex response was observed, since cell response occurred only after several days under nutrient restriction. We hypothesize that at first there was enough amount of nutrients within the central microchamber, and only after few days this concentration dropped below a threshold, which impelled the cells to migrate. As a consequence, the process for the creation and establishment of nutrient starvation and cell reaction could be better studied in future experiments using the developed microsystem and protocol. While modifying initial cell density or specific nutrient concentrations (glucose, etc.) within the central microchamber could lead to different responses and even to the generation of cellular autophagy and necrotic regions.

Conclusion

In this paper, we present a designed based protocol for hydrogel confinement within microfluidic devices made of hydrophilic materials. Therefore, it can be suitable for any hydrophilic material by a simple adjustment of the chamber design, facilitating the use

through wider channels require lower pressure compared to narrower ones in hydrophobic microdevices, the hydrogel will preferably fill the central chamber rather than blocking the lateral microchannels protected by the micropillars. Although PDMS-based microdevices are normally rendered hydrophilic in order to covalently bond them with glass coverslips or between two separated PDMS layers, this change is reversible. After some time (commonly from few hours up to few days), PDMS hydrophobicity is recovered and hydrogel confinement using pillars is achieved as explained above (Huang et al., 2009). However, in hydrophilic materials (contact angle is lower than 90°), pressure barriers have negative values, favoring spontaneous filling of narrower microchannels. Therefore, pillars on hydrophilic microdevices are unable to prevent leakage as they do in PDMS microdevices and specific changes on pillar shape must be performed.

The design presented in this paper represents an alternative strategy for microdevice filling based on geometric configurations known as capillary valves (Leu and Chang, 2004; Cho et al., 2007; Zimmermann et al., 2007; Chen et al., 2008). These valves are created by an abrupt change in microdevice wall direction. When interface reaches this abrupt change in wall direction, curvature of new materials such as polystyrene, COC, or SU-8. The developed protocol was tested with a hydrophilic SU-8-based microdevice, and produced positive results independent of the injected solution. After hydrogel confinement, biocompatibility of the microdevice was confirmed, showing cell viabilities >90% in all cases. Under restricted conditions, and due to cell metabolism, a strong migratory process toward the nutrient source was induced. Many other reports have shown previously examples of chemotaxis, but in such articles chemotaxis conditions are externally imposed. This is the first example in which the chemotaxis is generated naturally by cells within a microfluidic device, causing a self-induced migratory process. This self-induced process showed a dynamic very different compared with other imposed chemotaxis assays. Thus, combining the described methodology, with sensor integration within the SU-8 based microdevices, precise monitoring of metabolites could be done in real time during biological process like tumor growth and invasion.

Author Contributions

Conception and design: IO and LF. Protocol and methodology development: JA, RM, GL, MM, MA, JB, IO, and LF. Acquisition data: JA, MM, and GL. Manuscript writing and reviewing: JA, RM, IO, LF, and MD.

Acknowledgments

This work has been supported by the National Research Program of Spain, project "dpi2011-28262-c04-01" and "bes-2012-059940." JA studentship was provided by the Aragon government.

References

Ali, M. Y., Chuang, C. Y., and Saif, M. T. (2014). Reprogramming cellular phenotype by soft collagen gels. *Soft Matter* 10, 8829–8837. doi:10.1039/c4sm01602e

Altuna, A., Menendez de la Prida, L., Bellistri, E., Gabriel, G., Guimerà, A., Berganzo, J., et al. (2012). SU-8 based microprobes with integrated planar electrodes for enhanced neural depth recording. *Biosens. Bioelectron.* 37, 1–5. doi:10.1016/j.bios.2012.03.039

Altuna, A., Bellistri, E., Cid, E., Aivar, P., Gal, B., Berganzo, J., et al. (2013). SU-8 based microprobes for simultaneous neural depth recording and drug delivery in the brain. *Lab. Chip* 13, 1422–1430. doi:10.1039/c3lc41364k

Antoni, D., Burckel, H., Josset, E., and Noel, G. (2015). Three-dimensional cell culture: a breakthrough in vivo. *Int. J. Mol. Sci.* 16, 5517–5527. doi:10.3390/ijms16035517

Baker, B. M., and Chen, C. S. (2012). Deconstructing the third dimension: how 3D culture microenvironments alter cellular cues. *J. Cell. Sci.* 125, 3015–3024. doi:10.1242/jcs.079509

Bhattacharjee, N., Li, N., Keenan, T. M., and Folch, A. (2010). A neuron-benign microfluidic gradient generator for studying the response of mammalian neurons towards axon guidance factors. *Integr. Biol. (Camb)* 2, 669–679. doi:10.1039/c0ib00038h

Blanco, F., Agirregabiria, M., Garcia, J., Berganzo, J., Tijero, M., and Arroyo, M. (2004). Novel three-dimensional embedded SU-8 microchannels fabricated using a low temperature full wafer adhesive bonding. *J. Micromech. Microeng.* 14, 1047–1056. doi:10.1088/0960-1317/14/7/027

Breslin, S., and O'Driscoll, L. (2013). Three-dimensional cell culture: the missing link in drug discovery. *Drug Discov. Today* 18, 240–249. doi:10.1016/j.drudis.2012.10.003

Calvo, V., Ezkerra, A., Elizalde, J., Fernández, L., Berganzo, J., Mayora, K., et al. (2011). A highly integrated vertical SU8 valve for stepwise in-series reactions. *J. Micromech. Microeng.* 21, 065037. doi:10.1088/0960-1317/21/6/065037

Castaño-Álvarez, M., Fernández-Abedul, M. T., Costa-García, A., Agirregabiria, M., Fernández, L., Ruano-López, J., et al. (2009). Fabrication of SU-8 based microchip electrophoresis with integrated electrochemical detection for neurotransmitters. *Talanta* 80, 24–30. doi:10.1016/j.talanta.2009.05.049

Chen, J., Huang, P., and Lin, M. (2008). Analysis and experiment of capillary valves for microfluidics on a rotating disk. *Microfluid. Nanofluidics* 4, 427–437. doi:10.1007/s10404-007-0196-x

Chen, Y. A., King, A. D., Shih, H. C., Peng, C. C., Wu, C. Y., Liao, W. H., et al. (2011). Generation of oxygen gradients in microfluidic devices for cell culture using spatially confined chemical reactions. *Lab. Chip* 11, 3626–3633. doi:10.1039/c1lc20325h

Cho, H., Kim, H., Kang, J., and Kim, T. (2007). How the capillary burst microvalve works. *J. Colloid Interface Sci.* 306, 379–385. doi:10.1016/j.jcis.2006.10.077

Duval, D., González-Guerrero, A., Dante, S., Osmond, J., Monge, R., Fernández, L., et al. (2012). Nanophotonic lab-on-a-chip platforms including novel bimodal interferometers, microfluidics and grating couplers. *Lab. Chip* 12, 1987–1994. doi:10.1039/c2lc40054e

El-Ali, J., Sorger, P. K., and Jensen, K. F. (2006). Cells on chips. *Nature* 442, 403–411. doi:10.1038/nature05063

Ezkerra, A., Fernández, L., Mayora, K., and Ruano-López, J. (2007). Fabrication of SU-8 free-standing structures embedded in microchannels for microfluidic control. *J. Micromech. Microeng.* 17, 2264–2271. doi:10.1088/0960-1317/17/11/013

Ezkerra, A., Fernandez, L. J., Mayora, K., and Ruano-Lopez, J. M. (2011). SU8 diaphragm micropump with monolithically integrated cantilever check valves. *Lab. Chip* 11, 3320–3325. doi:10.1039/c1lc20324j

Falasca, M., Raimondi, C., and Maffucci, T. (2011). Boyden chamber. *Methods Mol. Biol.* 769, 87–95. doi:10.1007/978-1-61779-207-6_7

Farahat, W. A., Wood, L. B., Zervantonakis, I. K., Schor, A., Ong, S., Neal, D., et al. (2012). Ensemble analysis of angiogenic growth in three-dimensional microfluidic cell cultures. *PLoS ONE* 7:e37333. doi:10.1371/journal.pone.0037333

Fernández, L., Altuna, A., Tijero, M., Gabriel, G., Villa, R., Rodríguez, M., et al. (2009). Study of functional viability of SU-8-based microneedles for neural applications. *J. Micromech. Microeng.* 19, 025007. doi:10.1088/0960-1317/19/2/025007

Franco, S. J., and Muller, U. (2011). Extracellular matrix functions during neuronal migration and lamination in the mammalian central nervous system. *Dev. Neurobiol.* 71, 889–900. doi:10.1002/dneu.20946

Funamoto, K., Zervantonakis, I. K., Liu, Y., Ochs, C. J., Kim, C., and Kamm, R. D. (2012). A novel microfluidic platform for high-resolution imaging of a three-dimensional cell culture under a controlled hypoxic environment. *Lab. Chip* 12, 4855–4863. doi:10.1039/c2lc40306d

Gerhardt, H., and Betsholtz, C. (2005). "How do endothelial cells orientate?" in *Angiogenesis mechanisms*, Chap. 1. eds M. Clauss and G. Breier (Switzerland: Birkhäuser Basel), 3–15. doi:10.1007/3-7643-7311-3_1

González, I., Fernández, L., Gómez, T., Berganzo, J., Soto, J., and Carrato, A. (2012). A polymeric chip for micromanipulation and particle sorting by ultrasounds based on a multilayer configuration. *Sens. Actuators B Chem.* 144, 310–317. doi:10.1016/j.snb.2009.10.042

Guo, X., Elliott, C. G., Li, Z., Xu, Y., Hamilton, D. W., and Guan, J. (2012). Creating 3D angiogenic growth factor gradients in fibrous constructs to guide fast angiogenesis. *Biomacromolecules* 13, 3262–3271. doi:10.1021/bm301029a

Huang, C. P., Lu, J., Seon, H., Lee, A. P., Flanagan, L. A., Kim, H. Y., et al. (2009). Engineering microscale cellular niches for three-dimensional multicellular co-cultures. *Lab. Chip* 9, 1740–1748. doi:10.1039/b818401a

Huang, G., Wang, L., Wang, S., Han, Y., Wu, J., Zhang, Q., et al. (2012). Engineering three-dimensional cell mechanical microenvironment with hydrogels. *Biofabrication* 4, 042001. doi:10.1088/1758-5082/4/4/042001

Jeon, J. S., Chung, S., Kamm, R. D., and Charest, J. L. (2011). Hot embossing for fabrication of a microfluidic 3D cell culture platform. *Biomed. Microdevices* 13, 325–333. doi:10.1007/s10544-010-9496-0

Jha, A. K., Jackson, W. M., and Healy, K. E. (2014). Controlling osteogenic stem cell differentiation via soft bioinspired hydrogels. *PLoS ONE* 9:e98640. doi:10.1371/journal.pone.0098640

Jokinen, V., Suvanto, P., and Franssila, S. (2012). Oxygen and nitrogen plasma hydrophilization and hydrophobic recovery of polymers. *Biomicrofluidics* 6, 16501–1650110. doi:10.1063/1.3673251

Keenan, T. M., and Folch, A. (2008). Biomolecular gradients in cell culture systems. *Lab. Chip* 8, 34–57. doi:10.1039/B711887B

Kim, D. H., Hossain, M. A., Kim, M. Y., Kim, J. A., Yoon, J. H., Suh, H. S., et al. (2013a). A novel resveratrol analogue, HS-1793, inhibits hypoxia-induced HIF-1alpha and VEGF expression, and migration in human prostate cancer cells. *Int. J. Oncol.* 43, 1915–1924. doi:10.3892/ijo.2013.2116

Kim, B. J., Hannanta-anan, P., Chau, M., Kim, Y. S., Swartz, M. A., and Wu, M. (2013b). Cooperative roles of SDF-1alpha and EGF gradients on tumor cell migration revealed by a robust 3D microfluidic model. *PLoS ONE* 8:e68422. doi:10.1371/journal.pone.0068422

Kim, S., Kim, H. J., and Jeon, N. L. (2010). Biological applications of microfluidic gradient devices. *Integr. Biol. (Camb)* 2, 584–603. doi:10.1039/c0ib00055h

Korpos, E., Wu, C., Song, J., Hallmann, R., and Sorokin, L. (2010). Role of the extracellular matrix in lymphocyte migration. *Cell Tissue Res.* 339, 47–57. doi:10.1007/s00441-009-0853-3

Kothapalli, C. R., van Veen, E., de Valence, S., Chung, S., Zervantonakis, I. K., Gertler, F. B., et al. (2011). A high-throughput microfluidic assay to study neurite response to growth factor gradients. *Lab. Chip* 11, 497–507. doi:10.1039/c0lc00240b

Kotzar, G., Freas, M., Abel, P., Fleischman, A., Roy, S., Zorman, C., et al. (2002). Evaluation of MEMS materials of construction for implantable medical devices. *Biomaterials* 23, 2737–2750. doi:10.1016/S0142-9612(02)00007-8

Lesher-Perez, S. C., Frampton, J. P., and Takayama, S. (2013). Microfluidic systems: a new toolbox for pluripotent stem cells. *Biotechnol. J.* 8, 180–191. doi:10.1002/biot.201200206

Leu, T., and Chang, P. (2004). Pressure barrier of capillary stop valves in micro sample separators. *Sens. Actuators A* 115, 508–515. doi:10.1016/j.sna.2004.02.036

Lorenz, H., Despont, M., Fahrni, M., LaBianca, N., Vettiger, P., and Renaud, P. (1997). SU-8: a low-cost negative resist for MEMS. *J. Micromech. Microeng.* 7, 121–124. doi:10.1088/0960-1317/7/3/010

Mousseau, Y., Leclers, D., Faucher-Durand, K., Cook-Moreau, J., Lia-Baldini, A. S., Rigaud, M., et al. (2007). Improved agarose gel assay for quantification of growth factor-induced cell motility. *Biotechniques* 43, 509–516. doi:10.2144/000112557

Muinonen-Martin, A. J., Veltman, D. M., Kalna, G., and Insall, R. H. (2010). An improved chamber for direct visualisation of chemotaxis. *PLoS ONE* 5:e15309. doi:10.1371/journal.pone.0015309

Nagelkerke, A., Bussink, J., Mujcic, H., Wouters, B. G., Lehmann, S., Sweep, F. C., et al. (2013). Hypoxia stimulates migration of breast cancer cells via the PERK/ATF4/LAMP3-arm of the unfolded protein response. *Breast Cancer Res.* 15, R2. doi:10.1186/bcr3373

Nemani, K. V., Moodie, K. L., Brennick, J. B., Su, A., and Gimi, B. (2013). In vitro and in vivo evaluation of SU-8 biocompatibility. *Mater. Sci. Eng. C Mater. Biol. Appl.* 33, 4453–4459. doi:10.1016/j.msec.2013.07.001

Ni, M., Tong, W. H., Choudhury, D., Rahim, N. A., Iliescu, C., and Yu, H. (2009). Cell culture on MEMS platforms: a review. *Int. J. Mol. Sci.* 10, 5411–5441. doi:10.3390/ijms10125411

Nieto, M., Frade, J. M., Sancho, D., Mellado, M., Martinez-A, C., and Sanchez-Madrid, F. (1997). Polarization of chemokine receptors to the leading edge during lymphocyte chemotaxis. *J. Exp. Med.* 186, 153–158. doi:10.1084/jem.186.1.153

Noman, M. Z., Messai, Y., Carre, T., Akalay, I., Meron, M., Janji, B., et al. (2011). Microenvironmental hypoxia orchestrating the cell stroma cross talk, tumor progression and antitumor response. *Crit. Rev. Immunol.* 31, 357–377. doi:10.1615/CritRevImmunol.v31.i5.10

Page, H., Flood, P., and Reynaud, E. G. (2013). Three-dimensional tissue cultures: current trends and beyond. *Cell Tissue Res.* 352, 123–131. doi:10.1007/s00441-012-1441-5

Raines, E. W. (2000). The extracellular matrix can regulate vascular cell migration, proliferation, and survival: relationships to vascular disease. *Int. J. Exp. Pathol.* 81, 173–182. doi:10.1046/j.1365-2613.2000.00155.x

Rigat-Brugarolas, L. G., Elizalde-Torrent, A., Bernabeu, M., De Niz, M., Martin-Jaular, L., Fernandez-Becerra, C., et al. (2014). A functional microengineered model of the human splenon-on-a-chip. *Lab. Chip* 14, 1715–1724. doi:10.1039/c3lc51449h

Roussos, E. T., Condeelis, J. S., and Patsialou, A. (2011). Chemotaxis in cancer. *Nat. Rev. Cancer* 11, 573–587. doi:10.1038/nrc3078

Safavieh, R., and Juncker, D. (2013). Capillarics: pre-programmed, self-powered microfluidic circuits built from capillary elements. *Lab. Chip* 13, 4180–4189. doi:10.1039/c3lc50691f

Savino, W., Mendes-Da-Cruz, D. A., Smaniotto, S., Silva-Monteiro, E., and Villa-Verde, D. M. (2004). Molecular mechanisms governing thymocyte migration: combined role of chemokines and extracellular matrix. *J. Leukoc. Biol.* 75, 951–961. doi:10.1189/jlb.1003455

Schumacher, J., Grodrian, A., Kremin, C., Hoffmann, M., and Metze, J. (2008). Hydrophobic coating of microfluidic chips structured by SU-8 polymer for segmented flow operation. *J. Micromech. Microeng.* 5, 055019. doi:10.1088/0960-1317/18/5/055019

Smalley, K. S., Lioni, M., and Herlyn, M. (2006). Life isn't flat: taking cancer biology to the next dimension. *In vitro Cell. Dev. Biol. Anim.* 42, 242–247. doi:10.1290/0604027.1

Song, J. W., Bazou, D., and Munn, L. L. (2012). Anastomosis of endothelial sprouts forms new vessels in a tissue analogue of angiogenesis. *Integr. Biol. (Camb)* 4, 857–862. doi:10.1039/c2ib20061a

Torrejon, K. Y., Pu, D., Bergkvist, M., Danias, J., Sharfstein, S. T., and Xie, Y. (2013). Recreating a human trabecular meshwork outflow system on microfabricated porous structures. *Biotechnol. Bioeng.* 110, 3205–3218. doi:10.1002/bit.24977

Verdoy, D., Barrenetxea, Z., Berganzo, J., Agirregabiria, M., Ruano-López, J., Marimón, J., et al. (2012). A novel real time micro PCR based point-of-care device for *Salmonella* detection in human clinical samples. *Biosens. Bioelectron.* 32, 259–265. doi:10.1016/j.bios.2011.12.032

Vilares, R., Hunter, C., Ugarte, I., Aranburu, I., Berganzo, J., Elizalde, J., et al. (2010). Fabrication and testing of a SU-8 thermal flow sensor. *Sens. Actuators B Chem.* 147, 411–417. doi:10.1016/j.snb.2010.03.054

Wells, R. G. (2008). The role of matrix stiffness in regulating cell behavior. *Hepatology* 47, 1394–1400. doi:10.1002/hep.22193

Zhang, Y., Liu, Q., Wang, F., Ling, E. A., Liu, S., Wang, L., et al. (2013). Melatonin antagonizes hypoxia-mediated glioblastoma cell migration and invasion via inhibition of HIF-1alpha. *J. Pineal Res.* 55, 121–130. doi:10.1111/jpi.12052

Zicha, D., Dunn, G., and Jones, G. (1997). Analyzing chemotaxis using the Dunn direct-viewing chamber. *Methods Mol. Biol.* 75, 449–457.

Zigmond, S. H. (1977). Ability of polymorphonuclear leukocytes to orient in gradients of chemotactic factors. *J. Cell Biol.* 75, 606–616. doi:10.1083/jcb.75.2.606

Zimmermann, M., Hunziker, P., and Delamarche, E. (2007). Valves for autonomous capillary systems. *Microfluid. Nanofluidics* 5, 395–402. doi:10.1039/c2lc41083d

Conflict of Interest Statement: The authors declare that the research was conducted in the absence of any commercial or financial relationships that could be construed as a potential conflict of interest.

17

Laser patterning of non-linear optical $Bi_2ZnB_2O_7$ crystal lines in glass

Taisuke Inoue, Xuyi Gao, Kenji Shinozaki, Tsuyoshi Honma and Takayuki Komatsu *

Department of Materials Science and Technology, Nagaoka University of Technology, Nagaoka, Japan

Bi_2O_3-based glasses are very attractive from the viewpoints of low-melting, high refractive index, and crystallization, and the research on their glasses and glass-ceramics is at the frontiers of glass science and technology. Non-linear optical $Bi_2ZnB_2O_7$ crystal lines with a high orientation were patterned in $3Sm_2O_3$-$30.3Bi_2O_3$-$33.3ZnO$-$33.3B_2O_3$ glass by using a laser-induced crystallization technique. It was confirmed from transmission electron microscope observations that crystals were formed in the inside of the glass, i.e., at the beneath of $4\,\mu b$ from the surface, although lasers ($Yb:YVO_4$ laser with a wavelength of 1080 nm) were focused at the glass surface. A new potential for optical device applications was added in Bi_2O_3-based glasses from the present study.

Keywords: bismuth based glass, non-linear optical crystal, laser patterning, crystal orientation, glass-ceramics

Edited by:
Ashutosh Goel,
Rutgers-The State University of
New Jersey, USA

Reviewed by:
Jeetendra Sehgal,
Sterlite Technologies Limited, India
Zhihua Yang,
Chinese Academy of Sciences, China

***Correspondence:**
Takayuki Komatsu,
Department of Materials Science and
Technology, Nagaoka University of
Technology, 1603-1
Kamitomioka-cho, Nagaoka
940-2188, Japan
komatsu@mst.nagaokaut.ac.jp

Introduction

Micro-fabrications in glassy materials have found increasingly more applications in optoelectronics, telecommunications, and photonic devices such as optical gratings and waveguides, and laser irradiation to glasses has received considerable attention as a new tool of spatially selected micro-fabrications in glasses. For instance, a permanent change of refractive index can be induced in Ge-doped SiO_2 optical fibers by using ultraviolet laser irradiations to produce Bragg gratings under suitable exposure conditions. Spatially selected patterning of crystals in glasses by laser irradiations, i.e., laser-induced crystallization, is also extremely attractive, which is a new method for the design and control of the crystallization of glasses in comparison with the conventional crystallization technique using an electric furnace, and also opens a new door in crystal growth and patterning engineering. The design and control of the crystallization of glasses by using lasers are at the frontiers of the glass science and technology.

The present authors group proposed new laser irradiation techniques for the patterning of crystals with dot, line, and two-dimensional planar shapes in glasses, i.e., the rare-earth atom heat (REAH) processing and transition metal atom heat (TMAH) processing (Sato et al., 2001; Honma et al., 2003, 2006; Komatsu et al., 2007; Komatsu and Honma, 2013), and various crystals such as ferroelectric $LiNbO_3$ (Honma and Komatsu, 2010; Komatsu et al., 2011), multi-ferroic $'$-$RE_2(MoO_4)_3$ (RE: rare-earth) (Tsukada et al., 2009; Suzuki et al., 2010), non-linear optical $Li_2Si_2O_5$ (Honma et al., 2008), oxyfluoride $BaAlBO_3F_2$ (Shinozaki et al., 2012), and fluoride CaF_2 (Shinozaki et al., 2013) have been patterned successfully in glasses. In the laser-induced crystallization, a steep temperature gradient is created in the laser irradiated local region and such a steep temperature gradient is moved along laser scanning direction, consequently providing the patterning of crystals with a high orientation. It is of very importance to apply the laser-induced crystallization technique to various glasses, and such studies will lead a more deep understanding of the crystallization of glasses and also the possibility of new device applications of crystallized glasses (glass-ceramics).

In this study, we focus our attention on the patterning of non-linear optical $Bi_2ZnB_2O_7$ crystal lines with a high orientation in Bi_2O_3–ZnO–B_2O_3 glasses by using a laser-induced crystallization technique. Recently, $Bi_2MB_2O_7$ crystals (M = Zn, Sr, and Ca) have been successfully synthesized through the crystallization of corresponding glasses (Majhi and Varma, 2008a,b; Kanenishi et al., 2012; Shanmugavelu and Kumar, 2012). In particular, $Bi_2ZnB_2O_7$ with an orthorhombic structure is of interest, because $Bi_2ZnB_2O_7$ exhibits excellent non-linear optical properties (Barbier et al., 2005) and is crystallized from $33.3Bi_2O_3$-$33.3ZnO$-$33.3B_2O_3$ glass corresponding to the stoichiometric composition of $Bi_2ZnB_2O_7$. The values of second-order optical non-linearities, $d_{31} = 0.911$ pm/V, $d_{32} = 3.083$ pm/V, and $d_{33} = 1.015$ pm/V, have been reported for $Bi_2ZnB_2O_7$ single crystal (Su et al., 2013). Very recently, transparent glass-ceramics containing $Bi_2ZnB_2O_7$ nanocrystals have been synthesized in $30Bi_2O_3$-$40ZnO$-$30B_2O_3$ glass, and it has been found that they exhibit a large third-order optical non-linearity (Zeng et al., 2014). Furthermore, $Bi_2ZnB_2O_7$ nanocrystals have been patterned in ZnO–Bi_2O_3–B_2O_3 glasses by irradiations of femtosecond laser (Liu et al., 2015). There has been, however, no report on the laser patterning of $Bi_2ZnB_2O_7$ crystal lines (not dots) with a high orientation in glasses so far. In a previous paper (Inoue et al., 2010), we examined the features of electronic polarizability (optical basicity) and chemical bonding state in Bi_2O_3–ZnO–B_2O_3 glasses from density, refractive index, Raman scattering spectrum, and X-ray photoelectron spectrum (XPS) measurements, and proposed the formation of B–O–Bi and B–O–Zn bridging bonds in the glass structure. In this paper, we demonstrated that $Bi_2ZnB_2O_7$ crystal lines with a high orientation were patterned in Bi_2O_3–ZnO–B_2O_3 glasses by laser irradiations for the first time. Glasses based on the Bi_2O_3–ZnO–B_2O_3 system have also received much attention as new low-melting and shielding glasses (Maeder, 2013), and from this point of view, the understanding of the crystallization behaviors of Bi_2O_3–ZnO–B_2O_3 glasses is extremely important. The laser patterning of functional crystals in Bi_2O_3-based glasses has a great potential in application of non-linear optical integrated devices (Liu et al., 2015), and is at the frontiers of glass science and technology in Bi_2O_3-based glasses.

Experimental Section

Glasses with the compositions of $33.3Bi_2O_3$-$33.3ZnO$-$33.3Bi_2O_3$ and xSm_2O_3-$(33.3-x)Bi_2O_3$-$33.3ZnO$-$33.3B_2O_3$ ($x = 1$, 3, and 5) were prepared by using a conventional melt quenching technique. Commercial powders of reagent grade ZnO, Bi_2O_3, H_3BO_3, and Sm_2O_3 were melted in a platinum crucible at 1000°C for 30 min in an electric furnace. The melts were poured onto an iron plate and pressed to a thickness of ~1.5 mm by another iron plate. Glass transition (T_g) and crystallization peak temperatures (T_p) were determined by using differential thermal analysis (DTA) at a heating rate of 10 K/min. The quenched glasses were annealed at T_g for 30 min to release internal stresses. Densities of glasses were determined with the Archimedes method using distilled water as an immersion liquid (Inoue et al., 2010). Refractive indices at a wavelength (λ) of 632.8 nm (He–Ne laser) were measured at room temperature with a prism coupler (Metricon Model

2010). The melt-quenched plate-shaped glasses were heat-treated at some temperatures in an electric furnace, i.e., at 450–530°C for 3 h, and the crystalline phases formed were identified by using X-ray diffraction (XRD) analysis at room temperature. Raman scattering spectra at RT for the glasses and heat-treated samples were measured with a laser microscope (Tokyo Instruments Co., Nanofinder) operated at Ar^+ laser (wavelength: $\lambda = 488$ nm). In this apparatus, the data below 250 cm^{-1} cannot be measured due to the use of an edge filter.

Glasses were mechanically polished to a mirror finish with CeO_2 powders. The glass was irradiated by cw Yb:YVO_4 fiber laser (beam shape: single mode and ± 1 nm bandwidth) with $\lambda = 1080$ nm using objective lens (magnification: 50 times, numerical aperture: NA = 0.8). The laser beam was unpolarized and the diameter of laser spot was 2~3 μm. Plate-shaped glasses with a thickness of ~ 1 mm were put on the stage and mechanically moved during laser irradiations to pattern crystals. The crystalline phase in the laser-irradiated part was examined from micro-Raman scattering spectrum measurements. Second harmonic generation (SHG) microscopic measurements were performed for the laser-irradiated part using a Q-switched Nd:YAG (yttrium aluminum garnet) laser with $\lambda = 1064$ nm as a fundamental light source. SHG emissions ($\lambda = 532$ nm) were measured, although the fundamental one must be isolated by a cut-off filter before reaching to the CCD (Charged Coupled Devices) for image detection (Fujiwara et al., 2003). The morphology of crystals in laser-patterned lines were examined by using a transmission electron microscope (TEM: FE-TEM, JEOL, JEM-2100F) operating at 200 kV. Thin foils for TEM observations were prepared with a focused Ga ion beam (FIB) (JEOL, JIB-4500) method.

Results and Discussion

Crystallization Behavior of $33.3Bi_2O_3$-$33.3ZnO$-$33.3B_2O_3$ Glass

Prior to the laser patterning of $Bi_2ZnB_2O_7$ crystal lines in Bi_2O_3–ZnO–B_2O_3 glasses, it is necessary to confirm the formation of $Bi_2ZnB_2O_7$ crystals in the crystallization of a given glass. Basically, the crystalline phase formed by laser irradiations in glasses is the same as that formed by heat treatments in an electric furnace. The glass with the composition of $33.3Bi_2O_3$-$33.3ZnO$-$33.3B_2O_3$ corresponding to the stoichiometric composition of the $Bi_2ZnB_2O_7$ crystalline phase is designated here as 111BiZnB glass.

Figure 1 shows the DTA patterns for the bulk and powdered samples of as-quenched 111BiZnB glasses. In both samples, endothermic dips due to the glass transition and exothermic peaks due to the crystallization are detected clearly, providing the values of $T_g = 395$°C and $T_p = 562$°C for the bulk sample and $T_g = 395$°C and $T_p = 551$°C for the powdered sample. It is noted that the crystallization peak temperature ($T_p = 562$°C) for the bulk sample is close to that ($T_p = 551$°C) for the powdered sample, implying the possibility of the so-called bulk crystallization in the inside of glass. In **Figure 1**, the optical microscope photograph for the base glass is included. The glass is colorless in the visible region. The density (d) and refractive index (n) of 111BiZnB glass at room temperature are $d = 6.53$ g/cm^3 and $n = 2.06$, respectively.

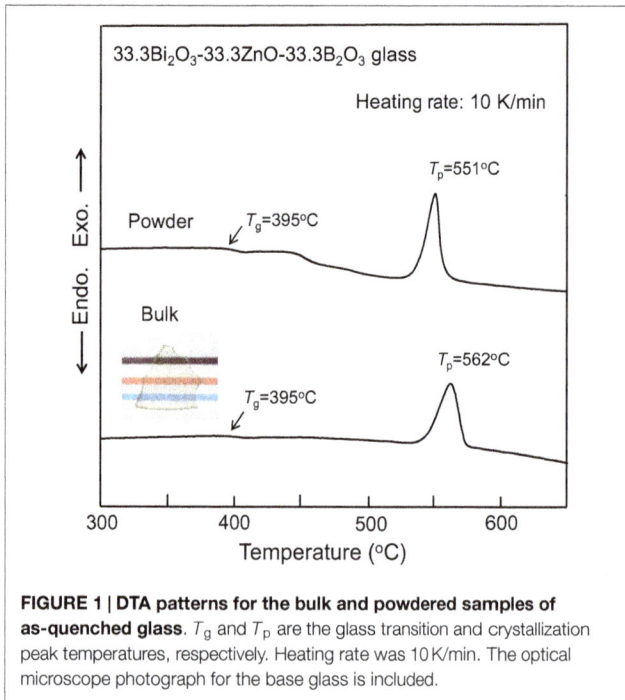

FIGURE 1 | DTA patterns for the bulk and powdered samples of as-quenched glass. T_g and T_p are the glass transition and crystallization peak temperatures, respectively. Heating rate was 10 K/min. The optical microscope photograph for the base glass is included.

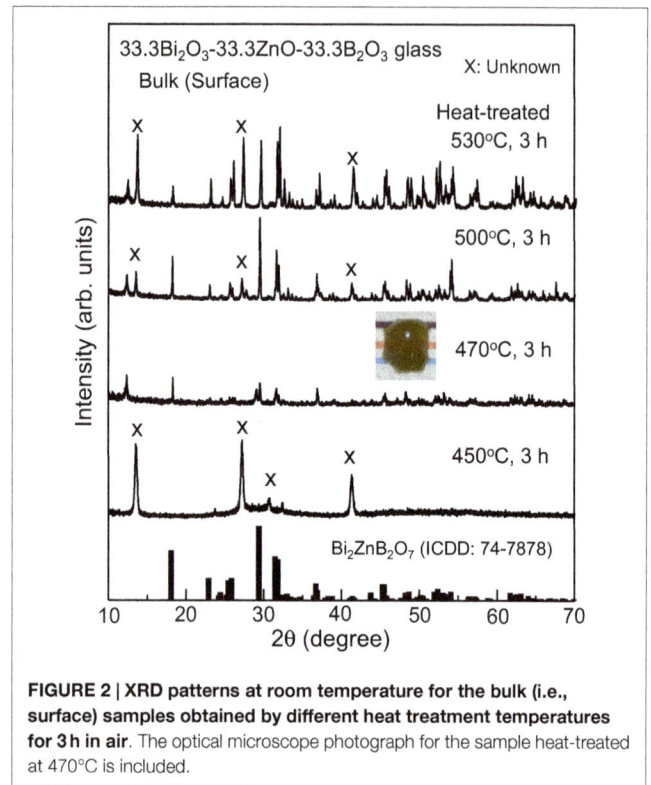

FIGURE 2 | XRD patterns at room temperature for the bulk (i.e., surface) samples obtained by different heat treatment temperatures for 3 h in air. The optical microscope photograph for the sample heat-treated at 470°C is included.

Figure 2 shows the XRD patterns at room temperature for the 111BiZnB samples obtained by different heat treatment temperatures for 3 h in air. In these XRD measurements, the bulk samples were used without any pulverizing and thus information on the crystalline phase being present at and near the glass surface is obtained. As seen in **Figure 2**, the formation of $Bi_2ZnB_2O_7$ crystals is confirmed in the samples heat-treated at 470, 500, and 530°C. Another crystalline phase is formed initially, which has not been identified (i.e., unknown crystalline phase) in our study. It has been reported that the -BiB_3O_6 crystalline phase is formed initially in the crystallization of 111BiZnB glass (Shanmugavelu and Kumar, 2012). In **Figure 2**, the optical microscope photograph for the sample heat-treated at 470°C is included. It is seen that the surface of the bulk sample is rounded. This behavior indicates that the softening/deformation of samples and crystallization occur almost simultaneously during heat treatments in 111BiZnB glass (Kanenishi et al., 2012). The high thermal stability against the crystallization in 111BiZnB glass, i.e., $\Delta T = T_p - T_g = 562°C - 395°C = 167°C$, would be one of the reasons for the sample deformation during heat treatments. As seen in **Figure 2**, the formation of unknown crystalline phase is detected in the sample heat-treated at 450°C from the XRD pattern; but in the sample heat-treated at 470°C, the formation of such an unknown crystalline phase is not confirmed. The crystallization at the glass surface is influenced by various factors such as surface quality, tips, cracks and scratches, foreign particles, and surrounding atmosphere in a given glass, i.e., the surface state (Müller et al., 2000). The shape of the sample heat-treated at 470°C was changed, i.e., the rounded shape (see **Figure 2**). This means that the surface state is largely changed in the heat treatment at 470°C, and thus the surface crystallization behavior would be influenced, providing the formation behavior of the unknown crystalline phase.

Figure 3 shows the XRD patterns at room temperature for the powdered samples, in which bulk plate-shape glasses were heat-treated and the samples were pulverized after heat treatments. **Figure 3** provides, therefore, information mainly on the crystalline phase present in the interior of samples. All XRD peaks in **Figure 3** are assigned to the $Bi_2ZnB_2O_7$ crystalline phase (ICDD: 74–7878), indicating that only the $Bi_2ZnB_2O_7$ crystalline phase is formed in all samples heat-treated at 450–530°C for 3 h in air. SHGs were confirmed in the crystallized samples (data are not shown here), indicating that the $Bi_2ZnB_2O_7$ crystalline phase formed in 111BiZnB glass exhibits the second-order optical non-linearity. **Figure 4** shows the Raman scattering spectra at room temperature for the base glass and crystallized (530°C, 3 h) samples. The glass shows very broad peaks being typical for glassy materials in the Raman scattering spectrum, i.e., at $\sim350, \sim600, \sim920, 1150 \sim 1400$, and $\sim1550 \, cm^{-1}$. From the Raman scattering spectra for various Bi_2O_3–ZnO–B_2O_3 glasses such as $30Bi_2O_3$-$30ZnO$-$40B_2O_3$, the formation of B–O–Bi and B–O–Zn bridging bonds in the glass structure has been proposed (Inoue et al., 2010). The broad peaks observed in **Figure 4** would be assigned as follows: (1) the band at $350 \, cm^{-1}$ can be connected with a symmetric stretching motion in Bi–O–Bi bonds between BiO_6 groups; (2) the band at $600 \, cm^{-1}$ could be assigned to symmetric stretching motions in B–O–Bi and B–O–Zn bonds; (3) the band at $920 \, cm^{-1}$ can be related mainly to asymmetric stretching motions in B–O–Bi and B–O–Zn bonds; (4) the bands at $1150 \sim 1400$ and $\sim1550 \, cm^{-1}$ can be assigned to an asymmetric stretching vibration of B–O bonds in BO_3 groups. On the other hand, the crystallized sample shows several sharp peaks at 347, 382, 497, 537, 577, 632, 699, and 741 cm^{-1} together

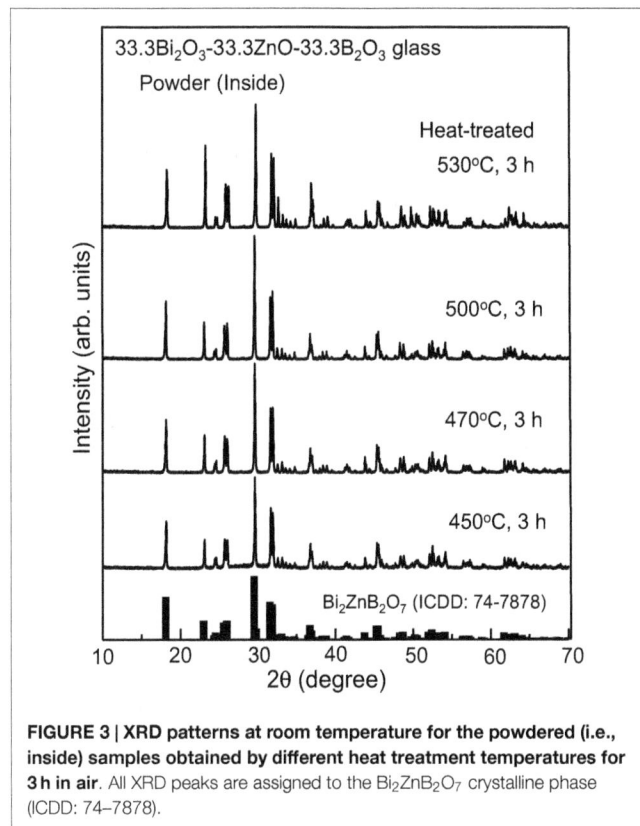

FIGURE 3 | XRD patterns at room temperature for the powdered (i.e., inside) samples obtained by different heat treatment temperatures for **3 h in air.** All XRD peaks are assigned to the $Bi_2ZnB_2O_7$ crystalline phase (ICDD: 74–7878).

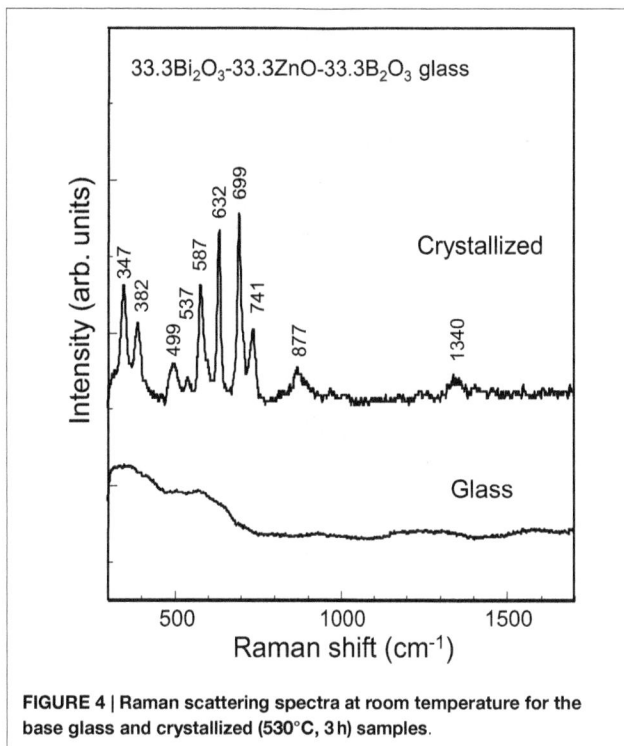

FIGURE 4 | Raman scattering spectra at room temperature for the base glass and crystallized (530°C, 3 h) samples.

with broad and weak peaks at 877, 1243, 1340, and 1548 cm^{-1}. Because $Bi_2ZnB_2O_7$ crystals are formed in the crystallized sample, it is obvious that the Raman scattering spectrum shown in **Figure 4** is for the $Bi_2ZnB_2O_7$ crystalline phase. In $Bi_2ZnB_2O_7$ with an orthorhombic structure ($a = 1.0827$ nm, $b = 1.10329$ nm, and $c = 0.48848$ nm) (Barbier et al., 2005), octahedral BiO_6 units, tetrahedral ZnO units, BO_3 and BO_4 units are present. ZnO_4 units and BO_3/BO_4 units are connected through a corner-shared bonding, i.e., the presence of B–O–Zn bonds, and are creating layers. The $Bi_2ZnB_2O_7$ crystalline phase is, therefore, constructed from the stacking of BiO_6, ZnO_4, BO_3, and BO_4 units. At this moment, the peaks shown in **Figure 4** have not been assigned in detail. However, the Raman scattering spectrum for $Bi_2ZnB_2O_7$ crystals would be used for the confirmation of the presence of $Bi_2ZnB_2O_7$ crystals in laser-patterned lines.

Laser Pattering of $Bi_2ZnB_2O_7$ Crystal Lines

In the laser-induced crystallization technique proposed and applied by our group (Komatsu et al., 2007; Komatsu and Honma, 2013), some amounts of rare-earth ions such as Sm^{3+} or transition metal ions such as Ni^{2+} must be included into the base glass for the absorption of laser. In this study, the glasses with the compositions of xSm_2O_3-$(33.3-x)Bi_2O_3$-$33.3ZnO$-$33.3B_2O_3$ ($x = 1, 3$, and 5) were prepared. The melt-quenched samples with these compositions were confirmed to be glasses from XRD and DTA measurements.

For the glass with $1Sm_2O_3$ ($x = 1$), it was found that laser irradiations with different conditions such as the laser powers of $P = 1.3$–1.5 W and the laser scanning speed of $S = 1.5$ μm/s

did not induce any crystallization. Furthermore, for the glass with $5Sm_2O_3$ ($x = 5$), it was difficult to pattern crystal lines with a homogeneous morphology. Here, therefore, the results for the glass with $3Sm_2O_3$ ($x = 3$), i.e., $3Sm_2O_3$-$30.3Bi_2O_3$-$33.3ZnO$-$33.3B_2O_3$ (designated here as 3Sm-111BiZnB glass) are presented. The 3Sm-111BiZnB glass has the values of $T_g = 410°C$ and $T_p = 553°C$. Because of the substitution of Sm_2O_3 for Bi_2O_3, the glass transition temperature is becoming high and the crystallization temperature is low compared with the 111BiZnB glass ($T_g = 395°C$ and $T_p = 562°C$) with no Sm_2O_3. The value of the optical absorption coefficient (α) at $\lambda = 1080$ nm (Yb:YVO_4 laser) was $\alpha = 10.5 cm^{-1}$ for 3Sm-111BiZnB glass. The reflection of light at the glass surface calculated using the Fresnel's equation was considered in the estimation of α (i.e., Lambert–Beer law).

Figure 5 shows the polarized optical microphotographs (POMs) for the samples obtained by laser irradiations. The laser scanning speed was fixed to $S = 1.5$ μm/s, and the laser power was changed in the range of $P = 1.30$–1.50 W. The laser was focused at the surface (not in the inside). The lines patterned with $P = 1.30$–1.45 W show a homogeneous distribution in color along the laser scanning direction, indicating that a stable structural change is taking place under these laser irradiation conditions. The morphology of the line patterned with $P = 1.50$ W is not homogeneous. **Figure 6** shows the micro-Raman scattering spectra at room temperature for the lines patterned by laser irradiations with $P = 1.35$ and 1.50 W. In both lines, several sharp peaks are observed and the positions of these peaks are almost similar to the $Bi_2ZnB_2O_7$ crystalline phase (**Figure 4**), indicating that the lines patterned with $P = 1.35$ and 1.50 W consist of $Bi_2ZnB_2O_7$ crystals. In **Figure 6**, the SHG microscopy images of these lines are presented, indicating that SHG emissions are founded to be

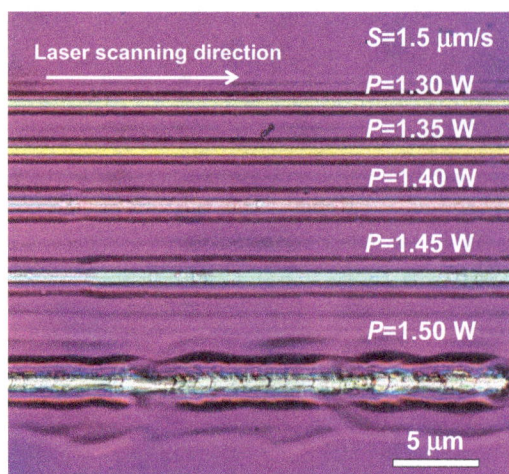

FIGURE 5 | Polarized optical microphotographs for the samples obtained by laser irradiations with the laser powers of $P = 1.3$–1.5 W and a laser scanning speed of $S = 1.5 \mu$m/s.

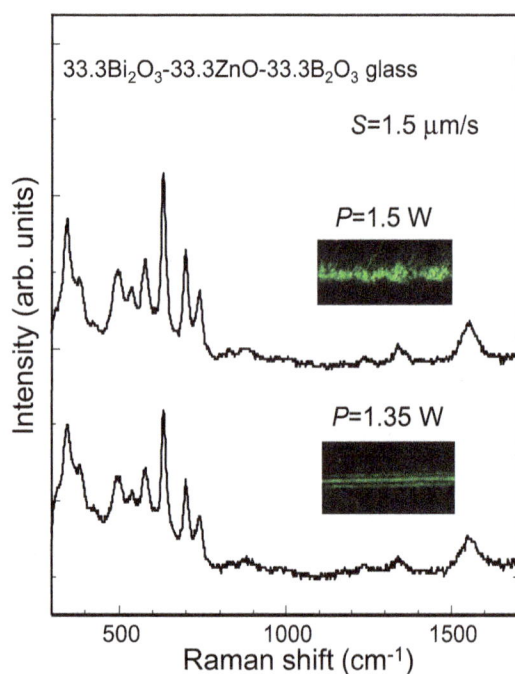

FIGURE 6 | Micro-Raman scattering spectra at room temperature for the lines patterned by laser irradiations with $P = 1.35$ and 1.50 W.

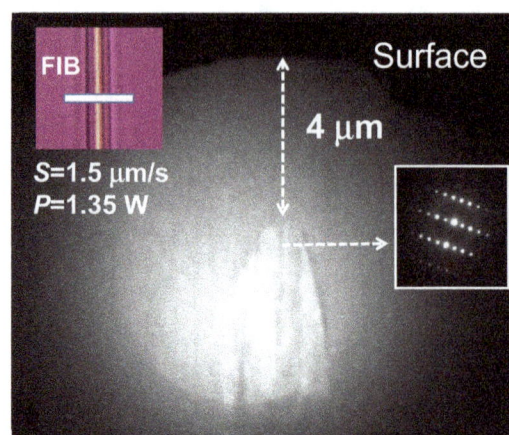

FIGURE 7 | A bright field image and selected area electron diffraction (SAED) pattern in TEM observations for the cross-section of the line patterned by laser irradiations with $S = 1.5 \mu$m/s and $P = 1.35$ W.

caused by only the crystal lines. Similar results were obtained for other lines patterned with $P = 1.30$, 1.40, and 1.45 W.

In order to examine the morphology and orientation of $Bi_2ZnB_2O_7$ crystals in the line patterned by laser irradiations with $S = 1.5 \mu$m/s and $P = 1.35$ W by using TEM observations, the line part was processed to a thin foil (~ 100 nm) shape using a FIB method. **Figure 7** shows the bright field image and selected area electron diffraction (SAED) pattern in TEM observations for the cross-section of the line. Almost a homogeneous bright field image is observed over the whole crystallized part. In the

SAED, the diffraction spots are observed clearly. These results suggest that $Bi_2ZnB_2O_7$ crystals in the cross-section (i.e., in the line) are highly oriented, and the line is not constructed from the assembly of $Bi_2ZnB_2O_7$ polycrystals. In order to determine definitely the orientation direction/quality of $Bi_2ZnB_2O_7$ crystals in the lines patterned, further TEM observations are required (Fan et al., 2012). In the present study, the laser focal position was the surface. However, as seen in **Figure 7**, $Bi_2ZnB_2O_7$ crystals were formed in the inside of the glass, i.e., at the beneath of 4μm from the surface. Usually, if the laser focal position is the surface of a given glass, crystals are basically formed at the glass surface [e.g. Komatsu et al. (2007) and Komatsu and Honma (2013)]. The crystal growth in the inside of the glass (**Figure 7**) observed in this study might be closely related to the high thermal stability (i.e., $\Delta T = T_p - T_g = 167^{\circ}$C) against the crystallization in the glass of $3Sm_2O_3$-$30.3Bi_2O_3$-$33.3ZnO$-$33.3B_2O_3$. As shown in **Figures 2** and **3**, although unknown crystals are formed initially at the glass surface, the main crystalline phase formed in the inside is the $Bi_2ZnB_2O_7$ crystalline phase. This would be one of the reasons for the formation of $Bi_2ZnB_2O_7$ crystals in the lines patterned by laser irradiations. Recently, non-linear optical β-BaB_2O_4 (β-BBO) crystal lines have been successfully patterned in the inside (up to the beneath of $\sim 200 \mu$m) of $8Sm_2O_3$-$42BaO$-$50B_2O_3$ glass by laser irradiations, in which the laser focal position was moved gradually from the surface to the inside (Nishii et al., 2015). The present study suggests that the patterning of $Bi_2ZnB_2O_7$ crystal lines would be possible even in the inside of more deep beneath around 100μm for Bi_2O_3–ZnO–B_2O_3 glasses by changing the laser focal position.

Frontiers of Laser Pattering of Crystals in Bi_2O_3-Based Glasses

Nowadays, it has been well recognized that Bi_2O_3 oxide is one of the important constituents for the design and control of optical, electrical, and thermal properties of oxide glasses, and numerous studies on Bi_2O_3-based glasses have been studied so far (Dumbaugh, 1986; Maeder, 2013). The most attractive features of Bi_2O_3 oxide are a large electronic polarizability and a weak single bond strength of Bi–O bonds (Dimitrov and Komatsu, 1999, 2002, 2005, 2010; Komatsu et al., 2010). Indeed, glasses with Bi_2O_3 exhibit large third order non-linear optical susceptibilities $^{(3)}$ of the order 10^{-12} esu (Terashima et al., 1997). Recently, Bi_2O_3–ZnO–B_2O_3 glasses have received much attention (Kim et al., 2007; Doweidar and Saddeek, 2009; Inoue et al., 2010; Hashimoto et al., 2011), in particular, because they are low melting glasses with high thermal stability and have been used as new sealing glasses instead of PbO-containing low melting glasses. The crystallization of Bi_2O_3-based glasses has also received much attention, because some functional crystals containing Bi_2O_3 are formed through the crystallization. For instance, high-T_c Bi_2O_3-based copper superconducting glass-ceramics have been synthesized through the crystallization of glasses with Bi_2O_3 (Komatsu et al., 1988, 1990). Non-linear optical $RE_xBi_{1-x}BO_3$ crystals (RE: Gd and Sm) showing strong SHGs have been also synthesized from RE_2O_3–Bi_2O_3–B_2O_3 glasses (Honma et al., 2002; Ihara et al., 2004; Koshiba et al., 2007). The design and control of new Bi_2O_3-based glasses and their crystallization are at the frontiers of the glass science and technology.

In this study, we demonstrated that non-linear optical $Bi_2ZnB_2O_7$ crystal lines with a homogeneous morphology and a high orientation are patterned in $3Sm_2O_3$-$30.3Bi_2O_3$-$33.3ZnO$-$33.3B_2O_3$ glass for the first time. Crystallized glasses (glass-ceramics) are composites of the glassy phase and crystalline phase, i.e., glass/crystal hybrid materials. In particular, the combination of glass and single crystal line would have a high potential for device applications. It is emphasized that a new potential for optical device applications is added in Bi_2O_3-based glasses from the present study. A large single crystal and crystal fibers of $Bi_2ZnB_2O_7$ have been fabricated in order to develop new solid state lasers with high transparency in the ultra-violet region

(Kozhayam et al., 2013; Su et al., 2013). The laser patterning of $Bi_2ZnB_2O_7$ crystal lines in glasses is a new and innovative approach for the single crystal growth technique. Very recently, non-linear optical -BBO crystal lines with bending and curved shapes have been patterned at the glass surface, and it has been demonstrated from birefringence imaging measurements that highly c-axis oriented -BBO crystals grow along laser scanning direction even if the laser scanning direction changes (Ogawa et al., 2013). As further studies, the patterning of bending and curved $Bi_2ZnB_2O_7$ crystal lines is strongly required in Bi_2O_3–ZnO–B_2O_3 glasses.

Conclusion

Non-linear optical $Bi_2ZnB_2O_7$ crystal lines with a high orientation were patterned in $3Sm_2O_3$-$30.3Bi_2O_3$-$33.3ZnO$-$33.3B_2O_3$ glass by using a laser-induced crystallization technique for the first time, in which continuous-wave Yb:YVO_4 fiber lasers with a wavelength of 1080 nm were irradiated at the glass surface. It was confirmed from transmission electron microscope observations that crystals were formed in the inside of the glass, i.e., at the beneath of $4\,\mu m$ from the surface. A new potential for optical device applications was added in Bi_2O_3-based glasses from the present study.

Acknowledgments

This work was supported from the Grant-in-Aid for Scientific Research from the Ministry of Education, Science, Sports, Culture and Technology, Japan (No. 23246114) and by Program for High Reliable Materials Design and Manufacturing in Nagaoka University of Technology.

References

Barbier, J., Penin, N., and Cranswick, L. M. (2005). Melilite-type borates $Bi_2ZnB_2O_7$ and $CaBiGaB_2O_7$. *Chem. Mater.* 17, 3130–3136. doi:10.1021/cm0503073

Dimitrov, V., and Komatsu, T. (1999). Electronic polarizability, optical basicity and nonlinear optical properties of oxide glasses. *J. Non Cryst. Solids* 249, 160–179. doi:10.1016/S0022-3093(99)00317-8

Dimitrov, V., and Komatsu, T. (2002). Classification of simple oxides: a polarizability approach. *J. Solid State Chem.* 163, 100–112. doi:10.1006/jssc.2001.9378

Dimitrov, V., and Komatsu, T. (2005). Classification of oxide glasses: a polarizability approach. *J. Solid State Chem.* 178, 831–846. doi:10.1016/j.jssc.2004.12.013

Dimitrov, V., and Komatsu, T. (2010). Average single bond strength and optical basicity of Bi_2O_3-B_2O_3 and Sb_2O_3-B_2O_3 glasses. *J. Non Cryst. Solids* 356, 258–262. doi:10.1016/j.jnoncrysol.2009.11.014

Doweidar, H., and Saddeek, Y. B. (2009). FTIR and ultrasonic investigations on modified bismuth borate glasses. *J. Non Cryst. Solids* 355, 348–354. doi:10.1016/j.jnoncrysol.2008.12.008

Dumbaugh, W. H. (1986). Heavy metal oxide glasses containing Bi_2O_3. *Phys. Chem. Glasses* 27, 119–123.

Fan, C., Poumellec, B., Lancry, M., He, X., Zeng, H., Erraji-Chahid, A., et al. (2012). Three-dimensional photoprecipitation of oriented $LiNbO_3$-like crystals in silica-based glass with femtosecond laser irradiation. *Opt. Lett.* 37, 2955–2957. doi:10.1364/OL.37.002955

Fujiwara, T., Sawada, T., Honma, T., Benino, Y., Komatsu, T., Takahashi, M., et al. (2003). Origin of intrinsic second-harmonic generation in crystallized GeO_2-SiO_2 glass films. *Jpn. J. Appl. Phys.* 42, 7326–7330. doi:10.1143/JJAP.42.7326

Hashimoto, T., Shimada, Y., Nasu, H., and Ishida, A. (2011). ZnO-Bi_2O_3-B_2O_3 glasses as molding glasses with high refractive indices and low coloration codes. *J. Am. Ceram. Soc.* 94, 2061–2066. doi:10.1111/j.1551-2916.2010.04383.x

Honma, T., Benino, Y., Fujiwara, T., Sato, R., and Komatsu, T. (2002). Optical nonlinear crystalline dot and line patterning in samarium bismuth borate glasses by YAG laser irradiation. *J. Ceram. Soc. Jpn.* 110, 398–402. doi:10.2109/jcersj.110.398

Honma, T., Benino, Y., Fujiwara, T., Komatsu, T., and Sato, R. (2003). Nonlinear optical crystal line writing in glass by yttrium aluminium garnet laser irradiation. *Appl. Phys. Lett.* 82, 892–894. doi:10.1063/1.1544059

Honma, T., Benino, Y., Fujiwara, T., and Komatsu, T. (2006). Transition metal atom heat processing for writing of crystal lines in glass. *Appl. Phys. Lett.* 88, 231105. doi:10.1063/1.2212272

Honma, T., and Komatsu, T. (2010). Patterning of two-dimensional planar lithium niobate architectures on glass surface by laser scanning. *Opt. Express* 18, 8019–8024. doi:10.1364/OE.18.008019

Honma, T., Nguyen, P. T., and Komatsu, T. (2008). Crystal growth behavior in CuO-doped lithium disilicate glasses by continuous-wave fiber laser irradiation. *J. Ceram. Soc. Jpn.* 116, 1314–1318. doi:10.2109/jcersj2.116.1314

Ihara, R., Honma, T., Benino, Y., and Komatsu, T. (2004). Second-order optical nonlinearities of metastable $BiBO_3$ phases in crystallized glasses. *Opt. Mater.* 27, 403–408. doi:10.1016/j.optmat.2004.07.002

Inoue, T., Honma, T., Dimitrov, V., and Komatsu, T. (2010). Approach to thermal properties and electronic polarizability from average single bond strength in ZnO-Bi_2O_3-B_2O_3 glasses. *J. Solid State Chem.* 183, 3078–3085. doi:10.1016/j.jssc.2010.10.027

Kanenishi, K., Sakida, S., Benino, Y., and Nanba, T. (2012). Surface crystallization of stoichiometric glass with $Bi_2ZnB_2O_7$ crystal using ultrasonic surface treatment followed by heat treatment. *J. Ceram. Soc. Jpn.* 120, 509–512. doi:10.2109/jcersj2. 120.509

Kim, B. S., Lim, E. S., Lee, J. H., and Kim, J. J. (2007). Effect of structure change on thermal and dielectric characteristics in low-temperature firing Bi_2O_3-B_2O_3-ZnO glasses. *J. Mater. Sci.* 42, 4260–4264. doi:10.1007/s10853-006-0638-y

Komatsu, T., Sato, R., Imai, K., Matusita, K., and Yamashita, T. (1988). High-T_c superconducting glass ceramics based on the Bi-Ca-Sr-Cu-O system. *Jpn. J. Appl. Phys.* 27, L550–L552. doi:10.1143/JJAP.27.L533

Komatsu, T., Hirose, C., Ohki, T., Sato, R., and Matusita, K. (1990). Preparation of Ag-coated superconducting $Bi_2Sr_2CaCu_2O_x$ glass ceramic fibers. *Appl. Phys. Lett.* 57, 183–185. doi:10.1063/1.103978

Komatsu, T., and Honma, T. (2013). Laser patterning and characterization of optical active crystals in glasses (review). *J. Asian Ceram. Soc.* 1, 9–16. doi:10.1016/j. jascer.2013.02.006

Komatsu, T., Ihara, R., Honma, T., Benino, Y., Sato, R., Kim, H. G., et al. (2007). Patterning of nonlinear optical crystals in glass by laser-induced crystallization. *J. Am. Ceram. Soc.* 90, 699–705. doi:10.1111/j.1551-2916.2006.01441.x

Komatsu, T., Ito, N., Honma, T., and Dimitrov, V. (2010). Electronic polarizability and its temperature dependence of Bi_2O_3-B_2O_3 glasses. *J. Non Cryst. Solids* 356, 2310–2314. doi:10.1016/j.jnoncrysol.2010.03.041

Komatsu, T., Koshiba, K., and Honma, T. (2011). Preferential growth orientation of laser-patterned $LiNbO_3$ crystals in lithium niobium silicate glass. *J. Solid State Chem.* 184, 411–418. doi:10.1016/j.jssc.2010.12.016

Koshiba, K., Honma, T., Benino, Y., and Komatsu, T. (2007). Patterning and morphology of nonlinear optical $Gd_xBi_{1-x}BO_3$ crystals in Gd_2O_3-Bi_2O_3-B_2O_3 glasses by YAG laser irradiation. *Appl. Phys. A* 89, 981–986. doi:10.1007/ s00339-007-4200-9

Kozhayam, N., Ferriol, M., Cochez, M., and Aillerie, M. (2013). Growth and characterization of bismuth zinc borate $Bi_2ZnB_2O_7$ crystal fibers by the micro-pulling down technique. *J. Cryst. Growth* 364, 51–56. doi:10.1016/j.jcrysgro.2012.11.059

Liu, Z., Zeng, H., Ji, X., Ren, J., Chen, G., Ye, J., et al. (2015). Formation of $Bi_2ZnB_2O_7$ nanocrystals in ZnO-Bi_2O_3-B_2O_3 glass induced by femtosecond laser. *J. Am. Ceram. Soc.* 98, 408–412. doi:10.1111/jace.13330

Maeder, T. (2013). Review of Bi_2O_3 based glasses for electronics and related applications. *Int. Mater. Rev.* 58, 3–40. doi:10.1179/1743280412Y.0000000010

Majhi, K., and Varma, K. B. R. (2008a). Structural, dielectric and optical properties of transparent glasses and glass-ceramics of $SrBi_2B_2O_7$. *J. Non Cryst. Solids* 354, 4543–4549. doi:10.1016/j.jnoncrysol.2008.06.010

Majhi, K., and Varma, K. B. R. (2008b). Structural, dielectric and optical properties of $CaBi_2B_2O_7$ glasses and glass-nanocrystal composites. *Mater. Chem. Phys.* 117, 494–499. doi:10.1016/j.matchemphys.2009.06.044

Müller, R., Zanotto, E. D., and Fokin, V. M. (2000). Surface crystallization of silicate glasses: nucleation sites and kinetics. *J. Non Cryst. Solids* 274, 208–231. doi:10.1016/S0022-3093(00)00214-3

Nishii, A., Shinozaki, K., Honma, T., and Komatsu, T. (2015). Morphology and orientation of α-BaB_2O_4 crystals patterned by laser in the inside of samarium

barium borate glass. *J. Solid State Chem.* 221, 145–151. doi:10.1016/j.jssc.2014. 09.031

Ogawa, K., Honma, T., and Komatsu, T. (2013). Birefringence imaging and orientation of laser patterned β-BaB_2O_4 crystals with bending and curved shapes in glass. *J. Solid State Chem.* 207, 6–12. doi:10.1016/j.jssc.2013.08.021

Sato, R., Benino, Y., Fujiwara, T., and Komatsu, T. (2001). YAG laser-induced crystalline dot patterning in samarium tellurite glasses. *J. Non Cryst. Solids* 289, 228–232. doi:10.1016/S0022-3093(01)00736-0

Shanmugavelu, B., and Kumar, V. V. R. K. (2012). Crystallization kinetics and phase transformation of bismuth zinc borate glass. *J. Am. Ceram. Soc.* 95, 2891–2898. doi:10.1111/j.1551-2916.2012.05342.x

Shinozaki, K., Honma, T., and Komatsu, T. (2012). New oxyfluoride glass with high fluorine content and laser patterning of nonlinear optical $BaAlBO_3F_2$ single crystal line. *J. Appl. Phys.* 112, 093506. doi:10.1063/1.4764326

Shinozaki, K., Noji, A., Honma, T., and Komatsu, T. (2013). Morphology and photoluminescence properties of Er^{3+}-doped CaF_2 nanocrystals patterned by laser irradiation in oxyfluoride glasses. *J. Fluor. Chem.* 145, 81–87. doi:10.1016/ j.jfluchem.2012.10.007

Su, X., Wang, Y., Yang, Z., Huang, X. C., Pan, S., Li, F., et al. (2013). Experimental and theoretical studies on the linear and nonlinear optical properties of $Bi_2ZnOB_2O_6$. *J. Phys. Chem. C.* 117, 14149–14157. doi:10.1021/jp4013448

Suzuki, F., Honma, T., and Komatsu, T. (2010). Origin of periodic domain structure in Er^{3+}-doped β'-$(Sm,Gd)_2(MoO_4)_3$ crystal lines patterned by laser irradiations in glasses. *J. Solid State Chem.* 183, 909–914. doi:10.1016/j.jssc.2010. 02.007

Terashima, K., Shimoto, T., and Yoko, T. (1997). Structure and nonlinear optical properties of PbO-Bi_2O_3-B_2O_3 glasses. *Phys. Chem. Glasses* 38, 211–217.

Tsukada, Y., Honma, T., and Komatsu, T. (2009). Self-organized periodic domain structure for second harmonic generation in ferroelastic β'-$(Sm,Gd)_2(MoO_4)_3$ crystal lines on glass surfaces. *Appl. Phys. Lett.* 94, 059901. doi:10.1063/1. 3076080

Zeng, H., Liu, Z., Jiang, Q., Li, B., Yang, C., Shang, Z., et al. (2014). Large third-order optical nonlinearity of ZnO-Bi_2O_3-B_2O_3 glass-ceramic containing $Bi_2ZnB_2O_7$ nanocrystals. *J. Euro. Ceram. Soc.* 34, 4383–4388. doi:10.1016/j.jeurceramsoc. 2014.06.031

Conflict of Interest Statement: The authors declare that the research was conducted in the absence of any commercial or financial relationships that could be construed as a potential conflict of interest.

The interaction of C_{60} on Si(111) 7×7 studied by supersonic molecular beams: interplay between precursor kinetic energy and substrate temperature in surface activated processes

Lucrezia Aversa[1]*, Simone Taioli[2,3,4], Marco Vittorio Nardi[5], Roberta Tatti[1], Roberto Verucchi[1] and Salvatore Iannotta[6]

[1] *Institute of Materials for Electronics and Magnetism (IMEM), National Research Council (CNR), Trento, Italy,* [2] *European Centre for Theoretical Studies in Nuclear Physics and Related Areas (ECT*), Bruno Kessler Foundation (FBK), Trento, Italy,* [3] *Trento Institute for Fundamental Physics and Applications (TIFPA), National Institute for Nuclear Physics (INFN), Trento, Italy,* [4] *Faculty of Mathematics and Physics, Charles University, Prague, Czech Republic,* [5] *Institut fur Physik, Humboldt-Universitat zu Berlin, Berlin, Germany,* [6] *Institute of Materials for Electronics and Magnetism (IMEM), National Research Council (CNR), Parma, Italy*

Edited by:
Alberto Corigliano,
Politecnico di Milano, Italy

Reviewed by:
Massimiliano Zingales,
Università degli Studi di Palermo, Italy
Shangchao Lin,
Florida State University, USA

*Correspondence:
Lucrezia Aversa,
Institute of Materials for Electronics and Magnetism (IMEM), CNR, Via Alla Cascata 56/C, Trento 38123, Italy
lucrezia.aversa@cnr.it

Buckminsterfullerene (C_{60}) is a molecule fully formed of carbon that can be used, owing to its electronic and mechanical properties, as "clean" precursor for the growth of carbon-based materials, ranging from π-conjugated systems (graphenes) to synthesized species, e.g., carbides such as silicon carbide (*SiC*). To this goal, C_{60} cage rupture is the main physical process that triggers material growth. Cage breaking can be obtained either thermally by heating up the substrate to high temperatures (630°C), after C_{60} physisorption, or kinetically by using supersonic molecular beam epitaxy techniques. In this work, aiming at demonstrating the growth of *SiC* thin films by C_{60} supersonic beams, we present the experimental investigation of C_{60} impacts on Si(111) 7×7 kept at 500°C for translational kinetic energies (KEs) ranging from 18 to 30 eV. The attained kinetically activated synthesis of *SiC* submonolayer films is probed by *in situ* surface electron spectroscopies (X-ray photoelectron spectroscopy and ultraviolet photoelectron spectroscopy). Furthermore, in these experimental conditions, the C_{60}-Si(111) 7×7 collision has been studied by computer simulations based on a tight-binding approximation to density-functional theory. Our theoretical and experimental findings point toward a kinetically driven growth of *SiC* on Si, where C_{60} precursor KE plays a crucial role, while temperature is relevant only after cage rupture to enhance Si and carbon reactivity. In particular, we observe a counterintuitive effect in which for low KE (below 22 eV), C_{60} bounces back without breaking more effectively at high temperature due to energy transfer from excited phonons. At higher KE ($22 < K < 30$ eV), for which cage rupture occurs, temperature enhances reactivity without playing a major role in the cage break. These results are in good agreement with *ab initio* molecular dynamics simulations. Supersonic molecular beam epitaxy is thus a technique able to drive material growth at low-temperature regime.

Keywords: fullerene, silicon carbide, thin film, surface dynamics, first-principle simulations

Introduction

The synthesis of carbon-based thin films and nanostructured compounds, such as carbides and graphene, on top of semiconductor or metal surfaces represents a serious challenge to the production of electronic devices and materials coating. Several methods, mainly based on the chemical vapor deposition (CVD) of organic molecules or molecular beam epitaxy (MBE), have been successfully developed to obtain large area carbon-based materials with the common feature of working at very high temperatures. For example, the low energy deposition of fullerene on silicon has been deeply studied with experimental approaches, mainly using standard techniques (Sakamoto et al., 1999; De Seta et al., 2000; Sanvitto et al., 2000; Balooch and Hamza, 1993). MBE, in particular, has shown to be a viable approach to silicon carbide (3C–SiC) synthesis at about 800°C, using fullerene (C_{60}) as carbon precursor and silicon as a growth substrate (Sanvitto et al., 2000; De Seta et al., 2000). The required high temperature, however, leads to structural defects at the nano- and micro-scale, mainly due to high lattice mismatch between SiC and Si and due to increasing thermal diffusion through the SiC film. Moreover, only amorphous Si films can be achieved at 800°C, and postdeposition thermal treatments at higher temperatures are necessary to improve crystallinity.

At room temperature, the most common process at the fullerene–surface interface, be it metallic or semiconductor, is represented by physisorption. Chemisorption is a viable way only on reconstructed silicon surfaces (both Si(111) 7×7 and Si(100) 2×1) due to the presence of several dangling bonds on the surface, the most reactive for fullerenes being the 7×7 reconstruction of Si(111). Furthermore, to induce cage breaking on the top of the surface, one needs to increase the substrate temperature. Finally, subsequent to rupture, the most common outcome on Si(111) 7×7 surfaces results in the formation of several new Si–C bonds up to the synthesis of silicon carbide (SiC) at 800°C.

To overcome the temperature issues, in a previous work (Verucchi et al., 2012), we reported the room temperature (about 20°C) synthesis of nanocrystalline 3C–SiC with fullerene on Si(111) 7×7 surface by the so-called supersonic molecular beam epitaxy (SuMBE) technique. Using this approach, the C_{60} translational kinetic energy (KE) can reach values up to 30–35 eV by aerodynamic acceleration due to isoentropic expansion in vacuum. We demonstrated that the cage rupture after impact with the silicon surface is driven by charge excitations to the excited levels of the system and the 3C–SiC synthesis is activated by the extra precursor KE content for both high and low (room T) substrate temperatures (Verucchi et al., 2002; Verucchi et al., 2012). At room temperature, where thermal contribution to observed processes is negligible, out-of-thermal-equilibrium conditions achieved by SuMBE induce chemical processes at the silicon surfaces, overcoming activation barriers toward the synthesis of ordered SiC. Moreover, by using C_{60} supersonic beams at 20 eV, we have shown that a polycrystalline 3C–SiC thick layer can be grown on Si(111) 7×7 at 750°C (Verucchi et al., 2002). This is due to the unquestionable ability of the high fullerene KE to induce cage breaking, synthesis of carbide as well as implement film structural order also in conditions where influence of temperature (and thus thermal equilibrium conditions) is significant.

In order to better identify the role played by temperature on SiC growth on Si by SuMBE, in this work, we present the experimental and *ab initio* investigations of supersonic C_{60} impacts on Si(111) 7×7 at temperature of 500°C, where MBE approach is not able to induce SiC formation (De Seta et al., 2000). This regime is intermediate between the room temperature experiments, which gave evidence of nanocrystalline SiC island formation, and the full synthesis regime at 750–800°C, where all the carbon precursors were transformed into an ordered uniform layer. Synthesis of SiC by SuMBE is thus expected also at 500°C, but interplay between fullerene KE and substrate temperature can be better addressed.

Using photoelectron spectroscopy and low-energy electron diffraction (LEED), we explore the chemical, structural, and electronic changes induced by varying the supersonic C_{60} beam KE at fixed substrate temperature. Experimental results point toward a SiC growth model driven by fullerene KE and where, surprisingly, temperature effects do not affect critically the cage rupture. Furthermore, aiming at modeling the interaction between C_{60} and Si surface for several initial KEs and different temperatures, in this work, we also show first-principle simulations based on the density-functional tight-binding (DFTB) method (Matthew et al., 1989; Elstner et al., 1998; Frauenheim et al., 2002) within the Born-Oppenheimer approximation. Classical molecular dynamics simulations of C_{60} impacts on Si have been already carried out by Averback (Hu et al., 2000) using a lower accuracy description of the atomic interaction through a Tersoff potential (Tersoff, 1988). However, classical molecular dynamics is not suitable for reactive problems out of thermal equilibrium, where bonds break and form as in our case. At variance, a DFTB-based approach has been recently applied to study high KE impact dynamics of fullerenes on graphite (Galli and Mauri, 1994) and room temperature growth of SiC (Taioli et al., 2013), displaying an accuracy comparable to density-functional theory (DFT).

Gaining insights on the chemical–physical mechanisms involved in SiC epitaxy by SuMBE and understanding their dependence on externally tunable parameters, such as the C_{60} beam KE and the substrate temperature, is at the basis to control and improve SiC synthesis even at room temperature and to transfer this technique to produce other materials (Taioli, 2014). Indeed, the study of the C_{60} cage breaking dynamics on metallic or semiconductor surfaces can reveal new paths toward the epitaxial growth of materials by our SuMBE approach directly on substrates already employed in the electronic device production, thus suitable for very large scale integration production.

Materials and Methods

Experiments have been performed at the IMEM-CNR laboratory in Trento in a tailored deposition *in situ* characterization facility operating in ultra-high vacuum (UHV). A detailed description of the SuMBE deposition technique can be found in the study by Milani and Iannotta (1999). On the computational side, DFTB calculations have been performed using the DFTB + code (Aradi et al., 2007; Elstner et al., 1998) within the self-consistent charge framework that leads to an improved description of the Coulomb interaction between atomic partial charges.

Surface and Structural Characterization

Chemical and electronic surface properties have been probed by means of X-ray photoelectron spectroscopy (XPS) and ultraviolet photoelectron spectroscopy (UPS). The UHV chamber is equipped with a CLAM2 Electron Hemisperical Analyzer, an MgK X-ray source, and a helium discharge lamp, thus the excitation photon for XPS is at 1253.6 eV and for UPS the HeI at 21.22 eV. The total resolution allowed by the analyzer is 0.95 eV for XPS and 0.1 eV for UPS. Spectra were acquired at normal electron acceptance geometry. Core-level binding energies (BEs) have been calculated using as a reference the Au surface after sputter cleaning, i.e., the 4f 7/2 Au level at 84.0 eV, while in UPS the spectra are referred to the Fermi level of a gold foil in electric contact with the sample. Core levels have been analyzed by Voigt lineshape deconvolution after background subtraction by a Shirley function (Hüfner, 1995). The typical precision for peak energy positioning is ± 0.05 eV, uncertainty for full width at half maximum (FWHM) is less than 5% and for area evaluation is about 2.5%. Surface structural characterization has been performed by LEED operated at 50 eV.

Thin Film Deposition

Substrates for C_{60} deposition were obtained from a Si(111) wafer (resistivity 1.20×10^{-4} m) cleaned by a modified Shiraki procedure (Verucchi et al., 2002). The silicon oxide film was then removed in vacuum by several annealing cycles, until a complete Si(111) 7×7 surface reconstruction. The entire procedure was checked by analyzing LEED images, and no traces of oxygen and carbon were detected from XPS analysis. Submonolayer (sub-ML) C_{60} films were deposited on clean and reconstructed Si(111) 7×7 surfaces kept at 500°C, using different precursor KE ranging from 18 to 30 eV. The deposition times were adjusted in order to reach a similar thickness for all the films for ease of comparison, with an average coverage of 0.5 ML ($\pm$ 0.1 ML). The C_{60} supersonic source is made of two coaxial quartz capillary tubes, resistively heated by a tantalum foil. The experiments were carried out using a free-jet expansions of H_2 in which the seeded fullerene particles are highly diluted (about 10^{-3} in concentration) and can reach a KE up to 30 eV, with an average growth rate on silicon of about 0.1 Å/min having the supersonic beam perpendicular to the substrate. Fullerene KEs of about 18, 22, 25, and 30 eV have been employed, as measured from time of flight analysis (Verucchi et al., 2002).

Theory

DFTB is based on a second-order expansion of the density appearing in the full DFT approach to electronic structure calculations. In this approach, the energy of a system of atoms is expressed as a sum of tight-binding-like matrix elements, a Coulomb interaction, and a repulsive pair-potential. The parameters appearing in the formulae expressing these contributions are evaluated using high-level electronic structure methods and are highly transferable to different physical and chemical environments (Slater–Koster parameters).

The advantage of using DFTB with respect to other *ab initio* methods, such as DFT, is due to the computational cost of this approach to electronic structure that is about two orders of magnitude cheaper than the corresponding full DFT calculation (Garberoglio and Taioli, 2012). As a result of this substantial speed gain, DFTB may be used (i) to investigate much larger systems than those accessible by DFT, (ii) to follow their dynamics for much longer timescales, and (iii) makes it possible to perform several tests by tuning the main parameters affecting the impact at an affordable computational cost.

In our simulations, the Si substrate is represented by eight silicon bilayers cleaved along the (111) direction.

According to the dimer adatom stacking fault model (DAS) (Takayanagi et al., 1985), only the first silicon bilayer participates into the superficial reconstruction, resulting in a 7×7 reconstructed surface. This reconstructed surface, represented in **Figure 1**, turns out to be optimal for our applications, showing high reactivity due to the presence of 19 dangling bonds on the surface (12 adatoms, six rest atoms, and one atom at the center of the hole in the unitary cell) and to its metallic character (Cepek et al., 1999). The computational supercell used in the impact calculations is hexagonal and, after optimization of both atomic positions and lattice vectors, measures 23.21 Å along the short diagonal and 26.79 Å along the large one. Fullerene was separately optimized and initially placed at 5 Å distance on the top of the silicon substrate and the cell size along the collision direction, perpendicular to the surface plane, was set to 50 Å to avoid spurious interaction among periodic images. The Brillouin zone was sampled at the Γ-point only, due to the large number of atoms in the unitary cell (848). Atomic interactions between chemical species were treated by the semirelativistic, self-consistent charge Slater–Koster parameter set "matsci-0-3" (Frenzel et al., 2009). To help and enhance convergence of band structure calculations, we employed a room temperature Fermi smearing of the electronic density. Molecular dynamics simulations were performed in the micro-canonical ensemble (NVE), setting the time step to 1 fs to enforce total energy conservation and each simulation lasted 2 ps.

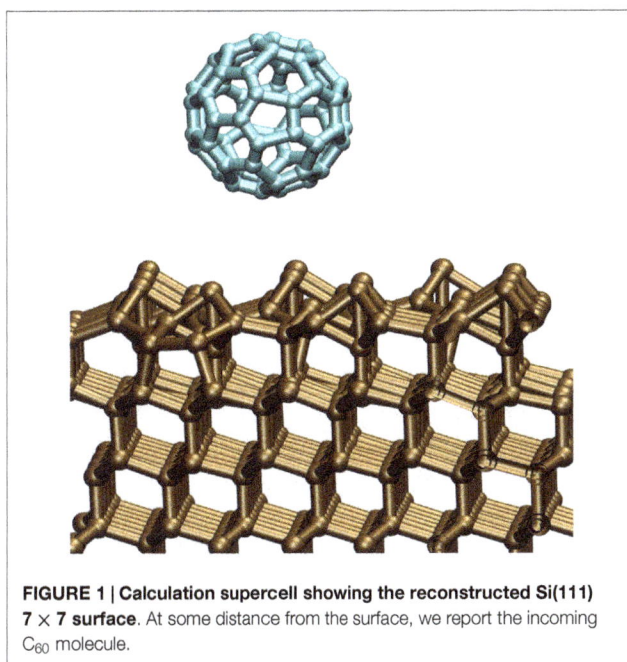

FIGURE 1 | Calculation supercell showing the reconstructed Si(111) 7 × 7 surface. At some distance from the surface, we report the incoming C_{60} molecule.

Finally, the last silicon bilayer was kept fixed during the dynamic evolution to simulate the presence of a silicon bulk.

As a second step of our analysis, we increased the substrate temperature up to 500°C to gain some insight on the role of temperature on SiC growth obtained via SuMBE. In particular, our goal was to find out the temperature effect on C_{60} cage mechanical stability.

Results

Photoemission

High-resolution C1s core levels for thin films deposited at increasing beam KE's are shown in **Figure 2**. They are all characterized by the presence of a main peak that is not fully symmetric but has a weak shoulder at lower BEs whose intensity increases significantly with beam KE. The lineshape analysis of core levels has been carried out by introducing the main component at 285.2 eV (FWHM of 1.1 eV) and up to two more features at low BE (with fixed FWHM of 1.0 and 0.9 eV), in particular at 284.1 and 283.2 eV. Peak positions do not change as the fullerene KE

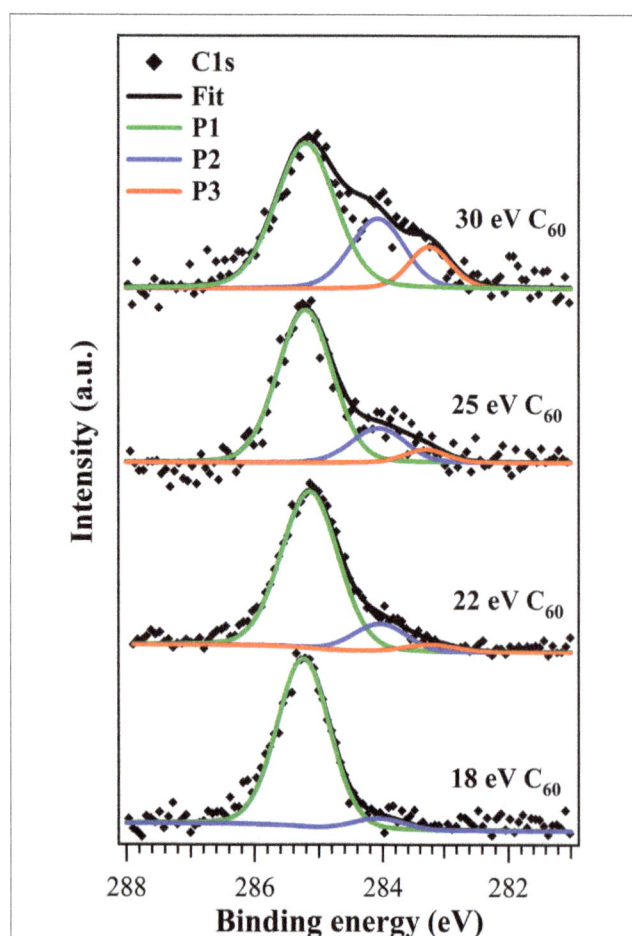

FIGURE 2 | C1s core level spectra (normalized in height) of submonolayer films grown at C_{60} kinetic energies of 18, 22, 25, and 30 eV. Green component corresponds to physisorbed fullerenes, blue component to chemisorbed cages, and red component to broken fullerenes and formation of silicon carbides.

increases. Previous experiments (Verucchi et al., 2012; Taioli et al., 2013) on the impact of supersonic fullerene on silicon at room and higher temperatures have shown that the molecule can undergo different paths, going from physisorption with intact cages to chemisorption, cage rupture and formation of new compounds, such as SiC, whose intensity increases with the precursor KE. The main peak at higher BE, P1, reflects the presence of unperturbed and physisorbed fullerenes, while the peaks at 284.1 and 283.2 eV can be identified as related, respectively, to chemisorbed species and to the formation of SiC occurring after cage breaking. The higher is the C_{60} KE, the more intense is the contribution from "totally reacted" molecules (see **Table 1**), i.e., the fraction of fullerenes leading to new species on the surface. The contribution from physisorbed molecules is always the main one. This is quite surprising, since the 30 eV precursor KE is the same as used in SuMBE room temperature experiments (Verucchi et al., 2012), where physisorbed species are <30 and 50% of the total C1s area at coverages of 0.30 and 0.65 ML.

Chemisorbed fullerenes (peak at 284.1 eV) are present for all films and increase with precursor KE, while SiC-related species (peak at 283.2 eV) are present only from the C_{60} KE of 22 eV and significantly increases only at 30 eV. The peak at 283.2 eV has a BE between that of 3C–SiC (Sakamoto et al., 1999; Mélinon et al., 1998) and non-stoichiometric SiC (Liu et al., 1995), thus it can be attributed to SiCs showing low structural order. The impossibility to achieve reliable results from the analysis of the Si2p core level (due to the prevailing and masking contribution coming from the bulk silicon on all surface components) makes any comment on the carbide stoichiometry purely speculative.

The valence band photoemission is reported in **Figure 3**, together with reference spectra for comparison from a clean silicon surface, a thick fullerene film and a completely synthesized SiC film by means of SuMBE at higher substrate temperatures. Si(111) 7 × 7 valence band is characterized by weak and broad structures, the one at about 0.9 eV related to the surface states, and a metallic character evidenced by the presence of signal at about 0 eV. C_{60} thick film shows intense and different bands, representing the different π and molecular orbitals. In particular, we found at about 2.0 eV the highest occupied molecular orbital (HOMO). At low beam KE, the features related to the clean silicon surface completely disappear, while the HOMO and the other molecular C_{60} bands are clearly visible. This is expected since the π and bands are reported to be fully developed (Ohno et al., 1991) also at this low coverage (less than 1 ML). There are, however, significant discrepancies between data found in literature, in particular with the work of Sakamoto et al. (1999). HOMO and HOMO-1 molecular levels (at 2.2 and 3.5 eV) are not fully developed and slightly shifted toward higher BEs with respect to the bulk organic film (2.0 and 3.3 eV). Moreover, the

TABLE 1 | Peak area% ($\pm$3) on total C1s emission.

C_{60} KE (eV)	% P1 (285.2 eV)	% P2 (284.1 eV)	% P3 (283.2 eV)
18	93	7	0
22	82	14	4
25	77	18	5
30	60	27	13

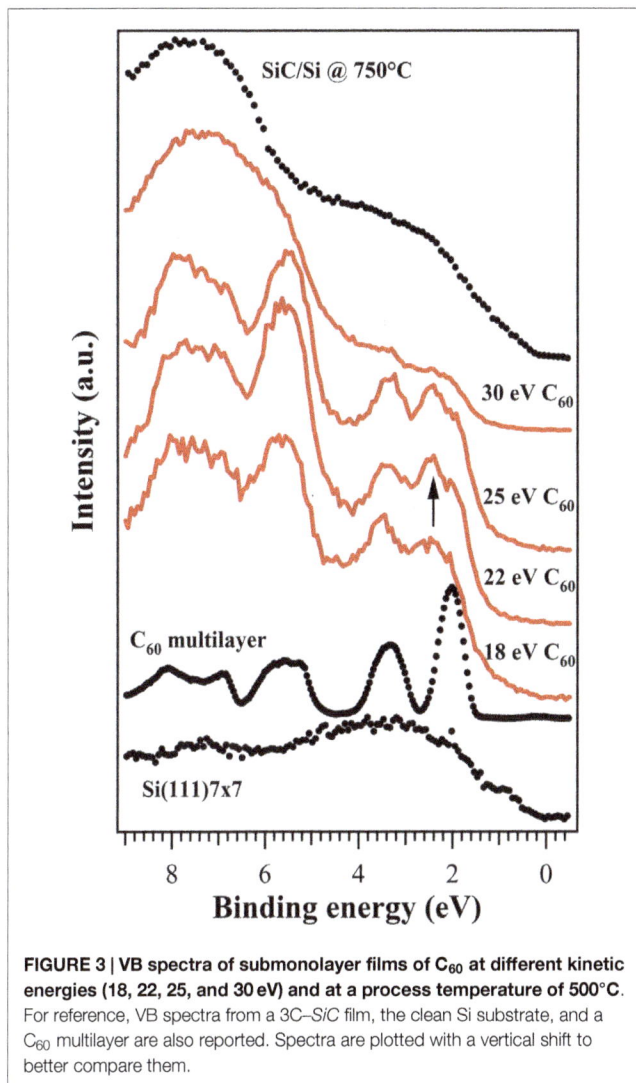

FIGURE 3 | VB spectra of submonolayer films of C$_{60}$ at different kinetic energies (18, 22, 25, and 30 eV) and at a process temperature of 500°C. For reference, VB spectra from a 3C–SiC film, the clean Si substrate, and a C$_{60}$ multilayer are also reported. Spectra are plotted with a vertical shift to better compare them.

There is a fair agreement between XPS and UPS data, suggesting that a C$_{60}$ KE of 22 eV highly improves chemisorption processes, while the cage rupture and SiC synthesis is at least partially achieved at 30 eV, where formation of defected carbides probably occurs.

LEED

From a structural standpoint, the interaction of C$_{60}$ with silicon surface at 500°C leads to quite a few changes, but not to the appearance of new extraspots related to the formation of ordered compounds. **Figure 4** shows the LEED patterns of Si(111) clean surface and of the system after C$_{60}$ impact at the lowest and highest KE, respectively, taken for a primary electron energy of 50 eV. The clean surface is characterized by the well-known 7×7 reconstruction pattern of Si(111) (**Figure 4A**), with bright spots within a high contrast background, typical of a clean and ordered surface. Upon low KE impact of fullerene, the reconstruction pattern disappears and only the 1×1 hexagonal pattern is visible (**Figure 4B**), with spots having lower brightness and sharpness. At highest energy impacts, the LEED pattern undergoes further modifications, with hexagonal pattern losing spots and becoming more similar to a triangular one. The extra spots (Verucchi et al., 2002) related to the formation of ordered 3C–SiC islands do not appear, a result that is in complete agreement with both XPS and UPS results, where formation of a not fully developed and defected carbide is clearly detectable.

Theory

The first step of our computational analysis is the simulation of the impacts with the silicon surface kept at room temperature. Depending on the initial KE (K_{C60}), we were able to identify three regimes: (i) for $K_{C60} < 75$ eV surface penetration does not occur with the cage undergoing simple distortion; (ii) between 100 and 150 eV, we found strong C$_{60}$ cage distortion and surface penetration; (iii) above 175 eV, the cage starts to break up and Si–C bond formation occurs with small surface damaging.

Thus, by using DFTB to describe interatomic forces in the C$_{60}$-silicon surface collision, the minimum impact KE to obtain heavy damage of the silicon surface and fragmentation of the C$_{60}$ cage is predicted to be well above 200 eV, slightly below Averback's findings. Movies of the full trajectory for 50 and 175 eV initial KE can be found in the Videos S1 and S2 in Supplementary Material.

We point out that there is an important discrepancy between simulations and experiments in the assessment of the precursor KE inducing cage break. In this regard, it is clear that DFTB is missing a comprehensive explanation of the cage break. The reason of this discrepancy was already identified and thoroughly discussed in Taioli et al. (2013). In that analysis, non-adiabatic effects and, specifically, the failure of the Born–Oppenheimer approximation was found responsible for this important discrepancy between simulation and experiments to determine cage rupture KE threshold. In this regard, due to very fast collisional times for the KE under investigations, electronic and nuclear motion do not decouple, as assumed within the framework of the Born–Oppenheimer approximation, as electrons do not relax fast enough to the ground state of the instantaneous positions of the nuclei. Thus, electronic and nuclear dynamics should be treated on an equal quantum mechanical footing.

bands in the 5–8 eV energy region loose some of their distinctive characteristics, most notably the shoulder at 5.2 eV disappears and the bands at 6.9 and 8.1 eV are no longer clearly resolved. For precursor KEs of 22 and 25 eV, a new peak appears in the UPS spectra between the HOMO and HOMO-1 structures (see arrow in **Figure 3**), probably is present also in film at 18 eV but very weak. A similar peak has been actually observed and identified as related to the strong chemical interaction with the surface (Sakamoto et al., 1999; De Seta et al., 2000). In the highest KE case, the valence band structure changes strongly and only a small residual of the original molecular π structure can be found in the 0–5 eV energy range. At BE of about 6–8 eV, the σ bands disappear and a broad structure is formed, reminding of the analogous structure of not-crystalline SiC (Aversa et al., 2003) in which the mixed C2p-Si3s orbitals ("sp-like" band in cubic SiC) are predominant with respect to the p-like C and Si orbitals at about 3 eV (Mélinon et al., 1998). This suggests that SiC synthesis is clearly achieved at 30 eV C$_{60}$ KE, even if we are dealing with a partially formed carbide, as demonstrated by the differences with the SiC valence band in **Figure 3** (top curve).

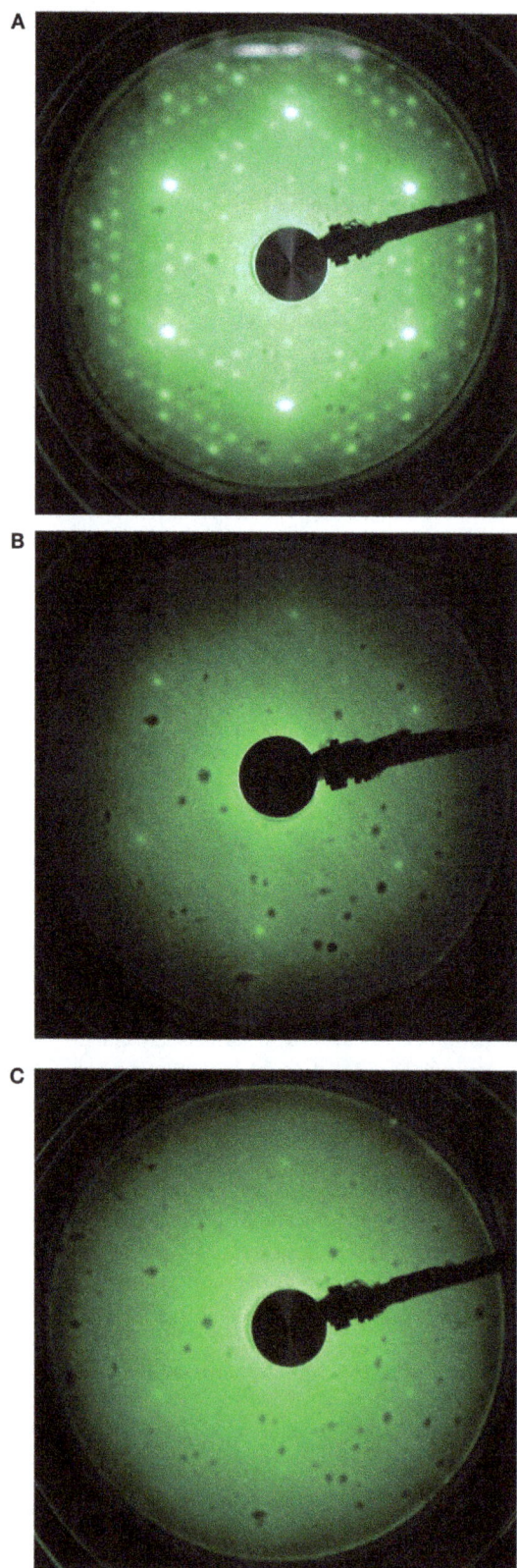

FIGURE 4 | LEED patterns (at 50 eV) of (A) a clean Si(111) 7 × 7 surface, (B) a submonolayer film of C_{60} at KE of 18 eV, and (C) a submonolayer film of C_{60} at KE of 30 eV.

However, in this work, our goal is primarily to investigate the role that temperature might play in the *SiC* growth by SuMBE. While we underline that a proper non-adiabatic quantum treatment of the collision processes, intertwining electronic and nuclear motion, should be used, in this analysis of temperature effects on growth this approach can be ruled out by two factors, namely physics and computational cost. First, the effect of temperature on the electrons is provided by the Fermi–Dirac distribution. In this statistics, temperature appears in the exponential factor at denominator and plays the role of "smearing" or increasing the electronic population in the excited states of the system. However, a temperature enhancement of 800 K corresponds to about 0.06 eV. The energetic distance between electronic orbitals close to the Fermi level in fullerene is much higher (of the order of 1 eV, corresponding to about 12000 K). Thus, the effect of a small temperature increase, such as that investigated in our work, on the electronic motion can be safely neglected, while nuclear motion is quite sensitive to temperature effects. Second, the computational cost associated with non-adiabatic simulations, at least two orders of magnitude larger than using DFTB, is unaffordable for screening the C_{60}-silicon impact for several KEs and different temperatures. Thus, the use of DFTB is totally justified to investigate temperature effects in the case of high KE fullerene impacts on a silicon surface.

MD simulations (Zhang et al., 1993; Kim and Tománek, 1994) indeed show that a break may occur if the total internal energy stored in the C_{60} cage is between 30 and 40 eV. The internal energy is defined as the total KE of the fullerene bouncing back after the impact minus the center of mass KE. A temperature increase may help to destabilize fullerene by exciting substrate phonons, which can transfer energy to C_{60}, eventually resulting in a lower cage breaking KE.

Thus, we performed a temperature dependent investigation of the internal KE after the impact, when fullerene bounces and leaves the silicon surface, in the same range of initial KEs previously examined at a temperature of 500°C. The system was thermostated to this temperature by a Nose–Hoover thermostat for 1 ps to reach thermal equilibrium and then NVE ensemble with a Fermi smearing corresponding to 500°C was adopted during the impact molecular dynamics run.

Surprisingly, the results point toward a not significant effect of temperature on the internal KE for all the impinging KEs.

The results are reported in **Figure 5** and demonstrate that the fullerene internal energy for the impinging KEs under investigation (from 25 to 175 eV) at room temperature conditions ranges from 13 to 28 eV (blue diamonds). For the assessment of the internal KE, we chose a molecular dynamics step corresponding to a maximum of the oscillating total KE immediately after the collision. A substrate temperature increase to 500°C (red diamonds) of course results into a higher final internal KE with respect to the room temperature case (blue diamonds) due to enhanced thermal fluctuations. However, this increase is not such to cause the rupture in the 25–175 eV energy range as much of the energy is spent by fullerene to increase its translational KE for leaving the surface, as shown by the evolution of the center of mass KE of C_{60} after impact (see blue and red circles in **Figure 5**).

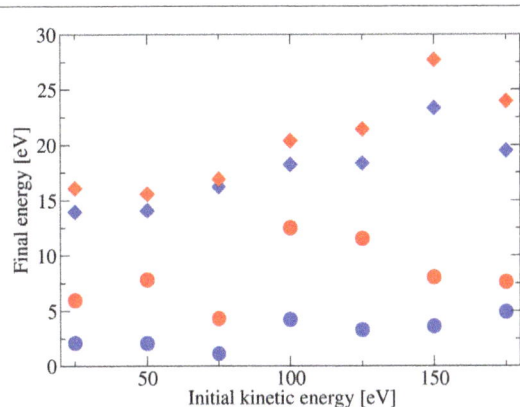

FIGURE 5 | Fullerene total internal kinetic energies after bouncing back from the Si(111) 7 × 7 surface at room temperature (blue diamonds) and 500°C (red diamonds) as a function of the initial C$_{60}$ kinetic energy. Furthermore, we show the center of mass kinetic energies for both 20°C (blue circles) and 500°C (red circles) substrate temperatures.

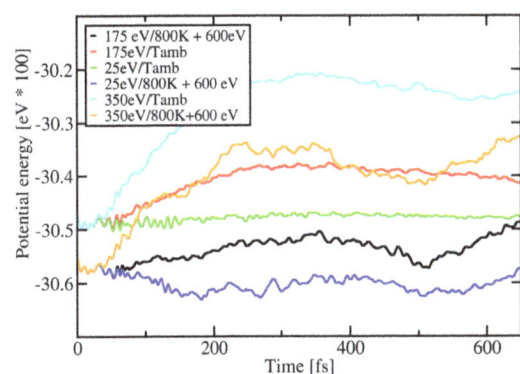

FIGURE 6 | Total interaction potential energy (eV*100) as a function of time (fs) for different kinetic energies (eV) and temperatures (K). The potential energy curves for 800 K are shifted up in energy by 600 eV for the sake of visibility.

We point out that the interaction potential energy of C$_{60}$ as a function of the distance from the silicon surface is not a quantity at our disposal (just the total interaction potential energy is, as shown below). This analysis could be in principle performed using classical molecular mechanics, where all the interactions terms are analytically known and are singly computable within the reach of this theory. However, in the conditions at which C$_{60}$–Si collision occurs, we are abundantly out of thermodynamic equilibrium with bonds breaking and forming. Thus, we are forced to use a "first-principles" technique, explicitly including electronic motion in a self-consistent way, to have a reliable description of the high KE impact. In DFTB, single terms of the potential energies are not a direct output, particularly as a function of the distance between C$_{60}$ and the silicon surface. In principle, one could perform a very expensive analysis to obtain this quantity, consisting in repeating all calculations at different KEs and temperatures as follows. Indeed, one could calculate separately the interaction potential energy of the isolated C$_{60}$ and silicon slab at each position, keeping fixed the atomic coordinates as by simulations of C$_{60}$ approaching

the surface. Finally, one could subtract the sum of these two separate contributions from the total interaction potential energy of the real interacting system at all distances, thus obtaining the C$_{60}$–silicon slab interaction potential energy as a function of the distance. Such a calculation is of course unfeasible due to the computational scaling of our approach.

The plot in **Figure 6** represents the total interaction potential energy, a quantity computable by DFTB, for three different KEs and two different temperatures as a function of computing time, which is somehow proportional to the distance of C$_{60}$ from the silicon surface (see **Figure 1**). In this plot, the potential energy curves for 800 K are shifted up in energy by 600 eV for the sake of visibility. From **Figure 6** one can clearly see that, at the point where C$_{60}$ starts experiencing the interaction with the silicon surface (below 100 fs), the total interaction potential energy changes abruptly only in the case of high KE (350 eV, orange and cyan lines for the cases of ambient temperature and 800 K, respectively). For this KE, C$_{60}$ is completely fragmented. Furthermore, in this high KE regime, the difference between the total interaction potential energy at ambient temperature and 800 K is clearly visible at all times. At variance, for low KE (25 eV, green and violet curve for the cases of ambient temperature and 800 K, respectively), where C$_{60}$ is just physisorbed on the silicon surface, the total interaction potential energy shows little change with time and is almost independently of temperature. Finally, at intermediate regimes (175 eV, red and black curve for the cases of ambient temperature and 800 K, respectively), where C$_{60}$ is partially fragmented, the change of the total interaction potential energy again seems to be due more to kinetic effects than to temperature, however showing larger fluctuations with increasing temperatures with respect to the ambient temperature impact.

To further enrich our discussion about temperature effects on *SiC* growth by SuMBE, we performed the calculation of the mean squared displacement (MSD) as a function of impinging C$_{60}$ KE and time for different temperatures. This quantity provides of course a measure of the dispersion or diffusion of carbon atoms of the C$_{60}$ molecule. These results are sketched in **Figures 7A,B**. In **Figure 7A**, we report, in particular, the behavior of the MSD of C$_{60}$ carbon atoms at ambient (black curve) and 800 K (red curve). From this figure, one safely concludes that temperature effects are relevant only at high KE regimes, when the cage is fragmented (350 eV) or partially broken (175 and 260 eV).

In **Figure 7B**, MSD is sketched as a function of time (or, which is the same, of C$_{60}$ approaching the silicon surface). Again, MSD is almost constant or slightly increasing at low KE regimes (25 and 175 eV) while is divergent at 350 eV, meaning that C$_{60}$ is completely fragmented. Furthermore, MSD is larger at higher temperature with a sharp increase at long times. This analysis clearly points toward a cage break model that is almost independent on temperature with kinetic effects playing a major role. Temperature effects become important only after cage rupture.

Discussion

Experimental results of C$_{60}$ supersonic impact on Si(111) 7 × 7 at 500°C show a picture completely different from what was observed using the same deposition technique at higher and lower

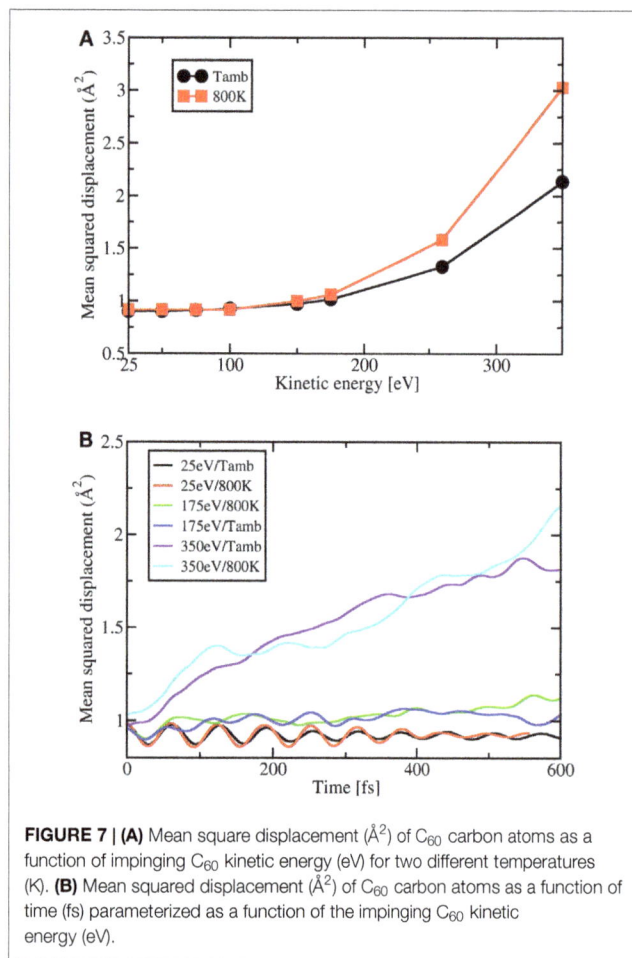

FIGURE 7 | (A) Mean square displacement (Å^2) of C_{60} carbon atoms as a function of impinging C_{60} kinetic energy (eV) for two different temperatures (K). **(B)** Mean squared displacement (Å^2) of C_{60} carbon atoms as a function of time (fs) parameterized as a function of the impinging C_{60} kinetic energy (eV).

substrate temperatures, evidencing an unexpected mechanism of interaction between the precursors and the surface. From previous experiments (Verucchi et al., 2002, 2012), as well as theoretical calculations (Taioli et al., 2013), one finds a clear indication of a kinetically driven growth mechanism leading to sub- or thick SiC layer formation already at room temperature, with temperature factor not playing a major role. Raising the silicon substrate temperature to 500°C should generally result in a higher silicon–carbon reactivity, with increasing carbide percentage over physisorbed fullerene. Core level and valence band analysis (**Figure 2**) point toward a completely different situation with respect to our expectations. Indeed, the great majority of species present on the surface after impact are unperturbed C_{60} up to high KE, whereby a reactivity increase is found in terms of percentage of chemisorbed fullerene and carbide formation. Valence band spectra suggest the same behavior, with molecular bands having characteristics more similar to physisorbed fullerene with C_{60} cage breaking and carbide synthesis clearly evident only for the 30 eV KE case. Nevertheless, we achieve formation of defected SiC and not ordered cubic 3C–SiC, as observed in previous experiments at 750–800°C (Verucchi et al., 2002) and room temperature (Verucchi et al., 2012).

At 500°C, C_{60} KE is crucial, with a threshold for SiC synthesis and cage breaking identified at about 30 eV. However, it is clear that chemical reactivity is different from both 750°C and room temperature cases. This may represent a surprising result, assuming that the higher the substrate temperature the easier should be the cage breaking. Indeed our calculations clearly indicate that the internal energy for a fixed value of the C_{60} initial KE is slightly higher if temperature is raised up to 500°C (see **Figure 5**). This trend is confirmed for all KEs investigated. However, our computational analysis shows that a major part of both the increasing KE and temperature is spent to eject more effectively the fullerene at the expense of an enhanced reactivity. This can be clearly seen in the low and intermediate (below 150 eV) KE regimes, where increasing the temperature corresponds to enhance proportionally the center of mass KE of the C_{60} bouncing back from the surface instead of the internal energy of the system, responsible for the cage break. A similar effect was observed in the collision of SF_6 from GaSe surface where an uptake of energy was identified (Boschetti et al., 1992). Additionally, at high KEs, when fullerene gets trapped within the silicon substrate and thus do not bounce back, temperature effects start to be more effective in transferring energy from silicon to fullerene vibrational modes, as reported in **Figure 5**, in perfect agreement with our experimental results. However, this is not enough to justify cage opening. Thus, both calculations and experimental results clearly point toward a kinetically driven fullerene cage breaking and SiC synthesis, with a KE threshold not critically affected by temperature below the well-known value of 750°C (Verucchi et al., 2002).

In conclusion, we have analyzed in details the impacts of energetic supersonic C_{60} on the reconstructed Si(111) 7×7 surface at 500°C by means of experiments and *ab initio* simulations. Surface chemical and structural characterization carried out at increasing fullerene KE (18 eV to 30 eV) has shown the presence of physisorbed, chemisorbed, and reacted molecules, the latter leading by the formation of SiC, probably defected cubic carbide or non-stoichiometric species. Compared to previous experiments with substrate kept at room temperature, we do not find an enhancement in reactivity with increasing temperature. DFTB calculations reveal that the excess of initial KE is spent more effectively to increase the translational KE (center of mass KE) and substrate temperature helps this process via phonon transfer, at least in the low and intermediate KE regime. Only for higher KE, when cage rupture occurs, the reactivity starts to be relevant, as shown in our MSD analysis, demonstrating that this process is kinetically driven. While our discussion has been limited to the case of SiC growth, we believe that SuMBE may represent a more general approach to grow materials in the low-temperature regime even on metals, notably nickel and copper, in which carbide formation may compete energetically with π-conjugate structures, such as graphene (Batzill, 2012; Lu et al., 2011).

Acknowledgments

The research leading to these results has received funding from the European Union Seventh Framework Programme under grant agreement no. 604391 Graphene Flagship and from project Super-Car (CMM-FBK, Fondazione Bruno Kessler). ST acknowledges support from the Istituto Nazionale di Fisica Nucleare through the

"Supercalcolo" agreement with Fondazione Bruno Kessler, from the European Science Foundation under the INTELBIOMAT Exchange Grant "Interdisciplinary Approaches to Functional Electronic and Biological Materials" and the Bruno Kessler Foundation for providing economical support through the "research mobility scheme" under which this work has been accomplished. Finally, ST gratefully acknowledges the Institute of Advanced Studies in Bologna for the support given under his ISA research

fellowship, and the high-performance computing service Archer in UK where impact calculations were performed.

References

Aradi, B., Hourahine, B., and Frauenheim, T. (2007). DFTB+, a sparse matrix-based implementation of the DFTB method. *J. Phys. Chem. A* 111, 5678–5684. doi:10.1021/jp070186p

Aversa, L., Verucchi, R., Boschetti, A., Podesta, A., Milani, P., and Iannotta, S. (2003). Fullerene freejets-based synthesis of silicon carbide. *Mater. Sci. Eng. B* 101, 169–173. doi:10.1016/S0921-5107(02)00703-1

Balooch, M., and Hamza, A. V. (1993). Observation of C_{60} cage opening on Si(111) 7x7. *Appl. Phys. Lett.* 63, 150–152. doi:10.1063/1.110382

Batzill M. (2012). The surface science of graphene. *Surf. Sci. Rep.* 67, 83–115. doi:10.1016/j.surfrep.2011.12.001

Boschetti, A., Cagol, A., Corradi, C., Jacobs, R., Mazzola, M., and Iannotta, S. (1992). Energy transfer processes and molecular degrees of freedom in the collision of SF_6 molecules with the GaSe(001) surface. *Chem. Phys.* 163, 179–191. doi:10.1016/0301-0104(92)87102-F

Cepek, C., Schiavuta, P., Sancrotti, M., and Pedio, M. (1999). Photoemission study of C_{60}/Si(111) adsorption as a function of coverage and annealing temperature. *Phys. Rev. B* 60, 2068–2073. doi:10.1103/PhysRevB.60.2068

De Seta, M., Tomozeiu, N., Sanvitto, D., and Evangelisti, F. (2000). SiC formation on Si(100) via C_{60} precursors. *Surf. Sci.* 460, 203–213. doi:10.1016/S0039-6028(00)00533-1

Elstner, M., Porezag, D., Jungnickel, G., Elsner, J., Haugk, M., Frauenheim, T., et al. (1998). Self-consistent-charge density-functional tight-binding method for simulations of complex materials properties. *Phys. Rev. B* 58, 7260–7268. doi:10.1103/PhysRevB.58.7260

Frauenheim, T., Seifert, G., Elstner, M., Niehaus, T., Köhler, C., Amkreutz, M., et al. (2002). Atomistic simulations of complex materials: ground-state and excited-state properties. *J. Phys. Condens. Mater.* 14, 3015–3047. doi:10.1088/0953-8984/14/11/313

Frenzel, J., Oliveira, A. F., Jardillier, N., Heine, T., and Seifert, G. (2009). *Semi-Relativistic, Self-Consistent Charge Slater-Koster Tables for Density-Functional Based Tight-Binding (DFTB) for Materials Science Simulations*. TU-Dresden 2004-2009.

Galli, G., and Mauri, F. (1994). Large scale quantum simulations: C_{60} impacts on a semiconducting surface. *Phys. Rev. Lett.* 73, 3471–3474.

Garberoglio, G., and Taioli, S. (2012). Modeling flexibility in metal – organic frameworks: comparison between density-functional tight-binding and universal force field approaches for bonded interactions. *Microporous Mesoporous Mater.* 163, 215–220. doi:10.1016/j.micromeso.2012.07.026

Hu, X., Albe, K., and Averback, R. S. (2000). Molecular-dynamics simulations of energetic C_{60} impacts on (2×1)-(100) silicon. *J. Appl. Phys.* 88, 49–54. doi:10.1063/1.373622

Hüfner, S. (1995). *Photoelectron Spectroscopy*. Berlin: Springer-Verlag.

Kim, S. G., and Tománek, D. (1994). Melting the fullerenes: a molecular dynamics study. *Phys. Rev. Lett.* 72, 2418–2421. doi:10.1103/PhysRevLett.72.2418

Liu, C. C., Lee, C., Cheng, K. L., Cheng, H. C., and Yew, T. R. (1995). Effect of SiH_4/CH_4 flow ratio on the growth of -SiC on Si by electron cyclotron resonance chemical vapor deposition at 500°C. *Appl. Phys. Lett.* 66, 168–170. doi:10.1063/1.113552

Lu, J., Yeo, P. S. E., Gan, C. K., Wu, P., and Loh, K. P. (2011). Transforming C_{60} molecules into graphene quantum dots. *Nat. Nanotechnol.* 6, 247. doi:10.1038/NNANO.2011.30

Matthew, W., Foulkes, C., and Haydock, R. (1989). Tight-binding models and density-functional theory. *Phys. Rev. B* 39, 12520–12536.

Mélinon, P., Kéghélian, P., Perez, A., Ray, C., Lermé, J., Pellarin, M., et al. (1998). Nanostructured SiC films obtained by neutral-cluster depositions. *Phys. Rev. B* 58, 16481–16490. doi:10.1103/PhysRevB.58.16481

Milani, P., and Iannotta, S. (1999). *Cluster Beam Synthesis of Nano-Structured Materials*. Berlin: Springer-Verlag.

Ohno, T. R., Chen, Y., Harvey, S. E., Kroll, G. H., Weaver, J. H., Haufler, R. E., et al. (1991). C_{60} bonding and energy-level alignment on metal and semiconductor surfaces. *Phys. Rev. B* 44, 13747–13755. doi:10.1103/PhysRevB.44.13747

Sakamoto, K., Kondo, D., Ushimi, Y., Harada, M., Kimura, A., Kakizaki, A., et al. (1999). Temperature dependence of the electronic structure of C_{60} films adsorbed on Si(001)-(2x1) and Si(111)-(7x7) surfaces. *Phys. Rev. B* 60, 2579–2591. doi:10.1103/PhysRevB.60.2579

Sanvitto, D., De Seta, M., and Evangelisti, F. (2000). Growth of thin C_{60} films on hydrogenated Si(100) surfaces. *Surf. Sci.* 452, 191–197. doi:10.1016/S0039-6028(00)00321-6

Taioli, S. (2014). Computational study of graphene growth on copper by first-principles and kinetic Monte Carlo calculations. *J. Mol. Model* 20, 2260. doi:10.1007/s00894-014-2260-2

Taioli, S., Garberoglio, G., Simonucci, S., a Beccara, S., Aversa, L., Nardi, M. V., et al. (2013). Non-adiabatic ab initio molecular dynamics of supersonic beam epitaxy of silicon carbide at room temperature. *J. Chem. Phys.* 138, 044701. doi:10.1063/1.4774376

Takayanagi, K., Tanishiro, Y., Takahashi, M., and Takahashi, S. (1985). Structural analysis of Si(111)-7×7 by UHV-transmission electron diffraction and microscopy. *J. Vac. Sci. Technol. A* 3, 1502–1506. doi:10.1016/j.ultramic.2013.04.005

Tersoff, J. (1988). New empirical approach for the structure and energy of covalent systems. *Phys. Rev. B.* 37, 6991–7000. doi:10.1103/PhysRevB.37.6991

Verucchi, R., Aversa, L., Ciullo, G., Podesta', A., Milani, P., and Iannotta, S. (2002). SiC film growth on Si(111) by supersonic beams of C_{60}. *Eur. Phys. J. B* 26, 509–514. doi:10.1140/epjb/e20020120

Verucchi, R., Aversa, L., Nardi, M. V., Taioli, S., a Beccara, S., Alfè, D., et al. (2012). Epitaxy of nanocrystalline silicon carbide on Si(111) at room temperature. *J. Am. Chem. Soc.* 134, 17400–17403. doi:10.1021/ja307804v

Zhang, B. L., Wang, C. Z., Chan, C. T., and Ho, K. M. (1993). Thermal disintegration of carbon fullerenes. *Phys. Rev. B* 48, 11381–11384.

Conflict of Interest Statement: The authors declare that the research was conducted in the absence of any commercial or financial relationships that could be construed as a potential conflict of interest.

19

New creep-resistant cast alloys with improved oxidation resistance in water vapor at 650–800°C

Sebastien Dryepondt*, Bruce A. Pint and Philip J. Maziasz

Materials Science and Technology Division, Oak Ridge National Laboratory, Oak Ridge, TN, USA

Edited by:
Her-Hsiung Huang,
National Yang-Ming University, Taiwan

Reviewed by:
Ole Øystein Knudsen,
SINTEF and NTNU, Norway
Guang-Ling Song,
Oak Ridge National Laboratory, USA

*Correspondence:
Sebastien Dryepondt,
Materials Science and Technology
Division, Oak Ridge National
Laboratory, 1 Bethel Valley Road,
MS6156, Oak Ridge, TN 37831, USA
dryepondtsn@ornl.gov

Cast stainless steel CF8C-Plus (19wt%Cr/12%Ni) has excellent creep properties, but limited oxidation resistance above 700°C in environments containing H_2O. One strategy to improve the alloy oxidation performance is to increase the Cr and Ni concentration. Two new alloys, with, respectively, 21wt%Cr–15wt%Ni and 22wt%Cr–17.5wt%Ni were therefore developed and their long-term oxidation behaviors in humid air were compared with the oxidation behavior of five other cast alloys. At 650°C and 700°C, all the alloys formed internal Cr-rich nodules, and outer nodules or layers rich in Fe and Ni, but they grew a protective Cr-rich inner layer over time. At 750°C, the lower alloyed steels such as CF8C-Plus showed large metal losses, but the two new alloys still exhibited a protective oxidation behavior. The 21Cr–15Ni alloy was severely oxidized in locations at 800°C, but that was not the case for the 22Cr–17.5Ni alloy. Therefore, the two new modified alloys represent a potential operating temperature gain of, respectively, 50°C and 100°C in aggressive environments compared with the CF8C-Plus alloy.

Keywords: cast steels, oxidation, water vapor, CF8C, protective scale, creep

Introduction

Increasing the operating temperature and pressure of internal combustion engines is a straightforward solution to improve the engine efficiency and decrease air pollutant emission. There is therefore a growing need for new cost-effective alloys with enhanced mechanical properties and oxidation resistance at 650°C and above (Davis et al., 2012). CF8C-Plus (CF8CP) is a cast stainless steel with improved high-temperature creep strength that was developed in 2002 by ORNL in collaboration with Caterpillar. The composition of the alloy is based on the standard CF8C steel, which is a cast version of the wrought type 347 alloy, but with the addition of Mn and N to form strengthening Nb-nano-carbonitride, and careful balancing of the other alloying elements to avoid the formation of the detrimental ferrite phase (Shingledecker et al., 2005, 2006; Evans et al., 2010). The alloy won in 2003 an R&D100 award and is commercially used to fabricate diesel exhaust components (Maziasz and Pint, 2010).

Creep properties of CF8CP steel can be further improved as is found in some stainless steels by adding Cu to form nanoscale Cu precipitates (Sawaragi and Hirano, 1990; Swindeman et al., 1990; Wu et al., 2005; Chi et al., 2012; Ha and Jung, 2012). The new version of CF8CP, called CF8CWCu, with ~3% of Cu and ~1% of W for solution strengthening, has demonstrated even better creep properties, similar to more expensive Ni-based alloys such as alloy 625 and 617 (Shingledecker et al., 2005; Maziasz and Pint, 2011). However, the oxidation resistance of these two alloys is degraded in the presence of water vapor (Maziasz and Pint, 2011), which will limit their use in aggressive atmospheres such as exhaust gas and steam. The effect of water vapor on the oxidation of ferritic and austenitic steels has been the focus of many studies (Quadakkers et al., 2009), and several mechanisms have been proposed to explain the detrimental impact of H_2O, including higher Cr_2O_3 growth rate (Henry et al., 2000; Hultquist et al., 2000), formation and evaporation of oxy-hydroxide (Asteman et al., 2000; Schutze et al., 2004; Young and Pint, 2006), and enhanced internal oxidation of Cr (Essuman et al., 2008; Quadakkers et al., 2009).

One straightforward route to improve oxidation resistance is to increase the level of Cr in the alloy to better promote the formation of an external Cr_2O_3 scale (Peraldi and Pint, 2004; Saunders et al., 2008; Meier et al., 2010), but increasing the Cr content alone could have a detrimental impact on the alloy mechanical properties. However, Peraldi and Pint demonstrated with a series of Fe–Cr–Ni model alloys that increasing Ni content can also improve oxidation performance in water vapor (Peraldi and Pint, 2004). Two new cast alloys have therefore been developed based on the CF8CWCu composition, with significant increases in Cr and Ni, to 21 wt% Cr and 15 wt% Ni for one alloy, and to 22 wt% Cr and 17.5 wt% Ni for the other alloy, compared with ~19 wt% Cr and 12 wt% Ni for the base CF8CP steel. Tensile and creep testing were conducted on the two newly developed alloys, and the results indicate that the mechanical properties of these alloys are modestly lower than alloy CF8CWCu, but still better than CF8CP and suitable for high-temperature structural applications (Unocic et al., submitted). The goal of this paper is to compare the long-term oxidation behavior in humid air at temperature ranging from 650°C to 800°C of these two new alloys with five other cast alloys, CF8C, CF10M (cast type 316 stainless steel), CF8CP, and two heats of CF8CWCu. The effect of Ni and Cr concentrations on the alloys oxidation performance is discussed in detail.

Experimental Procedure

Seven alloys were cast by Stainless Foundry and Engineering Inc., including three commercial cast alloys, CF8C, CF8C-Plus (CF8CP), and CF10M, two versions of the CF8CWCu alloy (different batches), and the two new modified alloys with higher levels of Cr and Ni compared to CF8CWCu, alloy 21Cr–15Ni and alloy 22Cr–17.5Ni. The composition of these alloys, as measured by inductively coupled plasma (ICP), combustion, and inert gas fusion (IGF) analyses, are given in **Table 1**.

Rectangular oxidation coupons, ~20 mm × 10 mm × 2 mm were machined from plates or Kiel blocks, polished to a 600 grit surface finish, and cleaned in an ultrasonic bath with acetone, and then methanol. Cyclic oxidation testing was conducted at 650°C, 700°C, 750°C, and 800°C for 1000 and 5000 h in air + 10% Vol. H_2O using four horizontal tube furnaces with sealed alumina chambers. Distilled water was atomized above its condensation temperature into dry air flowing at a rate of 850 cc/min (1.6–1.9/cm/s). Every 100 h, the oxidation coupons were removed from the furnace and weighted using a Mettler model XP205 balance with an accuracy of ±0.04 mg (~0.01 mg/cm²).

Specimens were cross-sectioned after exposure using conventional metallography techniques and observed by optical microscopy. Cu-plating was used to protect the oxide scale during specimen preparation. For selected specimens, the composition of the oxide layer and the alloy beneath the scale was measured using a JEOL JXA-8200 electron probe microanalyzer (EPMA) equipped with five vertical spectrometers. Only one CF8C and one CF10M oxidation coupons were exposed at each temperature, and the microstructure of these specimens was not characterized.

Results

Oxidation at 650°C

Specimen mass gains for the seven cast alloys exposed in air + 10% H_2O at 650°C and 700°C are displayed in **Figure 1**. At 650°C, the CF8CP, CF10M, CF8CWCu-1, and CF8CWCu-2 alloys all exhibit first a transient oxidation stage for ~500 h with faster oxidation rates, before reaching a steady slow-growing oxidation stage. The mass gains reached after the transient stage were slightly different for the two coupons of the same alloy, but the oxidation behaviors were similar. For the 21Cr and 22Cr alloys, an incubation period was observed for, respectively, ~1000 and 2000 h, with very low mass gain, followed by a nearly linear increase of the mass gain. For the 21Cr alloy, the mass gain rate decreased after 3000 h, to a rate similar to the rates measured for the other alloys. Alloy CF8C was the only alloy showing a constant decrease of the specimen mass for the 5000 h of exposure at 650°C (final mass loss of −2.9 mg/cm²), and spallation of the oxide scale was visually observed.

Cross-section images of the oxidation coupons after 1000 and 5000 h of exposure are shown in **Figure 2**. For the CF8CP, CF8CWCu-1, and CF8CWCu-2 alloys, an inhomogeneous ~10 μm bi-layer scale was observed after 1000 h, with thinner areas often related to interdendritic regions from the cast alloy microstructure. After exposure for 5000 h, large inner nodules were observed for alloys CF8CP and CF8CWCu-1, up to ~20 μm deep. For alloy CF8CWCu-2, the scale thickness increased only

---

TABLE 1 | Composition of the cast alloys measured by inductively coupled plasma (ICP), combustion, and inert gas fusion (IGF) analyses.

Alloy	Fe	Cr	Ni	Mn	Mo	Cu	Si	Nb	W	C	N
CF8C	67.6	18.7	10.4	0.7	0.28	0.21	1.23	0.47	0.02	0.02	0.08
CF10M	63.6	20.6	10.2	1	2.38	0.38	1.14	0.06	0.02	0.023	0.07
CF8CP	61.1	19.3	12.8	3.9	0.36	0.5	0.57	0.95	0.1	0.09	0.22
CF8CWCu-1	60.1	19.2	11.8	3	0.29	2.84	0.6	0.77	0.96	0.089	0.21
CF8CWCu-2	58.4	19.6	12.3	3.5	0.3	2.96	0.53	0.92	1.1	0.08	0.24
21Cr–15Ni	54.9	20.7	15.2	3.4	0.35	2.61	0.3	0.92	1.06	0.121	0.19
22Cr–17.5Ni	51	22.1	17.4	3.7	0.35	2.59	0.46	0.91	1.01	0.089	0.2

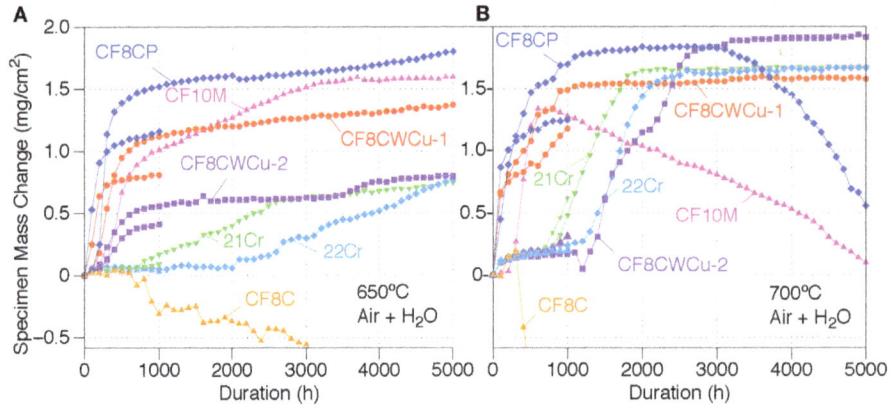

FIGURE 1 | Mass change curves for alloys CF8C, CF8CP, CF10M, CF8CWCu-1, CF8CWCu-2, 21Cr–15Ni, and 22Cr–17.5Ni exposed in air + 10% H_2O for 1000 and 5000 h at (A) 650°C and (B) 700°C.

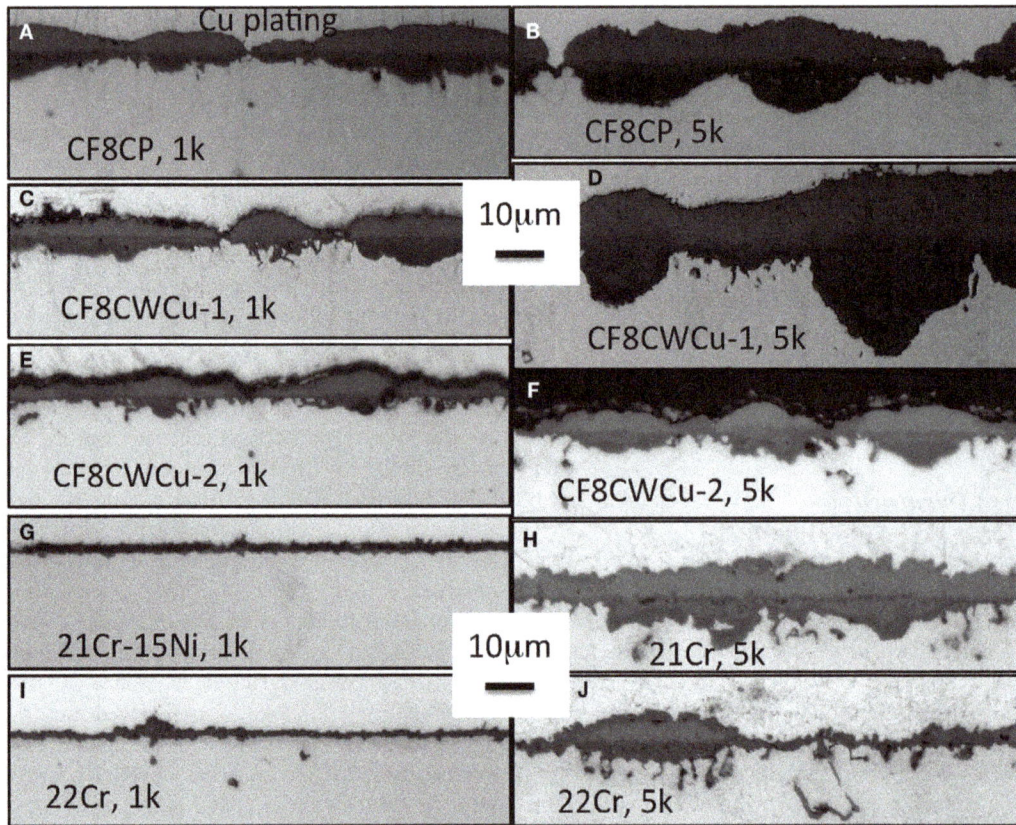

FIGURE 2 | Cross-section optical micrographs after exposure in air + 10% H_2O at 650°C, (A) CF8CP, 1000 h, (B) CF8CP, 5000 h, (C) CF8CWCu-1, 1000 h, (D) CF8CWCu-1, 5000 h, (E) CF8CWCu-2, 1000 h, (F) CF8CWCu-2, 5000 h, (G) 21Cr-15Ni, 1000 h, (H) 21Cr-15Ni, 5000 h (I) 22Cr-17.5Ni, 1000 h, and (J) 22Cr-17.5Ni, 5000 h.

slightly from 1000 to 5000 h of exposure. For the 21Cr and 22Cr alloys, the thin oxide scale observed after 1000 h is consistent with the very low mass gain measured at the beginning of the test. However, small nodules appeared all over the specimens after 5000 h of exposure, **Figures 2H,J**, which explains the higher reaction rate later in the 5000 h experiment.

Oxidation at 700°C

As can be seen in **Figure 1B**, a fast initial oxidation stage for alloy CF8CP and alloy CF8CWCu-1 at 700°C was again followed by a slow-growing steady-state stage. A third oxidation stage was observed for alloy CF8CP after ~3000 h, with constant mass losses. Alloy CF10M showed the same initial fast oxidation stage followed by a constant mass decrease for the remaining 4300 h. The CF8C coupon initially gained some weight, but then started to lose weight rapidly after only 400 h (final mass loss of $-10.5 \, mg/cm^2$), again because of scale spallation.

In the case of alloys, 21Cr, CF8CWCu-1, and 22Cr, slow mass gains again were initially observed, followed by a fast acceleration of the oxidation rate for ~1000 h to reach a new steady oxidation stage suggesting slow growth of the scale. The cross-section micrographs displayed in **Figure 3** show oxide scales at 700°C that are very consistent with the 700°C mass gains data. After 1000 h, a very thin oxide scale was observed for alloy CF8CWCu-2 and 22Cr, because the two alloys were still in the slow growth oxidation stage. A slightly thicker oxide scale was observed for the 21Cr alloy because the accelerated oxidation stage had already started. A significantly thicker oxide scale was already present for alloy CF8CP and CF8CWCu-1 after 1000 h, with deep internal nodules. After 5000 h, a thick duplex oxide was observed for all the alloys, with a wavy alloy/scale interface and a more uniform outer layer. Scale spallation was visually noticed for the CF8CP specimens

exposed for 5000 h, consistent with the mass loss exhibited in **Figure 1B**.

The EPMA elemental maps presented in **Figure 4** show that the oxide scale formed on alloy CF8CP was composed of three distinct layers; an outer Fe-rich oxide layer, a thin intermediate (Fe,Ni) layer consistent with spinel, and an inner layer of Cr-rich nodules with small additions of Si. Areas where the oxide scale is thinner are again related to the interdendritic zone of the alloy intersecting the surface, and it is easily identified by the characteristic enrichments in Mo and Nb. Traces of Mn were detected at the very outer surface of the oxide scale and in the inner Cr layer.

Similar chemical maps are displayed in **Figure 5**, this time for alloy 22Cr–17.5Ni after only 1000 h. The thin protective scale formed after 1000 h consists of an Cr–Mn–Si rich zone consistent with spinel or multi-layer structure (Mn-rich outer layer, Cr-rich oxide, and SiO_2 at metal–scale interface). Below the oxide scale, a zone depleted in Cr and Mn has formed that appears to be rich in Fe and Ni. This depleted zone is in general very thin,

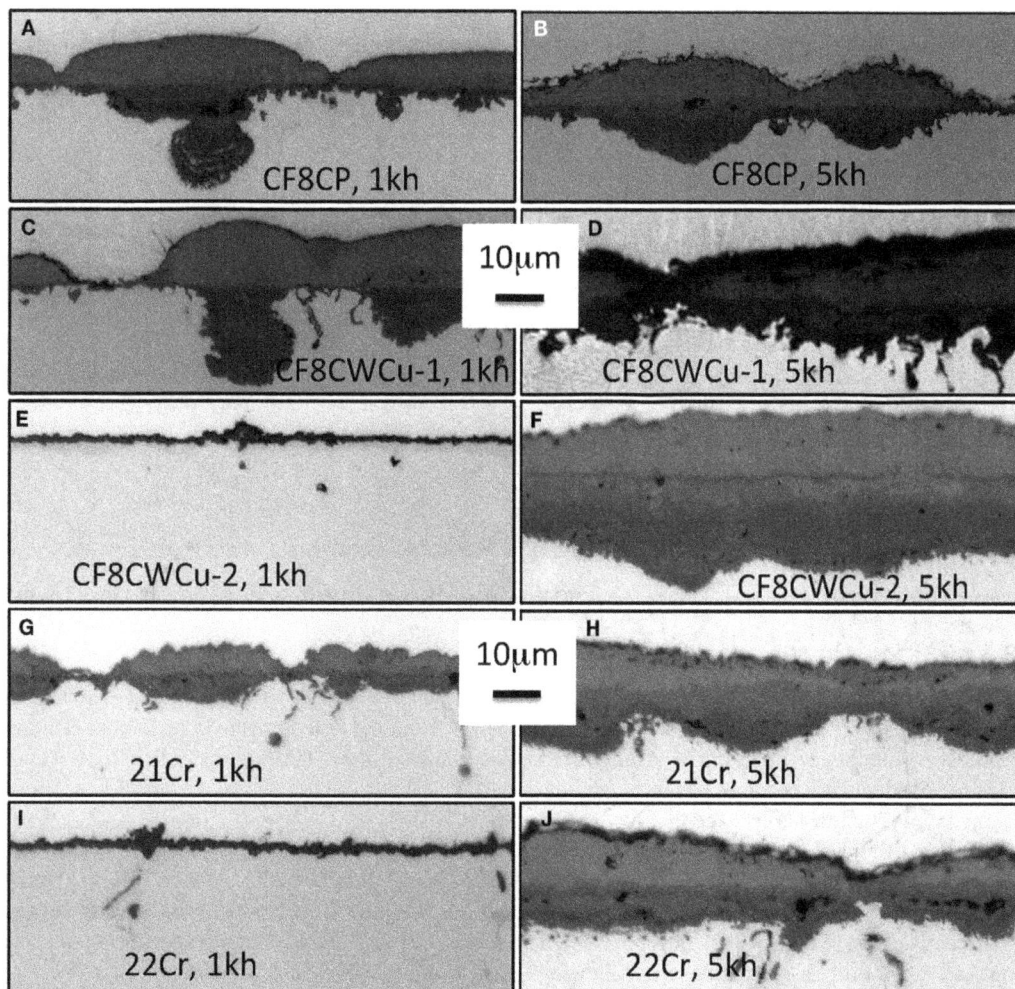

FIGURE 3 | Cross-section optical micrographs after exposure in air + 10% H_2O at 700°C, (A) CF8CP, 1000 h, (B) CF8CP, 5000 h, (C) CF8CWCu-1, 1000 h, (D) CF8CWCu-1, 5000 h, (E) CF8CWCu-2, 1000 h, (F) CF8CWCu-2, 5000 h, (G) 21Cr–15Ni, 1000 h, (H) 21Cr–15Ni, 5000 h, (I) 22Cr–17.5Ni, 1000 h, and (J) 22Cr–17.5Ni, 5000 h.

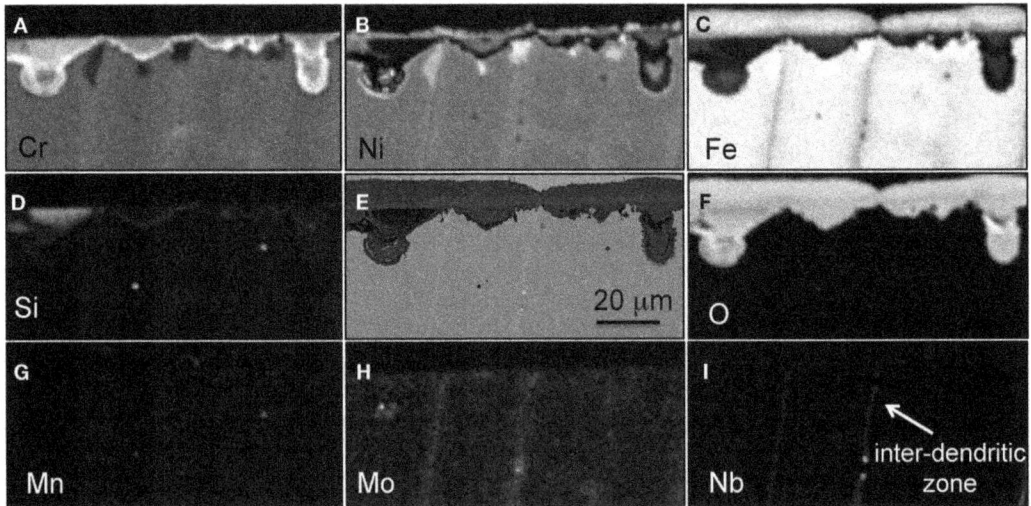

FIGURE 4 | EPMA mapping for alloy CF8CP after exposure for 1000 h in air + 10% H_2O at 700°C, (A) Cr, (B) Ni, (C) Fe, (D) Si, (E) SEM picture, (F) O, (G) Mn, (H) Mo, and (I) Nb.

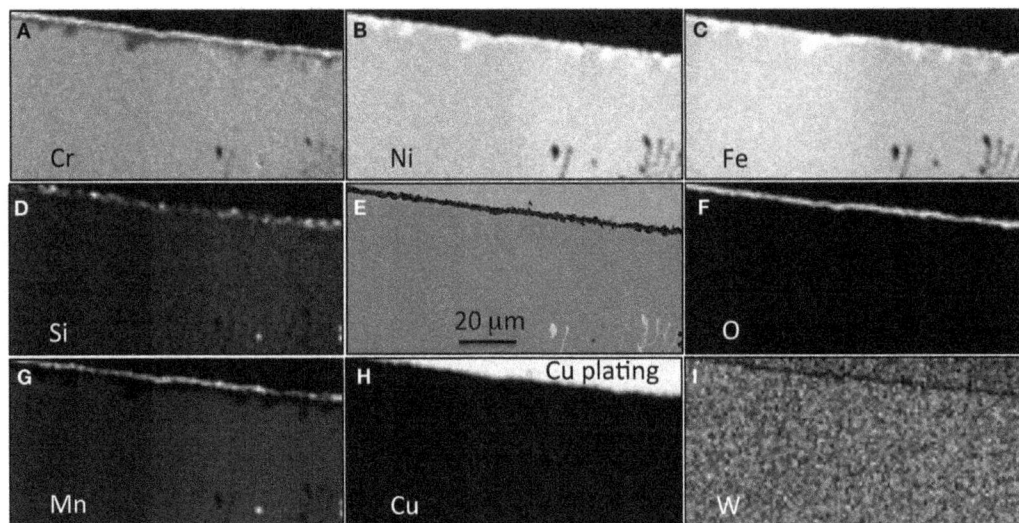

FIGURE 5 | EPMA mapping for alloy 22Cr–17.5Ni after exposure for 1000 h in air + 10% H_2O at 700°C, (A) Cr, (B) Ni, (C) Fe, (D) Si, (E) SEM picture, (F) O, (G) Mn, (H) Cu, and (I) W.

<7 μm, but larger local areas are observed, most likely because of the original chemical heterogeneity of the cast alloy. The Cr depletion is accelerated by the loss of Cr from the formation of oxy-hydroxide and once the depletion becomes large enough, this layer will be consumed, resulting in a faster reaction rate.

After exposure for 5000 h at 700°C, in **Figure 6**, four different oxide layers were present at the surface of alloy 22Cr: an outer Fe-rich layer (1), a thin and flat Cr-rich oxide layer that may be a remnant of the Cr-rich oxide seen in **Figure 5** (2), internal Ni-rich nodule-like regions (3), and a thin Cr-rich oxide layer following the nodule outlines (4). Si has segregated mostly in the two Cr-rich layers. Mn was also present in the Cr-rich layers, but Mn was also found at the oxide scale surface in the Fe-rich oxide layer. Again, the metal alloy beneath the second inner Cr-rich oxide

layer was depleted in Cr and Mn and enriched in Ni and Fe. It is worth noting that elongated precipitates rich in Cr, Mo, and Si have formed after 5000 h in the bulk of the 22Cr alloy.

Oxidation at 750°C

The mass changes for the CF8C, CF10M, CF8CP, CF8CW, CF8CWM, 21Cr, and 22Cr alloys at 750°C in air + 10% H_2O are shown in **Figure 7**. Significant mass losses were observed for alloys CF8C and CF8CP, reaching, respectively, −11 and −55 mg/cm² after only 1000 h. Such rapid mass loss is typically associated with massive scale spallation. The oxidation behaviors of the CF8CWCu-1 and CF8CWCu-2 alloys were clearly superior, with four different oxidation stages suggested by the mass change data. Small mass gains were measured at the beginning, followed by

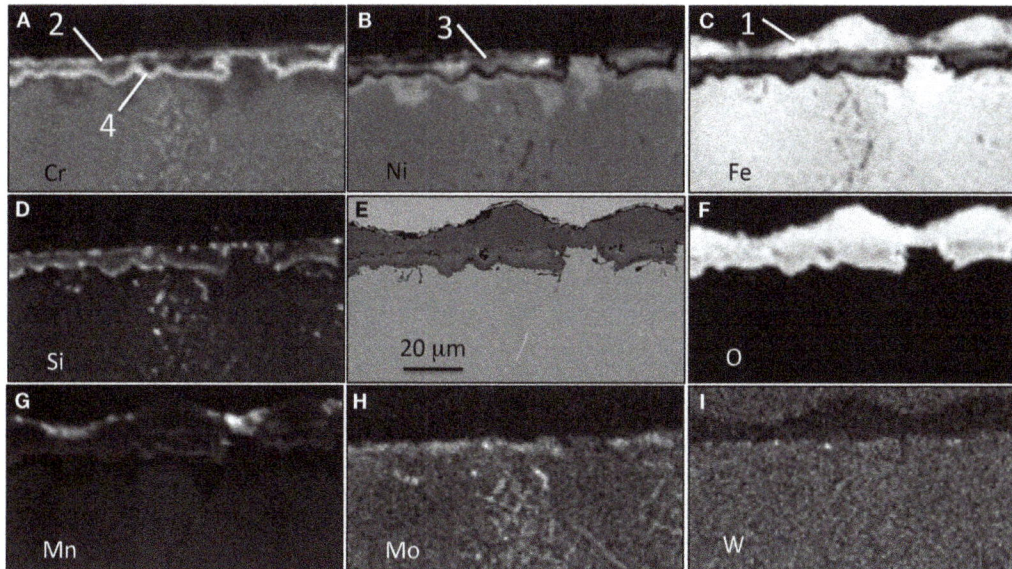

FIGURE 6 | EPMA mapping for alloy 22Cr–17.5Ni after exposure for 5000 h in air + 10% H₂O at 700°C, (A) Cr, (B) Ni, (C) Fe, (D) Si, (E) SEM picture, (F) O, (G) Mn, (H) Mo, and (I) W. Four different oxide layers are labels.

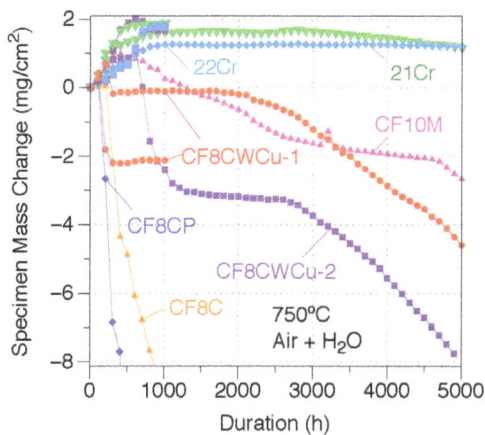

FIGURE 7 | Mass change curves for alloys CF8C, CF8CP, CF10M, CF8CWCu-1, CF8CWCu-2, 21Cr–15Ni, and 22Cr–17.5Ni exposed in air + 10% H₂O at 750°C for 1000 and/or 5000 h.

sudden mass losses. The specimen weights then remained the same for up to ~2500 h, and progressive mass decreases were observed during the final oxidation stage. In contrast, both the 21Cr and 22Cr alloys exhibited relatively protective oxidation behaviors at 750°C, with small mass gains for the first ~1000 h and subsequent slow mass loss for the remaining ~4000 h, consistent with small amounts of scale spallation or Cr oxy-hydroxide evaporation. As will be explained later, localized spallation is likely to be predominant. The CF10M alloy behaved in a similar manner, but the specimen loss rate was significantly higher.

Cross-section micrographs of the oxidation coupons exposed for 1000 and 5000 h at 750°C are presented in **Figure 8**. The CF8CP alloy exhibited a thick porous oxide scale after 1000 h composed of deep inner nodules and an inhomogeneous outer

layer with numerous cracks and obvious signs of spallation. The bi-layer oxide scale formed on alloy CF8CWCu-2 after 1000 h appeared similar to what was observed after 5000 h of exposure at 700°C. After 5000 h, the oxide scale was quite inhomogeneous, with many areas with large inner nodules below a thick outer layer, but locally a relatively thin oxide scale was still present, as seen in **Figure 8C**. Spallation of the outer layer was often observed in the areas with deep internal oxide nodules.

In the case of alloy, 21Cr and 22Cr, a ~15 μm multi-layer scale was mainly observed after 1000 h at 750°C, but a thin ~5 μm thick scale was also present locally (see white arrow in **Figure 8E**). After 5000 h of exposure, a slightly thicker multi-layer scale grew all over the specimen. Black arrows in **Figures 8D,F** highlight the presence of small voids for alloys CF8CWCu-2, 21Cr, and 22Cr observed beneath the inner oxide scale after exposure for 5000 h.

Oxidation at 800°C

The mass gain curves for the different alloys at 800°C are shown in **Figure 9**. For alloys CF8C, CF8CP, and CF8CWCu-1, rapid mass losses were measured and only 1000 h experiments were conducted. The two CF8CWCu-2 coupons gained weight during the first 100 h cycle and then lost weight for the remainder of the test, reaching −17 mg/cm² for the coupon tested for 5000 h. The oxidation resistance of alloy 21Cr was superior, with only a small mass gain for the first ~1000 h. However, the specimen exposed for a longer duration started to lose mass nearly linearly after 1000 h with a final mass loss of ~ −8 mg/cm² after 5000 h. Only alloy 22Cr exhibited a protective oxidation behavior at 800°C, with again small mass gains for the first ~1000 h, and then slow mass losses down to only ~ −0.37 mg/cm² after 5000 h.

Optical plan views and cross-sections of alloys CF8CWCu-2, 21Cr, and 22Cr after exposure for 5000 h at 800°C are presented in **Figure 10**. The oxidation degradation was clearly not

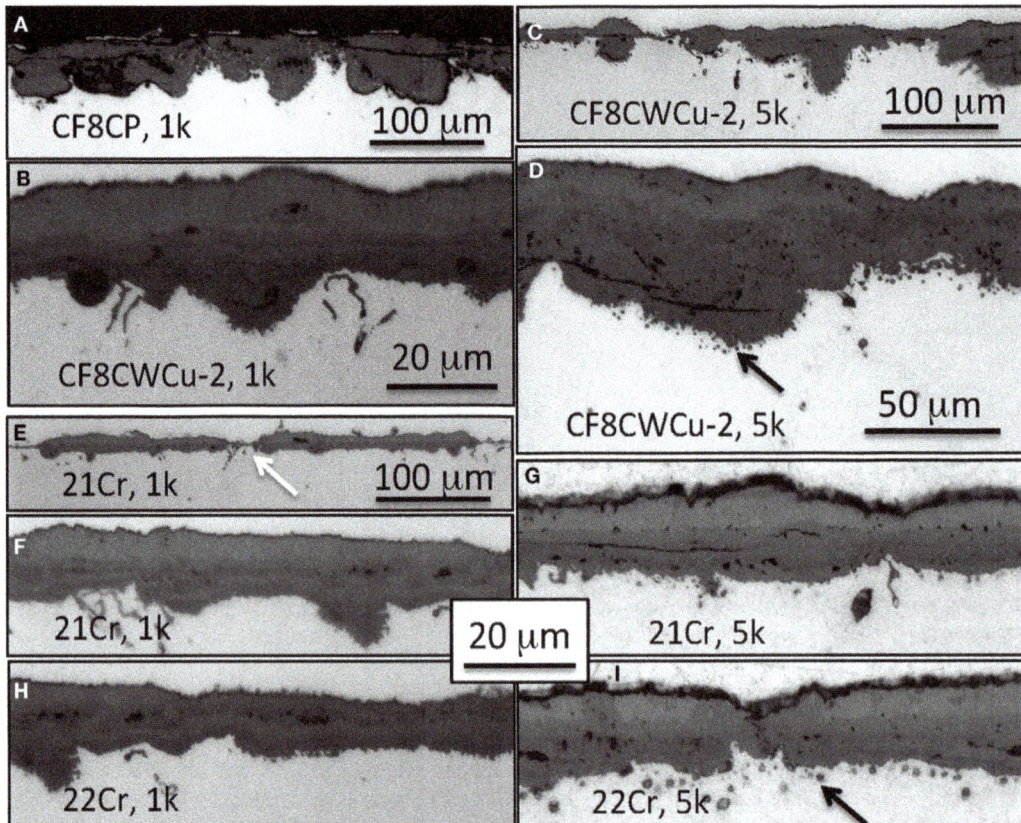

FIGURE 8 | Cross-section optical micrographs after exposure in air + 10% H_2O at 750°C, (A) CF8CP, 1000 h, (B) CF8CWCu-2, 1000 h, (C,D) CF8CWCu-2, 5000 h, (E,F) 21Cr–15Ni, 1000 h, (G) 21Cr–15Ni, 5000 h (H) 22Cr–17.5Ni, 1000 h, and (I) 22Cr–17.5Ni, 5000 h.

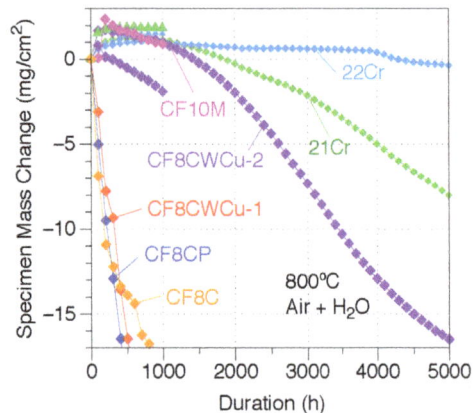

FIGURE 9 | Mass change curves for alloys CF8C, CF8CP, CF10M, CF8CWCu-1, CF8CWCu-2, 21Cr–15Ni, and 22Cr–17.5Ni exposed in air0 + 10% H_2O at 800°C for 1000 h for alloys CF8C and CF10M and 1000 and 5000 h for all the other alloys.

homogeneous for the CF8CWCu-2 and 21Cr alloys, and red arrows highlight areas with nodule formation and spallation.

Details of the scale cross-section structures are given in **Figure 11** for alloys CF8CP, CF8CWCu-1, CF8CWCu-2, 21Cr, and 22Cr. After 1000 h at 800°C, a nodular inner subscale was observed for alloys CF8CP and CF8CWCu-1, beneath a thick inner oxide layer. Most of the outer oxide layer had spalled away. **Figure 11C** clearly shows that after 1000 h at 800°C, the oxide scale at the surface of the CF8CWCu-2 sample was inhomogeneous, with some areas having a relatively thin oxide scale, while other areas exhibited large internal oxide nodules.

After 5000 h, the observations were similar, but now the nodular oxide areas grew in size and number. **Figures 11E,F** show, respectively, a thin oxide scale and a region of thick internal subscale, the external layer having spalled away. The thin scale after 5000 h looks very similar to the scale observed after 1000 h, in a similar low-corrosion area (**Figure 11D**). For the 21Cr specimens, the scale grown after 1000 h at 800°C was relatively homogeneous with a thickness varying from 10 to 20 µm (**Figures 11G,H**). However, after 5000 h, internal nodules became larger, **Figure 11I**, and, in some high-corrosion areas, very deep nodules and spallation of the outer scale were observed, similar to the observation for the CF8CWCu-2 specimen. The oxide scale formed on alloy 22Cr was similar to the scale grown on alloy 21Cr, both after 1000 and 5000 h, except that no high-corrosion/nodular oxide areas were observed (**Figures 11J–L**).

EPMA elemental maps for the CF8CP specimen exposed for 1000 h at 800°C are shown in **Figure 12**. A 25 µm thick Cr-rich inner oxide was observed. In this particular region, it appeared that there was a region of newly formed inner oxide with a lower Cr content and a center rich in Ni and Fe, likely because of metal

FIGURE 10 | Optical images of specimens exposed for 5000 h at 800°C in air + H₂O, (A) top view, alloy CF8CWCu-2, (B) top view, alloy 21Cr, (C) top view, alloy 22Cr, (D) Cross-section, alloy CF8CWCu-2, (E) Cross-section, alloy 21Cr, and (F) Cross-section, alloy 22Cr. Red arrows highlight areas with localized high rate corrosion attack.

entrapment. Note that where the inner layer is well-established, the adjacent metal is depleted in Cr and enriched in Ni. To the right where the new oxide nodule has formed, less Cr depletion is evident. Silicon was randomly distributed in the oxide scale and no Mn was detected. The outer layer containing Fe and Ni oxides had certainly spalled away. Similar elemental maps are presented in **Figure 13**, this time for alloy CF8CWCu-2, in a typical lower-corrosion area. Four or five different layers seem to be present: an outer discontinuous (Ni, Mn) spinel layer with a small amount of Cr (1), an Fe-rich layer (2), a Cr-rich layer containing some Fe (3), an Ni-rich layer with again some Fe (4), and then finally a Cr-rich inner layer (5). Silicon and Mn were present in the Cr and Ni layers. A very similar multi-layered scale was observed at the CF8CWCu-2 surface after exposure for 5000 h at 800°C, **Figure 14**. The only difference is the absence of an intermediate Cr-rich layer, but Cr was detected in the Ni-rich layer. EDX mapping of the scale grown on alloy 21Cr revealed a very similar oxide scale structure.

Discussion

Loss of Protective Oxidation Behavior

It is well known that the selective oxidation of Cr in ferritic and austenitic steels to form a protective chromia (or Cr-rich) scale and maintain it long term requires two conditions. First, the Cr content in the alloy must be sufficient to ensure the formation of a continuous external scale and avoid internal oxidation (Wagner, 1952). Second, the diffusion of Cr from the bulk of the alloy must be sufficient to maintain the scale and avoid significant depletion of Cr beneath the scale. Recall that in the presence of water vapor, the Cr consumption includes Cr lost due to the formation of volatile $CrO_2(OH)_2$, which likely explains why this

wet air environment is more aggressive than a similar test in dry or ambient air. The alloy composition, temperature, and oxidizing atmosphere can significantly affect the level of Cr required to meet these two criteria (Peraldi and Pint, 2004; Meier et al., 2010; Mu et al., 2013). At 650°C and 700°C, only the 21Cr and 22Cr alloys were able to form a (Cr,Mn)-rich layer during the first few hundred hours as shown in **Figure 1**. The exact structure of this thin scale has not been determined yet, but it was much more protective than the Fe and Ni-rich scale formed on the other alloys. Alloy CF8CWCu-2 formed a Cr-rich scale at 700°C but not at 650°C and alloys CF8CP and CF8CWCu formed non-protective mixed oxide layers. However, the formation of this mixed oxide layer may have sufficiently decreased the partial pressure of oxygen (PO₂) at the alloy surface to promote the formation of a continuous and more protective inner Cr-rich oxide layer (**Figure 4**), resulting in a significant oxidation rate decrease after 500–1000 h (**Figures 1 and 3**). The formation of the Cr-rich oxide layer was also likely facilitated by a local increase of the Cr content at the alloy surface due to the fast consumption of Fe and Ni in the alloy to form the Fe- and Ni-rich scales. A third factor is that the Fe- and Ni-rich layer may inhibit formation of the $CrO_2(OH)_2$, thus reducing the flux of Cr needed to maintain the Cr-rich scale. Previously, it was shown that the Mn-rich oxide does not inhibit the loss of Cr from alloy foils of Fe- and Ni-base chromia-formers, i.e., Cr loss in wet air exposures was independent of alloy Mn content (Pint, 2005).

For the 21Cr and 22Cr alloys, the consumption of Cr and Mn to form a protective (Cr,Mn) oxide layer resulted in a ~5 μm alloy depleted zone in Cr and Mn after 1000 h at 700°C (**Figure 5**, 22Cr alloy). After longer exposure, the Cr content at the surface dropped below the critical content required for the (Cr,Mn)-rich layer to form, and an outer Fe-rich layer (**Figure 6**, layer 1) and an inner Ni-rich subscale layer (layer 3) grew, respectively,

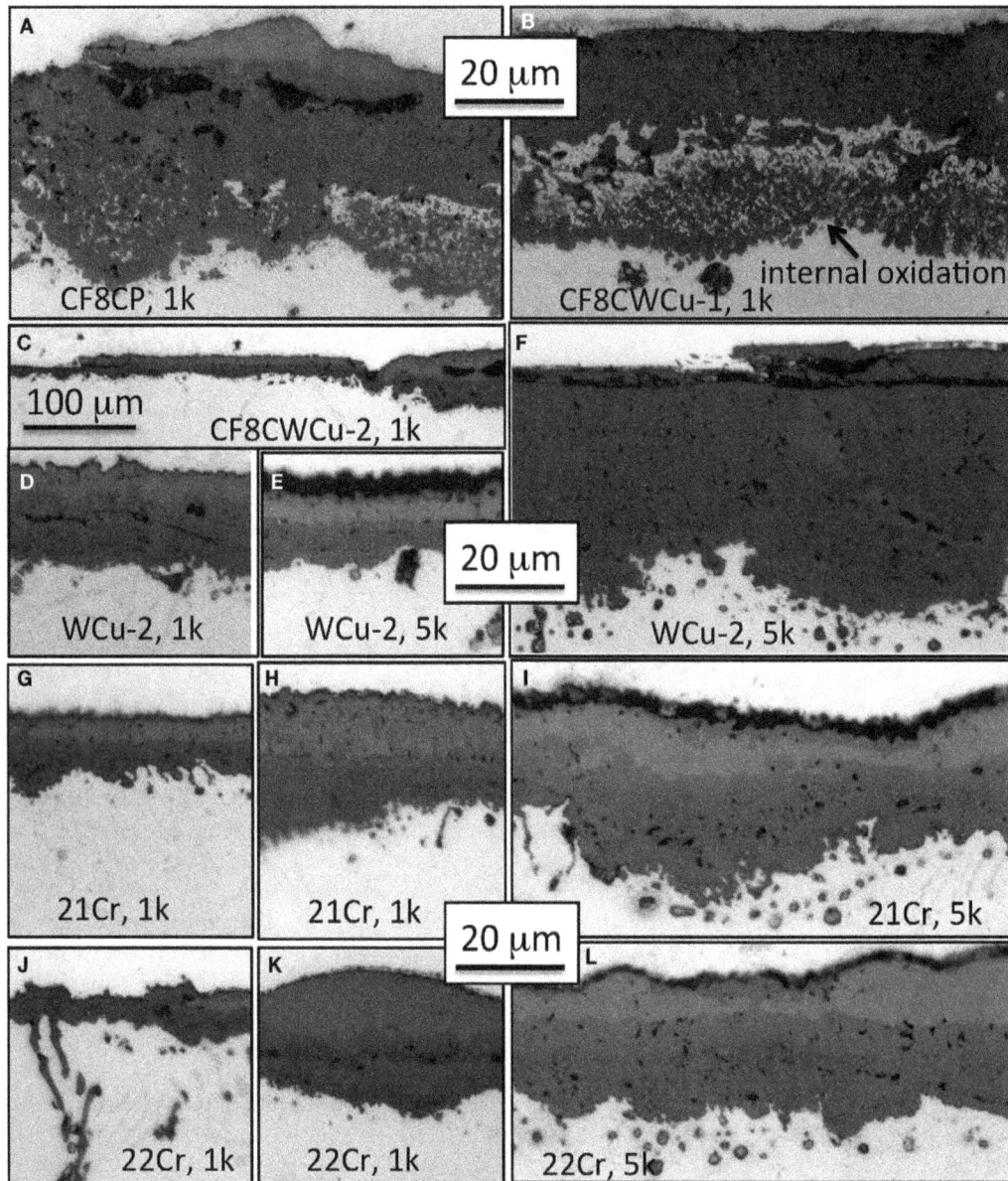

FIGURE 11 | Cross-section optical micrographs after exposure in air + 10%H₂O at 800°C, (A) CF8CP alloy for 1000 h, (B) CF8CWCu-1 alloy for 1000 h, (C,D) CF8CWCu-2 alloy for 1000 h, **(E,F) CF8CWCu-2 alloy for 5000 h, (G,H) 21Cr–15Ni alloy for 1000 h, (I) 21Cr–15Ni alloy for 5000 h, (J,K) 22Cr–17.5Ni alloy for 1000 h, and (L) 22Cr–17.5Ni alloy for 5000 h.**

above and below the original Cr-rich layer (layer 2). This fast consumption of Fe and Ni led in return to a relative increase of the Cr concentration at the alloy surface, above the threshold to form a Cr-rich layer, so that a second inner Cr-rich layer formed beneath the Ni-rich inner subscale (**Figure 6**, layer 4). This succession of more protective/less protective oxidation stages can be clearly seen in the mass gain curves in **Figure 1**. The formation of the second (Cr,Mn,Si)-rich layer led again to a depleted zone at the specimen surface, which could result, for exposure time long enough to decrease the Cr content at the surface below the Cr threshold, again in the formation of Fe and Ni-rich oxide scales. It is not clear if the accelerated consumption of Cr due to the formation and evaporation of $CrO_2(OH)_2$ is hindered by the presence of the outer

Fe-rich oxide layer. Similar depleted zones have been observed in austenitic steel superheater tubes (types 347H and 304H) after 4–30 kh in coal-fired boilers at ~550°C (Pint et al., Unpublished results)[1].

At 650°C and 700°C, the diffusion of Cr and Mn from the bulk to the alloy surface appears to be the limiting factor for the continuous growth of a more protective Cr-rich oxide scale. One reason some alloys showed more attack at 650°C than at 700°C is that faster Cr diffusion at the higher temperature may enable maintenance of the Cr-rich scale at long times.

---

[1] Pint, B. A., Leonard, D., and Shingledecker, J. (2015). Unpublished results.

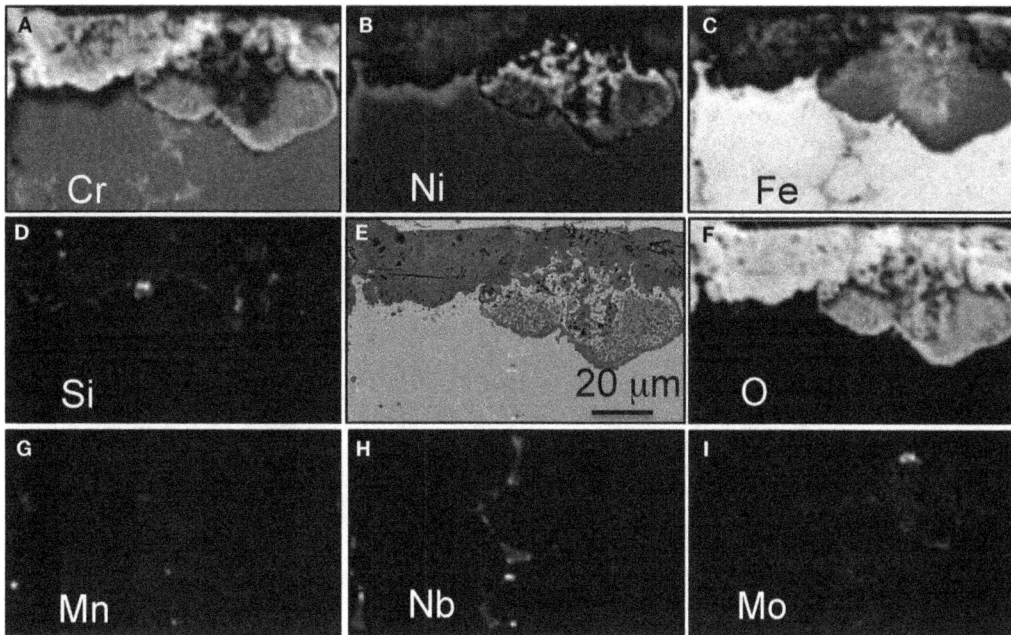

FIGURE 12 | EPMA mapping for alloy CF8CP exposed for 1000 h in air + 10% H_2O at 800°C, (A) Cr, (B) Ni, (C) Fe, (D) Si, (E) SEM picture, (F) O, (G) Mn, (H) Nd, and (I) Mo.

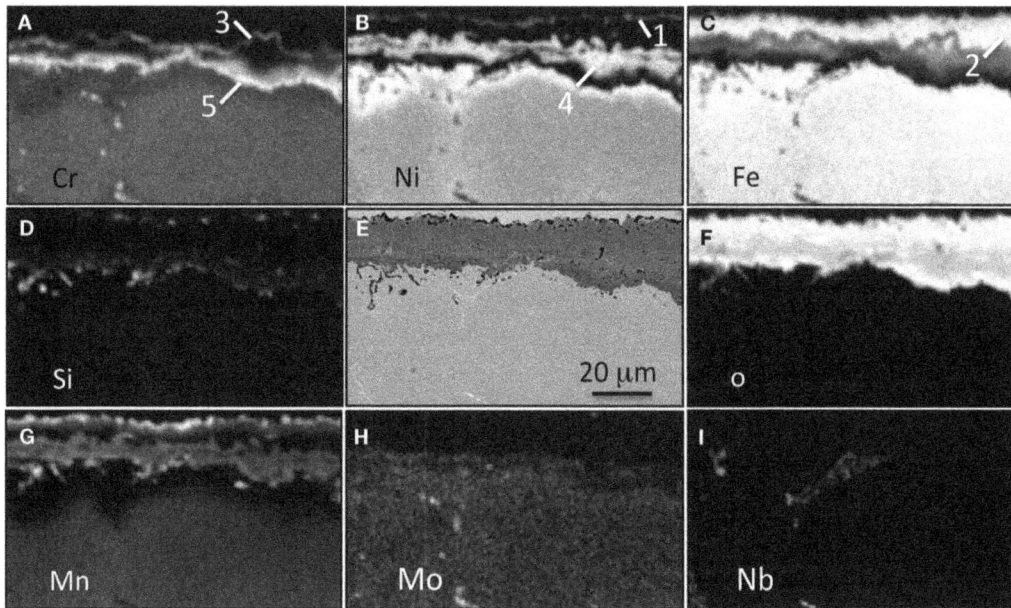

FIGURE 13 | EPMA mapping for alloy CF8CWCu-2 exposed for 1000 h in air + 10% H_2O at 800°C, (A) Cr, (B) Ni, (C) Fe, (D) Si, (E) SEM picture, (F) O, (G) Mn, (H) Mo, and (I) Nb. Five different oxide scales are labels.

For alloy CF8CP at 750°C and alloys CF8CP and CF8CWCu-1 at 800°C, an inner subscale was observed beneath large oxide nodules (**Figure 12**). The outer oxide layers rich in Fe and Ni were poorly adherent and spalled frequently during cooling. Therefore, the level of Cr in these alloys was insufficient to form the more protective Cr-rich layer. For alloy CF8CWCu-1 at 750°C and alloys CF8CWCu-2, 21Cr, and 22Cr at 750°C and 800°C, multi-layered oxide scales were observed after 1000 h, but an inner, more protective Cr-rich layer (**Figure 13**, layer 5) was able to form in most cases. However, massive nodule formation caused by local accelerated oxidation attack was observed (**Figures 10** and **11F**) for alloys CF8CWCu-1 and CF8CWCu-2 at 750°C, and the same alloys plus the 21Cr alloy at 800°C. The significant mass losses measured for all these alloys

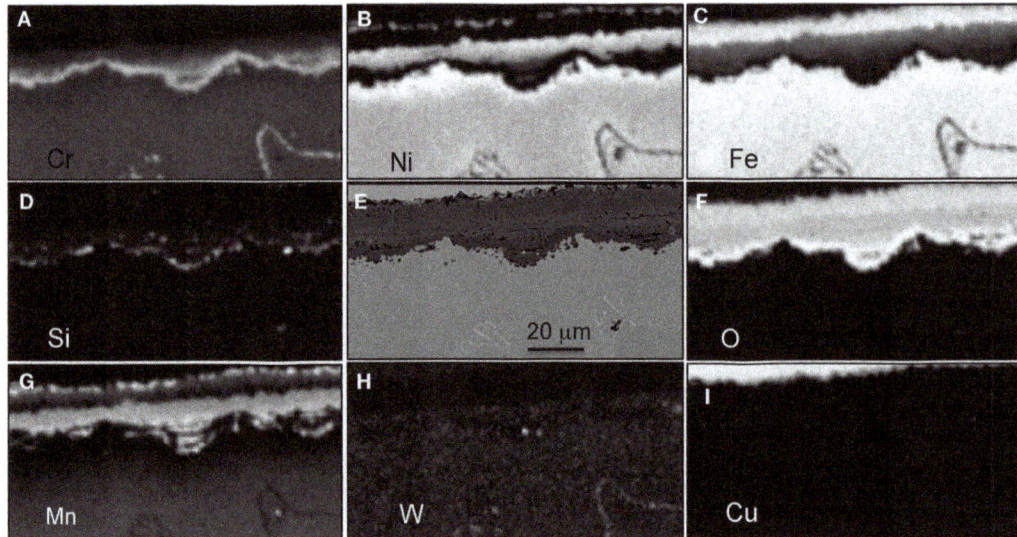

FIGURE 14 | EPMA mapping for alloy CF8CWCu-2 exposed for 5000 h in air + 10%H$_2$O at 800°C, (A) Cr, (B) Ni, (C) Fe, (D) Si, (E) SEM picture, (F) O, (G) Mn, (H) W, and (I) Cu.

(**Figures 7** and **11**) are related to the spallation occurring in these large nodular areas.

Effect of Alloy Composition on Oxidation Resistance

A summary of the specimen mass changes after 1000 or 5000 h in the 650–800°C range is shown in **Figure 15**. Considering low mass gain or low mass loss as acceptable after 5000 h for the desired application, the maximum operating temperature is ~650°C for alloy CF8C, 700°C for alloy CF8CP, between 700°C and 750°C for alloy CF10M, CF8CWCu-1, and CF8CWCu-2, between 750°C and 800°C for alloy 21Cr and ~800°C for alloy 22Cr.

The CF8C alloy did not perform as well as the other alloys because of its lower Cr and Ni contents, 18.7 and 10.4 wt%, respectively. Peraldi and Pint (2004) conducted 1 h cycle experiments at 650°C and 800°C in water vapor and showed, indeed, that increasing the Cr content from 12 to 20 wt% and/or the Ni content from 10% to 30% was beneficial for the alloy oxidation behavior. Similar results were shown for variations of Cr and Ni within the 347H specification, both higher Cr and higher Ni improved oxidation resistance (Pint and Dryepondt, 2014). Alloy CF10M has a similar composition as CF8C except for the addition of Mo and a higher level of Cr. Mo was not observed in the various oxide scales (**Figures 6** and **13**) and does not seem to play a role in the oxidation process, but the higher level of Cr, 20.6 wt%, is the likely reason for the better performance of alloy CF10M up to 750°C. The better oxidation behavior of alloys CF8CP, CF8CWCu-1, and CF8CWCu-2 compared with CF8C can be attributed to the higher levels of Ni and Mn. However, the reason for the superior performance of the WCu alloys compared with the CF8CP alloy is not clear. The only composition difference is the addition of Cu and W, but these two elements do not seem to participate in the oxidation process (**Figures 5** and **14**). The beneficial effect of Ni on oxidation resistance is not clearly understood, and Peraldi and Pint (2004) enumerated several potential

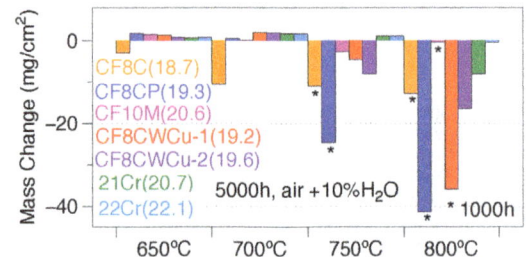

FIGURE 15 | Specimen mass changes for alloys CF8C, CF8CP, CF10M, CF8CWCu-1, CF8CWCu-2, 21Cr–15Ni, and 22Cr–17.5Ni exposed in air + 10%H$_2$O at 650–800°C for 1000 h or 5000 h. The Cr content of the alloys is indicated in parentheses.

explanations, including an effect of Ni on oxygen solubility, or on the activity of Cr and/or Fe at the oxide/alloy interface. Another plausible explanation is the stabilization of the austenite phase by Ni. In that case, the Cu addition is likely to play a similar austenite stabilizer role, which potentially enhances oxidation resistance. It is worth also noting that although both Mo and W are commonly used to strengthen alloys by solid solution, these two elements are chemically quite different. W is soluble in the austenite phase, is slow diffusing compared with Mo and Cr, and has a clear stabilizing effect on M$_{23}$C$_6$ carbides. W and Mo are, therefore, likely to have distinct effects on alloy oxidation resistance. Prior work with Fe–13at%Cr–5at% (W or Mo) alloys at 900°C (no H$_2$O) found that both Mo and W additions improved oxidation resistance. Normally a 13%Cr alloy could not form a chromia scale at 900°C, but both W and Mo enabled formation (Pint, Unpublished results)[2]. Thus, there is some potential for refractory metal additions to have a beneficial effect on oxidation resistance.

Another explanation for the better oxidation behavior of CF8CWCu alloys is that these alloys are more homogeneous than

---

[2]Pint, B. A. (2007). Unpublished results.

the CF8CP alloy. These cast alloys do not require any heat treatment after fabrication for optimum creep performance, but the chemical inhomogeneity between the dendritic and interdendritic structure may affect the alloy oxidation, as can be seen for example in **Figure 5**. In the 700–750°C range, small local concentration variations could result in a transition from protective to non-protective oxidation behavior. Material inhomogeneity could also explain why the CF8CWCu-2 is more oxidation resistant at 800°C compared with alloy CF8CWCu-1, although the compositions of the two alloys are similar.

As expected, the higher levels of Cr and Ni in the two new alloys resulted in an improvement of the oxidation performance compared with the base CF8CWCu alloys. CF10M and the 21Cr alloys have the same Cr content but the higher content for the 21Cr alloy of austenite-stabilizing elements such as Ni, Mn, Cu, and N, likely explains the superior oxidation behavior of the 21Cr alloy. The best performing alloy was the 22Cr alloy, with only moderate mass change at 800°C and accelerated corrosion attack only at the specimen corners. The increase to 22Cr and 17.5Ni resulted, therefore, in a potential increase of ~100°C for the alloy operating temperature from the standpoint of oxidation resistance compared with the standard CF8CP, going from ~700°C to 800°C.

Conclusion

Three commercial cast alloys, CF8C, CF10M, and CF8C-Plus, and four newly developed cast stainless steels were tested in air + 10% H_2O to simulate exhaust gas environments, at temperatures ranging from 650°C to 800°C for a duration of up to 5000 h. At 650°C and 700°C, a change in the oxidation behavior from protective to non-protective or non-protective to protective was observed due to depletion or enrichment beneath the oxide scale of elements such as Cr and Mn. For long-term exposure at 650° and 700°C, all the alloys, except for the CF8C, exhibited a relatively protective oxidation behavior due to the formation of an inner Cr, Mn-rich scale. Above 700°C, the alloys with the lower Cr and Ni contents, ~19 wt% Cr and less than 13 wt% Ni, started to locally form massive Fe-rich oxide nodules, and these alloys lost weight very rapidly when the outer portion of these nodules spalled. Increasing the Cr and Ni contents to, respectively, ~21 wt% Cr and 15 wt% Ni was sufficient to allow for the formation of an inner protective Cr-rich oxide layer at 750°C, but local highly corroded areas were observed at 800°C. This issue was solved by increasing the Cr and Ni level to 22 wt% Cr and 17.5 wt% Ni, and this newly developed cast alloy represents a potential gain of ~100°C in service temperature compared with the commercial CF8C-Plus alloy for operation in aggressive environments.

Acknowledgments

The authors would like to thank M. Stephens for his help with the oxidation exposures, T. Lowe, H. Longmire, and T. Jordan for the metallography work, and D. Leonard for conducting the microprobe measurements. They want also to acknowledge M. Brady for reviewing the manuscript. This research was sponsored by the U.S. Department of Energy, Office of Energy Efficiency and Renewable Energy, Advanced Manufacturing Office.

References

Asteman, H., Svensson, J. E., Norell, M., and Johansson, L. G. (2000). Influence of water vapor and flow rate on the high-temperature oxidation of 304L; effect of chromium oxide hydroxide evaporation. *Oxid. Met.* 54, 11–26. doi:10.1023/A:1004642310974

Chi, C., Yu, H., Dong, J., Liu, W., Cheng, S., and Liu, Z. (2012). The precipitation strengthening behavior of Cu-rich phase in Nb contained advanced Fe-Cr-Ni austenitic heat resistant steel for USC power plant application. *Prog. Nat. Sci.* 22, 175–185. doi:10.1016/j.pnsc.2012.05.002

Davis, P. B., Schutte, C. L., and Gibbs, J. L. (2012). *Progress Report for Propulsion Materials*. Washington, DC: US Department of Energy, Energy Efficiency and Renewable Energy.

Essuman, E., Meier, G. H., Zurek, J., Hansel, M., and Quadakkers, W. J. (2008). The effect of water vapor on selective oxidation. *Oxid. Met.* 69, 143–162. doi:10.1007/s11085-007-9090-x

Evans, N. D., Maziasz, P. J., Shingledecker, J. P., and Pollard, M. J. (2010). Structure and composition of nanometer-sized nitrides in a creep-resistant cast austenitic alloy. *Met. Trans.* 41, 3032–3041. doi:10.1007/s11661-010-0321-4

Ha, V. T., and Jung, W. S. (2012). Evolution of precipitate phases during long-term isothermal aging at 1083 K (810°C) in a new precipitation-strengthened heat-resistant austenitic stainless steel. *Metall. Mater. Trans.* 43A, 3366–3378. doi:10.1007/s11661-012-1150-4

Henry, S., Mougin, J., Wouters, Y., Petit, J. P., and Galerie, A. (2000). Characterization of chromia scales grown on pure chromium in different oxidizing atmospheres. *Mater. High Temp.* 17, 231. doi:10.1179/mht.2000.17.2.008

Hultquist, G., Tveten, B., and Hornlund, E. (2000). Hydrogen in chromium: influence on the high-temperature oxidation kinetics in H_2O, oxide-growth mechanisms, and scale adherence. *Oxid. Met.* 54, 1. doi:10.1023/A:1004610626903

Maziasz, P. J., and Pint, B. A. (2010). "High temperature performance of cast CF8C-Plus austenitic stainless steel," in *Proceedings of the ASME Turbo Expo 2010*, Glasgow.

Maziasz, P. J., and Pint, B. A. (2011). High temperature performance of cast CF8C-plus austenitic stainless steel. *J. Eng. Gas Turbine Power* 133, 1–5. doi:10.1115/1.4002828

Meier, G. H., Jung, K., Mu, N., Yanar, N. M., Pettit, F. S., Abellan, J. P., et al. (2010). Effect of alloy composition and exposure conditions on the selective oxidation behavior of ferritic Fe–Cr and Fe–Cr–X alloys. *Oxid. Met.* 74, 319–340. doi:10.1007/s11085-010-9215-5

Mu, N., Jung, K., Yanar, N. M., Pettit, F. S., Holcomb, G. R., Howard, B. H., et al. (2013). The effects of water vapor and hydrogen on the high-temperature oxidation of alloys. *Oxid. Met.* 79, 461–472. doi:10.1007/s11085-012-9349-8

Peraldi, R., and Pint, B. A. (2004). Effect of Cr and Ni contents on the oxidation behavior of ferritic and austenitic model alloys in air with water vapor. *Oxid. Met.* 61, 463–483. doi:10.1023/B:OXID.0000032334.75463.da

Pint, B. A. (2005). "The effect of water vapor on Cr depletion in advanced recuperator alloys," in *Proceedings of ASME Turbo Expo 2005* (Reno-Tahoe, NV: ASME), 1–8.

Pint, B. A., and Dryepondt, S. (2014). "Effect of alloy composition and surface engineering on steam oxidation resistance," in *Proc. 7th Inter. Conf. on Advances in Materials Technology for Fossil Power Plants*, eds D. Gandy and J. Shingledecker (Materials Park, OH: ASM International), 803–814.

Quadakkers, W. J., Zurek, J., and Hansel, M. (2009). Effect of water vapor on high-temperature oxidation of FeCr alloys. *JOM* 61, 44. doi:10.1007/s11837-009-0102-y

Saunders, S. R. J., Monteiro, M., and Rizzo, F. (2008). The oxidation behavior of metals and alloys at high temperatures in atmospheres containing water vapour: a review. *Prog. Mater. Sci.* 53, 775–837. doi:10.1016/j.pmatsci.2007.11.001

Sawaragi, Y., and Hirano, S. (1990). "Development of a new 18-8 austenitic steel (0.1C-18Cr-9Ni-3Cu-Nb, N) with high elevated temperature strength for fossil fired boilers," in *American Society of Mechanical Engineers, Pressure Vessels and Piping Division (Publication) PVP*, Vol. 201 (New York: ASME (American Society of Mechanical Engineers)), 141–146.

Schutze, M., Schorr, M., Renusch, D. P., and Vossen, J. P. T. (2004). The role of alloy composition, environment and stresses for the oxidation resistance of modern 9% Cr steels for fossil power stations. *Mater. Res.* 7, 111–123. doi:10.1590/S1516-14392004000100016

Shingledecker, J. P., Maziasz, P. J., Evans, N. D., and Pollard, M. J. (2005). "Alloy additions for improved creep-rupture properties of a cast austenitic alloy," in *Creep Deformation and Fracture, Design, and Life Extension*, eds R. S. Mishra, J. C. Earthman, S. V. Raj, and R. Viswanathan (Warrendale, PA: MS&T Publications), 129–138.

Shingledecker, J. P., Maziasz, P. J., Evans, N. D., Santella, M. L., and Pollard, M. J. (2006). CF8C plus: a new high temperature austenitic casting alloy for advanced power systems. *Energy Mater.* 1, 25–32. doi:10.1179/174892306X99697

Swindeman, R. W., Bolling, E., and Li, C. Y. (1990). "Copper-bearing stainless steels for high temperature applications," in *Proceedings of the ASME 2013 Pressure Vessels and Piping Conference* (Nashville, TN), 165–167.

Wagner, C. (1952). Theoritical analysis of the diffusion processes determining the oxidation rate of alloys. *J. Electrochem. Soc.* 99, 369–380. doi:10.1149/1.2779605

Wu, Q., Vasudevan, V. K., Shingledecker, J. P., and Swindeman, R.W. (2005). "Microstructure characterization of advanced boiler materials for ultra supercritical coal power plants," *Proceedings Fourth International Conference on Advances in Materials Technology for Fossil Power Plants* (Materials Park, OH: ASM-International), 748–761.

Young, D. J., and Pint, B. A. (2006). Chromium volatilization rates from Cr_2O_3 scales into flowing gases containing water vapor. *Oxid. Met.* 66, 137–153. doi:10.1007/s11085-006-9030-1

Conflict of Interest Statement: The Review Editor Guangling Song declares that, despite being affiliated to the same institution as the authors, the review process was handled objectively and no conflict of interest exists. The authors declare that the research was conducted in the absence of any commercial or financial relationships that could be construed as a potential conflict of interest.

20

Nanostructured colloidal particles by confined self-assembly of block copolymers in evaporative droplets

*Minsoo P. Kim and Gi-Ra Yi**

School of Chemical Engineering, Sungkyunkwan University, Suwon, South Korea

Block copolymers (BCPs) can create various morphology by self-assembly in bulk or film. Recently, using BCPs in confined geometries such as thin film (one-dimension), cylindrical template (two-dimension), or emulsion droplet (three-dimension), nanostructured BCP particles have been prepared, in which unique nanostructures of the BCP are formed via solvent annealing process and can be controlled depending on molecular weight ratio and interaction parameter of the BCPs, and droplet size. Moreover, by tuning interfacial property of the BCP particles, anisotropic particles with unique nanostructures have been prepared. Furthermore, for practical application such as drug delivery system, sensor, self-healing, metamaterial, and optoelectronic device, functional nanoparticles can be incorporated inside BCP particles. In this article, we summarize recent progress on the production of structured BCP particles and composite particles with metallic nanoparticles.

Keywords: block copolymer particle, nanostructure, hybrid particle

Edited by:
P. Davide Cozzoli,
University of Salento, Italy
Reviewed by:
M. Gabriella Santonicola,
Sapienza University of Rome, Italy
Markus Gallei,
Technische Universität Darmstadt,
Germany
Jiguang Liu,
Beijing Institute, China
Nathaniel Alexander Lynd,
University of Texas at Austin, USA

*Correspondence:
Gi-Ra Yi,
School of Chemical Engineering,
Sungkyunkwan University, 2066,
Seobu-ro, Jangan-gu, Suwon-si,
Gyeonggi-do 440-746, South Korea
yigira@skku.edu

What is Block Copolymer Particles?

Block copolymers (BCPs) have been widely used as templates for the nanostructured materials, since the BCPs can self-assemble to a variety of nanoscale morphologies like spheres, cylinders, and lamellar depending on their composition, volume fraction and molecular weight (Bates and Fredrickson, 1990; Hawker and Russell, 2005; Xiong et al., 2009; Mai and Eisenberg, 2012). The key to the use of BCPs for the complex nanostructures is to control the orientation and lateral order of the microdomains in the thin films. For instance, the use of solvent vapor can induce orientation of the BCP nanostructures (Malenfant et al., 2007). Kim and Libera (1998) carried out the effect of solvent on the morphological evolution of BCPs in the thin-film, resulting in perpendicular orientation of asymmetric poly(styrene-*block*-butadiene-*block*-styrene) (SBS) after solvent evaporation (Kim and Libera, 1998). Cavicchi and Russell (2007) reported a significant improvement on the lateral ordering of poly(isoprene-*block*-lactide) (PI-*b*-PLA) by solvent annealing in the casting film (Cavicchi and Russell, 2007). Besides the solvent annealing method, the orientation of BCP nanostructures can be also controlled by changing the surface properties of the substrate or the gap between the two substrates (Huang et al., 1998). For example, the periodic patterned substrates have been used to enhance or control the lateral order of microdomains in BCPs (**Figure 1A**) (Bita et al., 2008). Furthermore, various frustrated nanostructures inside confined geometry could be developed as shown in **Figures 1B–D** (Yu et al., 2007, 2006; Li et al., 2013). For one-dimensional (1D) confined geometry, it has been well known that perforated lamella phase can be formed instead of gyroid or bicontinuous phase (**Figure 1B**) (Koneripalli et al., 1995; Nikoubashman et al., 2014; Zhang et al., 2015a). In the case of cylindrical pores in anodized alumina oxide (AAO) membranes, various helical

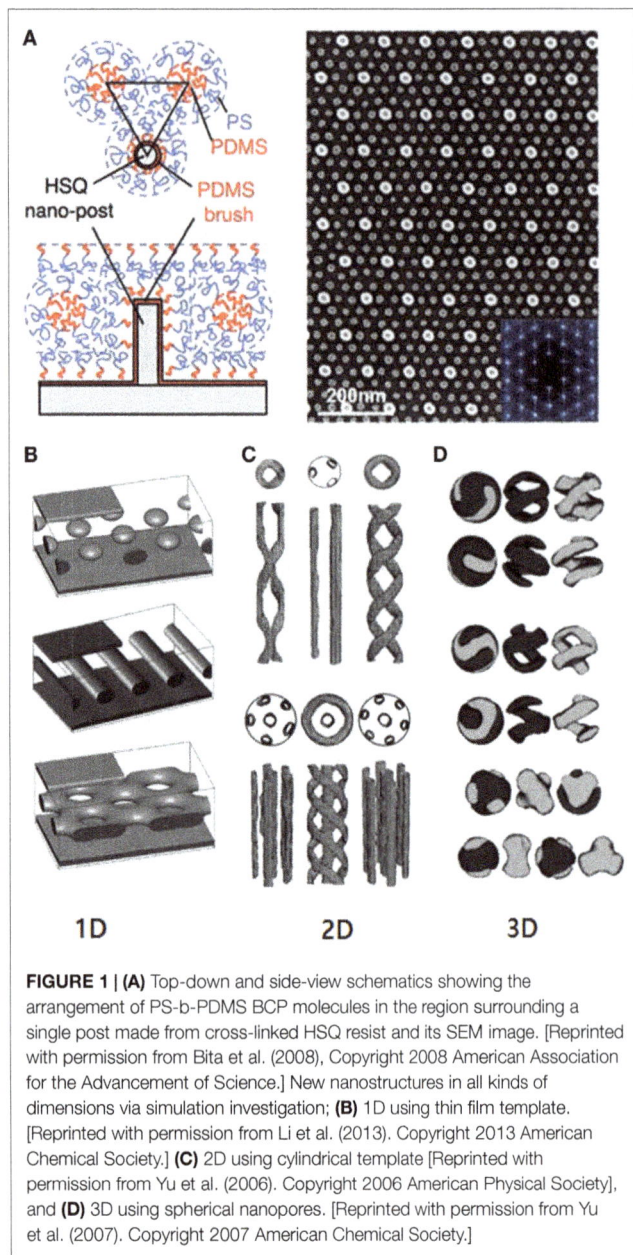

FIGURE 1 | (A) Top-down and side-view schematics showing the arrangement of PS-b-PDMS BCP molecules in the region surrounding a single post made from cross-linked HSQ resist and its SEM image. [Reprinted with permission from Bita et al. (2008), Copyright 2008 American Association for the Advancement of Science.] New nanostructures in all kinds of dimensions via simulation investigation; **(B)** 1D using thin film template. [Reprinted with permission from Li et al. (2013). Copyright 2013 American Chemical Society.] **(C)** 2D using cylindrical template [Reprinted with permission from Yu et al. (2006). Copyright 2006 American Physical Society], and **(D)** 3D using spherical nanopores. [Reprinted with permission from Yu et al. (2007). Copyright 2007 American Chemical Society.]

production of impact-resistant polymers, composite materials, catalytic supports, coatings, and adhesives (Sundberg and Durant, 2003). For instance, dyed multi-shell particles have been used in high-density 3D optical data storage and security data encryption (Pham et al., 2004). Most of core-shell structured colloids have been employed in bio-application such as drug delivery and biosensor (Otsuka et al., 2003; Wang et al., 2011; Robb et al., 2012; Yu et al., 2014; Louage et al., 2015). The researchers have prepared the colloidal particles composed of the assemblies of spherical polymeric micelles, which have been employed to study stimuli-reactive delivery or sensor system. In addition, hollow particles have been used in bio-application such as self-healing. Mohwald group reported that hollow polymer particle with onion-like nanoshell structure can be employed in bio-related application such as nanocapsule and self-healing (Sukhorukov et al., 2005; Grigoriev et al., 2009; Delcea et al., 2011; Latnikova et al., 2012). Structured spheres with periodic radial variation of the refractive index have attracted much attention due to their potential uses as spherical dielectric resonators in cavity quantum electrodynamics, optical switches, and limiters (Babin et al., 2003; Gourevich et al., 2006; Petukhova et al., 2008; Zhang et al., 2008a; Burlak et al., 2012). Interestingly, by tuning the surface property of the BCP particles, ellipsoid-structured particles from spherical particles can be created (Deng et al., 2013; Klinger et al., 2014). The shape-controlled particles can be applied in the fields such as multi-responsive materials, optical device, and biomedical delivery device.

Furthermore, incorporating functional materials such as metal or semiconductor into the nanostructured BCP microsphere can produce novel inorganic nanostructures with new characteristics, which depend on their spatial and size distribution (Jeon et al., 2009; Connal et al., 2012; Ku et al., 2013). The composite BCP microspheres have been prepared by selective interaction between one block of the BCP and polymer-capped nanoparticles or infiltration of inorganic precursors into the one block. For example, Ku et al. (2013) have fabricated hybrid BCP-quantum dot (QD) microspheres by spatially isolating different-colored QDs into core-blocks of the BCP micelles, which can control Förster resonance energy transfer (FRET) between QDs, resulting in formation of multicolor emission colloidal particles.

This review focuses mainly on how BCP colloidal particles could be formed with unique internal structures and then describes how nanoparticles could be incorporated or decorated. Finally, we will discuss remaining issues or problems for future applications.

Synthesis of Structured Block Copolymer Particles

Structured BCP particles can be prepared synthetically using mini-emulsion polymerization method (Landfester, 2001a, 2009;Rao and Geckeler, 2011). In the method, two different kinds of monomers are dispersed in aqueous media with surfactants to stabilize the interface between the monomers and water. Polymerization is performed by an initiator dispersed in the aqueous phase, resulting in formation of random BCPs. More effective method to fabricate the particles has been

structures were observed and they have been used as confining geometries (Lambooy et al., 1994; Dauphas et al., 2004; Wu et al., 2004; Xiang et al., 2004; Maiz et al., 2014; Hou et al., 2015). According to the theoretical prediction, more unusual morphologies like single helices, double helices, and meshes could be produced as shown in **Figure 1C** (Li and Wickham, 2006; Li et al., 2006; Yang et al., 2012; Bae et al., 2013; Xu et al., 2014; He et al., 2015; Zhang et al., 2015b; Zuo et al., 2015).

Likewise, in the case of three-dimensional (3D) confining geometries such as spheres, multi-shell, or onion-like structures or other morphologies as shown in **Figure 1D** could be developed (Yu et al., 2007). The nanostructures has been proved simulationally (Yu et al., 2007; Hao et al., 2014) and experimentally (Higuchi et al., 2012, 2008; Xu et al., 2015). These are practically important in a broad range of applications including the developed using radical polymerization (Smulders and Monteiro, 2004; Landfester, 2009; Wang and Zhang, 2012). For example, poly(*n*-butylacrylate-*block*-styrene) (PnBA-*b*-PS) particles were prepared by radical emulsion polymerization using a structural control agent (e.g., 1,1-diphenylethylene) or reversible addition-fragmentation chain-transfer (RAFT) agent, which is dispersed in the aqueous solution. Landfester et al. demonstrated that the particles, which have core-shell or hollow structures, can be used in the bio-application such as nanocapsules for drug delivery or self-healing materials (Landfester, 2001b, 2006; Landfester and Mailander, 2013; Lv et al., 2013; Vimalanandan et al., 2013). Moreover, hollow structured particles can be produced by layer-by-layer (LBL) assembly on colloidal particles as a template (Pastoriza-Santos et al., 2001; Coustet et al., 2014). The LBL technique is based on the consecutive assembly of oppositely charged polymer layers on a preformed charged spherical colloidal particle. After LBL assembly, the cores of the particles can be removed by selective dissolution or calcination, followed by formation of hollow particles. Mohwald et al. reported that the hollow particles were prepared by LBL assembly using charged polymer such as poly(diallyldimethylammonium chloride), poly(allylamine hydrochloride), or poly(styrene sulfonate) onto the charged silica or PS particles (Caruso et al., 1998a,b; Caruso and Mohwald, 1999; Shchukin and Mohwald, 2006). After removal of the cores in the particles, the hollow structured particles were obtained, which can be applied as biomaterials such as drug delivery carriers or self-healing (Peyratout and Dahne, 2004; Ma et al., 2005; Rivera Gil et al., 2008; Delcea et al., 2011; Latnikova et al., 2012; Yan et al., 2014). Although the structured particles prepared by mini-emulsion polymerization or LBL technique have narrow size distribution, the methods have been limited to control the inner nanostructure of the BCP particles.

On the other hand, BCP particles can be prepared directly from polymer solution by filling lithographically prepared voids and removing the template such as AAO template (Wu et al., 2004; Wang et al., 2008; Mei et al., 2011; Mei and Jin, 2013; Chu et al., 2014). Generally, a template is immersed in polymer solution during a constant time and then the template is picked up from the solution, resulting in formation of BCP nanotubes. After annealing of the BCP nanotubes, the template is selectively removed by etchant such as NaOH to release the nanotubes. Some groups demonstrated that the BCP particles were prepared by the templating method, which had spherical shape after exposure to neutral solvent for the BCP. Furthermore, mesoporous-structured BCP particles were fabricated by swelling of the BCP after exposure to a selective solvent which prefers one of the BCP (Mei et al., 2011; Mei and Jin, 2013).

Emulsion-encapsulation and solvent-evaporation method have been widely used for BCP particles, in which the polymer solution was dispersed in a non-solvent and the organic solvent inside emulsion was evaporated. Similarly, Thomas et al. (1990) experimentally demonstrated the formation of BCP particle using aerosol droplets as the confined geometry. Generally, BCP solution is atomized with an air jet atomizer, resulting in production of aerosol droplets. In nitrogen carrier gas, the droplets are transported to a heated tubular reactor where they are dried and heated for several seconds. After the droplets are cooled, the dried BCP particles are finally collected. The method using the aerosol droplets would produce small particles ranging from submicrometers to a few micrometers after completely removing the solvent from the droplets (Lu et al., 1999). Later then, various structured BCP particles were prepared from aerosol droplets where poly(styrene-*block*-4-vinylpyridine) (PS-*b*-P4VP) or triblock copolymer with poly(3-triethoxysilyl)propyl block were included (Zhang et al., 2008a,b, 2010a; Rahikkala et al., 2013). However, the nanostructures inside the BCP particles could not reach to equilibrium state or significantly overlapped between the neighboring domains because of fast removal of the solvent from the aerosol droplets.

In case of using emulsion droplets as confining geometries (**Figure 2A**), well-defined BCP particles can be prepared by the evaporationinduced microphase separation of BCPs. BCPs are dissolved in organic solvent, which is not miscible to water and then dispersed in aqueous solution containing surfactant. After slowly solvent evaporation from the droplets, the structured BCP particles are produced. When symmetric BCPs were assembled, onion-like structure was observed and other morphologies were also obtained by adding homopolymers as shown in **Figure 2B** (Jeon et al., 2007). The microdomains of spheres, cylinders, and lamellae were structured in concentric arrangements due to the external spherical confinement provided by the emulsion droplets. Yu et al. (2007) reported the morphologies in BCP particles by Monte-Carlo simulations. It was shown that the morphologies in the BCP particles were dependent on the equilibrium spacing of the BCP structures in the bulk films (L_0) and the diameters of the microspheres (D) as shown in **Figure 2C**. Moreover, interfacial tension was controlled using a mixture of surfactants, for example, poly(styrene-*block*-ethyleneoxide) (PS-*b*-PEO) and poly(butadiene-*block*-ethylene oxide) (PB-*b*-PEO) (Jeon et al., 2008). As shown in **Figure 3**, single type of surfactant produced onion-like spherical particles, but mixture of surfactants produced ellipsoidal particles with stacked lamellae nanostructures.

Another method to fabricate the structured BCP particles directly from polymer solution is using re-precipitation (Yabu et al., 2005; Li et al., 2010; Yabu, 2013). The polymer solution is slowly added to a large amount of poor solvent for the polymer, which is miscible with the good solvent. After evaporation of good solvent into the polymer solution, structured BCP particles can be produced. To selectively evaporate the good solvent, its boiling point should be lower than that of the poor solvent. For example, Higuchi et al. prepared the nanostructured BCP particles by controlled precipitation from a tetrahydrofuran (THF)-water mixture (Higuchi et al., 2008). Through gradual evaporation of THF in the solvent mixture with symmetric poly(styrene-*block*-isoprene) (PS-*b*-PI), various nanostructures in the particles were produced.

Generally, the nanostructures of BCP in thin film have been developed during thermal annealing (Bates and Fredrickson, 1990; Winey et al., 1994; Park et al., 2003; Tang et al., 2004; Hawker and Russell, 2005; Xu et al., 2005; Ahn et al., 2009; Marencic and Register, 2010; Lee et al., 2011; Bai et al., 2013) above glass transition temperature (T_g) or solvent annealing (Li et al., 1996; Zhao et al., 2004; Kim et al., 2006; Bang et al., 2007; Cavicchi and Russell, 2007; Jung and Ross, 2007; Malenfant et al., 2007; Zhang

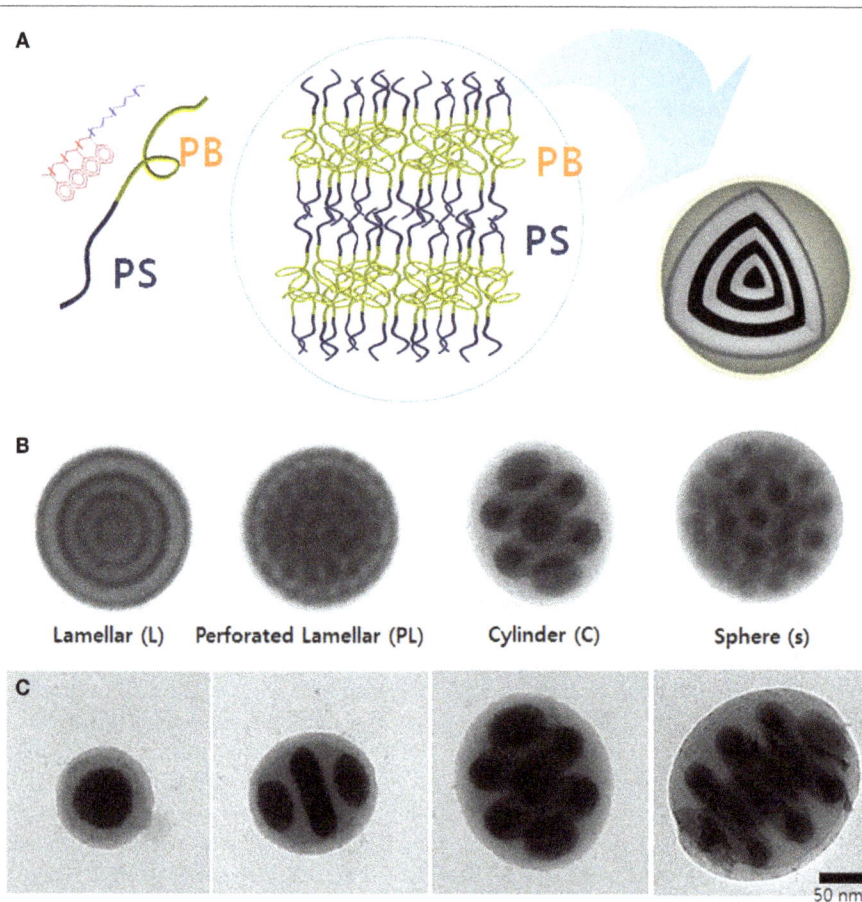

FIGURE 2 | 3D confinement effect of BCPs in emulsion droplet.
(A) Schematic illustration of preparation of PS-b-PB BCP particles with
nanostructures in emulsion droplets. Morphological transition of the BCP
particles depending on (B) volume fraction of the BCP and (C) droplet size.
[Reprinted with permission from Jeon et al. (2007). Copyright 2007 American
Chemical Society.]

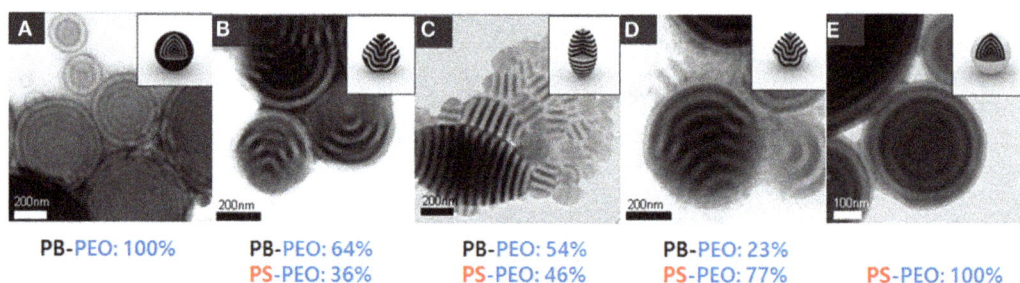

FIGURE 3 | TEM images of PS-b-PB colloidal particles depending
on the volume ratio (f_s) of PS block in the mixed surfactant of
PS-b-PEO and PB-b-PEO: (A) particles with onion-like
morphology for $f_s = 0$. (B) "Tulip-bulb" particles for $f_s = 0.36$.
(C) Prolate particles for $f_s = 0.46$. (D) "Tulip-bulb" particles with inverted
structure of the particles in (B) for $f_s = 0.77$. (E) Onion-like particles for
$f_s = 1$. [Reprinted with permission from Jeon et al. (2008). Copyright
2008 Wiley-VCH.]

et al., 2007; Chang and Lo, 2011; Chu et al., 2014; Kao et al., 2014) in the presence of solvent below T_g. Solvent annealing uses solvent vapor to increase chain mobility of BCP unlike thermal annealing in which the film are heated above T_g. Similarly, the nanostructures in the BCP particles can be accomplished by solvent annealing method (Li et al., 2010; Chi et al., 2011; Deng et al., 2013; Yabu, 2013; Jin and Fan, 2014) because the particles have been prepared from polymer solution. The morphological evolution of the BCP particle during solvent annealing has been reported theoretically and experimentally. Chi et al. studied solvent-induced self-assembly of BCPs under confinement via simulation (Chi et al., 2011). The self-assembled internal morphologies such as stacked lamellae, hexagonally packed cylinders, stacked toroids, or helices were predicted as functions of solvent-polymer interaction and the monomer concentration. Experimentally, Russell group demonstrated solvent-driven morphology of BCP particles using re-precipitation method from THF/water mixture (Li et al., 2010). They used a symmetric PS-*b*-PI BCP and chloroform for evolution of the BCP nanostructures. Since chloroform is a good solvent for both blocks, it can swell and anneal the BCP particles, resulting in formation of onion-like nanostructures. Upon the extended annealing, the nanostructures are evolved to cylindrical morphology and finally core-shell spherical structure.

Recently, microwave annealing approach have been developed for rapidly inducing self-assembly of BCP, by which highly ordered patterns were achieved within 3 min (Zhang et al., 2010b; Mokarian-Tabari et al., 2014). The method is based on simultaneously solvent-driven and thermal-driven mechanism by adsorption of microwave radiation into BCP chains, resulting in generation of BCP nanostructures in minutes. The nanostructure of BCP particles can be also created by microwave annealing method. Higuchi et al. examined morphological transitions of PS-*b*-PI BCP particles depending on experimental conditions such as temperature and time during microwave annealing (Higuchi et al., 2013). Even though the BCP particles are dispersed in aqueous solution, which has a higher dielectric constant and absorb most of microwave radiation, a small amount of the radiation can evolve the BCP nanostructures.

In addition, it is possible to change the particle shape from sphere to ellipsoid by tuning surface property of the BCP particles. Jintao group prepared reversible transformed BCP particles by solvent-adsorption annealing (Deng et al., 2013). In the article, poly(vinyl alcohol) (PVA) was used as a surfactant, which is a critical role in determining the particle shape, and chloroform, which is a good solvent for both blocks was used to swell and anneal the ellipsoid particles to spherical particles with onion-like structures. Moreover, they reported to fabricate polymeric Janus particles with hierarchically structures by phase separation of PS-*b*-P4VP and homopolymer poly(methyl methacrylate) (PMMA) binary blend (Deng et al., 2014). The Janus particles were tuned by varying copolymer composition, blending ratio, solvent selectivity, and particle size and reversibly transformed by solvent-adsorption process. Hawker group developed a new strategy to prepare ellipsoid BCP particles using modified surfactants (Jang et al., 2013; Klinger et al., 2014). Jang et al. (2013, 2011) prepared PS-*b*-PI terminated Au-core/Pt-shell nanoparticles as a surfactant for poly(styrene-*block*-2-vinylpyridine) PS-*b*-P2VP droplets. Since the nanoparticles preferentially adsorb at the emulsion interface and modify the interfacial interactions between PS-*b*-P2VP domains, the BCP droplets can be transformed to ellipsoid particle with striped nanostructure. Besides the nanoparticle surfactant, they chemically modified cetyltrimethylammonium bromide (CTAB) to hydroxyl-terminated CTAB (CTAB-OH). Although CTAB has a good affinity to PS, CTAB-OH prefer to P2VP chains, which can be produced onion-structured particles with P2VP as the outermost layer. When the fraction ratio of CTAB-OH and CTAB was between 0.7 and 0.8, ellipsoid particle was prepared with an internal nanostructure of stacked lamellae morphology.

Briefly, those factors to control the morphology in the BCP particles can be summarized as follows; the effects of the molecular weight and concentration of the BCP solution, the effect of the ratio of the particle diameter to the BCP structure period (D/L_0), the effect of the ratio of the good solvent to the poor solvent for the BCP on the microphase separation within the microspheres, and the effect of the interfacial tension between BCP solution and surfactants.

Composite Block Copolymer Particles

Block copolymer particles can be hybridized with functional inorganic, semiconductor, or metallic materials, to produce composite or hybrid BCP particles. There have been several different strategies for such a hybrid BCP particles, which can be categorized into chemical approach (MacLachlan et al., 2000; Manners, 2001, 2007; Korczagin et al., 2004; Whittell et al., 2011; Hudson et al., 2014; Zoetebier et al., 2015) using polymerization of metal-containing monomers and physical approach (Spatz et al., 1998; Kane et al., 1999; Tsutsumi et al., 1999; Lopes and Jaeger, 2001; Sohn and Seo, 2001; Boontongkong and Cohen, 2002; Bockstaller et al., 2003; Misner et al., 2003; Chiu et al., 2005; Kim et al., 2005; Yoo et al., 2010; Jang et al., 2011; Lim et al., 2011; Paek et al., 2011; Janczewski et al., 2014) such as *ex situ* and *in situ* approaches. The chemical approach is to use metal-containing monomer units. After living free-radical polymerization or ring-opening polymerization of the unit, metallopolymer can be produced. For example, Manner group reported a synthetic method to produce organo-metallic polymer. They performed ring-opening polymerization of spirocyclic silaferrocenophane as a monomer, resulting in formation of cross-linked Fe-containing polymer (MacLachlan et al., 2000). Moreover, they prepared fluorescent nanorod particles by micelle assembly of poly(ferrocenyldimethylsilane-*block*-(dimethylsiloxane-*random*-methylvinylsiloxane)) (PFS-*b*-(PDMS-*r*-PMVS)), which was synthesized by anionic polymerization (Hudson et al., 2014). The vinyl groups were then functionalized by photoinitiated hydrothiolation (thiolene-click reactions) with thiol reagents, providing primary hydroxyl or amino groups to which functional dyes may be attached. Vancso group reported a method to synthesize poly(ferrocenylsilane-*block*-methacrylate) (PFS-*b*-PMMA) by combined living radical polymerization. PFS was synthesized by anionic polymerization and PMMA was sequentially synthesized by ATRP, followed by production of PFS-*b*-PMMA (Korczagin et al., 2004). Although the chemical approach is to use inorganic precursor-containing monomers, it remains a challenge that the particles should be reduced as metal-polymer particles.

On the other hand, the physical approach to achieve hybrid BCPs is to use interaction between one block of BCPs and polymer ligands attached on the surface of NPs (*ex situ* method) or attraction between one block of BCPs and inorganic precursor (*in situ* method). The *ex situ* method is based on non-covalent interaction (e.g., van der Waals, – interactions or hydrogen bonding) or covalent linkage between each block and surface ligand of the NPs. For example, Bockstaller et al. (2003) reported alkanethiol-capped Au NPs localized at the interface between a symmetric poly(styrene-*block*-ethylene copropylene) (PS-*b*-PEP). Kim et al. (2005) studied PS-coated Au NPs self-assembled within

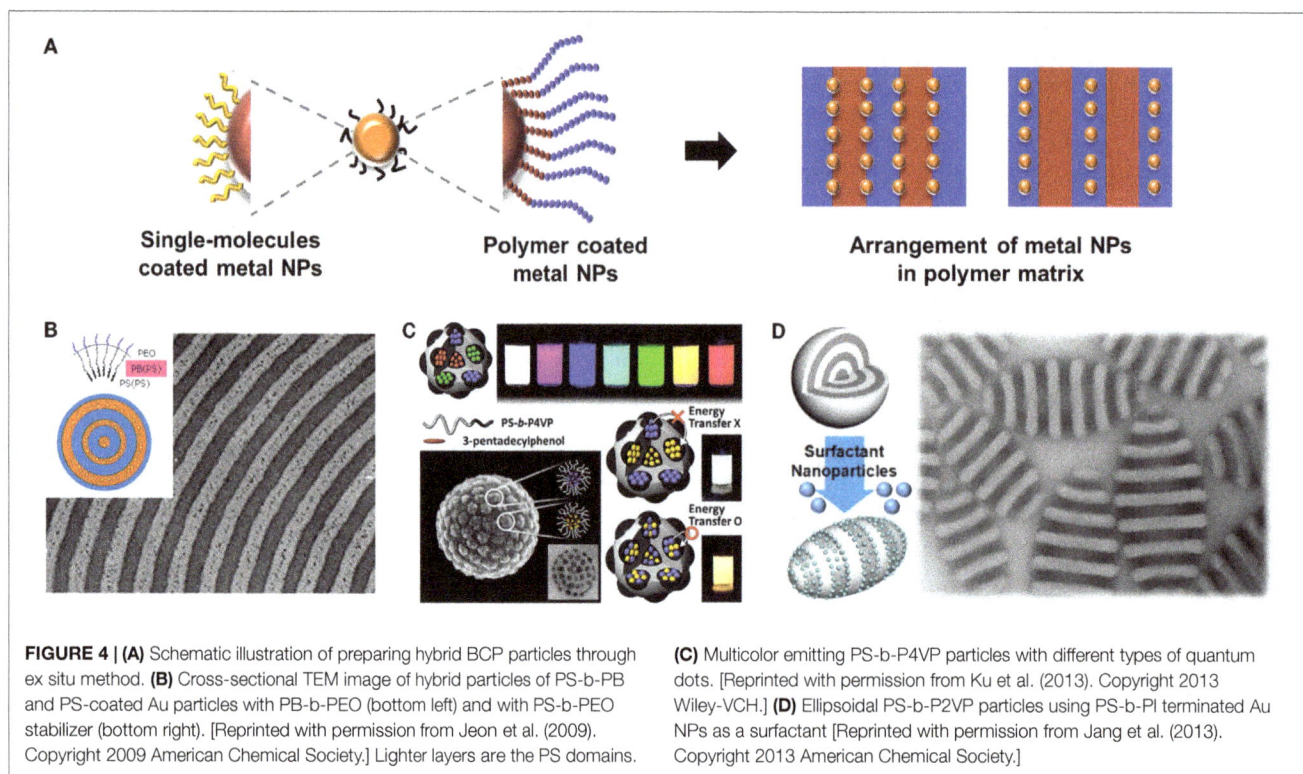

FIGURE 4 | (A) Schematic illustration of preparing hybrid BCP particles through ex situ method. **(B)** Cross-sectional TEM image of hybrid particles of PS-b-PB and PS-coated Au particles with PB-b-PEO stabilizer (bottom left) and with PS-b-PEO stabilizer (bottom right). [Reprinted with permission from Jeon et al. (2009). Copyright 2009 American Chemical Society.] Lighter layers are the PS domains. **(C)** Multicolor emitting PS-b-P4VP particles with different types of quantum dots. [Reprinted with permission from Ku et al. (2013). Copyright 2013 Wiley-VCH.] **(D)** Ellipsoidal PS-b-P2VP particles using PS-b-PI terminated Au NPs as a surfactant [Reprinted with permission from Jang et al. (2013). Copyright 2013 American Chemical Society.]

PS-b-P2VP. They found that the addition of NPs increased the effective volume fraction of the PS block and thus induced a lamellar-to-cylindrical phase transition. However, due to the steric hindrance between the core-shell type NPs and BCP chains, the density of NPs within the templates is expected to be much lower compared to the *in situ* approach. In addition, if the surface modification of NPs is necessary for such precise control, the intrinsic properties of NP can be deteriorated by increase in defect sites on the surface of NPs. The *in situ* method is based on electrostatic attraction between BCP chains and inorganic precursor. Generally, this approach has used Au, Pt, Pd, and Ag as inorganic precursors and poly(styrene-*block*-acrylic acid) (PS-b-PAA), and PS-b-P2VP as BCPs. For example, Chai et al. reported to fabricate linear metallic patterns using PS-b-P2VP (Chai et al., 2007). Since 2VP units of P2VP have been protonated under acidic condition, anionic metallic precursor (e.g., $AuCl_4^-$, $PdCl_4^{2-}$, $PtCl_4^{2-}$) interacted with the protonated 2VP units by electrostatic attraction. After their self-assembly on silicon substrate and metallization through chemical reduction, metallic patterns were produced, which can be extended toward increasing the conductance of the metallic nanowires and investigating the response of wire resistance to the environment for chemical and biosensing applications. In this section, we focus on *ex situ* and *in situ* approaches to produce hybrid BCP particles with nanostructures.

Ex Situ Method

The *ex situ* approach exploits the cooperative self-organization of pre-formed nanoparticles (NPs) and BCPs. The strategy for incorporating and controlling the location of NPs in a BCP or polymer blend involves tuning the surface properties of the NPs by functional polymer ligands, which play a critical role in particle location (**Figure 4**). The *ex situ* method of controlling the structural characteristics of the sequestered component uses cooperative self-organization of pre-made NPs and BCPs. Through this method, NPs with desired size and shape coated with oligomers or polymers can be incorporated in the polymer matrix. Once phase separation of the copolymer has occurred, the NPs can be confined in one of the segregated domains. Some have studied the control of the position in polymer matrix after coating the surface of the NPs using organic molecules (Zhao et al., 2009; Li et al., 2011). For example, Jeon et al. (2009) have prepared hybrid onionlike PS-b-PB particles containing PS-coated Au NPs. Since Au NPs were sufficiently covered with PS ligands, all Au NPs were located near the center of PS domain as shown in **Figure 4B**. Ku et al. (2013) have fabricated multicolor emitting hybrid PS-b-P4VP particles using QDs; white-color emitting particles were produced when two different types of QDs were independently incorporated into isolated PS-b-P4VP micelles, while orange-color emitting particles when the QDs were concurrently incorporated into the same micelles (**Figure 4C**). Furthermore, the NPs incorporated into the BCP particles could be used as a surfactant, which resulted in tuning the particle shapes. Jang et al. (2013) have reported that the spherical BCP particles could be changed to ellipsoidal particles when polymer-coated Au NPs were used as a surfactant which tunes the interfacial tension of the particles (**Figure 4D**). In addition, Ku et al. (2014, 2015) have investigated the effect of NP size and aspect-ratio (AR) to control the interfacial tension of the BCP particles. They prepared oleylamine-capped Au NPs with different size or oleic acid/oleylamine-capped CuPt nanorods with different AR, which was used as a surfactant for the BCP particles. Using supramolecular assembly between the prepared NPs and the BCP domain into the emulsion droplets, the authors

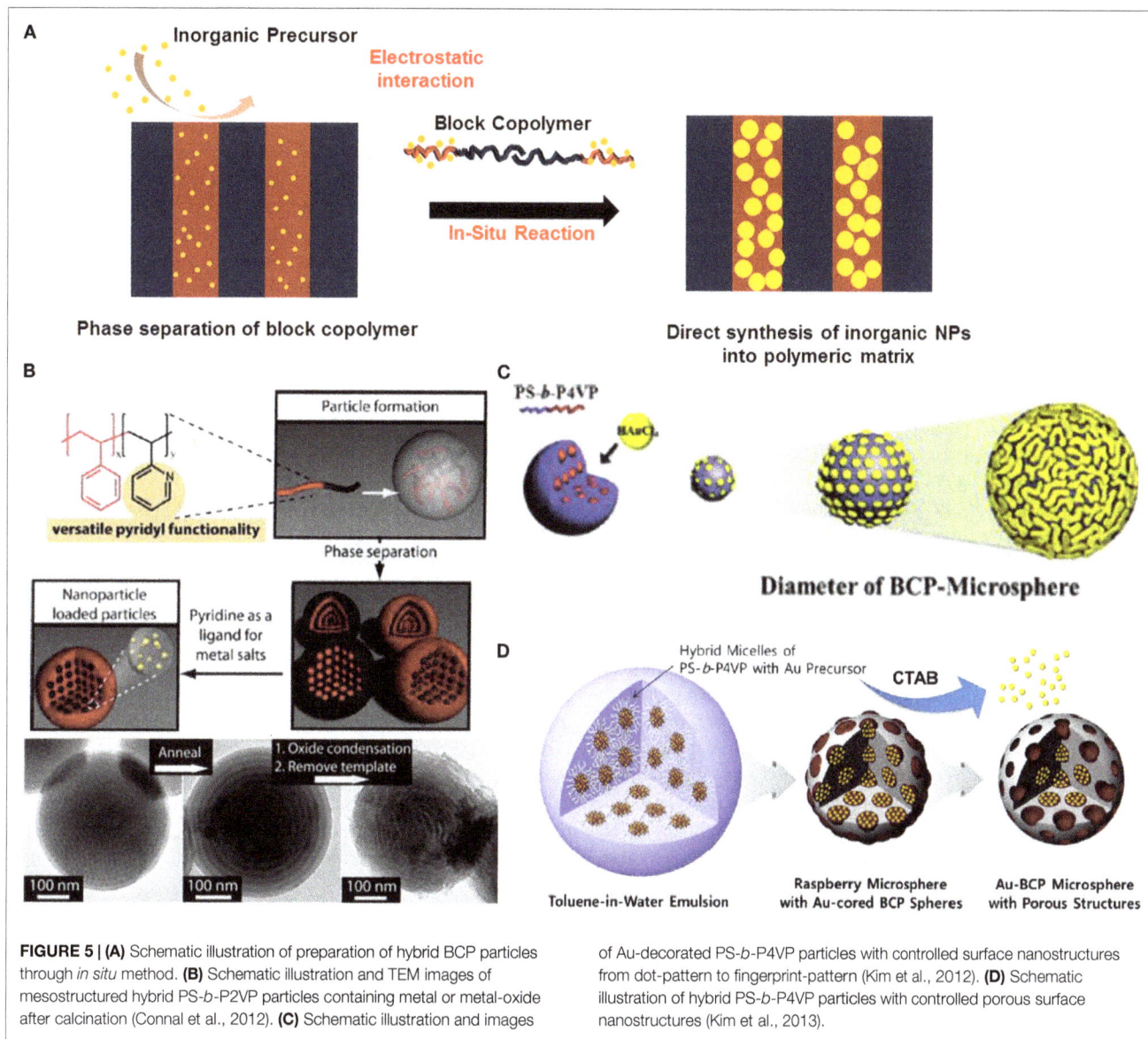

FIGURE 5 | (A) Schematic illustration of preparation of hybrid BCP particles through *in situ* method. **(B)** Schematic illustration and TEM images of mesostructured hybrid PS-*b*-P2VP particles containing metal or metal-oxide after calcination (Connal et al., 2012). **(C)** Schematic illustration and images of Au-decorated PS-*b*-P4VP particles with controlled surface nanostructures from dot-pattern to fingerprint-pattern (Kim et al., 2012). **(D)** Schematic illustration of hybrid PS-*b*-P4VP particles with controlled porous surface nanostructures (Kim et al., 2013).

demonstrated that convex-lens shaped PS-*b*-P4VP particles with defect-free and vertically ordered porous channels were created by tuning the interfacial tension.

Since the *ex situ* approach utilizes pre-made NPs that have desired size, shape, and uniformity, this approach is suitable for several applications that require inorganic and/or metallic NP with well controlled size and uniformity, for example, optoelectronic device (Ameri et al., 2009; Leventis et al., 2010; Yang et al., 2011; Liu, 2014), responsive sensor (Paek et al., 2011, 2014; Lee et al., 2013a; Yang et al., 2013), and catalyst (Campelo et al., 2009). Furthermore, the location of NPs within BCP matrix can be easily and precisely controlled by tailoring the surface properties of the NPs.

In situ Method

Another simple concept to incorporate the NPs into polymeric nanostructures is *in situ* method, followed by a chemical reduction step to obtain metallic nanoparticles (**Figure 5**). The *in situ*

approach is that NPs are directly synthesized within a BCP domain from metal precursors.

In that case, a specific and strong interaction between the precursors and the BCPs is required to guarantee the incorporation of metal precursors into the domain. In the *in situ* approach, preformed micelles of BCPs containing metal precursors are used as nanoreactors to synthesize NPs selectively in BCPs. Due to its chemical affinity, the salt selectively infiltrates the hydrophilic copolymer domain. The NPs are then form selectively within the precursor-loaded domains upon reduction. The targeted types of application mainly depend on the properties and functionality of the hybrid materials, which are determined by their chemical composition as well as morphological and structural parameters, for example, catalyst (Shastri and Schwank, 1985a,b; Wan et al., 1999; Bonnemann et al., 2000; Guczi et al., 2003), sensing device (Lee et al., 2013b; Choi et al., 2014), and metamaterials (Soukoulis and Wegener, 2011; Ramahi et al., 2012; Sheikholeslami et al., 2013).

Among the BCPs, polystyrene-*b*-poly (vinylpyridine) (PS-*b*-PVP) BCP has been widely used for preparing hybrid materials. This is because the nitrogen atoms in PVP block have unpaired electrons which interact strongly with metal precursor (Hayward et al., 2005; Chai et al., 2007). For example, Connal et al. (2012) have fabricated hybrid PS-*b*-P2VP particles by selectively infiltrating Au precursors (HAuCl$_4$) into the P2VP domain. In addition, they prepared mesoporous metal-oxide particles from PS-*b*-P2VP particles containing metal-oxide precursors after removing the BCP template by calcination. Furthermore, hybrid BCP particles with controlled surface nanoastructures also have been reported (Kim et al., 2012, 2013). The prepared spherical BCP particles using spherical PS-*b*-P4VP micelles were investigated the surface nanostructures from dot pattern to fingerprint pattern depending on the size of the BCP particles after infiltrating Au precursors into the BCP particles. In addition, we prepared dot-pattern hybrid BCP particle using spherical PS-*b*-P4VP micelles with Au precursors. After removing Au precursor into the P4VP domain, we produced hybrid BCP particles with porous surface which was controlled by tuning the amount of the Au precursors into the P4VP unit.

Summary and Outlook

Using BCPs in confined geometries such as aerosol or emulsion droplets, many nanostructured BCP particles have been prepared after solvent annealing process. They show a variety of new nanostructures due to confinement effect and interfacial properties. According to the theoretical prediction, however, much more nanostructures could be developed by more precise control of particle size and interfacial properties. Complex architecture of BCP such as star shape or multi-BCPs would also be applied into these BCP particles. Recently, these colloidal particles were functionalized by combination of various nanoparticles. Nanoparticles were mixed with BCP before particles formation (*ex situ* method) or precursors for nanoparticles were added after preparing BCP particles (*in situ* method). As an alternative route, nanopores were produced by removing selectively one of blocks, using molecular

templates or selective swelling process, which were then backfilled with other functional materials. Due to their unique nanostructure, these particles would be useful in various nanophotonic applications when the particles were functionalized with optically active materials. Particularly, these would be the new types of metamaterials if one of the phases is replaced with metallic materials. In addition, when the particles were transformed into nanoporous materials, it would also be useful for catalytic or sensor applications due to their high surface-to-volume ration and unusual well-defined pore structures. Although the nanostructured particles have been used in many applications such as drug delivery system, sensing device, self-healing, and optoelectronic device, it remains a challenge to scale-up monodisperse structured particles, which enable to investigate systematically the optical and electrical properties of the BCP particles with novel nanostructures.

Until now, the structured BCP particles have been prepared by high-mechanical force such as homogenizer, or sonicator, resulting in a broad range of the particle size. Polydisperse particles with different morphology and chemical heterogeneity can result in poor reproducibility and a variable property of the particles per batch. Some groups reported to fabricate uniform BCP particles using microfluidic method by which the particle diameters are ranged from hundreds nanometer to tens micrometer (Wang et al., 2010; Abate et al., 2011; Kuehne and Weitz, 2011), while other groups reported to prepare the structured BCP particles using membrane device (Tanaka et al., 2009; Connal et al., 2012; Deng et al., 2013), which enable a large-scale production of the particles but broader size distribution of the particles than that using microfluidic method. Therefore, it remains a challenge to produce large-scale and monodisperse structured BCP particles at sub-micrometer scale for practical application as mentioned above.

Acknowledgments

We acknowledge financial support by Korean NRF grant (no. 2010-0029409, NRF-2014R1A2A2A01006628).

References

Abate, A. R., Kutsovsky, M., Seiffert, S., Windbergs, M., Pinto, L. F. V., Rotem, A., et al. (2011). Synthesis of monodisperse microparticles from non-newtonian polymer solutions with microfluidic devices. *Adv. Mater.* 23, 1757–1760. doi:10.1002/adma.201004275

Ahn, H., Ryu, D. Y., Kim, Y., Kwon, K. W., Lee, J., and Cho, J. (2009). Phase behavior of polystyrene-b-poly(methyl methacrylate) diblock copolymer. *Macromolecules* 42, 7897–7902. doi:10.1021/la702745w

Ameri, T., Dennler, G., Lungenschmied, C., and Brabec, C. J. (2009). Organic tandem solar cells: a review. *Energy Environ. Sci.* 2, 347–363. doi:10.1039/b817952b

Babin, V., Garstecki, P., and Holyst, R. (2003). Multiple photonic band gaps in the structures composed of core-shell particles. *J. Appl. Phys.* 94, 4244–4247. doi:10.1063/1.1604932

Bae, D., Jeon, G., Jinnai, H., Huh, J., and Kim, J. K. (2013). Arrangement of block copolymer microdomains confined inside hemispherical cavities. *Macromolecules* 46, 5301–5307. doi:10.1021/ma4009324

Bai, P., Kim, M. I., and Xu, T. (2013). Thermally controlled morphologies in a block copolymer supramolecule via nonreversible order-order transitions. *Macromolecules* 46, 5531–5537. doi:10.1021/ma401033w

Bang, J., Kim, B. J., Stein, G. E., Russell, T. P., Li, X., Wang, J., et al. (2007). Effect of humidity on the ordering of PEO-based copolymer thin films. *Macromolecules* 40, 7019–7025. doi:10.1021/ma0710737

Bates, F. S., and Fredrickson, G. H. (1990). Block copolymer thermodynamics – theory and experiment. *Annu. Rev. Phys. Chem.* 41, 525–557. doi:10.1146/annurev.pc.41.100190.002521

Bita, I., Yang, J. K. W., Jung, Y. S., Ross, C. A., Thomas, E. L., and Berggren, K. K. (2008). Graphoepitaxy of self-assembled block copolymers on two-dimensional periodic patterned templates. *Science* 321, 939–943. doi:10.1126/science.1159352

Bockstaller, M. R., Lapetnikov, Y., Margel, S., and Thomas, E. L. (2003). Size-selective organization of enthalpic compatibilized nanocrystals in ternary block copolymer/particle mixtures. *J. Am. Chem. Soc.* 125, 5276–5277. doi:10.1021/ja034523t

Bonnemann, H., Brinkmann, R., Britz, P., Endruschat, U., Mortel, R., Paulus, U. A., et al. (2000). Nanoscopic Pt-bimetal colloids as precursors for PEM fuel cell catalysts. *J. New. Mater. Electrochem. Syst.* 3, 199–206.

Boontongkong, Y., and Cohen, R. E. (2002). Cavitated block copolymer micellar thin films: lateral arrays of open nanoreactors. *Macromolecules* 35, 3647–3652. doi:10.1021/ma0117357

Burlak, G., Diaz-de-Anda, A., and Zamudio-Lara, A. (2012). The narrow transmission peaks and field confinement produced by defects in a multilayered microsphere. *Opt. Commun.* 285, 1542–1549. doi:10.1016/j.optcom.2011.11.101

Campelo, J. M., Luna, D., Luque, R., Marinas, J. M., and Romero, A. A. (2009). Sustainable preparation of supported metal nanoparticles and their applications in catalysis. *ChemSusChem* 2, 18–45. doi:10.1002/cssc.200800227

Caruso, F., Caruso, R. A., and Mohwald, H. (1998a). Nanoengineering of inorganic and hybrid hollow spheres by colloidal templating. *Science* 282, 1111–1114. doi:10.1126/science.282.5391.1111

Caruso, F., Lichtenfeld, H., Giersig, M., and Mohwald, H. (1998b). Electrostatic self-assembly of silica nanoparticle – polyelectrolyte multilayers on polystyrene latex particles. *J. Am. Chem. Soc.* 120, 8523–8524. doi:10.1021/ja9815024

Caruso, F., and Mohwald, H. (1999). Preparation and characterization of ordered nanoparticle and polymer composite multilayers on colloids. *Langmuir* 15, 8276–8281. doi:10.1021/la990426v

Cavicchi, K. A., and Russell, T. P. (2007). Solvent annealed thin films of asymmetric polyisoprene-polylactide diblock copolymers. *Macromolecules* 40, 1181–1186. doi:10.1021/ma061163w

Chai, J., Wang, D., Fan, X. N., and Buriak, J. M. (2007). Assembly of aligned linear metallic patterns on silicon. *Nat. Nanotechnol.* 2, 500–506. doi:10.1038/nnano.2007.227

Chang, C. C., and Lo, C. T. (2011). Effect of particles on the structure of solvent-annealed block copolymer/nanoparticle composite thin film. *J. Phys. Chem. B* 115, 2485–2493. doi:10.1021/jp109321j

Chi, P., Wang, Z., Li, B. H., and Shi, A. C. (2011). Soft confinement-induced morphologies of diblock copolymers. *Langmuir* 27, 11683–11689. doi:10.1021/la202448c

Chiu, J. J., Kim, B. J., Kramer, E. J., and Pine, D. J. (2005). Control of nanoparticle location in block copolymers. *J. Am. Chem. Soc.* 127, 5036–5037. doi:10.1021/ja050376i

Choi, S. J., Kim, M. P., Lee, S. J., Kim, B. J., and Kim, I. D. (2014). Facile Au catalyst loading on the inner shell of hollow SnO$_2$ spheres using Au-decorated block copolymer sphere templates and their selective H$_2$S sensing characteristics. *Nanoscale* 6, 11898–11903. doi:10.1039/c4nr03706e

Chu, C. J., Chung, P. Y., Chi, M. H., Kao, Y. H., and Chen, J. T. (2014). Three-dimensional block copolymer nanostructures by the solvent-annealing-induced wetting in anodic aluminum oxide templates. *Macromol. Rapid Commun.* 35, 1598–1605. doi:10.1002/marc.201400222

Connal, L. A., Lynd, N. A., Robb, M. J., See, K. A., Jang, S. G., Spruell, J. M., et al. (2012). Mesostructured block copolymer nanoparticles: versatile templates for hybrid inorganic/organic nanostructures. *Chem. Mater.* 24, 4036–4042. doi:10.1021/cm3011524

Coustet, M., Irigoyen, J., Garcia, T. A., Murray, R. A., Romero, G., Cortizo, M. S., et al. (2014). Layer-by-layer assembly of polymersomes and polyelectrolytes on planar surfaces and microsized colloidal particles. *J. Colloid Interface Sci.* 421, 132–140. doi:10.1016/j.jcis.2014.01.038

Dauphas, N., van Zuilen, M., Wadhwa, M., Davis, A. M., Marty, B., and Janney, P. E. (2004). Clues from Fe isotope variations on the origin of early Archean BIFs from Greenland. *Science* 306, 2077–2080. doi:10.1126/science.1104639

Delcea, M., Mohwald, H., and Skirtach, A. G. (2011). Stimuli-responsive LbL capsules and nanoshells for drug delivery. *Adv. Drug Deliv. Rev.* 63, 730–747. doi:10.1016/j.addr.2011.03.010

Deng, R. H., Liang, F. X., Li, W. K., Yang, Z. Z., and Zhu, J. T. (2013). Reversible transformation of nanostructured polymer particles. *Macromolecules* 46, 7012–7017. doi:10.1021/ma401398h

Deng, R. H., Liu, S. Q., Liang, F. X., Wang, K., Zhu, J. T., and Yang, Z. Z. (2014). Polymeric Janus particles with hierarchical structures. *Macromolecules* 47, 3701–3707. doi:10.1021/ma500331w

Gourevich, I., Field, L. M., Wei, Z. X., Paquet, C., Petukhova, A., Alteheld, A., et al. (2006). Polymer multilayer particles: a route to spherical dielectric resonators. *Macromolecules* 39, 1449–1454. doi:10.1021/ma052167o

Grigoriev, D. O., Kohler, K., Skorb, E., Shchukin, D. G., and Mohwald, H. (2009). Polyelectrolyte complexes as a "smart" depot for self-healing anticorrosion coatings. *Soft Matter* 5, 1426–1432. doi:10.1039/b815147d

Guczi, L., Beck, A., Horvath, A., Koppany, Z., Stefler, G., Frey, K., et al. (2003). AuPd bimetallic nanoparticles on TiO$_2$: XRD, TEM, *in situ* EXAFS studies and catalytic activity in CO oxidation. *J. Mol. Catal. A Chem.* 204, 545–552. doi:10.1016/S1381-1169(03)00337-6

Hao, Q. H., Miao, B., Song, Q. G., Niu, X. H., and Liu, T. J. (2014). Phase behaviors of sphere-forming triblock copolymers confined in nanopores: a dynamic density functional theory study. *Polymer* 55, 4281–4288. doi:10.1016/j.polymer.2014.06.062

Hawker, C. J., and Russell, T. P. (2005). Block copolymer lithography: merging "bottom-up" with "top-down" processes. *MRS Bull.* 30, 952–966. doi:10.1557/mrs2005.249

Hayward, R. C., Chmelka, B. F., and Kramer, E. J. (2005). Crosslinked Poly(styrene)-Block-Poly(2-vinylpyridine) thin films as swellable templates for mesostructured silica and titania. *Adv. Mater.* 17, 2591–2595. doi:10.1002/adma.200500334

He, X., Zou, Z., Kan, D., and Liang, H. (2015). Self-assembly of diblock copolymer confined in an array-structure space. *J. Chem. Phys.* 142, 101912. doi:10.1063/1.4907532

Higuchi, T., Motoyoshi, K., Sugimori, H., Jinnai, H., Yabu, H., and Shimomura, M. (2012). Three-dimensional observation of confined phase-separated structures in block copolymer nanoparticles. *Soft Matter* 8, 3791–3797. doi:10.1039/c2sm07139h

Higuchi, T., Shimomura, M., and Yabu, H. (2013). Reorientation of microphase-separated structures in water-suspended block copolymer nanoparticles through microwave annealing. *Macromolecules* 46, 4064–4068. doi:10.1021/ma400620v

Higuchi, T., Tajima, A., Motoyoshi, K., Yabu, H., and Shimomura, M. (2008). Frustrated phases of block copolymers in nanoparticles. *Angew. Chem. Int. Ed.* 47, 8044–8046. doi:10.1002/anie.200803003

Hou, P. L., Fan, H. L., and Jin, Z. X. (2015). Spiral and mesoporous block polymer nanofibers generated in confined nanochannels. *Macromolecules* 48, 272–278. doi:10.1021/ma501933s

Huang, E., Russell, T. P., Harrison, C., Chaikin, P. M., Register, R. A., Hawker, C. J., et al. (1998). Using surface active random copolymers to control the domain orientation in diblock copolymer thin films. *Macromolecules* 31, 7641–7650. doi:10.1021/ma980705+

Hudson, Z. M., Lunn, D. J., Winnik, M. A., and Manners, I. (2014). Colour-tunable fluorescent multiblock micelles. *Nat. Commun.* 5, 1–8. doi:10.1038/ncomms4372

Janczewski, D., Song, J., and Vancso, G. J. (2014). Colloidal, water soluble probes constructed with quantum dots and amphiphilic poly(ferrocenylsilane) for smart redox sensing. *Eur. Polym. J.* 54, 87–94. doi:10.1016/j.eurpolymj.2014.02.012

Jang, S. G., Audus, D. J., Klinger, D., Krogstad, D. V., Kim, B. J., Cameron, A., et al. (2013). Striped, ellipsoidal particles by controlled assembly of diblock copolymers. *J. Am. Chem. Soc.* 135, 6649–6657. doi:10.1021/ja4019447

Jang, S. G., Khan, A., Dimitriou, M. D., Kim, B. J., Lynd, N. A., Kramer, E. J., et al. (2011). Synthesis of thermally stable Au-Core/Pt-Shell nanoparticles and their segregation behavior in diblock copolymer mixtures. *Soft Matter* 7, 6255–6263. doi:10.1039/c1sm05223c

Jeon, S. J., Yang, S. M., Kim, B. J., Petrie, J. D., Jang, S. G., Kramer, E. J., et al. (2009). Hierarchically structured colloids of diblock copolymers and Au nanoparticles. *Chem. Mater.* 21, 3739–3741. doi:10.1021/cm9011124

Jeon, S. J., Yi, G. R., Koo, C. M., and Yang, S. M. (2007). Nanostructures inside colloidal particles of block copolymer/homopolymer blends. *Macromolecules* 40, 8430–8439. doi:10.1021/ma0712302

Jeon, S. J., Yi, G. R., and Yang, S. M. (2008). Cooperative assembly of block copolymers with deformable interfaces: toward nanostructured particles. *Adv. Mater.* 20, 4103–4108. doi:10.1002/adma.200801377

Jin, Z. X., and Fan, H. L. (2014). Self-assembly of nanostructured block copolymer nanoparticles. *Soft Matter* 10, 9212–9219. doi:10.1039/c4sm02064b

Jung, Y. S., and Ross, C. A. (2007). Orientation-controlled self-assembled nanolithography using a polystyrene-polydimethylsiloxane block copolymer. *Nano Lett.* 7, 2046–2050. doi:10.1021/nl070924l

Kane, R. S., Cohen, R. E., and Silbey, R. (1999). Synthesis of doped ZnS nanoclusters within block copolymer nanoreactors. *Chem. Mater.* 11, 90–93. doi:10.1021/cm980468p

Kao, J., Thorkelsson, K., Bai, P., Zhang, Z., Sun, C., and Xu, T. (2014). Rapid fabrication of hierarchically structured supramolecular nanocomposite thin films in one minute. *Nat. Commun.* 5, 1–8. doi:10.1038/ncomms5053

Kim, B. J., Chiu, J. J., Yi, G. R., Pine, D. J., and Kramer, E. J. (2005). Nanoparticle-induced phase transitions in diblock-copolymer films. *Adv. Mater.* 17, 2618–2622. doi:10.1002/adma.200500502

Kim, G., and Libera, M. (1998). Morphological development in solvent-cast polystyrene-polybutadiene-polystyrene (SBS) triblock copolymer thin films. *Macromolecules* 31, 2569–2577. doi:10.1021/ma971349i

Kim, M. P., Kang, D. J., Jung, D. W., Kannan, A. G., Kim, K. H., Ku, K. H., et al. (2012). Gold-decorated block copolymer microspheres with controlled surface nanostructures. *ACS Nano* 6, 2750–2757. doi:10.1021/nn300194z

Kim, M. P., Ku, K. H., Kim, H. J., Jang, S. G., Yi, G. R., and Kim, B. J. (2013). Surface intaglio nanostructures on microspheres of gold-cored block copolymer spheres. *Chem. Mater.* 25, 4416–4422. doi:10.1021/cm402868q

Kim, S. H., Misner, M. J., Yang, L., Gang, O., Ocko, B. M., and Russell, T. P. (2006). Salt complexation in block copolymer thin films. *Macromolecules* 39, 8473–8479. doi:10.1021/ma061170k

Klinger, D., Wang, C. X., Connal, L. A., Audus, D. J., Jang, S. G., Kraemer, S., et al. (2014). A facile synthesis of dynamic, shape-changing polymer particles. *Angew. Chem. Int. Ed.* 53, 7018–7022. doi:10.1002/anie.201400183

Koneripalli, N., Singh, N., Levicky, R., Bates, F. S., Gallagher, P. D., and Satija, S. K. (1995). Confined block-copolymer thin-films. *Macromolecules* 28, 2897–2904. doi:10.1021/ma00112a041

Korczagin, I., Hempenius, M. A., and Vancso, G. J. (2004). Poly(ferrocenyisilane-block-methacrylates) via sequential anionic and atom transfer radical polymerization. *Macromolecules* 37, 1686–1690. doi:10.1021/ma0358172

Ku, K. H., Kim, M. P., Paek, K., Shin, J. M., Chung, S., Jang, S. G., et al. (2013). Multicolor emission of hybrid block copolymer-quantum dot microspheres by controlled spatial isolation of quantum dots. *Small* 9, 2667–2672. doi:10.1002/smll.201202839

Ku, K. H., Shin, J. M., Kim, M. P., Lee, C. H., Seo, M. K., Yi, G. R., et al. (2014). Size-controlled nanoparticle-guided assembly of block copolymers for convex lens-shaped particles. *J. Am. Chem. Soc.* 136, 9982–9989. doi:10.1021/ja502075f

Ku, K. H., Yang, H., Shin, J. M., and Kim, B. J. (2015). Aspect ratio effect of nanorod surfactants on the shape and internal morphology of block copolymer particles. *J. Polym. Sci. A Polym. Chem.* 53, 188–192. doi:10.1002/pola.27333

Kuehne, A. J. C., and Weitz, D. A. (2011). Highly monodisperse conjugated polymer particles synthesized with drop-based microfluidics. *Chem. Commun.* 47, 12379–12381. doi:10.1039/c1cc14251h

Lambooy, P., Russell, T. P., Kellogg, G. J., Mayes, A. M., Gallagher, P. D., and Satija, S. K. (1994). Observed frustration in confined block-copolymers. *Phys. Rev. Lett.* 72, 2899–2902. doi:10.1103/PhysRevLett.72.2899

Landfester, K. (2001a). The generation of nanoparticles in miniemulsions. *Adv. Mater.* 13, 765–768. doi:10.1002/1521-4095(200105)13:10<765::AID-ADMA765>3.0.CO;2-F

Landfester, K. (2001b). Polyreactions in miniemulsions. *Macromol. Rapid Commun.* 22, 896–936. doi:10.1002/1521-3927(20010801)22:12<896::AID-MARC896>3.3.CO;2-I

Landfester, K. (2006). Synthesis of colloidal particles in miniemulsions. *Annu. Rev. Mater. Res.* 36, 231–279. doi:10.1146/annurev.matsci.36.032905.091025

Landfester, K. (2009). Miniemulsion polymerization and the structure of polymer and hybrid nanoparticles. *Angew. Chem. Int. Ed.* 48, 4488–4507. doi:10.1002/anie.200900723

Landfester, K., and Mailander, V. (2013). Nanocapsules with specific targeting and release properties using miniemulsion polymerization. *Expert Opin. Drug Deliv.* 10, 593–609. doi:10.1517/17425247.2013.772976

Latnikova, A., Grigoriev, D., Schenderlein, M., Mohwald, H., and Shchukin, D. A. (2012). New approach towards "active" self-healing coatings: exploitation of microgels. *Soft Matter* 8, 10837–10844. doi:10.1039/c2sm26100f

Lee, J., Ryu, D. Y., and Cho, J. (2011). Temperature-pressure superposition in SANS chi for an A-b-B diblock copolymer. *Macromolecules* 44, 2387–2391. doi:10.1021/ma102102x

Lee, S. Y., Kim, S. H., Kim, M. P., Jeon, H. C., Kang, H., Kim, H. J., et al. (2013a). Freestanding and arrayed nanoporous microcylinders for highly active 3D SERS substrate. *Chem. Mater.* 25, 2421–2426. doi:10.1021/cm400298e

Lee, J. P., Chen, D. C., Li, X. X., Yoo, S., Bottomley, L. A., El-Sayed, M. A., et al. (2013b). Well-organized raspberry-like Ag@Cu bimetal nanoparticles for highly reliable and reproducible surface-enhanced Raman scattering. *Nanoscale* 5, 11620–11624. doi:10.1039/c3nr03363e

Leventis, H. C., King, S. P., Sudlow, A., Hill, M. S., Molloy, K. C., and Haque, S. A. (2010). Nanostructured hybrid polymer-inorganic solar cell active layers formed by controllable in situ growth of semiconducting sulfide networks. *Nano Lett.* 10, 1253–1258. doi:10.1021/nl903787j

Li, L., Matsunaga, K., Zhu, J. T., Higuchi, T., Yabu, H., Shimomura, M., et al. (2010). Solvent-driven evolution of block copolymer morphology under 3D confinement. *Macromolecules* 43, 7807–7812. doi:10.1021/ma101529b

Li, L., Miesch, C., Sudeep, P. K., Balazs, A. C., Emrick, T., Russell, T. P., et al. (2011). Kinetically trapped co-continuous polymer morphologies through intraphase gelation of nanoparticles. *Nano Lett.* 11, 1997–2003. doi:10.1021/nl200366z

Li, W. H., Liu, M. J., Qiu, F., and Shi, A. C. (2013). Phase diagram of diblock copolymers confined in thin films. *J. Phys. Chem. B* 117, 5280–5288. doi:10.1021/jp309546q

Li, W. H., and Wickham, R. A. (2006). Self-assembled morphologies of a diblock copolymer melt confined in a cylindrical nanopore. *Macromolecules* 39, 8492–8498. doi:10.1021/ma052151y

Li, W. H., Wickham, R. A., and Garbary, R. A. (2006). Phase diagram for a diblock copolymer melt under cylindrical confinement. *Macromolecules* 39, 806–811. doi:10.1021/ma052151y

Li, Z., Zhao, W., Liu, Y., Rafailovich, M. H., Sokolov, J., Khougaz, K., et al. (1996). Self-ordering of diblock copolymers from solution. *J. Am. Chem. Soc.* 118, 10892–10893. doi:10.1021/ja961713d

Lim, J., Yang, H., Paek, K., Cho, C. H., Kim, S., Bang, J., et al. (2011). "Click" synthesis of thermally stable au nanoparticles with highly grafted polymer shell and control of their behavior in polymer matrix. *J. Polym. Sci. A Polym. Chem.* 49, 3464–3474. doi:10.1002/pola.24782

Liu, R. C. (2014). Hybrid organic/inorganic nanocomposites for photovoltaic cells. *Materials* 7, 2747–2771. doi:10.3390/ma7042747

Lopes, W. A., and Jaeger, H. M. (2001). Hierarchical self-assembly of metal nanostructures on diblock copolymer scaffolds. *Nature* 414, 735–738. doi:10.1038/414735a

Louage, B., Zhang, Q. L., Vanparijs, N., Voorhaar, L., Vande Casteele, S., Shi, Y., et al. (2015). Degradable ketal-based block copolymer nanoparticles for anticancer drug delivery: a systematic evaluation. *Biomacromolecules* 16, 336–350. doi:10.1021/bm5015409

Lu, Y. F., Fan, H. Y., Stump, A., Ward, T. L., Rieker, T., and Brinker, C. J. (1999). Aerosol-assisted self-assembly of mesostructured spherical nanoparticles. *Nature* 398, 223–226. doi:10.1038/18410

Lv, L. P., Zhao, Y., Vilbrandt, N., Gallei, M., Vimalanandan, A., Rohwerder, M., et al. (2013). Redox responsive release of hydrophobic self-healing agents from polyaniline capsules. *J. Am. Chem. Soc.* 135, 14198–14205. doi:10.1021/ja405279t

Ma, N., Zhang, H. Y., Song, B., Wang, Z. Q., and Zhang, X. (2005). Polymer micelles as building blocks for layer-by-layer assembly: an approach for incorporation and controlled release of water-insoluble dyes. *Chem. Mater.* 17, 5065–5069. doi:10.1021/cm051221c

MacLachlan, M. J., Ginzburg, M., Coombs, N., Coyle, T. W., Raju, N. P., Greedan, J. E., et al. (2000). Shaped ceramics with tunable magnetic properties from metal-containing polymers. *Science* 287, 1460–1463. doi:10.1126/science.287.5457.1460

Mai, Y. Y., and Eisenberg, A. (2012). Self-assembly of block copolymers. *Chem. Soc. Rev.* 41, 5969–5985. doi:10.1039/c2cs35115c

Maiz, J., Zhao, W., Gu, Y., Lawrence, J., Arbe, A., Alegria, A., et al. (2014). Dynamic study of polystyrene-block-Poly(4-Vinylpyridine) copolymer in bulk and confined in cylindrical nanopores. *Polymer* 55, 4057–4066. doi:10.1016/j.polymer.2014.05.042

Malenfant, P. R., Wan, J., Taylor, S. T., and Manoharan, M. (2007). Self-assembly of an organic-inorganic block copolymer for nano-ordered ceramics. *Nat. Nanotechnol.* 2, 43–46. doi:10.1038/nnano.2006.168

Manners, I. (2001). Materials science – putting metals into polymers. *Science* 294, 1664–1666. doi:10.1126/science.1066321

Manners, I. (2007). Block copolymers with functional inorganic blocks: living addition polymerization of inorganic monomers. *Angew. Chem. Int. Ed.* 46, 1565–1568. doi:10.1002/anie.200604503

Marencic, A. P., and Register, R. A. (2010). Controlling order in block copolymer thin films for nanopatterning applications. *Annu. Rev. Chem. Biomol. Eng.* 1, 277–297. doi:10.1146/annurev-chembioeng-073009-101007

Mei, S. L., Feng, X. D., and Jin, Z. X. (2011). Fabrication of polymer nanospheres based on Rayleigh instability in capillary channels. *Macromolecules* 44, 1615–1620. doi:10.1021/ma102573p

Mei, S. L., and Jin, Z. X. (2013). Mesoporous block-copolymer nanospheres prepared by selective swelling. *Small* 9, 322–329. doi:10.1002/smll.201201504

Misner, M. J., Skaff, H., Emrick, T., and Russell, T. P. (2003). Directed deposition of nanoparticles using diblock copolymer templates. *Adv. Mater.* 15, 221–224. doi:10.1002/adma.200390050

Mokarian-Tabari, P., Cummins, C., Rasappa, S., Simao, C., Torres, C. M. S., Holmes, J. D., et al. (2014). Study of the kinetics and mechanism of rapid self-assembly in block copolymer thin films during solvo-microwave annealing. *Langmuir* 30, 10728–10739. doi:10.1021/la503137q

Nikoubashman, A., Register, R. A., and Panagiotopoulos, A. Z. (2014). Sequential domain realignment driven by conformational asymmetry in block copolymer thin films. *Macromolecules* 47, 1193–1198. doi:10.1021/ma402526q

Okubo, M., Saito, N., Takekoh, R., and Kobayashi, H. (2005). Morphology of polystyrene/polystyrene-block-Poly(Methyl Methacrylate)/Poly(Methyl Methacrylate) composite particles. *Polymer* 46, 1151–1156. doi:10.1016/j.polymer.2004.11.057

Otsuka, H., Nagasaki, Y., and Kataoka, K. (2003). PEGylated nanoparticles for biological and pharmaceutical applications. *Adv. Drug Deliv. Rev.* 55, 403–419. doi:10.1016/S0169-409X(02)00226-0

Paek, K., Chung, S., Cho, C. H., and Kim, B. J. (2011). Fluorescent and pH-responsive diblock copolymer-coated core-shell CdSe/ZnS particles for a color-displaying, ratiometric pH sensor. *Chem. Commun.* 47, 10272–10274. doi:10.1039/c1cc13848k

Paek, K., Yang, H., Lee, J., Park, J., and Kim, B. J. (2014). Efficient colorimetric pH sensor based on responsive polymer-quantum dot integrated graphene oxide. *ACS Nano* 8, 2848–2856. doi:10.1021/nn406657b

Park, C., Yoon, J., and Thomas, E. L. (2003). Enabling nanotechnology with self assembled block copolymer patterns. *Polymer* 44, 6725–6760. doi:10.1016/j.polymer.2003.08.011

Pastoriza-Santos, I., Scholer, B., and Caruso, F. (2001). Core-shell colloids and hollow polyelectrolyte capsules based on diazoresins. *Adv. Funct. Mater.* 11, 122–128. doi:10.1002/1616-3028(200104)11:2<122::AID-ADFM122>3.3.CO;2-E

Petukhova, A., Paton, A. S., Wei, Z. X., Gourevich, I., Nair, S. V., Ruda, H. E., et al. (2008). Polymer multilayer microspheres loaded with semiconductor quantum dots. *Adv. Funct. Mater.* 18, 1961–1968. doi:10.1002/adfm.200701441

Peyratout, C. S., and Dahne, L. (2004). Tailor-made polyelectrolyte microcapsules: from multilayers to smart containers. *Angew. Chem. Int. Ed.* 43, 3762–3783. doi:10.1002/anie.200300568

Pham, H. H., Gourevich, I., Oh, J. K., Jonkman, J. E. N., and Kumacheva, E. A. (2004). Multidye nanostructured material for optical data storage and security data encryption. *Adv. Mater.* 16, 516–520. doi:10.1002/chem.201406384

Rahikkala, A., Soininen, A. J., Ruokolainen, J., Mezzenga, R., Raula, J., and Kauppinen, E. I. (2013). Self-assembly of PS-b-P4VP block copolymers of varying architectures in aerosol nanospheres. *Soft Matter* 9, 1492–1499. doi:10.1039/C2SM26913A

Ramahi, O. M., Almoneef, T. S., Alshareef, M., and Boybay, M. S. (2012). Metamaterial particles for electromagnetic energy harvesting. *Appl. Phys. Lett.* 101, 173903. doi:10.1063/1.4764054

Rao, J. P., and Geckeler, K. E. (2011). Polymer nanoparticles: preparation techniques and size-control parameters. *Prog. Polym. Sci.* 36, 887–913. doi:10.1016/j.progpolymsci.2011.01.001

Rivera Gil, P., del Mercato, L. L., del-Pino, P., Munoz-Javier, A., and Parak, W. J. (2008). Nanoparticle-modified polyelectrolyte capsules. *Nano Today* 3, 12–21. doi:10.1039/c4cp05231e

Robb, M. J., Connal, L. A., Lee, B. F., Lynd, N. A., and Hawker, C. J. (2012). Functional block copolymer nanoparticles: toward the next generation of delivery vehicles. *Polym. Chem.* 3, 1618–1628. doi:10.1039/c2py20131c

Shastri, A. G., and Schwank, J. (1985a). Metal dispersion of bimetallic catalysts via stepwise chemisorption and surface titration.1. Ru-Au/Sio2. *J. Catal.* 95, 271–283. doi:10.1016/0021-9517(85)90028-4

Shastri, A. G., and Schwank, J. (1985b). Metal dispersion of bimetallic catalysts via stepwise chemisorption and surface titration.2. Ru-Au/Mgo. *J. Catal.* 95, 284–288. doi:10.1016/0021-9517(85)90028-4

Shchukin, D. G., and Mohwald, H. (2006). Sonochemical nanosynthesis at the engineered interface of a cavitation microbubble. *Phys. Chem. Chem. Phys.* 8, 3496–3506. doi:10.1039/b606104d

Sheikholeslami, S. N., Alaeian, H., Koh, A. L., and Dionne, J. A. (2013). A Metafluid exhibiting strong optical magnetism. *Nano Lett.* 13, 4137–4141. doi:10.1021/nl401642z

Smulders, W., and Monteiro, M. J. (2004). Seeded emulsion polymerization of block copolymer core-shell nanoparticles with controlled particle size and molecular weight distribution using xanthate-based RAFT polymerization. *Macromolecules* 37, 4474–4483. doi:10.1021/ma049496l

Sohn, B. H., and Seo, B. H. (2001). Fabrication of the multilayered nanostructure of alternating polymers and gold nanoparticles with thin films of self-assembling diblock copolymers. *Chem. Mater.* 13, 1752–1757. doi:10.1021/cm000939j

Soukoulis, C. M., and Wegener, M. (2011). Past achievements and future challenges in the development of three-dimensional photonic metamaterials. *Nat. Photonics* 5, 523–530. doi:10.1038/nphoton.2011.154

Spatz, J., Mossmer, S., Moller, M., Kocher, M., Neher, D., and Wegner, G. (1998). Controlled mineralization and assembly of hydrolysis-based nanoparticles in organic solvents combining polymer micelles and microwave techniques. *Adv. Mater.* 10, 473–475. doi:10.1002/(SICI)1521-4095(199804)10:6<473::AID-ADMA473>3.0.CO;2-Q

Sukhorukov, G., Fery, A., and Mohwald, H. (2005). Intelligent micro- and nanocapsules. *Prog. Polym. Sci.* 30, 885–897. doi:10.1016/j.progpolymsci.2005.06.008

Sundberg, D. C., and Durant, Y. G. (2003). Latex particle morphology, fundamental aspects: a review. *Polym. React. Eng.* 11, 379–432. doi:10.1081/PRE-120024420

Tanaka, T., Saito, N., and Okubo, M. (2009). Control of layer thickness of onionlike multilayered composite polymer particles prepared by the solvent evaporation method. *Macromolecules* 42, 7423–7429. doi:10.1021/ma901100n

Tang, P., Qiu, F., Zhang, H. D., and Yang, Y. L. (2004). Morphology and phase diagram of complex block copolymers: ABC linear triblock copolymers. *J. Phys. Chem. B* 108, 8434–8438. doi:10.1021/jp037911q

Thomas, E. L., Reffner, J. R., and Bellare, J. (1990). A menagerie of interface structures in copolymer systems. *J. Phys. Paris* 51, C7363–C7374. doi:10.1051/jphyscol:1990736

Tsutsumi, K., Funaki, Y., Hirokawa, Y., and Hashimoto, T. (1999). Selective incorporation of palladium nanoparticles into microphase-separated domains of Poly(2-vinylpyridine)-block-polyisoprene. *Langmuir* 15, 5200–5203. doi:10.1021/la990246l

Vimalanandan, A., Lv, L. P., Tran, T. H., Landfester, K., Crespy, D., and Rohwerder, M. (2013). Redox-responsive self-healing for corrosion protection. *Adv. Mater.* 25, 6980–6984. doi:10.1002/adma.201302989

Wan, B. S., Liao, S. J., and Yu, D. R. (1999). Polymer-supported palladium-manganese bimetallic catalyst for the oxidative carbonylation of amines to carbamate esters. *Appl. Catal. A Gen.* 183, 81–84. doi:10.1016/S0926-860X(99)00062-9

Wang, C. W., Oskooei, A., Sinton, D., and Moffitt, M. G. (2010). Controlled self-assembly of quantum dot-block copolymer colloids in multiphase microfluidic reactors. *Langmuir* 26, 716–723. doi:10.1021/la902427r

Wang, W., and Zhang, Q. (2012). Synthesis of block copolymer poly (N-Butyl Acrylate)-b-polystyrene by DPE seeded emulsion polymerization with monodisperse latex particles and morphology of self-assembly film surface. *J. Colloid Interface Sci.* 374, 54–60. doi:10.1016/j.jcis.2012.01.030

Wang, Y., Gosele, U., and Steinhart, M. (2008). Mesoporous block copolymer nanorods by swelling-induced morphology reconstruction. *Nano Lett.* 8, 3548–3553. doi:10.1021/nl8022687

Wang, Y. J., Hosta-Rigau, L., Lomas, H., and Caruso, F. (2011). Nanostructured polymer assemblies formed at interfaces: applications from immobilization and encapsulation to stimuli-responsive release. *Phys. Chem. Chem. Phys.* 13, 4782–4801. doi:10.1039/c0cp02287j

Whittell, G. R., Hager, M. D., Schubert, U. S., and Manners, I. (2011). Functional soft materials from metallopolymers and metallosupramolecular polymers. *Nat. Mater.* 10, 176–188. doi:10.1038/nmat2966

Winey, K. I., Gobran, D. A., Xu, Z. D., Fetters, L. J., and Thomas, E. L. (1994). Compositional dependence of the order-disorder transition in diblock copolymers. *Macromolecules* 27, 2392–2397. doi:10.1021/ma00087a005

Wu, Y. Y., Cheng, G. S., Katsov, K., Sides, S. W., Wang, J. F., Tang, J., et al. (2004). Composite mesostructures by nano-confinement. *Nat. Mater.* 3, 816–822. doi:10.1038/nmat1230

Xiang, H. Q., Shin, K., Kim, T., Moon, S. I., McCarthy, T. J., and Russell, T. P. (2004). Block copolymers under cylindrical confinement. *Macromolecules* 37, 5660–5664. doi:10.1021/ma049299m

Xiong, X. P., Eckelt, J., Zhang, L., and Wolf, B. A. (2009). Thermodynamics of block copolymer solutions as compared with the corresponding homopolymer solutions: experiment and theory. *Macromolecules* 42, 8398–8405. doi:10.1021/ma9014615

Xu, J., Wang, K., Li, J., Zhou, H., Xie, X., and Zhu, J. (2015). ABC triblock copolymer particles with tunable shape and internal structure through 3D confined assembly. *Macromolecules* 48, 2628–2636. doi:10.1021/acs.macromol.5b00335

Xu, T., Hawker, C. J., and Russell, T. P. (2005). Interfacial interaction dependence of microdomain orientation in diblock copolymer thin films. *Macromolecules* 38, 2802–2805. doi:10.1021/ma048005u

Xu, Y. C., Li, W. H., Qiu, F., and Lin, Z. Q. (2014). Self-assembly of 21-arm star-like diblock copolymer in bulk and under cylindrical confinement. *Nanoscale* 6, 6844–6852. doi:10.1039/c4nr01275e

Yabu, H. (2013). Self-organized precipitation: an emerging method for preparation of unique polymer particles. *Polym. J.* 45, 261–268. doi:10.1038/pj.2012.151

Yabu, H., Higuchi, T., and Shimomura, M. (2005). Unique phase-separation structures of block-copolymer nanoparticles. *Adv. Mater.* 17, 2062–2065. doi:10.1002/adma.200500255

Yan, Y., Bjonmalm, M., and Caruso, F. (2014). Assembly of layer-by-layer particles and their interactions with biological systems. *Chem. Mater.* 26, 452–460. doi:10.1021/cm402126n

Yang, H., Paek, K., and Kim, B. J. (2013). Efficient temperature sensing platform based on fluorescent block copolymer-functionalized graphene oxide. *Nanoscale* 5, 5720–5724. doi:10.1039/c3nr01486j

Yang, J., You, J. B., Chen, C. C., Hsu, W. C., Tan, H. R., Zhang, X. W., et al. (2011). Plasmonic polymer tandem solar cell. *ACS Nano* 5, 6210–6217. doi:10.1021/nn202144b

Yang, R. Q., Li, B. H., and Shi, A. C. (2012). Phase behavior of binary blends of diblock copolymer/homopolymer confined in spherical nanopores. *Langmuir* 28, 1569–1578. doi:10.1021/la204449x

Yoo, M., Kim, S., Lim, J., Kramer, E. J., Hawker, C. J., Kim, B. J., et al. (2010). Facile synthesis of thermally stable core-shell gold nanoparticles via photocross-linkable polymeric ligands. *Macromolecules* 43, 3570–3575. doi:10.1021/ma1000145

Yu, B., Li, B. H., Jin, Q. H., Ding, D., and Shi, A. C. (2007). Self-assembly of symmetric diblock copolymers confined in spherical nanopores. *Macromolecules* 40, 9133–9142. doi:10.1021/ma071624t

Yu, B., Sun, P. C., Chen, T. H., Jin, Q. H., Ding, D. T., Li, B. H., et al. (2006). Confinement-induced novel morphologies of block copolymers. *Phys. Rev. Lett.* 96, 138306. doi:10.1103/PhysRevLett.96.138306

Yu, H. Z., Qiu, X. Y., Nunes, S. P., and Peinemann, K. V. (2014). Biomimetic block copolymer particles with gated nanopores and ultrahigh protein sorption capacity. *Nat. Commun.* 5, 1–10. doi:10.1038/ncomms5110

Zhang, K., Gao, L., Chen, Y. M., and Yang, Z. Z. (2008a). Onionlike spherical polymer composites with controlled dispersion of gold nanoclusters. *Chem. Mater.* 20, 23–25. doi:10.1021/cm7028118

Zhang, K., Yu, X. L., Gao, L., Chen, Y. M., and Yang, Z. Z. (2008b). Mesostructured spheres of organic/inorganic hybrid from gelable block copolymers and arched nano-objects thereof. *Langmuir* 24, 6542–6548. doi:10.1021/la800096w

Zhang, K., Gao, L., Chen, Y. M., and Yang, Z. Z. (2010a). Onion-like microspheres with tricomponent from gelable triblock copolymers. *J. Colloid Interface Sci.* 346, 48–53. doi:10.1016/j.jcis.2010.02.039

Zhang, X. J., Harris, K. D., Wu, N. L. Y., Murphy, J. N., and Buriak, J. M. (2010b). Fast assembly of ordered block copolymer nanostructures through microwave annealing. *ACS Nano* 4, 7021–7029. doi:10.1021/nn102387c

Zhang, M. F., Yang, L., Yurt, S., Misner, M. J., Chen, J. T., Coughlin, E. B., et al. (2007). Highly ordered nanoporous thin films from cleavable polystyrene-block-poly(ethylene oxide). *Adv. Mater.* 19, 1571–1576. doi:10.1002/adma.200602461

Zhang, X., Wang, L., Zhang, L., Lin, J., and Jiang, T. (2015a). Controllable hierarchical microstructures self-assembled from multiblock copolymers confined in thin films. *Langmuir* 31, 2533–2544. doi:10.1021/la503985u

Zhang, T., Deng, H., Yang, T., and Li, W. (2015b). Defective morphologies kinetically formed in diblock copolymers under the cylindrical confinement. *Polymer* 65, 168–174. doi:10.1016/j.polymer.2015.03.059

Zhao, B., Haasch, R. T., and MacLaren, S. (2004). Solvent-induced self-assembly of mixed Poly(methyl methacrylate)/polystyrene brushes on planar silica substrates: molecular weight effect. *J. Am. Chem. Soc.* 126, 6124–6134. doi:10.1021/ja049570f

Zhao, Y., Thorkelsson, K., Mastroianni, A. J., Schilling, T., Luther, J. M., Rancatore, B. J., et al. (2009). Small-molecule-directed nanoparticle assembly towards stimuli-responsive nanocomposites. *Nat. Mater.* 8, 979–985. doi:10.1038/nmat2565

Zoetebier, B., Hempenius, M. A., and Vancso, G. J. (2015). Redox-responsive organometallic hydrogels for in situ metal nanoparticle synthesis. *Chem. Commun.* 51, 636–639. doi:10.1039/c4cc06988a

Zuo, Y. X., Wang, G. Q., Yu, Y., Zuo, C. C., Shi, L. Y., Shi, F. L., et al. (2015). Phase behavior of copolymers confined in multi-walled nanotubes: insights from simulations. *Polymers (Basel)* 7, 120–133. doi:10.3390/polym7010120

Conflict of Interest Statement: The authors declare that the research was conducted in the absence of any commercial or financial relationships that could be construed as a potential conflict of interest.

Permissions

All chapters in this book were first published in FMATS, by Frontiers; hereby published with permission under the Creative Commons Attribution License or equivalent. Every chapter published in this book has been scrutinized by our experts. Their significance has been extensively debated. The topics covered herein carry significant findings which will fuel the growth of the discipline. They may even be implemented as practical applications or may be referred to as a beginning point for another development.

The contributors of this book come from diverse backgrounds, making this book a truly international effort. This book will bring forth new frontiers with its revolutionizing research information and detailed analysis of the nascent developments around the world.

We would like to thank all the contributing authors for lending their expertise to make the book truly unique. They have played a crucial role in the development of this book. Without their invaluable contributions this book wouldn't have been possible. They have made vital efforts to compile up to date information on the varied aspects of this subject to make this book a valuable addition to the collection of many professionals and students.

This book was conceptualized with the vision of imparting up-to-date information and advanced data in this field. To ensure the same, a matchless editorial board was set up. Every individual on the board went through rigorous rounds of assessment to prove their worth. After which they invested a large part of their time researching and compiling the most relevant data for our readers.

The editorial board has been involved in producing this book since its inception. They have spent rigorous hours researching and exploring the diverse topics which have resulted in the successful publishing of this book. They have passed on their knowledge of decades through this book. To expedite this challenging task, the publisher supported the team at every step. A small team of assistant editors was also appointed to further simplify the editing procedure and attain best results for the readers.

Apart from the editorial board, the designing team has also invested a significant amount of their time in understanding the subject and creating the most relevant covers. They scrutinized every image to scout for the most suitable representation of the subject and create an appropriate cover for the book.

The publishing team has been an ardent support to the editorial, designing and production team. Their endless efforts to recruit the best for this project, has resulted in the accomplishment of this book. They are a veteran in the field of academics and their pool of knowledge is as vast as their experience in printing. Their expertise and guidance has proved useful at every step. Their uncompromising quality standards have made this book an exceptional effort. Their encouragement from time to time has been an inspiration for everyone.

The publisher and the editorial board hope that this book will prove to be a valuable piece of knowledge for researchers, students, practitioners and scholars across the globe.

List of Contributors

Jiawei Wang, Zhanshi Yao and Andrew W. Poon
Photonic Device Laboratory, Department of Electronic and Computer Engineering, The Hong Kong University of Science and Technology, Hong Kong, China

Zeyuan Cao and Bingqing Wei
Department of Mechanical Engineering, University of Delaware, Newark, DE, USA

Yoshiaki Nishijima
Department of Electrical and Computer Engineering, Graduate School of Engineering, Yokohama National University, Yokohama, Japan,

Saulius Juodkazis
Centre for Micro-Photonics, Faculty of Science, Engineering and Technology, Swinburne University of Technology, Melbourne, VIC, Australia,
Melbourne Centre for Nanofabrication (MCN), Australian National Fabrication Facility, Clayton, VIC, Australia,
Center of Nanotechnology, King Abdulaziz University, Jeddah, Saudi Arabia

Chung Hae Park and Patricia Krawczak
Polymers and Composites Technology and Mechanical Engineering Department, Ecole Nationale Supérieure des Mines de Douai, Douai, France

Neal S. Gupta
Department of Science and Technology, Bryant University, Smithfield, RI, USA

Sangryun Lee and Seunghwa Ryu
Department of Mechanical Engineering, Korea Advanced Institute of Science and Technology, Daejeon, South Korea

Edith Mäder
Leibniz-Institut für Polymerforschung Dresden e. V., Dresden, Germany
Institut für Werkstoffwissenschaft, Technische Universität Dresden, Dresden, Germany

Jianwen Liu and Janett Hiller
Leibniz-Institut für Polymerforschung Dresden e. V., Dresden, Germany

Weibang Lu and Qingwen Li
Advanced Materials Division, Suzhou Institute of Nano-Tech and Nano-Bionics, Suzhou, China

Serge Zhandarov
"V. A. Bely" Metal-Polymer Research Institute, National Academy of Sciences of Belarus, Gomel, Belarus

Tsu-Wei Chou
Center for Composite Materials, University of Delaware, Newark, DE, USA

Richard Geiger, Thomas Zabel and Hans Sigg
Laboratory for Micro- and Nanotechnology, Paul Scherrer Institut, Villigen, Switzerland

Fenghua Zhang , Zhichun Zhang , Tianyang Zhou and Jinsong Leng
Centre for Composite Materials and Structures, Harbin Institute of Technology, Harbin, China,

Yanju Liu
Department of Astronautical Science and Mechanics, Harbin Institute of Technology, Harbin, China

Brahim Aïssa
Qatar Environment and Energy Research Institute (QEERI), Qatar Foundation, Doha, Qatar,
Department of Smart Materials and Sensors for Space Missions, MPB Technologies Inc., Montreal, QC, Canada

Nasir K. Memon, Adnan Ali and Marwan K. Khraisheh
Qatar Environment and Energy Research Institute (QEERI), Qatar Foundation, Doha, Qatar

Ane Altuna
microLIQUID, Mondragon, Spain,

Javier Berganzo
Microsystems Department, IK4-IKERLAN, Mondragon, Spain,

Luis J. Fernández
Centro de Investigación Biomédica en Red, Biomateriales y Nanomedicina (CIBER-BBN), Zaragoza, Spain
Group of Structural Mechanics and Materials Modelling (GEMM), Aragón Institute of Engineering Research (I3A), Universidad de Zaragoza, Zaragoza, Spain

Jie Cheng
Department of Biomedical Engineering, College of Health Science, Korea University, Seoul, South Korea

Sang-Hoon Lee
Department of Biomedical Engineering, College of Health Science, Korea University, Seoul, South Korea
Department of Bio-Convergence Engineering, College of Health Science, Korea University, Seoul, South Korea
KU-KIST Graduate School of Converging Science and Technology, Korea University, Seoul, South Korea

Jesus Marino Falcón and Idalina Vieira Aoki
Polytechnic School, University of São Paulo, São Paulo, Brazil,

Tiago Sawczen
Nanocorr – Aditivos Inteligentes e Soluções Contra Corrosão Ltd., Paraná, Brazil

Grahame Mackenzie and Alberto Diego-Taboada
Department of Chemistry, University of Hull, Hull, UK
Sporomex Ltd., Driffield, UK

Andrew N. Boa
Department of Chemistry, University of Hull, Hull, UK

Stephen L. Atkin
Weill Cornell Medical College Qatar, Doha, Qatar

Thozhukat Sathyapalan
Hull York Medical School, University of Hull, Hull, UK

Taisuke Inoue, Xuyi Gao, Kenji Shinozaki, Tsuyoshi Honma and Takayuki Komatsu
Department of Materials Science and Technology, Nagaoka University of Technology, Nagaoka, Japan

Sebastien Dryepondt, Bruce A. Pint and Philip J. Maziasz
Materials Science and Technology Division, Oak Ridge National Laboratory, Oak Ridge, TN, USA

Minsoo P. Kim and Gi-Ra Yi
School of Chemical Engineering, Sungkyunkwan University, Suwon, South Korea

www.ingramcontent.com/pod-product-compliance
Lightning Source LLC
Chambersburg PA
CBHW050439200326
41458CB00014B/4996